Enzyme-Based Computing Systems

Enzyme-Based Computing Systems

Evgeny Katz

WILEY-VCH

Author

Dr. Evgeny Katz
Clarkson University
Department of Chemistry
Clarkson Avenue 8
NY
United States

Cover
Image was kindly provided by
Dr. Evgeny Katz

Library of Congress Card No.:
applied for

British Library Cataloguing-in-Publication Data
A catalogue record for this book is available from the British Library.

Bibliographic information published by the Deutsche Nationalbibliothek
The Deutsche Nationalbibliothek lists this publication in the Deutsche Nationalbibliografie; detailed bibliographic data are available on the Internet at <http://dnb.d-nb.de>.

Print ISBN: 978-3-527-34570-0
ePDF ISBN: 978-3-527-81996-6
ePub ISBN: 978-3-527-81998-0
oBook ISBN: 978-3-527-81999-7

Cover Design Adam-Design, Weinheim, Germany
Typesetting SPi Global, Chennai, India
Printing and Binding Markono Print Media Pte Ltd, Singapore
Printed on acid-free paper

10 9 8 7 6 5 4 3 2 1

In the memory of Professor Vladimir Privman, my good friend and collaborator.

Contents

Preface

The use of biomolecular systems for processing information, performing logic operations, computational operations, and even automata performance is a rapidly developing research area. The entire field was named with the general buzzwords, "biomolecular computing" or "biocomputing." Exciting advances in the area include the use of various biomolecular systems including proteins/enzymes, DNA, RNA, DNAzymes, antigens/antibodies, and even whole biological (usually microbial) cells operating as "hardware" for unconventional computing. The present book concentrates on enzymatic systems, which involve biocatalytic reactions utilized for information processing (biocomputing). Extensive ongoing research in the enzyme-based biocomputing, mimicking Boolean logic gates, has been motivated by potential applications in biotechnology and medicine. Furthermore, novel biosensor concepts have been contemplated with multiple inputs processed biochemically before the final output is coupled to transducing electronic or optical systems. These applications have warranted recent emphasis on networking of enzyme logic gates. First few gate networks have been experimentally realized, including coupling, for instance, to signal-responsive electrodes for signal readout. In order to achieve scalable, stable network design and functioning, considerations of noise propagation and control have been initiated as a new research direction. Optimization of single enzyme-based gates for avoiding analog noise amplification has been explored, as were certain network optimization concepts. The book reviews and exemplifies these developments, as well as offers an outlook for possible future research foci. The latter include design and uses of non-Boolean network elements, e.g., filters, as well as other developments motivated by potential novel biosensor and biotechnology applications. The most important feature of the enzyme biocomputing systems is their operation in biochemical and even biological environment. Many different applications of these systems, in addition to unconventional computation, are feasible, while their biosensor/biomedical use is obviously one of the most important applications. Interfacing of biological systems with biosensors, "smart" signal-responsive materials, and bioelectronic devices is of highest importance for future developments in the area of biomolecular computing.

The various topics covered highlight key aspects and the future perspectives of the enzyme-based computing. The different topics addressed in this book will be of high interest to the interdisciplinary community active in the area of unconventional biocomputing. The readers can find additional complementary material on molecular [1] and biomolecular [2] computing published recently by Wiley-VCH. It is hoped that the book will be important and beneficial for researchers and students working in various areas related to biochemical computing, including biochemistry, materials science, computer science, and so on. Furthermore, the book is aimed to attract young scientists and introduce them to the field while providing newcomers with an enormous collection of literature references. I, indeed, hope that the book will spark the imagination of scientists to further develop the topic.

The text was carefully proofread, and the figures were meticulously redrawn and checked to eliminate possible typos, mistakes, and unclear meaning. Still because of the large volume and big number (230) figures, some problems may appear. If this happens, the readers are advised to go to the original publications following the references provided.

A significant amount of the discussed material has originated from the studies to which I have personally contributed. I am very grateful to all scientists, researchers, and students who have participated in this research and have made the achieved results possible.

I would like to conclude this preface by thanking my wife Nina for her support in every respect in the past 47 years. Without her help and support, it would not have been possible to complete this work.

Evgeny Katz

Potsdam, NY, USA
January 2019

References

1 Katz, E. (ed.) (2012). *Molecular and Supramolecular Information Processing: From Molecular Switches to Logic Systems.* Weinheim: Wiley-VCH.
2 Katz, E. (ed.) (2012). *Biomolecular Information Processing – From Logic Systems to Smart Sensors and Actuators.* Weinheim: Wiley-VCH.

Acknowledgment

Professor Vladimir Privman (1955–2018), Director of the Center for Quantum Device Technology and Robert A. Plane Professor of Physics with joint appointments in the Department of Chemistry and Biomolecular Science and Department of Electrical and Computer Engineering (Clarkson University, NY, USA), was a great contributor to the research area of enzyme computing, and this book would not be possible without his work. The majority of the material presented in this book includes contributions by Professor Privman.

His research interests spanned broad areas of advanced technology, including bio-inspired information processing, synthesis of colloids and nanoparticles, kinetics of surface processes at the nanoscale, physics of semiconductor devices, spintronics, quantum computing, statistical mechanics, chemical kinetics, and surface and polymer science.

Professor Privman began earning recognition early in his career, receiving the Petroleum Research Fund Young Investigator Award and the Clarkson University Graham Award for Young Faculty Excellence. He contributed to a wide range of scientific fields and was a lecturer or moderator at national and international conferences every year. He authored/coauthored over 280 research papers, major reviews, and books. He served on numerous boards of scientific journals, and national funding agencies, and received an American Physical Society Outstanding Referee Award. In 2005 he was named a fellow of the American Physical Society, which recognized his fundamental contributions and professional leadership in statistical physics; surface, colloid, and polymer science; and quantum information science. In 2010 he was named an International Academy, Research, and Industry Association (IARIA) fellow.

Over the past 10 years, Professor Privman has been among the key players in the unconventional computing field. Particularly noteworthy are his contributions to the integration of biomolecular computing and actuation, implementation of biochemical logical gates, biomolecular signal processing, networked enzymatic gates with filtering, associative memory based on enzymatic cascades, biochemical logic for drug release, biomolecular filters for signal separation, enzymatic systems for information processing, and digital biosensors. Professor Privman's contributions to quantum computing were in the evaluation of decoherence for quantum computing architectures, modeling of semiconductor spintronics, quantum control, nuclear spin-based memory and logic in quantum hall semiconductors, Hamiltonians for quantum computing, and three-spin XOR gate.

In 2005 Professor Privman edited the Special Issue containing papers from the 2004 IEEE Nanotechnology Council (NTC) Quantum Device Technology Workshop, which was held on 17–21 May 2004, in Clarkson University, Potsdam, NY. The contents of the issue demonstrated breakthroughs in several fields of novel materials and devices, including biochemical logical gates, styrene butadiene rubber nanocomposites, swarms of microscale nanorobots, robots for target therapies, biomolecular motors, magnetoresistive detection of nanoparticles, and self-assembly of quantum dots. In 2017 the *International Journal of Parallel, Emergent and Distributed Systems* (vol. 32, issue 1) published a special issue *Signal processing, biosensing, and computing with bio-inspired and biochemical systems* compiled and edited by Professor Privman. He presented the field of unconventional computing with diverse contributions such as reaction–diffusion chemistry implementation of neural networks, fluidic infrastructure for enzyme-based Boolean logic circuits, architectures of nano-biointerfaces, modeling of enzymatic signal processing, wireless sensor networks with biological cultures, biosensors and memristors in networks of plants, oscillator dynamics of slime mold, insulin biosensor, and biocomputing in forensic analysis.

Professor Privman was highly regarded by his peers and students. He was proud of his trainee's success and advancement and took an active role in mentoring undergraduate, graduate, postdoc, and senior researchers in several departments at Clarkson University. He enjoyed training and collaborating with scientists throughout the United States and internationally. His passing is a great loss to the scientific community.

Professor Vladimir Privman's Works on Biomolecular and Enzymatic Computing

1. A.V. Okhokhonin, S. Domanskyi, Y. Filipov, M. Gamella, A.N. Kozitsina, V. Privman, E. Katz, Biomolecular release from alginate-modified electrode triggered by chemical inputs processed through a biocatalytic cascade – Integration of biomolecular computing and actuation. *Electroanalysis* **2018**, *30*, 426–435.
2. M.L. Wood, S. Domanskyi, V. Privman, Design of high quality chemical XOR gates with noise reduction. *ChemPhysChem* **2017**, *18*, 1773–1781.
3. S. Domanskyi, V. Privman, Modeling and modifying response of biochemical processes for biocomputing and biosensing signal processing. Ch. 3 in: *Advances in Unconventional Computing, Vol. 2: Prototypes, Models and Algorithms*, pp. 61–83, edited by A. Adamatzky, Vol. 23 of Emergence, Complexity and Computation, Springer Nature, Basel, Switzerland, **2017**.
4. V. Privman, Theoretical modeling expressions for networked enzymatic signal processing steps as logic gates optimized by filtering. *Int. J. Parallel Emergent Distrib. Syst.* **2017**, *32*, 30–43.

5. Y. Filipov, S. Domanskyi, M.L. Wood, M. Gamella, V. Privman, E. Katz, Experimental realization of high quality biochemical XOR gate. *ChemPhysChem* **2017**, *18*, 2908–2915.
6. A. Verma, B.E. Fratto, V. Privman, E. Katz, Design of flow systems for improved networking and reduced noise in biomolecular signal processing in biocomputing and biosensing applications. *Sensors* (MDPI) **2016**, *16*, article No. 1042.
7. V. Privman, E. Katz, Can bio-inspired information processing steps be realized as synthetic biochemical processes? *Physica Status Solidi A* **2015**, *212*, 219–228.
8. E. Katz, V. Privman, O. Zavalov, Structure of feed-forward realizations with enzymatic processes. *Proceedings of The Eighth International Conference on Quantum, Nano/Bio, and Micro Technologies (ICQNM 2014)*, ThinkMind Online Publishing, Wilmington, DE, **2014**, pp. 22–27.
9. V. Privman, S. Domanskyi, S. Mailloux, Y. Holade, E. Katz, Kinetic model for a threshold filter in an enzymatic system for bioanalytical and biocomputing applications. *J. Phys. Chem. B* **2014**, *118*, 12435–12443.
10. V. Privman, S. Domanskyi, S. Mailloux, Y. Holade, E. Katz, Kinetic model for a threshold filter in an enzymatic system for bioanalytical and biocomputing applications. *J. Phys. Chem. B* **2014**, *118*, 12435–12443.
11. V. Privman, O. Zavalov, L. Halámková, F. Moseley, J. Halámek, E. Katz, Networked enzymatic logic gates with filtering: New theoretical modeling expressions and their experimental application. *J. Phys. Chem. B* **2013**, *117*, 14928–14939.
12. S. Bakshi, O. Zavalov, J. Halámek, V. Privman, E. Katz, Modularity of biochemical filtering for inducing sigmoid response in both inputs in an enzymatic AND gate. *J. Phys. Chem. B* **2013**, *117*, 9857–9865.
13. K. MacVittie, J. Halámek, V. Privman, E. Katz, A bioinspired associative memory system based on enzymatic cascades. *Chem. Commun.* **2013**, *49*, 6962–6964.
14. V. Privman, B.E. Fratto, O. Zavalov, J. Halámek, E. Katz, Enzymatic AND logic gate with sigmoid response induced by photochemically controlled oxidation of the output. *J. Phys. Chem. B* **2013**, *117*, 7559–7568.
15. O. Zavalov, V. Bocharova, J. Halámek, L. Halámková, S. Korkmaz, M.A. Arugula, S. Chinnapareddy, E. Katz, V. Privman, Two-input enzymatic logic gates made sigmoid by modifications of the biocatalytic reaction cascades. *Int. J. Unconventional Computing* **2012**, *8*, 347–365.
16. V. Bocharova, O. Zavalov, K. MacVittie, M.A. Arugula, N.V. Guz, M.E. Dokukin, J. Halámek, I. Sokolov, V. Privman, E. Katz, Biochemical logic approach to biomarker-activated drug release. *J. Mater. Chem.* **2012**, *22*, 19709–19717.
17. O. Zavalov, V. Bocharova, V. Privman, E. Katz, Enzyme-based logic: OR gate with double-sigmoid filter response. *J. Phys. Chem. B* **2012**, *116*, 9683–9689.
18. V. Bocharova, K. MacVittie, S. Chinnapareddy, J. Halámek, V. Privman, E. Katz, Realization of associative memory in an enzymatic process: Toward biomolecular networks with learning and unlearning functionalities. *J. Phys. Chem. Lett.* **2012**, *3*, 1234–1237.

19. J. Halámek, O. Zavalov, L. Halámková, S. Korkmaz, V. Privman, E. Katz, Enzyme-based logic analysis of biomarkers at physiological concentrations: AND gate with double-sigmoid "filter" response. *J. Phys. Chem. B* **2012**, *116*, 4457–4464.
20. S. Domanskyi, V. Privman, Design of digital response in enzyme-based bioanalytical systems for information processing applications. *J. Phys. Chem. B* **2012**, *116*, 13690–13695.
21. V. Privman, Approaches to control of noise in chemical and biochemical information and signal processing, Ch. 12 in: *Molecular and Supramolecular Information Processing. From Molecular Switches to Logic Systems*, E. Katz (Ed.), Wiley-VCH, Weinheim, **2012**, pp. 281–303.
22. J. Wang, J.R. Windmuller, P. Santosh, M.-C. Chuang, E. Katz, J. Halamek, V. Bocharova, M. Pita, V. Privman, Patent: Enzyme-Logic Biosensing, WO/2011/116151 (September 22, **2011**).
23. V. Privman, Control of noise in chemical and biochemical information processing. *Israel J. Chem.* **2011**, *51*, 118–131.
24. J. Halámek, J. Zhou, L. Halámková, V. Bocharova, V. Privman, J. Wang, E. Katz, Biomolecular filters for improved separation of output signals in enzyme logic systems applied to biomedical analysis. *Anal. Chem.* **2011**, *83*, 8383–8386.
25. J. Halámek, V. Bocharova, M.A. Arugula, G. Strack, V. Privman, E. Katz, Realization and properties of biochemical-computing biocatalytic XOR gate based on enzyme inhibition by a substrate. *J. Phys. Chem. B* **2011**, *115*, 9838–9845.
26. M. Pita, V. Privman, M.A. Arugula, D. Melnikov, V. Bocharova, E. Katz, Towards biochemical filter with sigmoidal response to pH changes: Buffered biocatalytic signal transduction. *PhysChemChemPhys* **2011**, *13*, 4507–4513.
27. V. Privman, Error-control and digitalization concepts for chemical and biomolecular information processing systems. *J. Comput. Theor. Nanosci.* **2011**, *8*, 490–502.
28. V. Pedrosa, D. Melnikov, M. Pita, J. Halámek, V. Privman, A. Simonian, E. Katz, Enzymatic logic gates with noise-reducing sigmoid response. *Int. J. Unconventional Computing* **2010**, *6*, 451–460.
29. V. Privman, J. Halámek, M.A. Arugula, D. Melnikov, V. Bocharova, E. Katz, Biochemical filter with sigmoidal response: Increasing the complexity of biomolecular logic. *J. Phys. Chem. B* **2010**, *114*, 14103–14109.
30. V. Privman, J. Zhou, J. Halámek, E. Katz, Realization and properties of biochemical-computing biocatalytic XOR gate based on signal change. *J. Phys. Chem. B* **2010**, *114*, 13601–13608.
31. D. Melnikov, G. Strack, J. Zhou, J.R. Windmiller, J. Halámek, V. Bocharova, M.-C. Chuang, P. Santhosh, V. Privman, J. Wang, E. Katz, Enzymatic AND logic gates operated under conditions characteristic of biomedical applications. *J. Phys. Chem. B* **2010**, *114*, 12166–12174.
32. E. Katz, V. Privman, J. Wang, Towards biosensing strategies based on biochemical logic systems. *The Fourth International Conference on Quantum, Nano and Micro Technologies (ICQNM 2010).* February 10–16, **2010** - St. Maarten, Netherlands Antilles. Proceedings. pp. 1–9.

33. E. Katz, V. Privman, Enzyme-based logic systems for information processing. *Chem. Soc. Rev.* **2010**, *39*, 1835–1857.
34. V. Privman, Biomolecular computing: Learning through play. *Nature Nanotechnol.* **2010**, *5*, 767–768.
35. V. Privman, V. Pedrosa, D. Melnikov, M. Pita, A. Simonian, E. Katz, Enzymatic AND-gate based on electrode-immobilized glucose-6-phosphate dehydrogenase: Towards digital biosensors and biochemical logic systems with low noise. *Biosens. Bioelectron.* **2009**, *25*, 695–701.
36. M.A. Arugula, J. Halámek, E. Katz, D. Melnikov, M. Pita, V. Privman, G. Strack, Optimization of enzymatic logic gates and networks for noise reduction and stability. *Second International Conference on Advances in Circuits, Electronics and Micro-Electronics*, Proceedings, IEEE Comp. Soc. Publ. (Los Alamitos, California) **2009**, 1–7.
37. D. Melnikov, G. Strack, M. Pita, V. Privman, E. Katz, Analog noise reduction in enzymatic logic gates. *J. Phys. Chem. B* **2009**, *113*, 10472–10479.
38. V. Privman, M.A. Arugula, J. Halámek, M. Pita, E. Katz, Network analysis of biochemical logic for noise reduction and stability: A system of three coupled enzymatic AND gates. *J. Phys. Chem. B* **2009**, *113*, 5301–5310.
39. V. Privman, G. Strack, D. Solenov, M. Pita, E. Katz, Optimization of enzymatic biochemical logic for noise reduction and scalability: How many biocomputing gates can be interconnected in a circuit? *J. Phys. Chem. B* **2008**, *112*, 11777–11784.
40. L. Fedichkin, E. Katz, V. Privman, Error correction and digitalization concepts in biochemical computing. *J. Comput. Theor. Nanoscience* **2008**, *5*, 36–43.

Vladimir Privman

List of Abbreviations

αAmy	α-amylase (enzyme)
βAmy	β-amylase (enzyme)
α-KTG	α-ketoglutaric acid
Δf	oscillation frequency change measured by QCM
λ	wavelength
λ_{max}	wavelength of maximum absorbance in optical spectra
Θ	angle of incident light beam (in SPR measurements)
2-OG	2-oxoglutarate
2-PGA	2-phosphoglyceric acid (or salt form)
3-oxo-C12-HSL	3-oxododecanoyl homoserine lactone (QS signaling molecule)
AA	African American (ethnic origin)
Abs	optical absorbance
ABT	abdominal trauma
ABTS	2,2′-azino-*bis*(3-ethylbenzothiazoline-6-sulfonic acid) (chromogenic substrate used to follow peroxidase activity)
$ABTS_{ox}$	oxidized ABTS (colored product)
Ac	acetic acid
AChE	acetylcholinesterase (enzyme)
AcP	acetyl phosphate
AcidP	acid phosphatase (enzyme)
ADH	alcohol dehydrogenase (enzyme)
ADP	adenosine 5′-diphosphate
AFM	atomic force microscope (microscopy)
Ala	alanine (amino acid)
Ald	acetaldehyde
ALT	alanine transaminase (enzyme)
AMG	amyloglucosidase (enzyme)
AND	AND Boolean logic gate
anti-DNP	anti-dinitrophenyl IgG polyclonal antibody
anti-NT	anti-nitrotyrosine IgG polyclonal antibody
AOx	alcohol oxidase (enzyme)
AP	alkaline phosphatase (enzyme)
APTES	(3-aminopropyl)triethoxysilane (silanizing agent for modification of electrodes and nanoparticles)

ArNHOH	oxidizable hydroxylamine (product of TNT biocatalytic reduction)
ArNO	nitroso compound (product of ArNHOH biocatalytic oxidation)
Asc	ascorbate
ASCII	American Standard Code for Information Interchange
ATM	automated teller machine (as an example of an electronic device with a keypad lock system)
ATP	adenosine 5′-triphosphate
BHQ2	Black Hole Quencher® (fluorescence quencher)
Bo	borrow digit (output signal in a half-subtractor)
BSA	bovine serum albumin
Bu	butyric acid
Bu-O-Et	ethyl butyrate ester
Bu-O-Me	methyl butyrate ester
C	carry digit (output signal in a half-adder)
C4-HSL	*N*-butanoyl-l-homoserine lactone (QS signaling molecule)
CA	Caucasian (ethnic origin)
CA	chronoamperometry
CaM	calmodulin
cAMP	cyclic adenosine monophosphate (a second messenger important in biological processes)
ChOx	choline oxidase (enzyme)
CK	creatine kinase (enzyme)
CoA	coenzyme A
CN	4-chloro-1-naphthol
CNOT	Controlled NOT (reversible logic gate)
CN-ox	CN insoluble oxidized product
CNT(s)	carbon nanotube(s)
Crt	creatine
CrtP	creatine phosphate
CSWAP	Controlled-Swap (logic gate)
D	delay (flip-flop memory)
D	difference digit (output signal in a half-subtractor)
DC	direct current
DCPIP	dichlorophenolindophenol ($DCPIP_{red}$ and $DCPIP_{ox}$ are reduced and oxidized forms of DCPIP, respectively; DCPIP also corresponds to the oxidized form)
DDC	diethyldithiocarbamate (product of DS reduction)
DFG	double Feynman gate (reversible logic gate)
DHA	dehydroascorbate (product of Asc oxidation)
Diaph	diaphorase (enzyme)
dmo-bpy	4,4′-dimethoxy-2,2′-bipyridine (ligand in the redox active complex: $Os(dmo\text{-}bpy)_2Cl$)
DNA	deoxyribonucleic acid
DNAzyme	deoxyribozyme (catalytically active DNA)
DNP	2,4-dinitrophenyl (used as an antigen for anti-DNP)

DNT	2,4-dinitrotoluene
DS	disulfiram
DTT	dithiothreitol
Dz	another abbreviation for DNAzyme
E	potential applied or measured in electrochemical experiments
E°	standard redox potential (derived from electrochemically reversible cyclic voltammogram)
EDC	1-ethyl-3-(3-dimethylaminopropyl)carbodiimide (carbodiimide coupling reagent)
EIS	electrolyte–insulator–semiconductor
ELISA	enzyme-linked immunosorbent assay
EN	enolase (enzyme)
Est	esterase (enzyme)
Et-O-Ac	ethyl acetate ester
EtOH	ethanol
f	oscillation frequency measured with QCM
FAM	fluorescein derivative used for labeling biomolecules
FET	field-effect transistor
FITC	fluorescein isothiocyanate (fluorescent label)
Frc	fructose
G6PDH	glucose 6-phosphate dehydrogenase (enzyme)
GDH	glucose dehydrogenase (enzyme)
Glc	glucose
Glc1P	glucose-1-phosphate
Glc6P	glucose-6-phosphate
Glc6PA	gluconate-6-phosphate acid (product of Glc6P oxidation)
GlcA	gluconic acid (product of glucose oxidation)
Glu	glutamate (amino acid, salt form)
GluOx	glutamate oxidase (enzyme)
GlutOx	glutathione oxidase (enzyme)
GOx	glucose oxidase (enzyme)
GR	glutathione reductase (enzyme)
GSH	glutathione (reduced form)
GSSG	glutathione (dimeric oxidized form)
HEPES	(4-(2-hydroxyethyl)-1-piperazineethanesulfonic acid) (buffer)
HK	hexokinase (enzyme)
HPLC	high-performance liquid chromatography
HRP	horseradish peroxidase (enzyme)
HRP-Ab	antibody labeled with HRP enzyme
HS	hemorrhagic shock
HAS	human serum albumin
i	current density produced by a biofuel cell on an external ohmic resistance
ID	Identity (YES) gate
INHIB	Inhibited Boolean logic gate

Inv	invertase (enzyme)
INV	Inverter (logic element)
I_p	peak current (measured with cyclic voltammetry)
IPTG	isopropyl β-D-thiogalactoside (artificial inducer in cellular regulating processes).
IR	infrared (light)
i_{sc}	short circuit current density produced by a biofuel cell on an external ohmic resistance
ITO	indium tin oxide (electrode)
JK	Jack Kilby (flip-flop memory)
Lac	lactate
LDH	lactate dehydrogenase (enzyme)
LI	liver injury
LOx	lactate oxidase (enzyme)
LSPR	localized surface plasmon resonance
Luc	luciferase (enzyme)
Lucif	luciferin
M13	calmodulin-binding peptide
Maj	majority logic gate
Mal	malate
Malt	maltose
MB	methylene blue (electron transfer mediator operating with GOx); MB_{ox} and MB_{red} are oxidized and reduced forms of MB, respectively
MDH	malate dehydrogenase (enzyme)
MHC I	MHC class I molecules are one of two primary classes of major histocompatibility complex molecules and are found on the cell surface of all nucleated cells in the bodies of jawed vertebrates
Min	minority logic gate
MMP2 and MMP7	matrix metalloproteinases (cancer biomarkers)
MNP(s)	magnetic nanoparticle(s)
MP-11	microperoxidase-11
MPh	maltose phosphorylase (enzyme)
MPAX	methyl paraoxon (acetylcholinesterase inhibitor; model nerve agent)
MWCNT(s)	multiwalled carbon nanotube(s)
NAD^+	nicotinamide adenine dinucleotide (oxidized form)
NADH	nicotinamide adenine dinucleotide (reduced form)
NADH-POx	NADH peroxidase (enzyme)
$NADP^+$	β-nicotinamide adenine dinucleotide phosphate oxidized
NADPH	β-nicotinamide adenine dinucleotide phosphate reduced
NAD(P)H	represent either NADH or NADPH
NAND	NOT–AND Boolean logic gate
NE	norepinephrine (catecholamine hormone neurotransmitter)
NHS	*N*-hydroxysuccinimide

NOR	NOT–OR Boolean logic gate
NOT	Inverted Identity Boolean logic gate
NP(s)	nanoparticle(s)
NRd	nitroreductase (enzyme)
NT	3-nitro-L-tyrosine (used as an antigen for anti-NT)
NXOR	NOT-Exclusive-OR Boolean logic gate
O.D.	optical density (in optical absorbance measurements)
QCM	quartz crystal microbalance
Q-F	oligonucleotide labeled with a fluorescent dye at one end and with a quencher at another end; F is a fluorescent dye; Q is a quencher
QS	quorum sensing
OPH	organophosphorous hydrolase (enzyme)
OR	OR Boolean logic gate
OS	oxidative stress
Q_t	initial (present) state of a flip-flop device
Q_{t+1}	next state of a flip-flop device
Q_{t+2}	next, next state of a flip-flop device
OxAc	oxaloacetate
O/W	oil-in-water Pickering emulsion
Qz6	Quasar 670 (fluorescent dye)
P2VP	poly(2-vinyl pyridine)
P4VP	poly(4-vinyl pyridine)
PAX	paraoxon (acetylcholinesterase inhibitor; model nerve agent)
PB	Prussian blue
PBSE	1-pyrenebutanoic acid succinimidyl ester (heterobifunctional reagent)
P.D.	power density produced by a biofuel cell on an external ohmic resistance
$P.D._{max}$	maximum power density produced by a biofuel cell on an external optimized ohmic resistance
PDH	pyruvate dehydrogenase (enzyme)
PDI	protein disulfide-isomerase (enzyme)
PEI	polyethyleneimine
PEO	poly(ethylene oxide)
PEP	phospho(enol)pyruvic acid (or phosphoenol pyruvate in the form of salt)
Pi	inorganic phosphate
PK	pyruvate kinase (enzyme)
pK_a	acid dissociation constant
PNP	*p*-nitrophenol
PNPP	*p*-nitrophenyl phosphate
POx	pyruvate oxidase (enzyme)
Ppy	polypyrrole
Ppy-ox	polypyrrole oxidized state
Ppy-red	polypyrrole reduced state

PQQ	pyrroloquinoline quinone
PQQ-GDH	PQQ-dependent glucose dehydrogenase (enzyme)
PS	polystyrene
Pyr	pyruvate
R	reset signal
R	reflectance measured by SPR
R	external load resistance connected to a biofuel cell
R_{cell}	ohmic resistance measured in a bulk solution in an electrochemical cell
RE	reference electrode
R_{et}	electron transfer resistance (measured by Faradaic impedance spectroscopy)
RI	radiation injury
RNA	ribonucleic acid
RNS	reactive nitrogen species
ROC	receiver operating characteristic
ROS	reactive oxygen species
S	set signal
S	sum digit (output signal in a half-adder)
SAND	single inversion AND (logic gate equivalent to NOT–AND operation, where inversion NOT is applied to one of the inputs)
SEM	scanning electron microscopy
SPE	screen-printed electrode
SPR	surface plasmon resonance
SR	set/reset (flip-flop memory)
STI	soft tissue injury
SWV	square wave voltammetry
T	toggle (flip-flop memory)
TBI	traumatic brain injury
t_g	gate time (time of reaction after which the gate response is measured)
TMB	3,3′,5,5′-tetramethylbenzidine (chromogenic substrate used to follow peroxidase activity)
TMB_{dox}	TMB double-oxidized product
TMB_{ox}	oxidized colored form of TMB
TMB_{red}	TMB reduced original state (the same as TMB)
TMB_{sox}	TMB single-oxidized product (the same as TMB_{ox})
TNT	trinitrotoluene (explosive)
Tris	2-amino-2-(hydroxymethyl)propane-1,3-diol (buffer)
UV	ultraviolet (light)
Ure	urease (enzyme)
V	voltage produced by a biofuel cell on an external ohmic resistance
V_a	alternative voltage applied between the conducting support and reference electrode of the EIS devise

V_{bias}	constant (bias) voltage applied between the conducting support and reference electrode of the EIS devise
V_{FB}	flat band voltage of the EIS device
V_{oc}	open circuit voltage produced by a biofuel cell
W/O	water-in-oil Pickering emulsion
XOR	Exclusive-OR Boolean logic gate
YES	Identity (ID) Boolean logic gate

1

Introduction

1.1 Motivation and Applications

Exponential development of computing systems based on silicon materials and binary algorithms formulated as the "Moore's law" [1] (Figure 1.1) is coming to the end being limited by further component miniaturization and by the speed of operation. Conceptually novel ideas are needed to break through these limitations. The quest for novel ideas in the information processing has resulted in several exciting directions in the general area of unconventional computing [2–4], including research in quantum computing [5] and biologically inspired molecular computing [6–9]. Molecular computing systems, generally motivated by mimicking natural biological information processing [10, 11], are not necessarily based on biomolecules and could be represented by synthetic molecules with signal-controlled switchable properties. Synthetic molecular systems and nano-species have been designed to mimic the operation of Boolean logic gates and demonstrate basic arithmetic functions and memory units. However, despite progress achieved in assembling synthetic molecular systems performing basic Boolean operations and simple computations, these systems have limited complexity, and further increase of their complexity is very challenging. A new advance in the development of molecular information systems has been achieved with the use of biomolecular species [12] (Figure 1.2) such as DNA/RNA [13–16], oligopeptides [17], proteins [18], enzymes [2, 19, 20], antigens/antibodies [21], and even whole biological cells/organisms [22–24] capable of operating in a biological environment [25], borrowing some ideas from systems biology [26]. The advantage of the biomolecular computing systems is their ability to be integrated in artificially designed complex reacting processes mimicking multistep information processing networks. These systems are still far away from the natural information processing in cells but are already much more complex than pure synthetic molecular systems. In fact, biochemical reactions are at the core of the mechanism of life itself, and therefore one could set rather ambitious expectations for how far can (bio)chemical reaction systems be scaled up in complexity, if not speed, for information processing. While in a long perspective a "biocomputer" might become a reality [27], particularly for some special applications, e.g., for solving complex combinatorial problems [28], potentially promising to have an advantage over silicon-based electronic computers

Enzyme-Based Computing Systems, First Edition. Evgeny Katz.

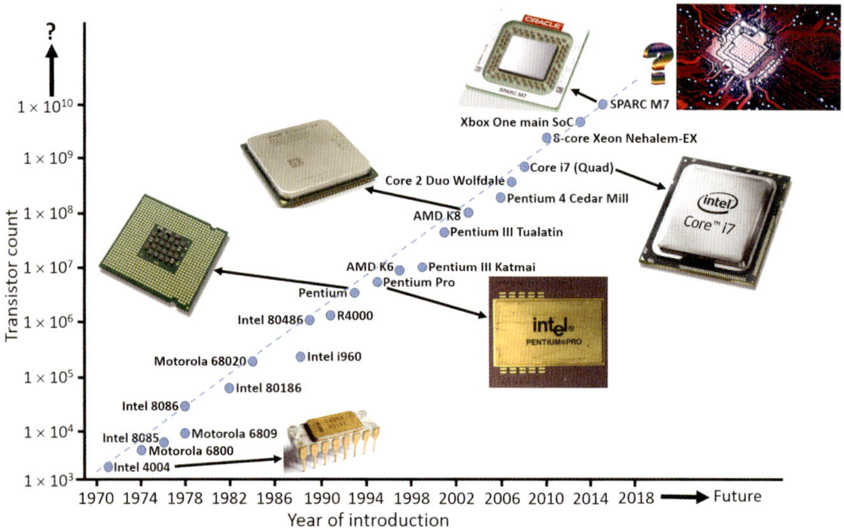

Figure 1.1 Moore's law – exponential increase of transistors on integrated circuit chips. (The plot shown in the figure is based on the data provided by Wikipedia: https://en.wikipedia.org/wiki/Moore%27s_law.) Source: Katz 2018 [2]. Adapted with permission of John Wiley and Sons.

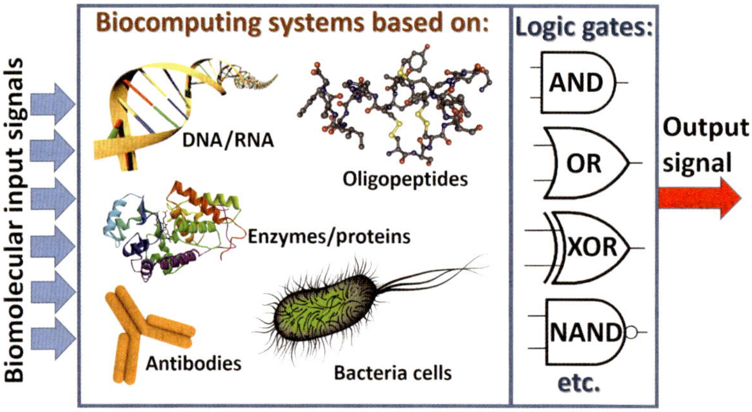

Figure 1.2 Biomolecular computing systems mimicking operation of different Boolean logic gates and circuitries can be based on various species including oligopeptides, enzymes/proteins, DNA/RNA, antibodies, and even whole biological (e.g., microbial) cells. Source: Katz 2018 [2]. Adapted with permission of John Wiley and Sons.

due to parallel computing performed by numerous biomolecular units, the present level of technology does not allow any practical use of biomolecular systems for real computational applications. For achieving any practical result soon, some other applications, different from making a biocomputer, should be considered using the (bio)molecular systems with a limited complexity. One of the immediate possible applications for molecular logic systems is a special kind of biosensing [29–31] where the multiple input signals are logically processed through chemical reactions resulting in YES/NO decisions in the binary (**0**,**1**) format. In this subarea of biomolecular logic systems, practical results are already possible at the present level of the system complexity, particularly for biomedical applications [32, 33]. Overall, the research in molecular/biomolecular information processing, which has been motivated originally to progress unconventional computing applications, is broadly developing to areas not directly related to computing in its narrow definition. This research is bringing us to novel areas in sensing/biosensing [29–31], switchable "smart" materials controlled by logically processed signals [34–36], bioelectronic devices (e.g., biofuel cells) controlled by external signals [37, 38], signal-controlled release processes [39–43], etc.

1.2 Enzyme-Based Logic Gates and Short Logic Circuits

While the major research efforts have been directed to the DNA-based computing systems [10, 13–15], mostly aiming at computing applications in their direct narrow definition [27, 44–50] and expecting acceleration of the computing process due to massively parallel data processing [28, 51], enzyme logic systems [19, 20] received smaller attention since they are less promising for real large-scale computational applications. Growing interest to the enzyme logic systems is based on their activation with physiologically relevant biomolecular signals (metabolites) appearing at physiological concentrations [52–54] allowing low-scale information processing for biomedical applications, such as binary (YES/NO format) biosensing [32, 33], signal-controlled materials, and implantable bioelectronic devices [55].

Enzyme-based logic gates are usually realized through relatively simple enzyme-catalyzed reactions [19, 20] (Figure 1.3). Rapid progress in enzyme-based information processing systems has resulted in the design of biocatalytic cascades mimicking various Boolean logic gates, including AND [52, 56–59], OR [59, 60], NAND [61], NOR [57, 61], CNOT [62], XOR [57, 59, 63–65], INHIB [57, 59], Identity [57], and Inverter [57] gates. In order to digitalize chemical processes, the reacting species considered as logic input signals were applied at two levels of their concentrations: their physical absence (zero concentration) was defined as logic **0** input, while logic **1** input was defined as experimentally optimized and conveniently high concentration, thus allowing significant separation in the produced output signals when inputs **0** and **1** were applied in different combinations. Depending on specific needs set by applications, the input signals were defined as variable concentrations of substrates and/or cofactors reacting with enzymes [57] or different concentrations of the biocatalytic enzymes added to a "soup" of substrates/cofactors being ready to react with the enzymes [59]. The non-variable part of the system was considered as a "machinery" operating with the variable input signals applied in various combinations. Multistep biocatalytic cascades activated by several

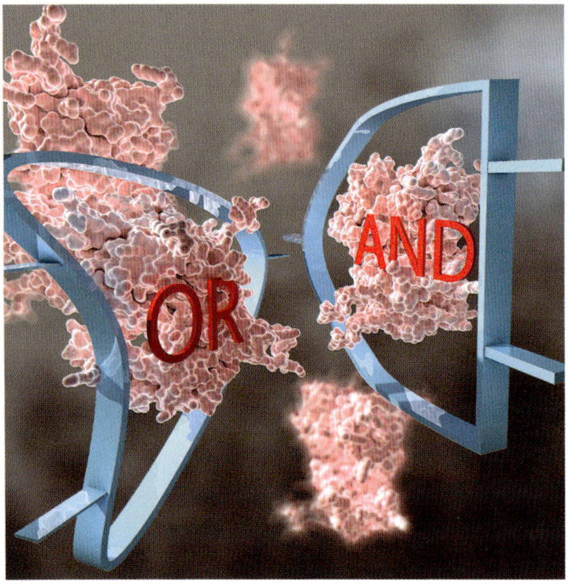

Figure 1.3 Enzyme-based Boolean logic gates – artistic vision. Source: Katz 2018 [2]. Adapted with permission of John Wiley and Sons.

(more than two) input signals have been assembled to mimic logic circuits composed of concatenated logic gates [66–69]. Various reaction cascades have been designed to mimic different combinations of concatenated logic gates; however, they usually do not include more than three or four logic steps. Due to noise formation and amplification through the reaction steps, the number of logic steps is limited, and theoretical estimation limits the system complexity by approximately 10 logic steps (which have never been realized experimentally in those enzyme-biocatalyzed reactions) [70]. Complex branched biocatalytic reactions realized in flow cell systems have been used to mimic operation of reversible logic gates, such as Feynman gate, Double Feynman gate, Toffoli gate, Peres gate, and Fredkin gate [71–74].

The following chapters present different logic systems based on the enzyme reactions, their optimization, and applications. While the designed systems demonstrated many different logic/computing processes, their operation provided only low-scale information processing, which is not sufficient for building a biomolecular computer. However, the designed systems have found important applications in various signal-controlled bioelectronic devices, biosensors, and stimuli-responsive materials. The research in the biomolecular computing, particularly using enzyme reactions for information processing, motivated initially by computational goals and expected to compete with silicon-based microelectronics, finally moved to signal-switchable devices processing a few signals in limited complexity processes. Therefore, the results obtained in the research area represent limited interest for pure computational applications but offer highly promising applications in bioelectronics, particularly operating in biological environment and being adaptive to biological processes.

References

1 Moore, G.E. (1998). *Proc. IEEE* 86: 82–85.
2 Katz, E. (2019). *ChemPhysChem* 20: 9–22.
3 Calude, C.S., Costa, J.F., Dershowitz, N. et al. (eds.) (2009). Unconventional computation. In: *Lecture Notes in Computer Science*, vol. 5715. Berlin: Springer.
4 Adamatzky, A. (ed.) (2017). *Advances in Unconventional Computing*, in series: Emergence, Complexity and Computation, 2 volumes. Switzerland: Springer.
5 Mermin, N.D. (2007). *Quantum Computer Science: An Introduction*. Cambridge: Cambridge University Press.
6 Szacilowski, K. (2012). *Infochemistry – Information Processing at the Nanoscale*. Chichester: Wiley.
7 de Silva, A.P. (2013). *Molecular Logic-Based Computation*. Cambridge: Royal Society of Chemistry.
8 Katz, E. (ed.) (2012). *Molecular and Supramolecular Information Processing – From Molecular Switches to Logic Systems*. Weinheim: Wiley-VCH.
9 Sienko, T., Adamatzky, A., Rambidi, N.G., and Conrad, M. (eds.) (2003). *Molecular Computing*. Cambridge, MA: MIT Press, USA.
10 Spitzer, N.C. and Sejnowski, T.J. (1997). *Science* 277: 1060–1061.
11 Kampfner, R.R. (1989). *BioSystems* 22: 223–230.
12 Katz, E. (ed.) (2012). *Biomolecular Computing – From Logic Systems to Smart Sensors and Actuators*. Weinheim: Wiley-VCH.
13 Stojanovic, M.N., Stefanovic, D., and Rudchenko, S. (2014). *Acc. Chem. Res.* 47: 1845–1852.
14 Stojanovic, M.N. and Stefanovic, D. (2011). *J. Comput. Theor. Nanosci.* 8: 434–440.
15 Ezziane, Z. (2006). *Nanotechnology* 17: R27–R39.
16 Xie, Z., Wroblewska, L., Prochazka, L. et al. (2011). *Science* 333: 1307–1311.
17 Ashkenasy, G., Dadon, Z., Alesebi, S. et al. (2011). *Isr. J. Chem.* 51: 106–117.
18 Unger, R. and Moult, J. (2006). *Proteins* 63: 53–64.
19 Katz, E. and Privman, V. (2010). *Chem. Soc. Rev.* 39: 1835–1857.
20 Katz, E. (2015). *Curr. Opin. Biotechnol.* 34: 202–208.
21 Halámek, J., Tam, T.K., Chinnapareddy, S. et al. (2010). *J. Phys. Chem. Lett.* 1: 973–977.
22 Rinaudo, K., Bleris, L., Maddamsetti, R. et al. (2007). *Nature Biotechnol.* 25: 795–801.
23 Arugula, M.A., Shroff, N., Katz, E., and He, Z. (2012). *Chem. Commun.* 48: 10174–10176.
24 Adamatzky, A. (2010). *Physarum Machines – Computers from Slime Mould*. London: World Scientific.
25 Kahan, M., Gil, B., Adar, R., and Shapiro, E. (2008). *Physica D* 237: 1165–1172.

26 Alon, U. (2006). *An Introduction to Systems Biology: Design Principles of Biological Circuits*. Boca Raton: Chapman & Hall/CRC.
27 Benenson, Y. (2009). *Mol. Biosyst.* 5: 675–685.
28 Adleman, L.M. (1994). *Science* 266: 1021–1024.
29 Katz, E., Wang, J., Privman, M., and Halámek, J. (2012). *Anal. Chem.* 84: 5463–5469.
30 Wang, J. and Katz, E. (2011). *Isr. J. Chem.* 51: 141–150.
31 Wang, J. and Katz, E. (2010). *Anal. Bioanal. Chem.* 398: 1591–1603.
32 Halámková, L., Halámek, J., Bocharova, V. et al. (2012). *Analyst* 137: 1768–1770.
33 Halámek, J., Windmiller, J.R., Zhou, J. et al. (2010). *Analyst* 135: 2249–2259.
34 Minko, S., Katz, E., Motornov, M. et al. (2011). *J. Comput. Theor. Nanoscience* 8: 356–364.
35 Tokarev, I., Gopishetty, V., Zhou, J. et al. (2009). *ACS Appl. Mater. Interfaces* 1: 532–536.
36 Pita, M., Minko, S., and Katz, E. (2009). *J. Mater. Sci.: Materials in Medicine* 20: 457–462.
37 Katz, E. and Minko, S. (2015). *Chem. Commun.* 51: 3493–3500.
38 Katz, E. (2010). *Electroanalysis* 22: 744–756.
39 Okhokhonin, A.V., Domanskyi, S., Filipov, Y. et al. (2018). *Electroanalysis* 30: 426–435.
40 Filipov, Y., Gamella, M., and Katz, E. (2018). *Electroanalysis* 30: 1281–1286.
41 Gamella, M., Privman, M., Bakshi, S. et al. (2017). *ChemPhysChem* 18: 1811–1821.
42 Gamella, M., Zakharchenko, A., Guz, N. et al. (2017). *Electroanalysis* 29: 398–408.
43 Katz, E., Pingarrón, J.M., Mailloux, S. et al. (2015). *J. Phys. Chem. Lett.* 6: 1340–1347.
44 Eghdami, H. and Darehmiraki, M. (2012). *Artificial Intelligence Rev.* 38: 223–235.
45 Xu, J. and Tan, G.J. (2007). *J. Comput. Theor. Nanosci.* 4: 1219–1230.
46 Zhang, Z.Z., Fan, C.H., and He, L. (2005). *Curr. Nanosci.* 1: 91–95.
47 Sarkar, M., Ghosal, P., and Mohanty, S.P. (2017). *IEEE Trans. Nanobiosci.* 16: 383–399.
48 Sarkar, M., Ghosal, P., and Mohanty, S.P. (2017). *Natural Comput.* 16: 463–472.
49 Gang, F. and Yan, Z. (2011). *J. Comput. Theor. Nanosci.* 8: 274–281.
50 Su, X.P. and Smith, L.M. (2004). *Nucleic Acids Res.* 32: 3115–3123.
51 Reif, J.H. (1999). *Algorithmica* 25: 142–175.
52 Halámek, J., Zavalov, O., Halámková, L. et al. (2012). *J. Phys. Chem. B* 116: 4457–4464.
53 Katz, E., Bocharova, V., and Privman, M. (2012). *J. Mater. Chem.* 22: 8171–8178.
54 Privman, M., Tam, T.K., Bocharova, V. et al. (2011). *ACS Appl. Mater. Interfaces* 3: 1620–1623.
55 Katz, E. (ed.) (2014). *Implantable Bioelectronics - Devices, Materials and Applications*. Weinheim: Wiley-VCH.

56 Bakshi, S., Zavalov, O., Halámek, J. et al. (2013). *J. Phys. Chem. B* 117: 9857–9865.
57 Baron, R., Lioubashevski, O., Katz, E. et al. (2006). *J. Phys. Chem. A* 110: 8548–8553.
58 Privman, V., Fratto, B.E., Zavalov, O. et al. (2013). *J. Phys. Chem. B* 117: 7559–7568.
59 Strack, G., Pita, M., Ornatska, M., and Katz, E. (2008). *ChemBioChem* 9: 1260–1266.
60 Zavalov, O., Bocharova, V., Privman, V., and Katz, E. (2012). *J. Phys. Chem. B* 116: 9683–9689.
61 Zhou, J., Arugula, M.A., Halámek, J. et al. (2009). *J. Phys. Chem. B* 113: 16065–16070.
62 Moseley, F., Halámek, J., Kramer, F. et al. (2014). *Analyst* 139: 1839–1842.
63 Halámek, J., Bocharova, V., Arugula, M.A. et al. (2011). *J. Phys. Chem. B* 115: 9838–9845.
64 Privman, V., Zhou, J., Halámek, J., and Katz, E. (2010). *J. Phys. Chem. B* 114: 13601–13608.
65 Filipov, Y., Domanskyi, S., Wood, M.L. et al. (2017). *ChemPhysChem* 18: 2908–2915.
66 Niazov, T., Baron, R., Katz, E. et al. (2006). *Proc. Natl. Acad. USA* 103: 17160–17163.
67 Privman, V., Arugula, M.A., Halámek, J. et al. (2009). *J. Phys. Chem. B* 113: 5301–5310.
68 Privman, V., Zavalov, O., Halámková, L. et al. (2013). *J. Phys. Chem. B* 117: 14928–14939.
69 Strack, G., Ornatska, M., Pita, M., and Katz, E. (2008). *J. Am. Chem. Soc.* 130: 4234–4235.
70 Privman, V., Strack, G., Solenov, D. et al. (2008). *J. Phys. Chem. B* 112: 11777–11784.
71 Fratto, B.E. and Katz, E. (2016). *ChemPhysChem* 17: 1046–1053.
72 Katz, E. and Fratto, B.E. (2017). Enzyme-based reversible logic gates operated in flow cells. In: *Advances in Unconventional Computing*, vol. 2, Chapter 2 (ed. A. Adamatzky), 29–59. Springer.
73 Fratto, B.E. and Katz, E. (2015). *ChemPhysChem* 16: 1405–1415.
74 Fratto, B.E., Guz, N., and Katz, E. (2015). *Parallel Processing Letters* 25 (01): 1540001.

2

Boolean Logic Gates Realized with Enzyme-Catalyzed Reactions: Unusual Look at Usual Chemical Reactions

2.1 General Introduction and Definitions

Chemical and biochemical reactions are not binary processes by their nature; thus in order to use them for mimicking binary computing processes, including realization of Boolean logic gates, they should be processed in a special way to allow binary operations. This artificial way to treat chemical processes has some similarity with electronics, where the electrical parameters (voltage and current) are also not naturally binary, but they can be processed as binary parameters [1]. The electronic parameters can be artificially converted to binary values by assigning **0** and **1** binary values to low and high electronic parameters. In chemical processes mimicking the electronics, similar assignments can be used considering low and high concentrations of chemical species as **0** and **1** binary values. This is quite easy for concentrations of (bio)chemical reacting species applied to model reacting system. Indeed, in the simplest approach, the initial reagents can be applied at physically zero concentration (meaning their absence) and experimentally convenient high concentrations selected to have reasonable reaction time scales and produced signal intensities. The situation when the reacting species are absent in the system is usually considered as the logic **0** value for the input signals. The experimentally optimized high concentrations corresponding to the logic input **1** can vary depending on the reaction kind and methods used to analyze the chemical changes produced. For this way to define the input signals, the output signals (the produced chemicals) are set by the reactions proceeding in the system. In most of cases the output **0** can be physically zero concentration since the reactions can be fully inhibited in the absence of the initial reactants. The output **1** can be optimized to have significant concentration above zero by selecting appropriate initial concentrations for the input reagents. Overall, in model systems, where the input signals are fully controlled by the researcher, the output **0** and **1** signals can be significantly different, thus allowing their easy experimental discrimination.

In some specific applications (e.g., in biomedical/biosensor usage), the low and high concentrations of the initial reagents corresponding to the logic values **0** and **1**, respectively, can be set by natural reasons. For example, the logic **0** and **1** inputs can be defined as normal physiological and abnormal pathological concentrations of the reacting species [2–4]. In this case the gap separating the logic inputs **0** and

Enzyme-Based Computing Systems, First Edition. Evgeny Katz.

1 may not be large, thus resulting in the output signals with a small difference. This can result in complications for their binary discrimination. This kind of bioanalytical logic systems is overviewed in Chapter 11. The present chapter of the book concentrates on examples of the most frequently designed model systems performing Boolean logic operations without pretending to have practical applications and being only illustrations for the biochemical realization of logic systems. For simplicity, the example model systems operate with the logic input **0** defined as the zero concentration of the corresponding species, as explained before.

While in general the (bio)chemical logic systems can be designed using various molecular species [5], the present chapter is concentrated on the enzyme-catalyzed reactions performing Boolean logic operations [6, 7]. The example systems operate with the binary (**0**,**1**) variable low molecular species (substrates, cofactors), while the biocatalytic species (i.e., enzymes) and some auxiliary reagents (promoters, salts, buffers, etc.) are preserved at their constant (non-variable) concentrations, thus being considered as the "machinery" part of the logic gates (Figure 2.1A). It should be noted that the "constant" concentrations of the "machinery" components correspond to their initial concentrations, while some of them (particularly for co-substrates) can change in course of the

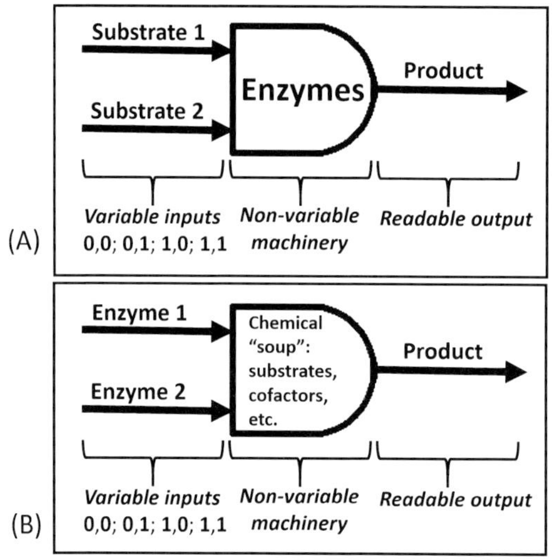

Figure 2.1 (A) The enzyme logic system with the non-variable "machinery" part composed of enzymes and the variable input signals represented by the corresponding substrates converted to the product through the enzyme-catalyzed process. The final product represents the output produced according to the logic "program" implemented in the biocatalytic system. (B) The enzyme logic system with the non-variable "machinery" part composed of a "soup" of substrates and other required species and the variable input signals represented by enzymes added to the mixed chemical composition. The enzyme-catalyzed reaction results in the product that represents the output produced according to the logic "program" implemented in the biocatalytic system. The present schematically shown examples operate with two input signals and one output signal produced in the course of the reaction process. In general, the number of the input and output signals may vary. Source: Katz 2018 [7]. Adapted with permission of John Wiley and Sons.

reaction processes. In the most studied experimental approaches, the enzyme systems carrying out Boolean logic operations were used in solutions, performing simple biocatalytic reactions [6, 7]. Recently, experiments have been reported with enzymes as the "gate machinery" immobilized on solid supports [8, 9] or at interfaces [10, 11] allowing their separation from the reacting species. It should be noted that this definition of the logic inputs and machinery part can be changed. The enzymes can operate as the logically variable inputs, and all other reactants (the "soup" of all species required for the biocatalytic reactions) can be considered as the non-variable machinery part [12] (Figure 2.1B). This approach might be particularly important when the enzyme inputs originate from natural sources, being, for example, biomarkers of some biological processes [13–15] (see Chapter 11 for details).

2.2 Fundamental Boolean Logic Operations Mimicked with Enzyme-Catalyzed Reactions

Complex logic networks performing sophisticated logic operations are described in Chapters 4 and 5. The present section concentrates on fundamental Boolean logic operations, processing two input signals and producing one output signal readable by optical and electrochemical means: Identity (YES) gate, Inverted Identity (NOT) gate, OR gate, NOR gate, XOR gate, NXOR gate, AND gate, NAND gate, and INHIB gate. The overviewed systems are only examples, and different enzyme reactions can be used to mimic the same logic operations [6, 7, 16, 17].

2.2.1 Identity (YES) Gate

This logic gate is the most trivial and the easiest in the chemical realization. Actually, any chemical reaction resulting in a chemical product appearing in the presence of a corresponding original substrate can be considered as the YES Boolean function. In the biochemical, enzyme-based, realization, the reaction should be catalyzed by an enzyme to yield a product in the presence of a substrate. Many different enzymatic reactions were reported for mimicking this simple logic operation [6, 7, 17]. For practical reasons, the product should be conveniently detected with a simple analytical tool, possibly optically or electrochemically. If other analytical methods are used [18], the possible reacting systems can be significantly broadened. The YES logic system exemplified in Figure 2.2A is based on the biocatalytic reduction of NAD^+ to NADH in the presence of lactate (Lac) as the reductive substrate. The reaction is catalyzed by lactate dehydrogenase (LDH). The biocatalytic production of NADH results in the optical absorbance increase at $\lambda = 340$ nm (absorbance maximum of NADH [19]), which is considered as the output signal (Figure 2.2B). The absorbance increase above a certain threshold is defined as the output signal **1**, which appears only in the presence of Lac (input **1**). Otherwise, in the absence of Lac (input **0**), the reaction cannot proceed, and the original NAD^+ is preserved with

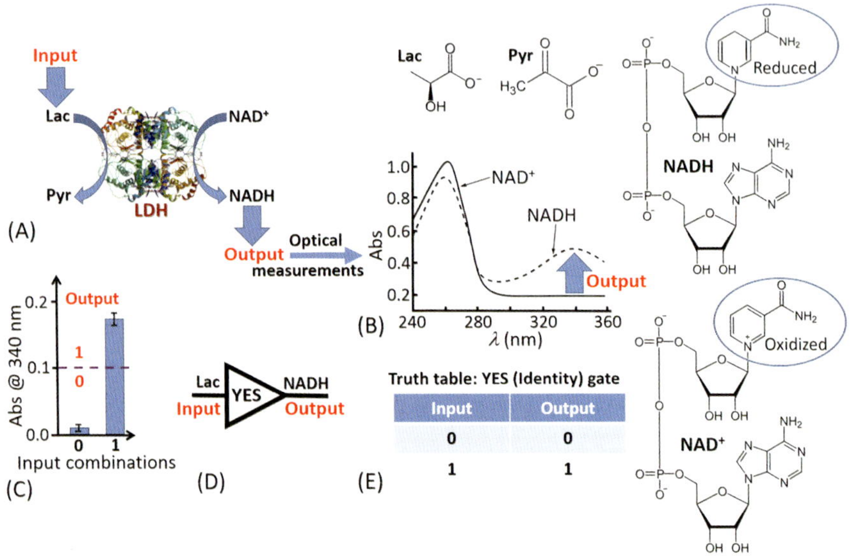

Input	Output
0	0
1	1

Figure 2.2 (A) The biocatalytic reaction mimicking the Identity (YES) gate. The NADH reaction product appears always when lactate (Lac) is added to the reaction solution. (B) The optical absorbance changes corresponding to the reduction of NAD^+ and formation of NADH in the course of the reaction catalyzed by lactate dehydrogenase (LDH). Note that absorbance at $\lambda = 340$ nm corresponding to NADH is increasing in the course of the reaction. (C) The bar chart showing the logic outputs **0** and **1** corresponding to the input signals **0** and **1** applied to the system. The dashed line separates the high and low absorbance values corresponding to **1** and **0** outputs. (D) The scheme of the Identity (YES) gate activated by Lac and producing NADH in response to the input signal. (E) The truth table corresponding to the YES gate. The Lac, pyruvate (Pyr), NAD^+, and NADH structures are shown at the right. Note that images of enzymes used in the schemes (here and in all figures in the book) may not correspond exactly to the enzymes used in the discussed systems – the images are used for illustrating purposes only. Source: Katz 2018 [7]. Adapted with permission of John Wiley and Sons.

a small absorbance at $\lambda = 340$ nm (output **0**) (Figure 2.2C). Therefore, the biocatalytic process resembles the Boolean YES (Identity) logic gate (Figure 2.2D,E). While the considered process is a standard bioanalytical assay, similar reactions mimicking the YES logic gate are highly important parts of very sophisticated logic systems, e.g., in realization of logically reversible Feynman gate, Double Feynman gate, Toffoli gate, and Peres gate [20–22] (see Chapter 5). These gates (including the YES part) have been utilized for transduction of logic signals from enzyme-based to DNA-based computing systems [23]. Overall, despite its simplicity, the YES gate should be considered as the important part of biocomputing systems.

2.2.2 Inverted Identity (NOT) Gate

The Inverted Identity (NOT) Boolean gate is essentially the same as the YES gate, but with the inverted output signal, thus producing the output **1** in response to the input **0** and *vice versa* (Figure 2.3). The NOT process can be realized with various enzymatic reactions [6, 7, 17] and the difference from the YES gate

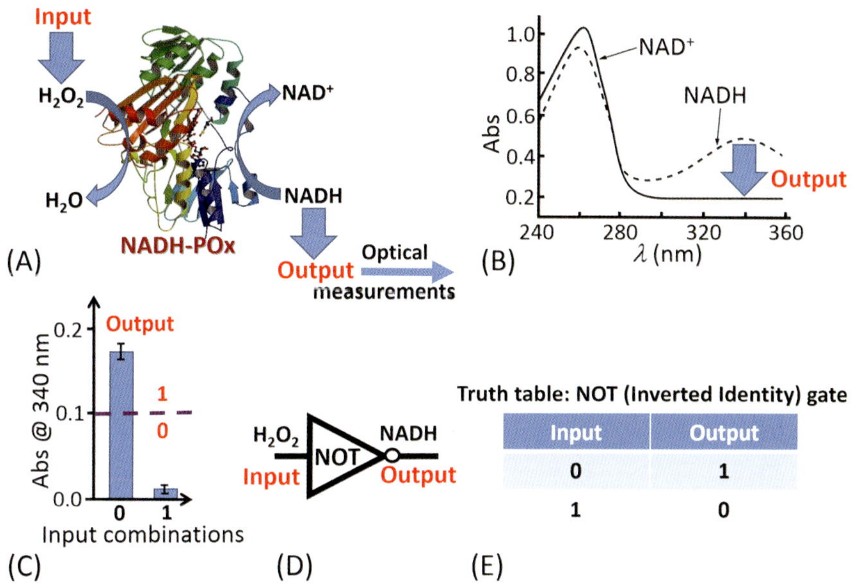

Figure 2.3 (A) The biocatalytic reaction mimicking the Inverted Identity (NOT) gate. The NADH consumption is observed always when H_2O_2 is added to the reaction solution. (B) The optical absorbance changes corresponding to the oxidation of NADH and formation of NAD^+ in the course of the reaction catalyzed by NADH peroxidase (NADH-POx). Note that absorbance at $\lambda = 340$ nm corresponding to NADH is decreasing in the course of the reaction. (C) The bar chart showing the logic outputs **0** and **1** corresponding to the input signals **1** and **0** applied to the system. The dashed line separates the high and low absorbance values corresponding to **1** and **0** outputs. (D) The scheme of the Inverted Identity (NOT) gate activated by H_2O_2 and consuming NADH in response to the input signal. (E) The truth table corresponding to the NOT gate. Source: Katz 2018 [7]. Adapted with permission of John Wiley and Sons.

might be only in the definition of the output signal. The system exemplified in Figure 2.3A results in the consumption of NADH (compared with Figure 2.2A where NADH is produced). When the output signal is defined as the absorbance at $\lambda = 340$ nm (corresponding to the NADH concentration), the output signal **1** appears in response to the input signal **0** (the absence of the oxidative substrate). On the other hand, in the presence of H_2O_2 (input **1**), NADH is oxidized to NAD^+ in the reaction catalyzed by NADH peroxidase (NADH-POx), thus resulting in the absorbance decrease (output **0**) (Figure 2.3B). The process resembles the logic function NOT (Figure 2.3C–E). This logic function, despite its simplicity, is a very important part of various biocomputing systems, e.g., operating as an inverter in a half-subtractor [17, 24].

2.2.3 OR Gate

The Boolean OR gate can be easily realized using two biocatalytic reactions proceeding in parallel and resulting in the same chemical product [6, 7, 17]. Figure 2.4A shows two biocatalytic reactions producing NADH that can be

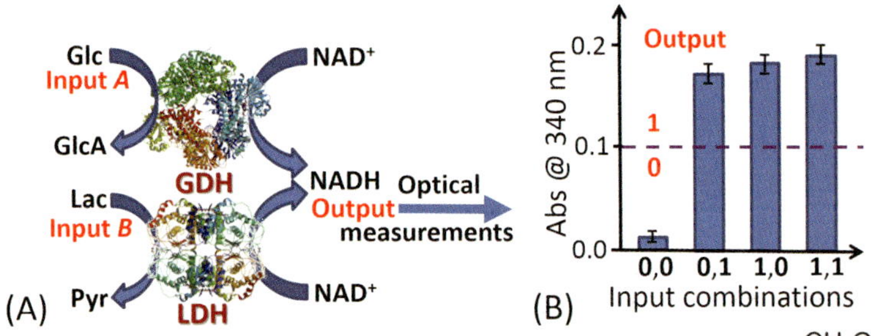

Truth table: OR gate

Input *A*	Input *B*	Output
0	0	0
0	1	1
1	0	1
1	1	1

(C)

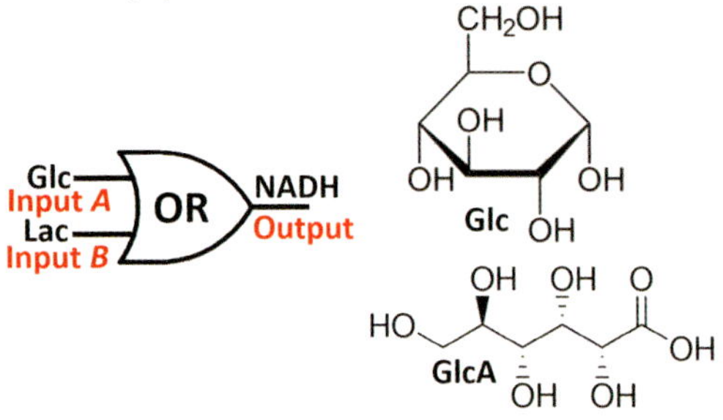

(D)

Figure 2.4 (A) Two parallel reactions catalyzed by glucose dehydrogenase (GDH) and lactate dehydrogenase (LDH) mimicking the OR logic gate. The NADH production is observed when either or both biocatalytic reactions activated. (B) The bar chart showing the logic outputs **0** and **1** (measured as the optical absorbance corresponding to the produced NADH) upon application of the input signals (Glc and Lac) in four combinations (**0**,**0**; **0**,**1**; **1**,**0**; **1**,**1**). The dashed line separates the high and low absorbance values corresponding to **1** and **0** outputs. (C) The truth table corresponding to the OR gate. (D) The scheme of the OR gate activated by Glc and Lac, then producing NADH in response to the input signals. The Glc and GlcA (gluconic acid, product of glucose oxidation) structures are shown at the right. Pyruvate (Pyr) is the product of Lac biocatalytic oxidation. Source: Katz 2018 [7]. Adapted with permission of John Wiley and Sons.

analyzed by measuring optical absorbance at $\lambda = 340$ nm. One reaction catalyzed by glucose dehydrogenase (GDH) reduces NAD^+ to NADH in the presence of glucose (Glc), Input *A*, while the second reaction catalyzed by lactate dehydrogenase (LDH) results in the same product in the presence of lactate (Lac), Input *B*. The optical absorbance corresponding to NADH produced increases in the presence of either or both reducing substrates (input combinations **0**,**1**; **1**,**0**; and **1**,**1**). The optical absorbance is defined as the output signal. The biocatalytic system should be optimized to yield similar absorbance values in the presence of either or both reducing substrates. The only situation when the absorbance does not increase is the absence of both reducing substrates (Glc and Lac, input combination **0**,**0**) (Figure 2.4B). The biocatalytic system based on two parallel reactions resembles the Boolean OR logic gate (Figure 2.4C,D). It should be noted that the enzyme-based OR logic gates are among the most frequently reported systems [6, 7, 25, 26], probably because of their simplicity. The OR gates can find various applications, e.g., in bioanalytical systems operating in YES/NO mode [27].

2.2.4 NOR Gate

The Boolean NOR logic gate is essentially the same as the OR gate, but with the inverted output signal, thus producing the output **1** in response to the absence of both input reagents (**0,0** input combination) and the output **0** in the presence of either or both inputs (**0,1**; **1,0**; and **1,1** combination). This logic response can be obtained when both input reagents inhibit the biocatalytic reaction, which is running only in their absence. Such design of the NOR logic gate is obviously possible, but not easy for realization. The largest disadvantage of this approach is the difficulty to produce the inhibiting species through other biocatalytic reactions, thus making incorporation of the NOR gate in a multi-gate logic network very challenging. On the other hand, another approach based on consumption of the output reagent rather than its production is very simple and convenient for integrating the NOR gate in a series of concatenated logic gates. Figure 2.5A shows two biocatalytic reactions operating in parallel and both resulting in consumption of NADPH (the cofactor similar to NADH but with an additional phosphate group). One reaction catalyzed by glutathione reductase (GR) and activated with glutathione oxidized (GSSG), Input *A*, resulted in oxidation of NADPH and decrease of the corresponding absorbance at $\lambda = 340$ nm. The second reaction catalyzed by diaphorase (Diaph) and activated with $[Fe(CN)_6]^{3-}$ resulted in the same output. Thus, the optical absorbance, being the final measurable output signal, was decreased (output **0**) when either or both input reagents were present (input signal combinations **0,1**; **1,0**; and **1,1**) and was preserved at the original high value (output **1**) when both input reagents were absent (input signal combination **0,0**) (Figure 2.5B). The logic function upon operation of this biocatalytic system resembles the Boolean NOR logic gate (Figure 2.5C,D). Most importantly, the present formulation of the biocatalytic NOR gate allows its easy integration in complex logic networks [28], e.g., operating as a multiplexer, and its use in biosensor-alert devices [29]. It should be also noted that NOR logic gate is a universal gate, which can be used to assemble logic circuits performing all other logic operations [30]. While this is mostly applicable to electronics but hardly possible in chemical realization because of the complexity of this approach, it is still a reason to attract interest to this logic gate.

2.2.5 XOR Gate

The Boolean XOR logic gate is one of the most important logic elements needed to design complex logic systems performing reversible logic operations [20–22] and arithmetic functions [24] (see also Chapters 5 and 9). Because of its importance, the XOR gate was realized in many different enzyme-based systems [17, 31–33]. Despite the high interest to the XOR gate, its realization in biocatalytic system is difficult and usually includes an "artificial" assumption – the output signal is frequently considered as the absolute value of the generated signal. This approach allows the easy realization of the XOR gate using two biocatalytic reactions proceeding in the opposite directions. Figure 2.6A shows an example system operating as a XOR gate and considering absolute value of the optical absorbance changes produced in the system as the output signal. One

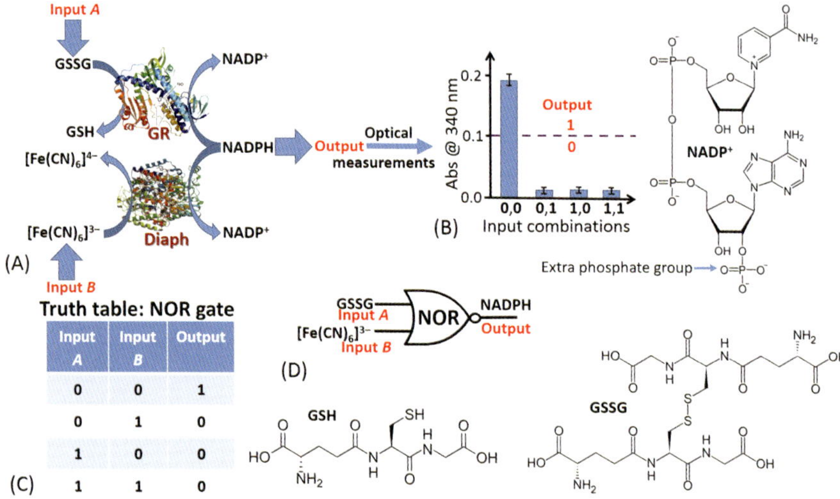

Input A	Input B	Output
0	0	1
0	1	0
1	0	0
1	1	0

Figure 2.5 (A) Two parallel reactions catalyzed by glutathione reductase (GR) and diaphorase (Diaph) mimicking the NOR logic gate. The NADPH consumption is observed when either or both biocatalytic reactions are activated. (B) The bar chart showing the logic outputs **0** and **1** (measured as the optical absorbance corresponding to the residual amount of NADPH) upon application of the input signals, glutathione oxidized (GSSG) and $[Fe(CN)_6]^{3-}$, in four combinations (**0**,**0**; **0**,**1**; **1**,**0**; **1**,**1**). The dashed line separates the high and low absorbance values corresponding to **1** and **0** outputs. (C) The truth table corresponding to the NOR gate. (D) The scheme of the NOR gate activated by GSSG and $[Fe(CN)_6]^{3-}$, then consuming NADPH in response to the input signals. The structures of GSH (reduced glutathione), GSSG (oxidized glutathione), and $NADP^+$ (oxidized form of the NADPH cofactor) are shown at the right. Source: Katz 2018 [7]. Adapted with permission of John Wiley and Sons.

reaction catalyzed by NADH peroxidase (NADH-POx) and activated by H_2O_2 (oxidative species; Input *A*) resulted in oxidation of NADH, thus decreasing its optical absorbance at $\lambda = 340$ nm. The second reaction catalyzed by lactate dehydrogenase (LDH) and activated by lactate (Lac) (reductive species, Input *B*) resulted in reduction of NAD^+, thus, yielding NADH and increasing its absorbance at $\lambda = 340$ nm. The optical absorbance was not changed (output **0**) in the absence of both reactive species (**0**,**0** input combination). Application of either input species (input combinations **0**,**1** and **1**,**0**) activated one of the reactions, thus increasing or decreasing the NADH concentration and its absorbance, respectively. Despite the fact that the chemical changes and the absorbance changes are different (opposite), both results can be considered as the same output **1** if the absolute value of the optical change is defined as the output signal [17, 32] (Figure 2.6B). The most complicated gate operation, which is critically important for the XOR logic operation, is a low response (output signal **0**) to the both inputs applied (**1**,**1** input combination). The both input concentrations should be optimized to compensate the chemical changes, thus keeping the NADH concentration and its absorbance almost unchanged. Overall, the balanced input concentrations (**0**,**0** and **1**,**1**) should not produce any changes in the system, while their unbalanced concentrations (**0**,**1**

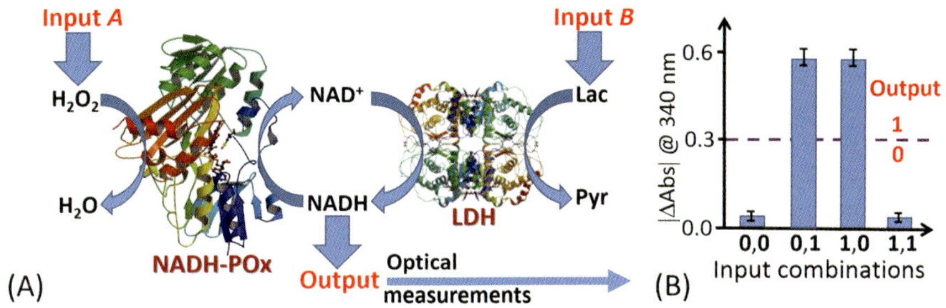

Truth table: XOR gate

Input *A*	Input *B*	Output
0	0	0
0	1	1
1	0	1
1	1	0

(C)

(D)

Figure 2.6 (A) Two reactions catalyzed by NADH peroxidase (NADH-POx) and lactate dehydrogenase (LDH) performing oxidation/reduction of NADH/NAD^+ mimicking an XOR logic gate. The output signal generated by the logic gate is the absolute value of the optical absorbance change measured at $\lambda = 340$ nm corresponding to the NADH concentration. (B) The bar chart showing the logic outputs **0** and **1** (measured as the absolute value of the optical absorbance change) upon application of the input signals H_2O_2 and lactate (Lac) in four combinations (**0**,**0**; **0**,**1**; **1**,**0**; **1**,**1**). The dashed line separates the high and low absorbance values corresponding to **1** and **0** outputs. (C) The truth table corresponding to the XOR gate. (D) The scheme of the XOR gate activated by H_2O_2 and Lac. Source: Katz 2018 [7]. Adapted with permission of John Wiley and Sons.

and **1**,**0**) should drive the reaction away from the original state, increasing or decreasing the absorbance changes – both changes being considered as the same output regardless the change direction. Overall, the biocatalytic cascade shown schematically in Figure 2.6A can mimic the logic XOR operation (Figure 2.6C,D) if Inputs *A* and *B* are properly optimized. The "artificial" assigning of the absolute value of the signal generated as the output signal has advantages and disadvantages. The advantage of this approach is the simplicity of the XOR gate realization, which can be realized using many reversible transformations driven by different enzymes in the opposite directions. The disadvantage is the limitation of the XOR gate use in concatenated logic gate networks (note that the same output signal **1** corresponds to increasing concentrations of chemically different species, NAD^+ and NADH in the present example). Indeed, the absolute value changes can be considered as the final signal in a cascade of reactions mimicking a network of concatenated logic gates; however, this gate cannot be extended to any other logic operation since two different chemical species (e.g., NAD^+ and NADH) cannot operate chemically in the same way even the logic output signal for them considered to be identical.

2.2.6 NXOR Gate

The Boolean NXOR gate (XOR with the inverted output signals) is a rather exotic system rarely realized in biocatalytic systems [34]. Indeed, the output signal **1** should be generated by the NXOR gate in the absence of any input (input combination **0**,**0**) and in the presence of both inputs (input combination **1**,**1**) (Figure 2.7F), which means that the output signal chemically produced by the system should not be dependent on the absence/presence of the input species. On the other hand, the presence of either input applied separately (input combinations **0**,**1** and **1**,**0**) should inhibit the biocatalytic process yielding a low output signal considered as the logic value **0**. In a simple way it is difficult to have two species inhibiting a biocatalytic reaction when they are applied separately, but being not effective inhibitors when applied together. Figure 2.7A offers one possible approach to realization of this process. The primary reaction catalyzed by horseradish peroxidase (HRP) in the presence of H_2O_2 results in oxidation of 2,2′-azino-*bis*(3-ethylbenzothiazoline-6-sulfonic acid) (ABTS) substrate, yielding a colored oxidized product ($ABTS_{ox}$) (Figure 2.7B) (see the ABTS and $ABTS_{ox}$ structures in Figure 2.8). It should be noted that all chemical species

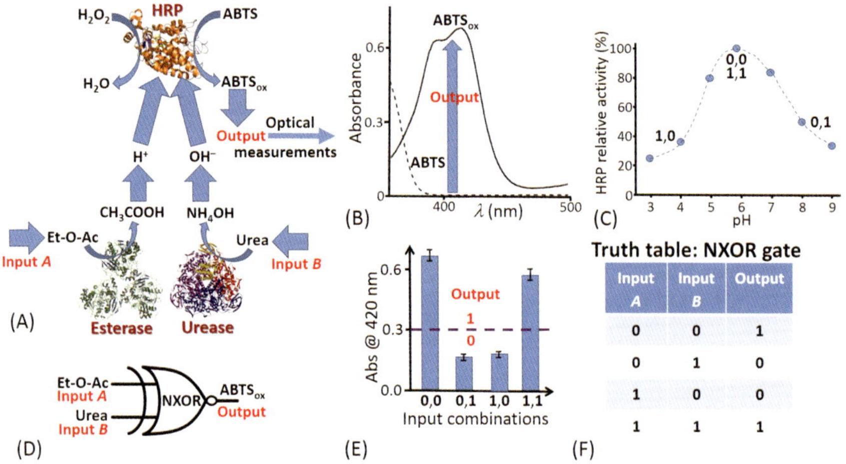

Figure 2.7 The system outlined mimics the features of the NXOR logic gate. (A) The reaction catalyzed by horseradish peroxidase (HRP) controlled by the *in situ* changed pH value. Two other reactions catalyzed by esterase and urease are used to decrease and increase the pH, respectively. The primary input signals, ethyl acetate (Et-O-Ac) and urea (Inputs *A* and *B*, respectively), are consumed by the esterase and urease in order to change the pH value. The final output measured optically corresponds to the oxidized form of ABTS ($ABTS_{ox}$) produced in the reaction catalyzed by HRP with different rates depending on the pH value. (B) Optical absorbance of ABTS (the original substrate added as the "machinery" part) and $ABTS_{ox}$ (the final output). (C) The pH-controlled rate of the HRP-catalyzed reaction. The reaction rates observed after applying the input signals in different combinations are shown on the plot. (D) The NXOR gate scheme. (E) The bar chart showing the optical absorbance measured at $\lambda = 420$ nm corresponding to the $ABTS_{ox}$ concentration after applying the input signals in different combinations for 20 minutes. The dashed line separates the high and low absorbance values corresponding to **1** and **0** outputs. (F) The truth table of the NXOR logic gate. Source: Katz 2018 [7]. Adapted with permission of John Wiley and Sons.

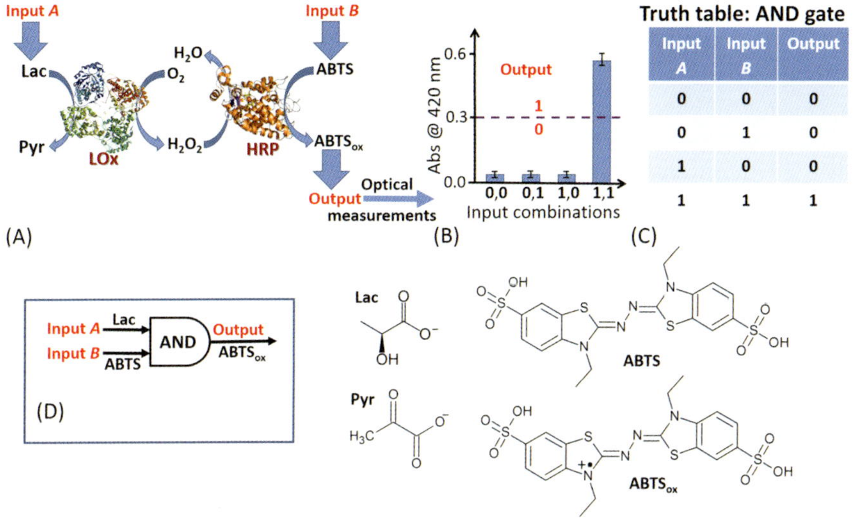

Figure 2.8 (A) Two consecutive reactions catalyzed by lactate oxidase (LOx) and horseradish peroxidase (HRP) mimicking the AND logic gate. The oxidation of ABTS (yielding the colored product $ABTS_{ox}$) proceeds only when both biocatalytic reactions are activated, thus requiring the presence of both input signals (Lac and ABTS) at their logic input **1** (**1**,**1** combination). (B) The optical absorbance (λ_{max} = 420 nm) corresponding to the biocatalytically produced $ABTS_{ox}$ measured after applying the input signals in different combinations, featuring the AND logic gate. The dashed line separates the high and low absorbance values corresponding to **1** and **0** outputs. (C) The truth table of the AND logic gate. (D) The scheme of the AND gate activated with two inputs (Lac and ABTS). Chemical structures of Lac, Pyr (oxidized product of Lac), ABTS, and $ABTS_{ox}$ are shown. Source: Katz 2018 [7]. Adapted with permission of John Wiley and Sons.

participating in this biocatalytic process are not variable inputs (non-variable "machinery" components); thus the reaction is running continuously in the absence of the logic inputs (**0**,**0** combination). However, as many other enzymes, HRP has an optimum pH value that allows the biocatalytic reaction to proceed with the maximum rate. Shifting the pH to lower (acidic) or higher (basic) values results in decreasing the enzyme activity, thus inhibiting the biocatalytic production of $ABTS_{ox}$, which is the final output signal (Figure 2.7C). The pH shift can be generated by two reactions catalyzed by esterase (Est) in the presence of an ester (ethyl acetate, Et-O-Ac, in the present example, Input *A*) and by urease (Ure) in the presence of urea (Input *B*). The former reaction results in the acidic pH due to hydrolytic decomposition of Et-O-Ac and production of acetic acid, while the latter reaction results in the basic pH due to hydrolytic decomposition of urea yielding ammonia. Therefore, the input combinations **1**,**0** and **0**,**1** result in the opposite pH changes, however, both resulting in the pH shift from the optimum value and both inhibiting the HRP reaction (Figure 2.7C). If the input concentrations (Et-O-Ac and urea) are properly optimized, application of both of them (input combination **1**,**1**) can neutralize each other, thus resulting in no pH changes, keeping HRP at the optimum pH value with the enzyme activity similar to the original value (Figure 2.7C). This allows logic operation expected for the NXOR gate (Figure 2.7D) with the experimentally measured output signals (Figure 2.7E) corresponding to the truth table characteristic of the NXOR gate (Figure 2.7F). It should be noted that the HRP activity dependent on the pH value is not a unique enzyme feature – similar bell-shaped function for enzyme activity on pH is often observed for many other enzymes [35–37], thus allowing similar NXOR design with the use of many enzymes producing different output species analyzed by various methods. The only requirement for such NXOR gate realization is a sharp enzyme activity change with the pH variation. Also, the optimum pH value should be located not far from the neutral pH, thus allowing the pH shift in both directions (to acidic and to basic pH values). The reactions biocatalyzed by Est and Ure have been already used for similar pH variation to realize set/reset functions in various bioelectronic switchable systems [9]. Some other enzymes capable of pH changes can be used instead of Est and Ure, thus allowing many biomolecular systems performing NXOR logic operation. The only limitation for the discussed design of the NXOR gate is the low buffer concentration in the solution to allow pH changes due to biocatalytic transformations. Obviously, the gate time (when the output signal is read) should be optimized to allow the different output signals before all of them reach saturation. Indeed, the pH shift generated by the input signals does not inhibit completely the biocatalytic reaction and only decreases the reaction rate; thus the output signal will reach the same saturated value for any pH, if the reaction time is long enough. The NXOR realization based on completely different concepts has been reported as well [34]; thus, the presently discussed system is only an example.

2.2.7 AND Gate

The Boolean AND logic gate is one of the most frequently designed systems (along with the OR gate, which is also very common), particularly realized with enzyme-catalyzed reactions [16, 17, 25, 38–44]. While the OR logic gate, discussed earlier, can be easily realized with two biocatalytic reactions operating in parallel, the AND gate is often designed to operate as a cascade of two consecutive biocatalytic processes (Figure 2.8A). Two input reactants (usually enzyme substrates) activate two steps in the process, and the final output product appears only when both steps are active (input combination **1**,**1**). Otherwise, when either or both input reactants are missing (input combinations **0**,**1**; **1**,**0**; and **0**,**0**), the biocatalytic cascade cannot be completed, and the final product is not produced. The example reaction cascade shown in Figure 2.8A operates in this way. Oxidation of lactate (Lac; Input *A*) results in generation of H_2O_2 in the reaction catalyzed by lactate oxidase (LOx). The *in situ* produced H_2O_2 oxidizes ABTS (Input *B*), resulting in the colored oxidized form of ABTS ($ABTS_{ox}$) observed optically ($\lambda_{max} = 420$ nm) as the final output signal (Figure 2.8B). The increased optical absorbance, corresponding to the formation of $ABTS_{ox}$, was measured only when both inputs (Lac and ABTS) appeared at their logic inputs **1**, thus mimicking the AND logic gate operation (Figure 2.8C,D). Numerous other biocatalytic processes based on different enzymes and being activated with various substrates can operate in the similar logic way. The number of the biocatalytic steps and number of enzymes involved can be different. In the simplest realization it could be a single enzyme activated with a substrate/co-substrate pair representing two input signals. In a different approach, one or both input signals can be represented by enzymes, while the mixture of substrates/co-substrates/cofactors can be a non-variable "machinery" part of the logic gate [12]. The enzyme-based AND logic systems have been extensively used for various biosensing systems, logically responding in the binary (YES/NO) way to the presence or absence of the analyzed species (see Chapter 11 for many examples).

2.2.8 NAND Gate

The NAND logic gate is essentially the same as the AND gate but with the inverted output signal. The inversion of the output signal might be performed after the AND logic operation in a special biocatalytic reaction [45]; however, in the easiest way the inversion can be done simply by changing the output signal definition. Instead of selecting the generated product as the output signal, the consumed (decreasing in its concentration) species can be defined as the output signal species. Figure 2.9A shows an example reaction cascade catalyzed by two enzymes and activated by two input signals. Importantly, the output signal (measured as the optical absorbance) was defined as the NADH concentration, which is decreasing upon the biocatalytic process. This definition of the output signal results in the signal inversion required for the NAND logic gate.

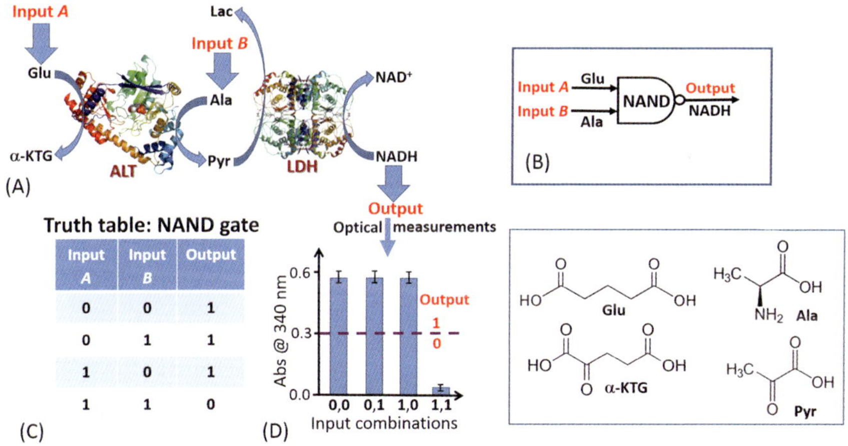

Figure 2.9 (A) Two consecutive reactions catalyzed by alanine transaminase (ALT) and lactate dehydrogenase (LDH) mimicking the NAND logic gate. The consumption of NADH (yielding the corresponding absorbance decrease at $\lambda_{max} = 340$ nm) proceeds only when both biocatalytic reactions are activated, thus requiring the presence of both input signals (Glu and Ala) at their logic input **1** (**1**,**1** combination). (B) The scheme of the NAND gate activated with two inputs (Glu and Ala). (C) The truth table of the NAND logic gate. (D) The optical absorbance ($\lambda_{max} = 340$ nm) corresponding to the residual amount of NADH not consumed by the biocatalytic reactions. The absorbance changes correspond to the NAND gate features. The dashed line separates the high and low absorbance values corresponding to **1** and **0** outputs. Chemical structures of Glu, α-KTG (oxidized product of Glu), Ala, and Pyr are shown. Source: Katz 2018 [7]. Adapted with permission of John Wiley and Sons.

Glutamate (Glu; Input *A*) is oxidized by alanine (Ala; Input *B*) in the reaction catalyzed by alanine transaminase (ALT) (Figure 2.9A). This biocatalytic reaction can be considered as the AND logic gate since it proceeds only in the presence of both reacting species (Glu and Ala) corresponding to the **1**,**1** logic input combination. The intermediate (not analyzed) product (pyruvate, Pyr) cannot be produced when either or both input reactants are missing (input combinations **0**,**1**; **1**,**0**; and **0**,**0**). When Pyr is produced in the first reaction step, it is reduced to lactate (Lac), resulting in the consumption (oxidation) of NADH in the second reaction catalyzed by lactate dehydrogenase (LDH). This step operates as a signal inverter – when the Pyr concentration is increasing in course of the first reaction, the NADH concentration is decreasing in the second reaction step. The whole step (AND inverter system) operates as the NAND logic gate (Figure 2.9B). The optically measured response, corresponding to the NADH concentration, mimics the NAND logic gate (Figure 2.9D), reproducing the truth table characteristic of the Boolean NAND logic function (Figure 2.9C). It should be noted that the enzyme-based NAND logic gate is a model of the universal gate, which theoretically can be used to construct all other Boolean gates. This way of constructing various gates using only NAND gates is important in electronics. However, it is hardly realized in biomolecular systems due to the logic circuitry complexity; indeed connecting together many NAND gates operating in concert is a very challenging goal, making this approach inconvenient. Still keeping in mind the fact that the NAND gate is a universal logic function [30], its formulation and study is an important part of the biomolecular computing.

2.2.9 INHIB Gate

INHIB (Inhibited) logic gate represents an interesting Boolean function where one of the inputs cancels (inhibits) the logic operation when the input appears at the logic input **1**. Figure 2.10A shows an example system mimicking the INHIB logic gate. Acetylcholine (Input *B*) is hydrolyzed to choline and acetic acid in the reaction catalyzed by acetylcholinesterase (AChE). Then the *in situ* produced choline is oxidized by oxygen in the reaction catalyzed by choline oxidase (ChOx), also generating H_2O_2. Finally, the produced H_2O_2 oxidizes ABTS in the reaction catalyzed by horseradish peroxidase (HRP), producing the colored $ABTS_{ox}$ detected optically as the final output signal. The optically measured output signal appears at the logic input **1** when Input *B* is applied at the logic input **1**. In the absence of acetylcholine (input **0**), the final product $ABTS_{ox}$ cannot be produced since the whole biocatalytic cascade is mute. The first biocatalytic reaction (hydrolysis of acetylcholine) can be inhibited by the addition of paraoxon (PAX), which is an effective inhibitor of AChE [46] (*note that PAX is toxic and requires very careful handling*). When PAX (Input *A*) is applied at the logic input **1**, it inhibits the biocatalytic transformations, resulting in no production of $ABTS_{ox}$ (output signal **0**). Overall, the final output signal (optical absorbance corresponding to the formation of $ABTS_{ox}$; $\lambda_{max} = 420$ nm) appears at the logic input **1** only when Input *B* (activating input) is **1** and Input *A* (inhibiting input) is **0** (Figure 2.10D). Therefore, the biocatalytic cascade shown in Figure 2.10A operates as the Boolean INHIB logic gate (Figure 2.10B),

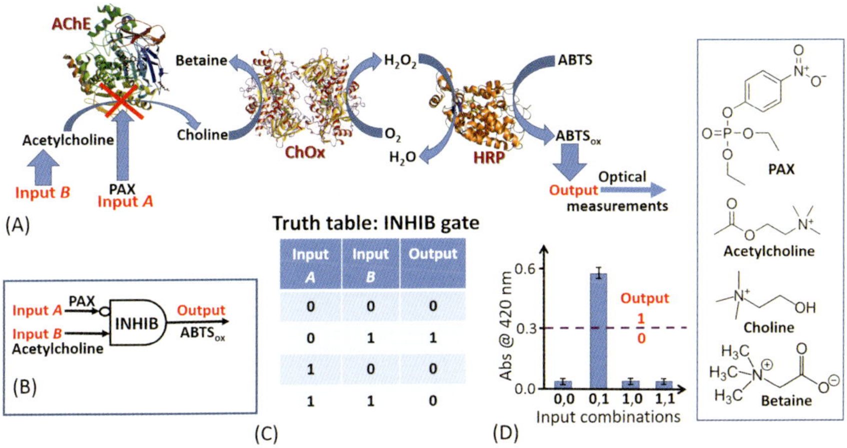

Figure 2.10 (A) The biocatalytic cascade operated with two input signals (PAX and acetylcholine) mimicking the INHIB logic gate. Acetylcholine is activating the biocatalytic cascade, resulting in the production of choline, then H_2O_2, and finally the colored product $ABTS_{ox}$. The PAX input inhibits the first enzyme (acetylcholinesterase, AChE), thus inhibiting the whole set of the reactions preventing formation of $ABTS_{ox}$. (B) The scheme of the INHIB logic gate with two input signals (PAX and acetylcholine) and $ABTS_{ox}$ as the output signal. (C) The truth table of the INHIB logic gate. (D) The optical absorbance (λ_{max} = 420 nm) corresponding to the biocatalytically produced $ABTS_{ox}$ measured after applying the input signals in different combinations, featuring the INHIB logic gate. The dashed line separates the high and low absorbance values corresponding to **1** and **0** outputs. The structures of paraoxon (PAX), acetylcholine, choline (hydrolyzed product of acetylcholine), and betaine (oxidation product of choline) are shown. Source: Katz 2018 [7]. Adapted with permission of John Wiley and Sons.

corresponding to the truth table of the INHIB logic operation (Figure 2.10C). It should be noted that the Boolean INHIB logic gate can be realized in a biocatalytic cascade without direct inhibition of an enzyme, but operating in a different way, still performing the same logic function [17]. Depending on which input results in the inhibition, the logic gate can be INHIB A or INHIB B (being inhibited by Input *A* or Input *B*, respectively). The example system shown in Figure 2.10 operates as the INHIB A gate.

2.2.10 Summary on the Basic Boolean Gates Realized with Enzyme Systems

The previous subsections concentrated on the design of enzyme-catalyzed reactions utilized to mimic fundamental Boolean logic gates: Identity (YES), Inverted Identity (NOT), OR, NOR, XOR, NXOR, AND, NAND, and INHIB, which are essential elements for constructing various biocomputing systems. The present example systems are discussed to highlight the conceptual points, rather than to collect all possible variants of such systems. The same Boolean logic gates can be realized with endless biochemical reactions using various enzymes, substrates, cofactors, activators, inhibitors, etc.

Newcomers to the research area who are not familiar with the use of enzymes for mimicking logic gates and particularly do not know about applications of these systems might be disappointed at this time. Indeed, the biocatalytic systems exemplified in the previous subsections are rather simple and standard biochemical reactions, and their consideration as binary-operating logic gates seems to be quite artificial for most of chemists. However, when continuously reading the book, the readers will realize that these simple logic gates can be assembled in sophisticated logic networks [47], particularly performing reversible logic functions [21]. Their optimization (through experimental and theoretical modeling) is not trivial and includes highly sophisticated procedures [33]. Also, the enzyme-catalyzed binary processes can be used in various biomolecular memory [48] and information security [49] devices. The overviewed logic gates have been interfaced with various modified electrodes [50, 51] and electronic chips [52] providing novel signal-switchable functionality. Finally, various binary biosensing systems generating answers in the YES/NO format have been designed for various biomedical [2], forensic [53], and biotechnological [54] applications with the use of similar Boolean logic gates. The readers are welcome to continue reading the book to get explanations about the importance and applications of the fundamental logic gates outlined in this section.

The following subsections will continue the discussion of the enzyme-based logic gates moving forward to more sophisticated systems with the higher complexity and additional functionalities.

2.3 Modular Design of NOR and NAND Logic Gates

At present all kinds of (bio)chemical logic gates, based on synthetic signal-switchable molecules [55–57] or natural biomolecules [5, 58], particularly based on enzyme-catalyzed reactions [6, 7], have different compositions and operate with different chemical reactions. In other words, all logic gates mimicking Boolean operations have different and unique design (see examples in Section 2.2). When they operate as stand-alone molecular systems, their operation might be fine, but connecting them in logic networks is complicated and often not possible. There are many challenging problems that have to be solved to allow easy assembling of (bio)molecular systems in multicomponent computing devices – one of them is modularity of the logic elements. The modularity approach means assembling of different logic gates from standard interchangeable units. The full realization of this approach is difficult, but the first attempts to achieve this goal are illustrated in this section.

It is known that in electronic systems NAND and NOR logic gates are universal and all other Boolean functions can be derived from their compositions [30]. Standard NAND/NOR gates are used in electronic integrated circuitries to perform all needed digital functions [59]. It would be an advantage for chemical computing systems to use this standard modular approach. Thus, the development of the NAND and NOR logic gates with modular exchangeable sub-units composed of enzymes performing biocatalytic reactions interconnected in cascades mimicking the logic operations is the important step to the modular design of biocomputing networks. It should be noted that all previously designed (bio)molecular NAND/NOR logic gates used unique designs without the modularity concept [60–69]. At the present stage the developed (bio)molecular NAND/NOR gates do not pretend to mimic all possible Boolean operations because of the complexity of the required networks. Still one should keep in mind that their potential use as universal gates might be possible in the future.

The modular design started [45] from AND/OR gate sub-units (Figure 2.11A,B), producing glucose as the output signals connected to an Inverter (INV) gate sub-unit (Figure 2.11C), accepting glucose as the input, and converting it to NADH being the final output. The AND/OR sub-units were interchangeable and connectable to the INV unit. Note that all parts of the biocomputing system have operated in a solution; thus, their integration was only functional, but not structural. The AND gate sub-unit was composed of maltose phosphorylase (MPh), biocatalyzing reaction of maltose and phosphate (Pi) (defined as Input *A* and Input *B*, respectively) to yield glucose (Glc) and glucose-1-phosphate (Glc1P) as a by-product. The experimentally optimized concentrations of maltose and phosphate were considered as logic inputs **1**, while their absence was defined as logic **0** inputs. The optimization of the input concentrations aimed at the production of the output signals (glucose concentrations) comparable with those generated by another interchangeable OR logic sub-unit described later. The glucose production was activated only in the presence of the both input signals (**1**,**1**), while it was inhibited in the absence of either of them or both (**0**,**1**; **1**,**0**; **0**,**0**), thus resembling the Boolean AND logic operation. The OR

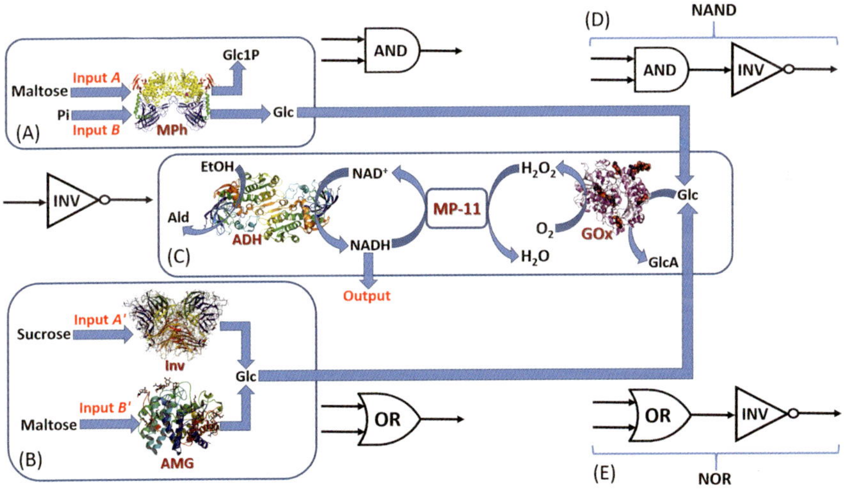

Figure 2.11 Modular design of the enzyme-based NAND and NOR gate. (A) The reaction activated with maltose and phosphate (Pi) and catalyzed by maltose phosphorylase (MPh) mimics the AND logic gate. (B) Two parallel reactions activated with sucrose and maltose and catalyzed by invertase (Inv) and amyloglucosidase (AMG), respectively, mimic the OR logic gate. (A–B) Both biocatalytic reactions produce glucose, but upon different combinations of the input signals. (C) The biocatalytic cascade consuming glucose produced in the preceding reactions and converting the glucose intermediate output to consumption of NADH considered as the final output – the cascade operates as an inverter, converting AND OR gates to NAND and NOR, respectively.

gate sub-unit was composed of amyloglucosidase (AMG) and invertase (Inv), biocatalyzing conversion of sucrose and maltose (defined as Input A' and Input B', respectively) to glucose. Similarly to the AND gate sub-unit, the optimized concentrations of maltose and sucrose were considered as logic inputs **1**, while their absence was defined as logic **0** inputs. The glucose production was activated by any of the parallel biocatalytic reaction or by both of them in the presence of the either or both input signals (**0,1**; **1,0**; **1,1**), while it was inhibited in the absence of the both inputs (**0,0**), thus resembling the Boolean OR logic operation. The INV sub-unit included alcohol dehydrogenase (ADH) to produce constantly NADH upon reaction of the included in the system NAD^+ and ethanol being the non-variable parts of the system "machinery." Another biocatalytic entity, microperoxidase-11 (MP-11), was included to reverse the NADH production, returning it to the initial oxidized state NAD^+ in the presence of H_2O_2 produced *in situ* by glucose oxidase (GOx) in the presence of glucose coming from the connected AND/OR gate sub-units. It should be noted that the catalytic activity of MP-11 was high enough to overwhelm the biocatalytic reduction process, resulting in the production of NADH. The optimization of the MP-11 concentration was performed for the specific H_2O_2 concentrations produced *in situ* by the preceding biocatalytic cascade. Finally, the NADH output signal was generated only in the absence of glucose incoming from the AND/OR logic sub-units, thus inverting the **1** signal of glucose to **0** signal of NADH and resembling the NAND/NOR logic when the INV is connected to the AND/OR sub-units, respectively.

Figure 2.12 shows the output signals of the NAND gate composed of the AND–INV sub-units for different combinations of the primary input signals (maltose and phosphate): **0,0**; **0,1**; **1,0**; and **1,1**. The output signals were recorded as the absorbance spectra of NADH being the final output signal of the gate. It should be noted that similar output signals could be derived from the electrochemical detection of NADH in the presence of multienzyme logic system [14]. The large absorbance, $\lambda_{max} = 340$ nm, corresponding to the *in situ* generated NADH was observed in the absence of the either input or both primary input signals (**0,1**; **1,0**; **0,0**) (Figure 2.12, curves a–c). The presence of the both input signal (**1,1**) inhibited the absorbance formation characteristic of NADH (Figure 2.12, curve d), thus resembling the NAND features. The absorbance values below 0.15 at $\lambda_{max} = 340$ nm were considered as digital **0** values, while the absorbance above 0.2 was defined as digital **1** output (Figure 2.12, inset). Similar to the paradigm used in electronics [59], the absorbance values between two thresholds were defined as digitally undefined. Figure 2.13 shows the output signals of the NOR gate composed of the OR–INV sub-units for different combinations of the primary input signals (maltose and sucrose): **0,0**; **0,1**; **1,0**; and **1,1**. The output signals were recorded as the absorbance spectra of NADH being the final output signal of the gate. The large absorbance, $\lambda_{max} = 340$ nm, corresponding to the *in situ* generated NADH was observed only in the absence of both primary input signals (**0,0**) (Figure 2.13, curve a), while the presence of either or both input signals (**0,1**; **1,0**; and **1,1**) inhibited the absorbance formation characteristic of NADH (Figure 2.13, curves b–d). The absorbance values below 0.15 at $\lambda_{max} = 340$ nm were considered as digital **0** values, while the absorbance

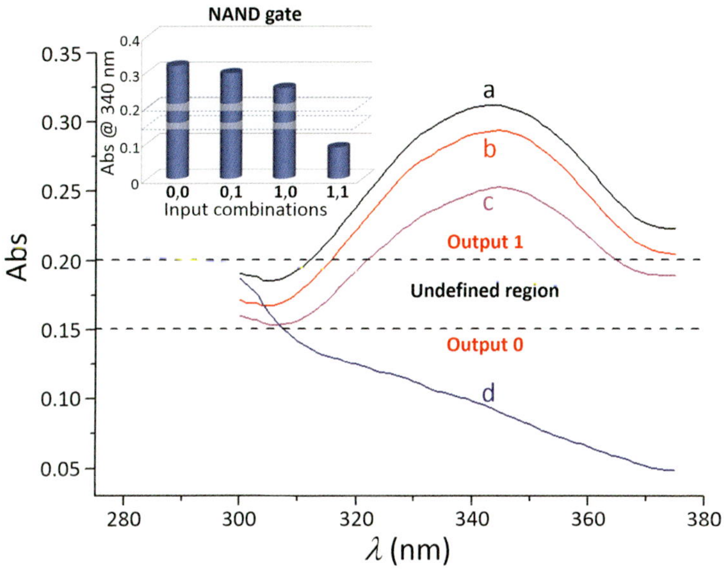

Figure 2.12 Absorbance spectra corresponding to the output signal of NADH generated by the enzyme-based NAND logic gate (see the gate scheme in Figure 2.11A,C) upon application of different combination of the input signals of maltose and phosphate: (a) **0**,**0**; (b) **0**,**1**; (c) **1**,**0**; (d) **1**,**1**. Inset: the output signal values measured at $\lambda_{max} = 340$ nm corresponding to the logic **1** when higher than 0.2 and **0** when lower than 0.15. The system response appearing between two threshold dashed lines is considered as undefined. Source: Zhou et al. 2009 [45]. Adapted with permission of American Chemical Society.

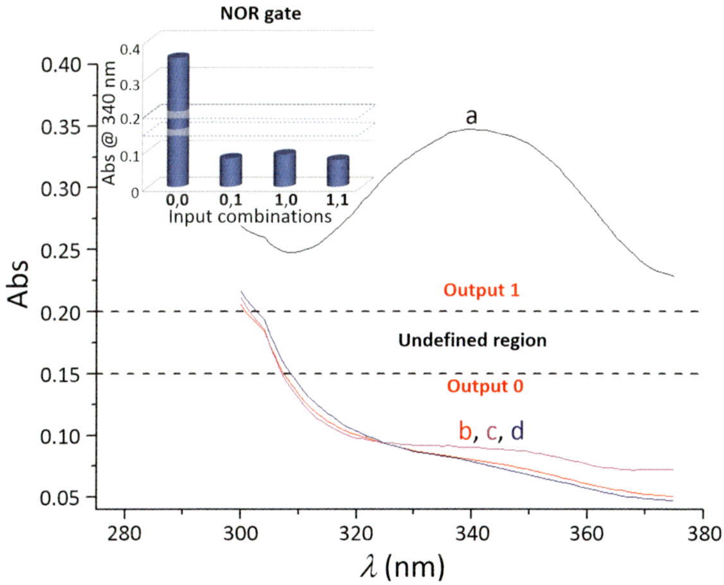

Figure 2.13 Absorbance spectra corresponding to the output signal of NADH generated by the enzyme-based NOR logic gate (see the gate scheme in Figure 2.11B,C) upon application of different combinations of the input signals of maltose and sucrose: (a) **0**,**0**; (b) **0**,**1**; (c) **1**,**0**; (d) **1**,**1**. Inset: The output signal values measured at $\lambda_{max} = 340$ nm corresponding to the logic **1** when higher than 0.2 and **0** when lower than 0.15. The system response appearing between two threshold dashed lines is considered as undefined. Source: Zhou et al. 2009 [45]. Adapted with permission of American Chemical Society.

above 0.2 was defined as digital **1** output (Figure 2.13, inset). The response pattern generated by the system was characteristic of the Boolean NOR logic gate (Figure 2.13, inset).

The systems with interchangeable AND/OR units connected to the same INV unit illustrated the modular design of NAND/NOR logic gates. Still in this very first attempt to design logic networks with the modular composition, all reacting elements were in the same solution. The next step in this research direction should obviously include physical separation of the reacting units, using, for example, flow cell systems similar to those discussed in Chapter 5. Microfluidic lab-on-a-chip systems will be even more appropriate when biocomputing systems with the modular design are brought to practical applications, being used, for example, as binary biosensing devices (see Chapter 11).

2.4 Majority and Minority Logic Gates

The present section overviews new enzyme systems mimicking Boolean gates performing "majority" and "minority" logic operations. In Boolean logic, the circuit with the majority function returns true (output **1**) only when more than 50% of its inputs are true (inputs **1**) and false otherwise; the minority function operates in the opposite way, being the complement of the majority function [70]. While, in general, the majority and minority logic gates can operate with n input signals ($n > 2$), their simplest realization includes three inputs. Therefore, the 3-input majority gate generates the logic output **1** (true) when any two or all three inputs have logic values **1**, and, otherwise, if more than one input has logic value **0**, the output is **0** (false). The 3-input minority gate generates the logic output **1** only when one or none of the inputs has a logic value of **1**, while any two or all inputs with the value **1** result in output **0**. These logic operations are considered highly important for electronic/computing systems [70–73]. Majority (voting) functions are used in fault-tolerant computing [74] and other applications, including synthesis of reversible logic circuits [75]. Overall, a three-input majority gate is one of the most basic logic gates; on the molecular level it can serve as a basic and versatile building block for constructing more complex circuits. However, until now, its biomolecular realization was reported mostly for DNA-based logic systems [76–83], with very few results for enzyme-catalyzed reactions [84]. The complimentary minority logic gate received even less attention in the area of molecular computing [84]. The new realization of the majority and minority gates [84] using enzyme catalytic cascades discussed in this section allows integration of these gates with other Boolean enzyme-based gates, yielding logic networks of high complexity with unique logic features.

Figure 2.14 shows the general composition of the enzyme-based 3-input majority gate with a following filter unit [84] (detailed discussion on the "filters" is offered in Chapter 3) and preceding AND gate selected as an example. The majority gate was realized with three reactions catalyzed by three NAD^+-dependent enzymes, glucose dehydrogenase (GDH), glucose-6-phosphate dehydrogenase (G6PDH), and alcohol dehydrogenase (ADH), operating in parallel. The biocatalytic reactions were activated by input signals *A*, *B*, and *C*, represented by

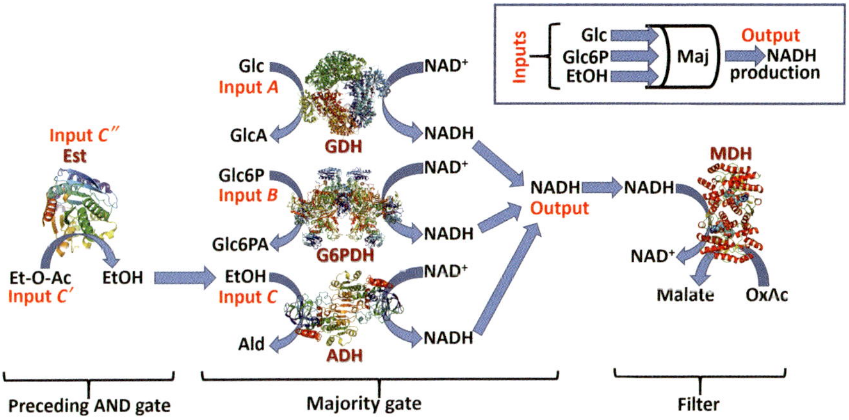

Figure 2.14 The 3-input majority gate mimicked with three parallel reactions – all producing the same NADH upon oxidation of glucose (Glc) catalyzed by glucose dehydrogenase (GDH), oxidation of glucose-6-phosphate catalyzed by glucose-6-phosphate dehydrogenase (G6PDH), and oxidation of ethanol (EtOH) catalyzed by alcohol dehydrogenase (ADH). The ethanol input was also produced *in situ* through the reaction catalyzed by esterase (Est) mimicking the preceding AND gate. The produced NADH was processed by the enzyme "filter" catalyzed by malate dehydrogenase (MDH). The filter operation resulted in better resolution between the **0** and **1** logic outputs produced by the majority gate. Source: Mailloux et al. 2014 [84]. Adapted with permission of American Chemical Society.

the corresponding substrates: glucose (Glc), glucose-6-phosphate (Glc6P), and ethanol (EtOH), respectively. Aiming at the simplest concept demonstration, logic value **0** for all input signals was defined as the absence of the substrates, while logic input **1** was defined as the experimentally convenient and optimized concentrations. All three reactions resulted in the production of NADH upon reduction of the cofactor NAD^+ and concomitant oxidation of the substrate inputs. The NADH formation was followed by absorbance measurements at $\lambda = 340$ nm, which is typical for NADH analysis [19] (Figure 2.15A). As expected for the parallel operating biocatalytic reactions, the NADH production was increased with an increasing number of reactions contributing to the process. In other words, in the absence of all substrates (meaning logic **0** for all inputs; **0,0,0** combination), all three enzymes were mute, resulting in no production of NADH (Figure 2.15A, curve a). Applying any one of the three input signals (Glc or Glc6P or EtOH) at its logic input **1** (**0,0,1**; **0,1,0**; **1,0,0** combinations) resulted in the activation of one biocatalytic reaction, thus resulting in NADH production (Figure 2.15A, curves b–d). Note that the enzyme activities and input concentrations were optimized to result in similar rates of NADH generation in all three biocatalytic processes. When two or all three inputs were applied at logic inputs **1** (**0,1,1**; **1,0,1**; **1,1,0**; **1,1,1** combinations), the rate of the NADH production was increased due to parallel operation of two or three biocatalytic reactions (Figure 2.15A, curves e–h). The biocatalytically produced and optically measured NADH was considered as the output signal with the low and high concentrations defined as logic **0** and **1**, respectively, where logic **0** corresponds to no NADH production or NADH generation by any single pathway, while

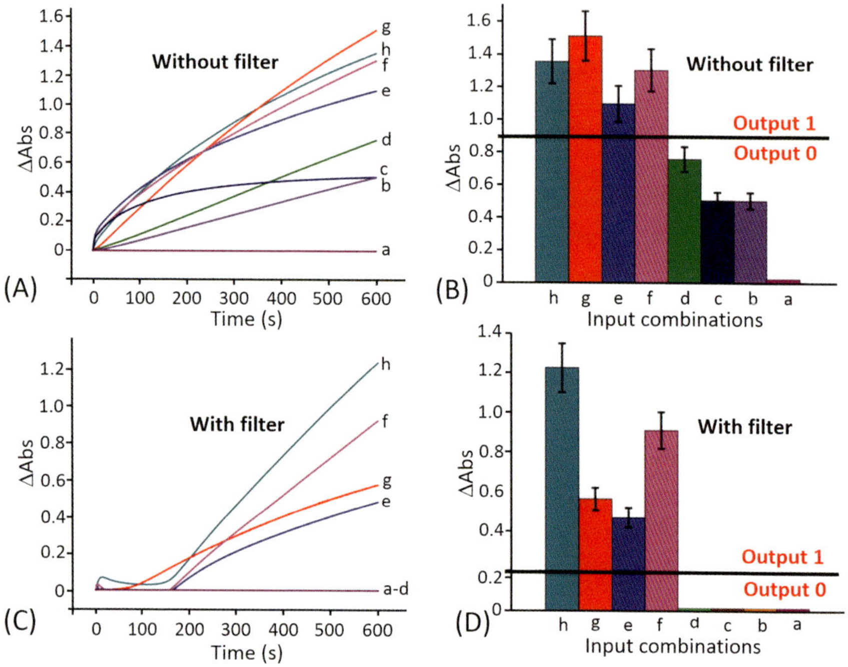

Figure 2.15 (A) Time-dependent absorbance changes corresponding to NADH production by the majority gate system (see Figure 2.14 for the scheme) activated by various combinations of the input signals. (B) The bar chart showing the optical absorbance of NADH measured at $\lambda = 340$ nm at the gate time of 600 seconds for various combinations of the input signals. (C) Time-dependent absorbance changes corresponding to NADH production by the majority gate system activated by various combinations of the input signals after processing the output signal through the filter – note better resolution between the **0** and **1** output signals. (D) The bar chart showing the optical absorbance of NADH measured at $\lambda = 340$ nm at the gate time of 600 s for various combinations of the input signals after processing the output signal through the filter. The input signal combinations were encoded in the following way: (a) **0,0,0**; (b) **0,1,0**; (c) **0,0,1**; (d) **1,0,0**; (e) **1,0,1**; (f) **0,1,1**; (g) **1,1,0**; (h) **1,1,1**. The input signals are shown in the following order: *A*, *B*, *C*. Input signals are defined in Figure 2.14. The horizontal solid lines separate the logic output values **0** and **1**. Source: Mailloux et al. 2014 [84]. Adapted with permission of American Chemical Society.

logic **1** corresponds to NADH production by any combination of two pathways or all three of them, following the definition of a majority gate [70]. While there is a clear difference between NADH production in one of the possible reaction pathways and in two/three of them operating simultaneously, the difference between physical values of logic **0** and **1** outputs is only ca. twofold at the gate time of 600 seconds (Figure 2.15B). The threshold separating **0** and **1** outputs is shown as a horizontal solid line in Figure 2.15B.

Recent research aiming at reducing noise in the operation of enzyme-based logic gates resulted in the formulation of "filter" systems consuming a part of the output signal, thus bringing the output to physical zero when it is produced in small concentrations [26, 38, 40, 41, 85] (see more details about "filters" used in biocatalytic reactions in Chapter 3). Applying this approach to the majority gate, a biocatalytic cascade partially consuming the produced NADH was assembled (Figure 2.14). Malate dehydrogenase (MDH) oxidized NADH produced by the reactions included in the majority gate with the concomitant reduction of oxaloacetate (OxAc) to malate. It should be noted that MDH can reversibly catalyze the reaction in both directions [86]. Thus, accumulation of malate eventually pushes the opposite reaction to occur, finally resulting in a steady state condition, limiting the extent of NADH oxidation [87]. The concentration of OxAc was experimentally optimized to consume the entire amount of NADH produced by any single reaction pathway in the majority gate, thus bringing the logic **0** output to its physically zero level (Figure 2.15C, curves a–d). This resulted in much better separation between the logic **0** and **1** values of the output signal (Figure 2.15D). In other words, before the filtering process, the NADH concentrations (measured as optical absorbance at $\lambda = 340$ nm) corresponding to the logic output values **0** and **1** were different by ca. twofold, both being much above physical zero. When the filter reaction was applied, the NADH concentration corresponding to logic **0** became physically zero, while the concentration corresponding to logic **1** was significantly above zero. The threshold separating the **0** and **1** outputs is shown as a solid line in Figure 2.15D. It should be noted that the mechanism of the filter operation is different from all previously studied biomolecular filters (see discussion in Chapter 3). The mechanistic consideration and theoretical modeling of this filter operation was reported elsewhere [88], being outside the scope of the present chapter.

The 3-input minority gate was realized similarly to the majority gate described above with three reactions biocatalyzed by three NAD^+-dependent enzymes, malate dehydrogenase (MDH), diaphorase (Diaph), and lactic dehydrogenase (LDH), operating in parallel (Figure 2.16). The biocatalytic reactions were activated by input signals *A*, *B*, and *C* represented by the corresponding substrates: oxaloacetate (OxAc), methylene blue (MB), and pyruvate (Pyr), respectively. Note that all the substrates were electron-accepting (oxidizing) species, which induce the biocatalytic oxidation of NADH. This pathway is opposite compared with the reactions used in the majority gate. Thus, all three reactions resulted in conversion of NADH to NAD^+ upon its biocatalytic oxidation and concomitant reduction of the substrate inputs. Logic values **1** for inputs *A*, *B*, and *C* were experimentally optimized, while the absence of the corresponding substrates was considered as logic value **0**. The output signal was defined as the optical absorbance of NADH ($\lambda_{max} = 340$ nm) (Figure 2.17A,B). As expected for the parallel operating biocatalytic reactions, the NADH consumption (conversion to NAD^+) was increased with the increasing number of reactions contributing to this process. In other words, in the absence of all substrates (meaning logic **0** value for all inputs, **0,0,0** combination), all three enzymes were mute, resulting in no consumption of NADH (Figure 2.17A, curve a). Applying any one of three input signals (OxAc, MB, or Pyr) at its logic input **1** (**0,0,1**; **0,1,0**; **1,0,0** input combinations) resulted in activation of one biocatalytic reaction, resulting in NADH consumption and decreasing optical absorbance (Figure 2.17A, curves b–d). Note that the enzyme activities and input concentrations were optimized to result in similar rates of NADH consumption in all three biocatalytic processes. When two or all three inputs were applied at logic input **1** (**0,1,1**; **1,0,1**;

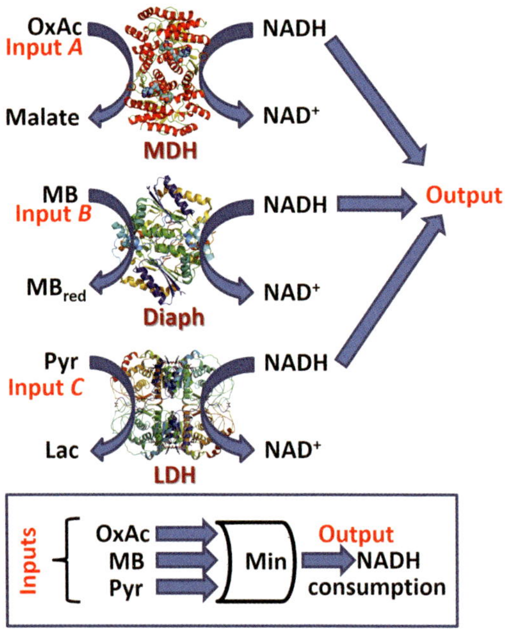

Figure 2.16 The enzyme-based 3-input minority gate mimicked with three parallel reactions – note that the reactions result in consumption (oxidation) of NADH, opposite to the reactions used for mimicking the majority gate. The system is activated with oxaloacetate (OxAc) reduced to malate in the reaction catalyzed by malate dehydrogenase (MDH), methylene blue (MB) reduced to MB_{red} in the reaction catalyzed by diaphorase (Diaph), and pyruvate (Pyr) reduced to lactate (Lac) in the reaction catalyzed by lactate dehydrogenase (LDH). Source: Mailloux et al. 2014 [84]. Adapted with permission of American Chemical Society.

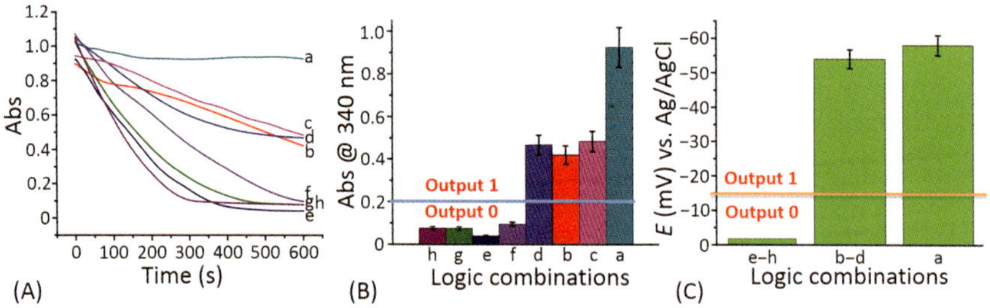

Figure 2.17 (A) Time-dependent absorbance changes corresponding to the NADH concentration upon operation of the minority gate (see Figure 2.16 for the scheme) activated by various combinations of the input signals. (B) The bar chart showing the optical absorbance of NADH measured at $\lambda = 340$ nm at the gate time of 600 seconds for various combinations of the input signals. (C) Bar chart showing the potentials produced on the PQQ-modified electrode in the presence of NADH upon operation of the minority gate after application of the input signals in different combinations. The gate time for the potential measurements was 600 seconds. The input signal combinations were encoded in the following way: (a) **0,0,0**; (b) **0,1,0**; (c) **0,0,1**; (d) **1,0,0**; (e) **1,0,1**; (f) **0,1,1**; (g) **1,1,0**; (h) **1,1,1**. The input signals are shown in the following order: *A*, *B*, *C*. Input signals are defined in Figure 2.16. The solid lines separate the logic output values **0** and **1**. Source: Mailloux et al. 2014 [84]. Adapted with permission of American Chemical Society.

1,1,0; **1,1,1** input combinations), the rate of NADH consumption was increased due to parallel operation of two or three biocatalytic reactions (Figure 2.17A, curves e–h). Figure 2.17B shows the optically measured output signals at the gate time of 600 seconds, where the low and high absorbance values correspond to logic **0** and **1** values, respectively. As expected for the minority function [70], the logic output value **1** was obtained only if one or none of the input signals appeared at their logic input **1**. In the case where more than one input was applied at the logic input **1**, the system returned output **0**. However, the physical values corresponding to the logic **1** output were spread over a broad range of optical absorbance (Figure 2.17B). In order to consolidate the physical measure of the output signal, the output signal was measured electrochemically as the potential electrocatalytically produced on the PQQ-modified electrode in the presence of NADH (see the detailed discussion on this kind of measurements in Section 6.2.2). The potentials measured upon operation of the minority gate demonstrated small deviations from the average values – all potentials corresponding to the logic input **1** were measured close to −50 mV, while the potentials corresponding to the logic value **0** were ca. 0 mV (measured vs. Ag/AgCl reference electrode) (Figure 2.17C), thus significantly improving the outcome from the logic operation and resulting in smaller noise. The improved output signals measured electrochemically can be explained by the logarithmic dependence of the measured potentials on the NADH concentration corresponding to the Nernst equation. The electrochemical readout of the output signals produced by the majority and minority gates allowed their use for logically controlled release of biomolecular species from modified electrode surfaces (this application of the logic systems is discussed in detail in Section 12.1).

Enzyme-based logic systems can be extended to multistep biocatalytic cascades mimicking concatenated logic gates [47, 49, 85, 89] (see the detailed discussion on the logic networks composed of many logic gates in Chapter 4). The connectivity of logic gates was also illustrated with the present majority gate [84]. An additional reaction biocatalytically converting ethyl acetate (Et-O-Ac), input signal C', to ethanol (EtOH) in the presence of esterase (Est), input signal C'', was added to produce EtOH *in situ* for its reaction in the majority gate (Figure 2.14). It should be noted that for experimental convenience the enzyme Est was considered as a logically variable input signal, similarly to the previously designed enzyme logic gates [12] (Figure 2.1B). Logic values **0** for both additional input signals were defined as their absence, while their logic inputs **1** were experimentally optimized to activate the majority gate. The EtOH production proceeded only in the presence of both input signals. In other words, when signals C' (Et-O-Ac) and C'' (Est) were both applied at their logic inputs **1**, EtOH was produced, thus mimicking an AND logic operation. The *in situ* produced EtOH (the output signal of this AND gate) was used in substitution of the input signal C in the majority gate. Note that this input appeared only when inputs C' and C'' both appeared at their **1** logic values. Overall, the concatenated AND-majority circuit generated output **1** in the following input $ABC'C''$ combinations: **1,1,1,1**; **1,1,1,0**; **1,1,0,1**; **0,1,1,1**; **1,0,1,1**; **1,1,0,0**. Figure 2.18 shows operation of the majority gate extended with the preceding AND gate and following filter system. While it was not demonstrated experimentally, it is quite clear that the majority and minority logic gates can be connected not only to the preceding but also to the following logic gates being activated with the NADH output produced in the majority/minority systems. Also, the 3-input biocatalytic system used for mimicking the majority gate can be applied for signal-controlled switching between different logic operations (see Section 2.5).

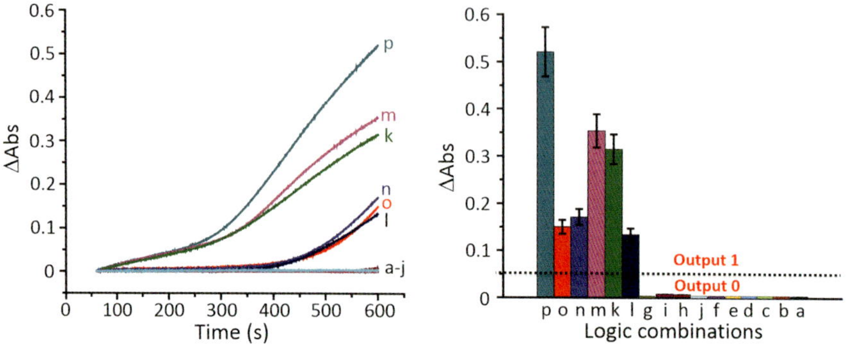

Figure 2.18 (A) Time-dependent absorbance changes corresponding to NADH production by the majority gate system integrated with the preceding AND gate and followed by the filter (see Figure 2.14 for the reaction scheme) activated by various combinations of the input signals. (B) The bar chart showing the optical absorbance of NADH measured at $\lambda = 340$ nm at the gate time of 600 seconds for various combinations of the input signals. The input signal combinations were encoded in the following way: (a) **0,0,0,0**; (b) **1,0,0,0**; (c) **0,1,0,0**; (d) **0,0,1,0**; (e) **0,0,0,1**; (f) **1,0,1,0**; (g) **0,0,1,1**; (h) **0,1,1,0**; (i) **0,1,0,1**; (j) **1,0,0,1**; (k) **0,1,1,1**; (l) **1,1,0,0**; (m) **1,0,1,1**; (n) **1,1,0,1**; (o) **1,1,1,0**; (p) **1,1,1,1**. The input signals are shown in the following order: *A*, *B*, *C*′, *C*″. Input signals are defined in Figure 2.14. The dashed line separates the logic output values **0** and **1**. Source: Mailloux et al. 2014 [84]. Adapted with permission of American Chemical Society.

2.5 Reconfigurable Logic Gates

For many applications, including bioanalytical signal processing [90], the logic operations should be reprogrammed without restructuring chemical systems and without changing the corresponding devices. This functionality may be achieved with the addition of switchable pathways activated on demand, resulting in processes performing different logic operations. While signal-responsive synthetic molecules or supramolecular/nanostructured systems operating as reconfigurable logic gates must be very sophisticated [91–94], biomolecular systems with the reconfigurable logic features [95, 96] might be much easier for realization, particularly using enzyme-catalyzed reactions [34, 84]. Two examples of enzyme-based reconfigurable logic systems switchable between different logic operations by a special external signal are presented in the next two sections. While the first example system operated in a one-pot solution [84], the second example was realized in a flow cell device with the reacting units located separately in different cells [34].

2.5.1 3-Input Logic Gates Switchable Between AND–OR Logic Functions Operating in a Solution

The 3-input gate, described in the previous section as the majority gate (Figure 2.14), can be programmed to act as an OR gate or an AND gate by fixing any one of the three inputs as a programming input. If the programming input is **0**, the two remaining inputs perform the AND logic operation. If the programming input is **1**, the OR operation is performed on the other two inputs.

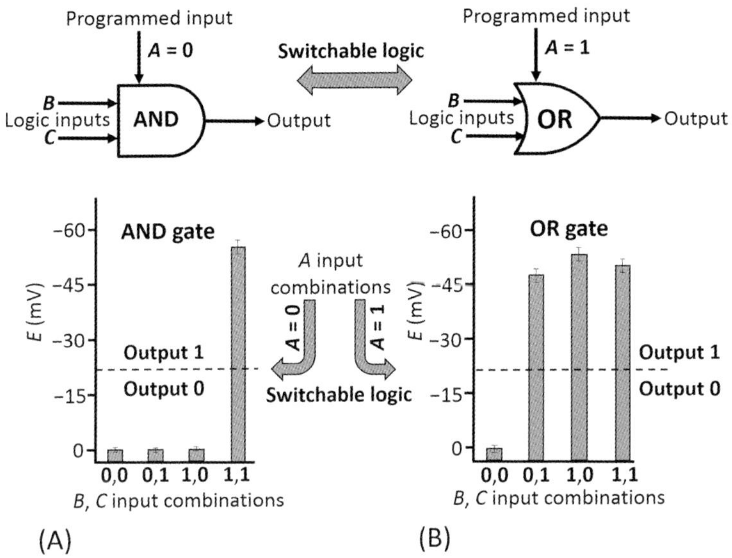

Figure 2.19 Switchable operation of the 3-input gate. The gate operates as AND or OR logic for Inputs *B* and *C* when Input *A* is reserved as the programming input and applied at the logic value **0** or **1**, respectively. The output signals are shown as the potentials generated on the PQQ-modified electrode in the presence of various combinations of input signals *B*, *C*: (a) **0,0**; (b) **0,1**; (c) **1,0**; (d) **1,1**. Input signals are defined in Figure 2.14.

In order to illustrate the switchable logic operation, Input *A* can be selected as the programmed input, while Inputs *B* and *C* were used as logic inputs. Figure 2.19 shows operation of the AND and OR logic gates realized with the programming Input *A* held at the logic values **0** and **1**, respectively. The output signal was defined as the potential measured on the PQQ-modified electrode, where the potential values of ca. −50 mV and 0 mV were considered as logic values **1** and **0**, respectively. Obviously, the selection of the input for operating as the programmed input depends on the specific application. In other words, any input from three input signals can be selected as the programmed input for switching between AND and OR operations for two other inputs.

2.5.2 Enzyme-Based Logic Gates Switchable Between OR, NXOR, and NAND Boolean Operations Realized in a Flow System

The present section outlines the logic processing of chemical input signals in the mode switchable between OR, NXOR, and NAND operations performed in a flow device with a modular design [34]. Figure 2.20A shows the biocatalytic cascade used for the realization of the switchable logic gates. First, two biocatalytic reactions operating in parallel resulted in the production of NADH. One reaction catalyzed by glucose dehydrogenase (GDH) resulted in oxidation of glucose (Glc) and concomitant reduction of NAD^+ to NADH. In another parallel reaction that was catalyzed by glucose-6-phosphate dehydrogenase (G6PDH), the oxidation of glucose-6-phosphate (Glc6P) occurred, also yielding NADH as the second product. Then NADH produced in the both reactions was oxidized back to NAD^+ in

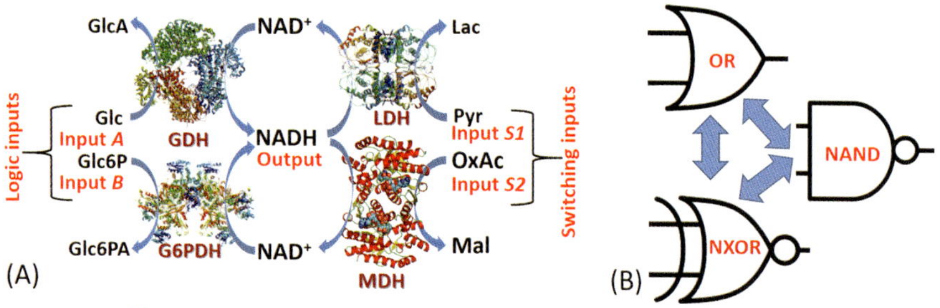

OR gate

Input A	Input B	Output
0	0	0
0	1	1
1	0	1
1	1	1

NXOR gate

Input A	Input B	Output
0	0	1
0	1	0
1	0	0
1	1	1

NAND gate

Input A	Input B	Output
0	0	1
0	1	1
1	0	1
1	1	0

Figure 2.20 (A) The biocatalytic cascade mimicking various Boolean logic gates: OR, NXOR, and NAND, depending on the logic value of the switching input (see the detailed explanation of the switchable gate operation in the text). (B) The switchable logic gates realized by the biocatalytic system. Bottom: Truth tables of the Boolean logic gates. Source: Fratto et al. 2014 [34]. Adapted with permission of Royal Society of Chemistry.

the presence of two enzymes, lactic dehydrogenase (LDH) and malate dehydrogenase (MDH), when activated by their substrates, pyruvate (Pyr) and oxaloacetic acid (OxAc), respectively.

It should be noted that the biocatalytic reactions contributing to the cascade were not always proceeding, but they were activated selectively upon addition of their respective substrates. Two of them, Glc and Glc6P, were considered as the logic inputs for operating logic gates (OR, NXOR, and NAND), while two others, Pyr and OxAc, were used as switching inputs to allow switching between three different logic operations (Figure 2.20B). As usual for biomolecular systems mimicking Boolean logic gates, the logic **0** values of the input signals were defined as the complete absence of the corresponding reactants. The logic inputs **1** for the input signals were optimized experimentally while accounting for dilution by all other pumped solutions (note that the reactions were performed in flow cells, as explained later) to yield the output signal with the comparable value for all combinations of the logic inputs. NAD^+ and NADH were always present in the initial solution as a part of the biomolecular "machinery" system. The output signal generated by the system was measured as the absolute value of the optical absorbance change at $\lambda = 340$ nm corresponding to the variation of the NADH concentration. This definition for the output signal has already been used frequently for XOR gates based on the enzymatic systems [32].

The biocatalytic system mimicking logic gate operations was assembled in a flow device shown in Figure 2.21. The enzymes were immobilized in the flow compartments allowing each enzyme to be immobilized in a different cell. The biocatalytic reactions proceeded step by step when the reacting solution was moved from one flow cell to another using a peristaltic pump. The logic inputs (Glc and Glc6P) were injected in all possible combinations (**0**,**0**; **0**,**1**; **1**,**0**; **1**,**1**) into

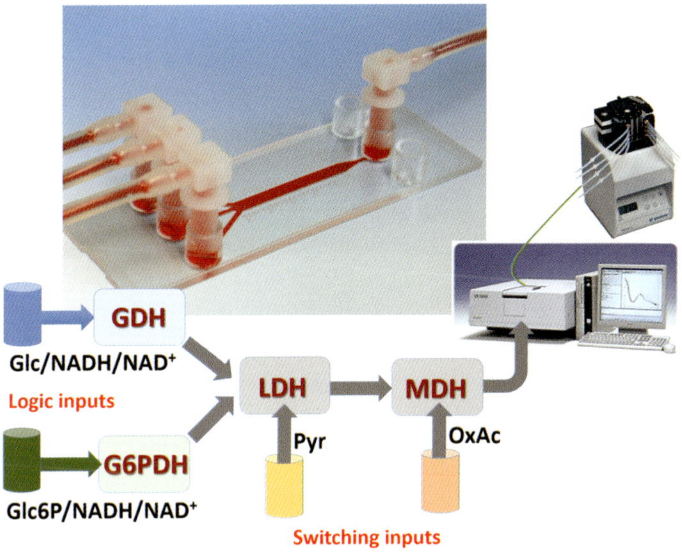

Figure 2.21 The assembled flow system used for the realization of the biocatalytic cascade operating as the switchable logic gates depending on the values of the switching signals. The red color in the flow cell does not correspond to the real experimental solutions. Source: Fratto et al. 2014 [34]. Adapted with permission of Royal Society of Chemistry.

the first flow cells operating in parallel (Figure 2.21). Both switching inputs (Pyr and OxAc) were applied at their logic value **0** (meaning the physical absence of these inputs) to the third and fourth flow cells to realize the OR gate operation on the logic inputs (Glc and Glc6P). When the first switching input (Pyr) was applied at its logic input **1** and the second switching input was at its **0** value, the NXOR gate operation was realized for the logic inputs. In order to realize the NAND gate operation for the logic inputs, both switching inputs were applied at their **1** value. The output signal in the form of optical changes corresponding to the NADH concentration was measured after the solution moved through all four flow cells, reaching a flow cuvette in the spectrophotometer (Figure 2.21). Application of various values (**0,0**; **1,0**; and **1,1**) of the switching inputs resulted in the reversible transition between three possible logic gates (OR, NXOR, and NAND, respectively) processing the logic inputs (Glc and Glc6P) (Figure 2.20B). The truth tables of these logic gates are shown in Figure 2.20, bottom.

Figure 2.22A shows the time-dependent output signals produced by the flow system in response to the various combinations of the input signals when the OR logic gate was realized upon application of the switching inputs in their **0,0** combination. Logic inputs (Glc and Glc6P) applied at their **0,0** combination resulted in no formation of NADH, thus keeping the absorbance unchanged (Figure 2.22A). When any of the logic inputs was applied at its **1** logic value (**0,1** or **1,0** combinations), the NADH concentration was increased, thus resulting in the absorbance increase (Figure 2.22A). The simultaneous application of the both input signals at their **1** value (**1,1** input combination) resulted in the double increase in the absorbance value due to the operation of the both parallel

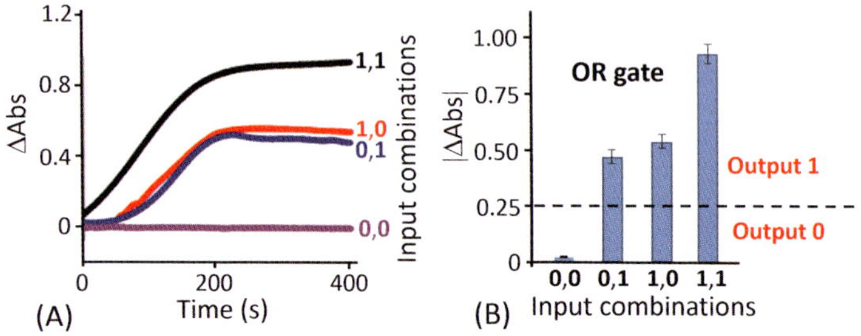

Figure 2.22 (A) Time-dependent outputs produced by the biocatalytic system operated in the OR gate mode with both switching inputs applied at their **0** value. (B) The bar chart showing the output values measured at the gate time of 400 seconds. The dashed line shows a threshold separating logic **0** and **1** output values. The curves and bars show the outputs generated with the following logic input value combinations: **0**,**0**; **0**,**1**; **1**,**0**; and **1**,**1**. Source: Fratto et al. 2014 [34]. Adapted with permission of Royal Society of Chemistry.

reactions (Figure 2.22A). Figure 2.22B represents the bar chart of the output signal values obtained at the gate time of 400 seconds.

Figure 2.23A shows the time-dependent output signals produced by the flow system in response to the various combinations of the input signals when the NXOR logic gate was realized upon application of the switching inputs in their **1**,**0** combination. When both logic inputs (Glc and Glc6P) were applied at their **0** value, the NADH production in the first reacting cells did not occur; however the first switching signal (Pyr) applied at its **1** value resulted in the consumption of NADH existing in the "machinery" solution, thus resulting in the overall decrease of the measured absorbance (Figure 2.22A). When any of the logic inputs was applied at its **1** value (**0**,**1** or **1**,**0** input combinations), the NADH formation in one of the parallel operating flow cells was compensated by the NADH oxidation in the next cell due to the presence of Pyr, thus resulting in no overall changes in the absorbance value (Figure 2.23A). Application of the both logic inputs at their **1** logic values (**1**,**1** input combination) resulted in the

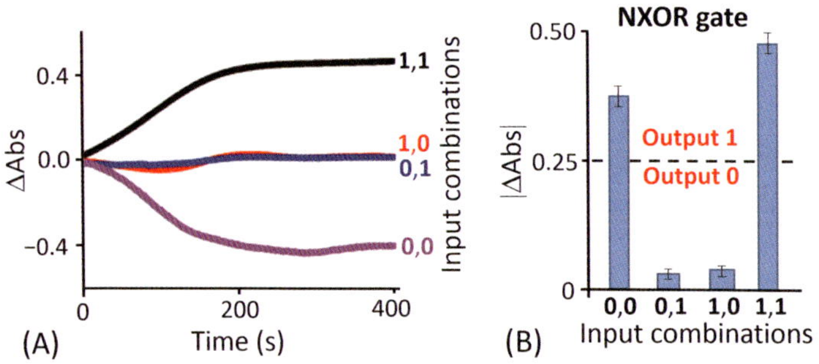

Figure 2.23 (A) Time-dependent outputs produced by the biocatalytic system operated in the NXOR gate mode with the switching inputs applied at **1**,**0** values. (B) The bar chart showing the output values measured at the gate time of 400 seconds. The dashed line shows a threshold separating logic **0** and **1** output values. The curves and bars show the outputs generated with the following logic input value combinations: **0**,**0**; **0**,**1**; **1**,**0**; and **1**,**1**. Source: Fratto et al. 2014 [34]. Adapted with permission of Royal Society of Chemistry.

NADH concentration increase, which cannot be compensated by the NADH oxidation in the presence of Pyr, thus resulting in the overall absorbance increase (Figure 2.23A). Figure 2.23B represents the bar chart of the output signal values obtained at the gate time of 400 seconds. It should be noted that the output signal was defined as the absolute value of the absorbance change. In other words, the output signal value was considered as the positive value regardless of the decreasing or increasing the absorbance due to the formation or consumption of NADH in the course of the biocatalytic reactions.

Figure 2.24A shows the time-dependent output signals produced by the flow system in response to the various combinations of the input signals when the NAND logic gate was realized upon application of the switching inputs in their **1,1** combination. When the logic input signals (Glc and Glc6P) were applied at their **0,0** logic combination, the NADH concentration was decreased due to the oxidative reactions activated by the presence of Pyr and Oxacet, thus resulting in the absorbance decrease (Figure 2.24A). When one of the logic inputs was applied at its **1** value (**0,1** or **1,0** logic combinations), the NADH production was overshot by the oxidative reactions activated by the both switching inputs applied at **1** value (Figure 2.24A). Finally, the NADH reduction upon application of the both logic inputs at their **1** value was compensated by the oxidation in the presence of Pyr and OxAc, thus keeping the absorbance value unchanged (Figure 2.24A). Figure 2.24B represents the bar chart of the output signal values obtained at the gate time of 400 seconds. Like before, the output signal was defined as the absolute value of the absorbance change.

The designed flow system mimicking various Boolean logic gates represents a new step in the development of biomolecular computing systems. It offers some programming features that allow reversible switching between different modes of logic operations. The system complexity has the potential to be increased by attaching additional flow cells to the input channels to perform various logic operations prior to the signal processing in the present device. Presently, the system in its current configuration does not allow its extension by attaching additional gates to the output channel. In other words, the present system can operate

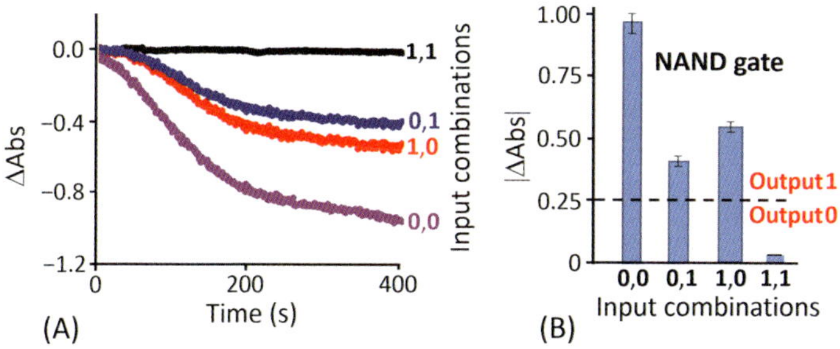

Figure 2.24 (A) Time-dependent outputs produced by the biocatalytic system operated in the NAND gate mode with the switching inputs applied at **1,1** values. (B) The bar chart showing the output values measured at the gate time of 400 s. The dashed line shows a threshold separating logic **0** and **1** output values. The curves and bars show the outputs generated with the following logic input value combinations: (a) **0,0**; (b) **0,1**; (c) **1,0**; and (d) **1,1**. Source: Fratto et al. 2014 [34]. Adapted with permission of Royal Society of Chemistry.

only as the end of the logic network. This problem is caused by the definition of the output signal as the absolute value of the absorbance change. In this definition the logic input **1** is the same for the increasing or decreasing of the NADH concentration, which is difficult to accept for the ensuing logic operations. This problem can be eliminated with another definition of the output signal, which is unambiguously related to the NADH concentration increase. This alternative output signal definition has already been realized in an XOR gate [31, 33]; however, its use in the present switchable system requires its redesign using different biocatalytic processes. Future progress in the design of signal-switchable and reprogrammed logic systems could benefit many bioanalytical systems, which are used for processing various biomarkers important for biomedical analysis [2–4].

2.6 Conclusions and Perspectives

The logic gates exemplified above were aimed at concept demonstration without any specific practical application. One can consider this research step as filling a toolbox with necessary components for future work. However, the question "What kind of applications is feasible?" is inevitable. This question also appears regarding the future of molecular computing using synthetic nonbiological systems [97]. The original idea of using molecular systems, particularly based on DNA molecules, for information processing was motivated by the possibility to perform massive parallel data processing allowing faster results, at least for some combinatorial problems [98]. Other research prospects included miniaturization of micro-size electronic elements to the size of single molecules [99]. Both possibilities are still under discussion, particularly for the DNA-based computing systems, as seen in the recent literature [100–102], but none of the expected advantages of (bio)molecular computing have been practically realized after more than 20 years of research, thus resulting in some disappointment. While this unfortunate lack of practical applicability is characteristic of all (bio)molecular computing/logic systems, it is even more typical for the enzyme logic gates overviewed in this chapter. Indeed, limited complexity and difficulties in assembling large information processing networks are in part what inhibit biomolecular systems from competing with electronic computers. If pure computational application of the biomolecular systems is not achievable, at least at the present level of technology, other potential applications have to be formulated to justify the continued research. One of the possible alternatives is a special kind of biosensing in which biomolecular logic systems process a number of biomarkers making binary decision in the binary form of YES/NO about specific biomedical problems [2]. Logic analysis of biomolecules could be performed intracellularly [103, 104], thus resulting in novel medical diagnostic tools. Combining logic analysis with chemical intervention (e.g., drug release or regulation of biochemical pathways) would result in the novel theranostic (combination of diagnostics with therapy) approach. When biosensing, particularly for biomedical applications, is considered as the primary aim of biomolecular logic systems, logic **0** and **1** values should be redefined [2]. Indeed, when biomolecular logic gates were designed for simple concept demonstration, arbitrary values were assigned to the input concentrations corresponding to logic **0** and **1**. Usually, logic **0** was considered as the complete absence of the chemical input and logic **1** was defined as a conveniently high concentration of the input to allow well-separated **0** and **1** values of the output signal. In the biosensor/bioanalytical applications, logic **0** input values should correspond to physiologically normal concentrations of biomarkers, and logic **1** input values should be defined as pathophysiologically elevated (or in some cases decreased) biomarker concentrations. In this case the biomolecular reactions will proceed at all combinations of inputs signals, even if all of them are applied at logic **0** values, thus resulting in much narrower gap separating output **1** and **0** values, making diagnostic conclusion YES/NO more difficult.

The following chapters in the book will address further research and advances in the enzyme logic gates formulated in this chapter. This will include the work on combining the enzyme logic gates in small logic networks, their optimization, and practical applications, particularly for designing binary biosensors with YES/NO output responses. Other advances and applications of the enzyme logic gates such as their interfacing with different transducers and bioelectronic devices are discussed in the following chapters as well. Overall, the fundamental logic gates formulated in this chapter are important elements of many systems with higher complexity discussed in the following chapters. The readers can select some of the other chapters for reading, while skipping others, depending on specific interest, but this chapter is highly important for understanding the following material, thus becoming a conceptually important part of the book.

References

1 Maini, A.K. (2007). *Digital Electronics: Principles, Devices and Applications*. Wiley.
2 Katz, E., Wang, J., Privman, M., and Halámek, J. (2012). *Anal. Chem.* 84: 5463–5469.
3 Wang, J. and Katz, E. (2010). *Anal. Bioanal. Chem.* 398: 1591–1603.
4 Wang, J. and Katz, E. (2011). *Isr. J. Chem.* 51: 141–150.
5 Katz, E. (ed.) (2012). *Biomolecular Information Processing – From Logic Systems to Smart Sensors and Actuators*. Weinheim: Wiley-VCH.
6 Katz, E. and Privman, V. (2010). *Chem. Soc. Rev.* 39: 1835–1857.
7 Katz, E. (2019). *ChemPhysChem* 20: 9–22.
8 Pita, M., Krämer, M., Zhou, J. et al. (2008). *ACS Nano* 2: 2160–2166.
9 Bocharova, V., Tam, T.K., Halámek, J. et al. (2010). *Chem. Commun.* 46: 2088–2090.
10 Pita, M. and Katz, E. (2008). *J. Am. Chem. Soc.* 130: 36–37.
11 Okhokhonin, A.V., Domanskyi, S., Filipov, Y. et al. (2018). *Electroanalysis* 30: 426–435.
12 Strack, G., Pita, M., Ornatska, M., and Katz, E. (2008). *ChemBioChem* 9: 1260–1266.
13 Halámková, L., Halámek, J., Bocharova, V. et al. (2012). *Analyst* 137: 1768–1770.

14 Manesh, K.M., Halámek, J., Pita, M. et al. (2009). *Biosens. Bioelectron.* 24: 3569–3574.
15 Halámek, J., Windmiller, J.R., Zhou, J. et al. (2010). *Analyst* 135: 2249–2259.
16 Baron, R., Lioubashevski, O., Katz, E. et al. (2006). *Org. Biomol. Chem.* 4: 989–991.
17 Baron, R., Lioubashevski, O., Katz, E. et al. (2006). *J. Phys. Chem. A* 110: 8548–8553.
18 Katz, E. (2017). *ChemPhysChem* 18: 1688–1713.
19 Passonneau, J.V. and Lowry, O.H. (1993, Chapter 1). *Enzymatic Analysis: A Practical Guide*, 3–22. Totowa: Humana.
20 Katz, E. and Fratto, B.E. (2017). Enzyme-based reversible logic gates operated in flow cells. In: *Advances in Unconventional Computing*, vol. 2, Chapter 2 (ed. A. Adamatzky), 29–59. Springer.
21 Fratto, B.E. and Katz, E. (2015). *ChemPhysChem* 16: 1405–1415.
22 Fratto, B.E., Guz, N., and Katz, E. (2015). *Parallel Processing Lett.* 25 (01): 1540001.
23 Guz, N., Fedotova, T.A., Fratto, B.E. et al. (2016). *ChemPhysChem* 17: 2247–2255.
24 Fratto, B.E., Lewer, J.M., and Katz, E. (2016). *ChemPhysChem* 17: 2210–2217.
25 Poghossian, A., Katz, E., and Schöning, M.J. (2015). *Chem. Commun.* 51: 6564–6567.
26 Zavalov, O., Bocharova, V., Privman, V., and Katz, E. (2012). *J. Phys. Chem. B* 116: 9683–9689.
27 von Maltzahn, G., Harris, T.J., Park, J.-H. et al. (2007). *J. Am. Chem. Soc.* 129: 6064–6065.
28 Arugula, M.A., Bocharova, V., Pita, M. et al. (2010). *J. Phys. Chem. B* 114: 5222–5226.
29 Chuang, M.-C., Windmiller, J.R., Santhosh, P. et al. (2011). *Chem. Commun.* 47: 3087–3089.
30 Kumar Sarkar, S., De, A.K., and Sarkar, S. (2015). *Foundation of Digital Electronics and Logic Design*. Singapore: Pan Stanford.
31 Halámek, J., Bocharova, V., Arugula, M.A. et al. (2011). *J. Phys. Chem. B* 115: 9838–9845.
32 Privman, V., Zhou, J., Halámek, J., and Katz, E. (2010). *J. Phys. Chem. B* 114: 13601–13608.
33 Filipov, Y., Domanskyi, S., Wood, M.L. et al. (2017). *ChemPhysChem* 18: 2908–2915.
34 Fratto, B.E., Roby, L.J., Guz, N., and Katz, E. (2014). *Chem. Commun.* 50: 12043–12046.
35 Herlet, J., Kornberger, P., Roessler, B. et al. (2017). *Biotechnol. Biofuels* 10: 234.
36 Prieto, M.A., Vazquez, J.A., and Murado, M.A. (2015). *Analyst* 140: 3587–3602.
37 Tipton, K.F. and Dixon, H.B.F. (1979). *Methods Enzymol.* 63: 183–234.
38 Bakshi, S., Zavalov, O., Halámek, J. et al. (2013). *J. Phys. Chem. B* 117: 9857–9865.

39 Gdor, E., Katz, E., and Mandler, D. (2013). *J. Phys. Chem. B* 117: 16058–16065.
40 Privman, V., Fratto, B.E., Zavalov, O. et al. (2013). *J. Phys. Chem. B* 117: 7559–7568.
41 Halámek, J., Zavalov, O., Halámková, L. et al. (2012). *J. Phys. Chem. B* 116: 4457–4464.
42 Melnikov, D., Strack, G., Zhou, J. et al. (2010). *J. Phys. Chem. B* 114: 12166–12174.
43 Privman, V., Pedrosa, V., Melnikov, D. et al. (2009). *Biosens. Bioelectron.* 25: 695–701.
44 Miyake, T., Josberger, E.E., Keene, S. et al. (2015). *APL Materials* 3: 014906.
45 Zhou, J., Arugula, M.A., Halámek, J. et al. (2009). *J. Phys. Chem. B* 113: 16065–16070.
46 Ray, A., Liu, J., Karanth, S. et al. (2009). *Toxicol. Appl. Pharmacol.* 236: 341–347.
47 Niazov, T., Baron, R., Katz, E. et al. (2006). *Proc. Natl. Acad. USA.* 103: 17160–17163.
48 MacVittie, K. and Katz, E. (2014). *Int. J. General Systems* 43: 722–739.
49 Strack, G., Ornatska, M., Pita, M., and Katz, E. (2008). *J. Am. Chem. Soc.* 130: 4234–4235.
50 Privman, M., Tam, T.K., Bocharova, V. et al. (2011). *ACS Appl. Mater. Interfaces* 3: 1620–1623.
51 Katz, E., Bocharova, V., and Privman, M. (2012). *J. Mater. Chem.* 22: 8171–8178.
52 Honarvarfard, E., Gamella, M., Poghossian, A. et al. (2017). *Applied Materials Today* 9: 266–270.
53 Kramer, F., Halámková, L., Poghossian, A. et al. (2013). *Analyst* 138: 6251–6257.
54 Katz, E. and Pita, M. (2009). *Chem. Eur. J.* 15: 12554–12564.
55 Szacilowski, K. (2012). *Infochemistry – Information Processing at the Nanoscale*. Chichester: Wiley.
56 de Silva, A.P. (2013). *Molecular Logic-Based Computation*. Cambridge: Royal Society of Chemistry.
57 Katz, E. (ed.) (2012). *Molecular and Supramolecular Information Processing – From Molecular Switches to Logic Systems*. Weinheim: Wiley-VCH.
58 Katz, E. (2015). *Curr. Opin. Biotechnol.* 34: 202–208.
59 Bostock, G. (1988). *Programmable Logic Devices. Technology and Applications*. New York, NY: McGraw-Hill.
60 De Silva, A.P., Dixon, I.M., Gunaratne, H.Q.N. et al. (1999). *J. Am. Chem. Soc.* 121: 1393–1394.
61 Straight, S.D., Liddell, P.A., Terazono, Y. et al. (2007). *Adv. Funct. Mater.* 17: 777–785.
62 Turfan, B. and Akkaya, E.U. (2002). *Org. Lett.* 4: 2857–2859.
63 Wang, Z., Zheng, G., and Lu, P. (2005). *Org. Lett.* 7: 3669–3672.
64 Baytekin, H.T. and Akkaya, E.U. (2000). *Org. Lett.* 2: 1725–1727.
65 Zong, G., Xiana, L., and Lua, G. (2007). *Tetrahedron Lett.* 48: 3891–3894.

66 Massey, M., Medintz, I.L., Ancona, M.G., and Algar, W.R. (2017). *ACS Sensors* 2: 1205–1214.
67 Chen, J.H. and Zeng, L.W. (2013). *Biosens. Bioelectron.* 42: 93–99.
68 Zhou, N., Windmiller, J.R., Valdés Ramírez, G. et al. (2011). *Anal. Chim. Acta* 703: 94–100.
69 Du, Y., Peng, P., and Li, T. (2018). *Chem. Commun.* 54: 6132–6135.
70 Zhang, R., Gupta, P., and Jha, N.K. (2007). *IEEE Trans. on Computer-Aided Design of Integrated Circuits and Systems* 26: 1233–1245.
71 Goldmann, M. and Karpinski, M. (1998). *SIAM J. Comput.* 27: 230–246.
72 Yang, X., Cai, L., and Kang, Q. (2012). *J. Comput. Theor. Nanosci.* 9: 621–625.
73 Imre, A., Csaba, G., Ji, L. et al. (2006). *Science* 311: 205–208.
74 Stroud, C.E. (1994). *IEEE Trans. VLSI Systems* 2: 516–521.
75 Yang, G., Hung, W.N.N., Song, X., and Perkowski, M. (2005). *Theoretical Computer Science* 334: 259–274.
76 Li, W., Yang, Y., Yan, H., and Liu, Y. (2013). *Nano Lett.* 13: 2980–2988.
77 Zhu, J., Zhang, L., Dong, S., and Wang, E. (2013). *ACS Nano* 7: 10211–10217.
78 Gao, J.T., Liu, Y.Q., Lin, X.D. et al. (2017). *Sci. Rep.* 7: 14014.
79 George, A.K. and Singh, H. (2017). *Micro Nano Lett.* 12: 143–146.
80 Zhu, Q., Gao, Y., and Li, Y.H. (2016). *ChemistrySelect* 1: 4554–4559.
81 Yang, J., Song, Z.C., Liu, S. et al. (2016). *ACS Appl. Mater. Interfaces* 8: 22451–22456.
82 Zhou, C.Y., Liu, D.L., Wu, C.T. et al. (2016). *Nanoscale* 8: 17524–17531.
83 Fan, D.Q., Wang, K., Zhu, J.B. et al. (2015). *Chem. Sci.* 6: 1973–1978.
84 Mailloux, S., Guz, N., Zakharchenko, A. et al. (2014). *J. Phys. Chem. B* 118: 6775–6784.
85 Privman, V., Zavalov, O., Halámková, L. et al. (2013). *J. Phys. Chem. B* 117: 14928–14939.
86 Silverstein, E. and Sulebele, G. (1969). *Biochemistry* 8: 2543–2550.
87 Mailloux, S., Zavalov, O., Guz, N. et al. (2014). *Biomater. Sci.* 2: 184–191.
88 Privman, V., Domanskyi, S., Mailloux, S. et al. (2014). *J. Phys. Chem. B* 118: 12435–12443.
89 Privman, V., Arugula, M.A., Halámek, J. et al. (2009). *J. Phys. Chem. B* 113: 5301–5310.
90 Halámek, J., Bocharova, V., Chinnapareddy, S. et al. (2010). *Mol. Biosyst.* 6: 2554–2560.
91 Sun, W., Xu, C.H., Zhu, Z. et al. (2008). *J. Phys. Chem. C* 112: 16973–16983.
92 de Silva, A.P., Dobbin, C.M., Vancea, T.P., and Wannalerse, B. (2009). *Chem. Commun.* 1386–1388.
93 Budyka, M.F., Potashova, N.I., Gavrishova, T.N., and Lee, V.M. (2009). *J. Mater. Chem.* 19: 7721–7724.
94 Li, Z.X., Liao, L.Y., Sun, W. et al. (2008). *J. Phys. Chem. C* 112: 5190–5196.
95 He, K.Y., Li, Y., Xiang, B.B. et al. (2015). *Chem. Sci.* 6: 3556–3564.
96 Buckhout-White, S., Claussen, J.C., Melinger, J.S. et al. (2014). *RSC Adv.* 4: 48860–48871.

97 Pischel, U., Andreasson, J., Gust, D., and Pais, V.F. (2013). *ChemPhysChem* 14: 28–46.
98 Adleman, L.M. (1994). *Science* 266: 1021–1024.
99 Schmidt, K.A., Henkel, C.V., Rozenberg, G., and Spaink, H.P. (2004). *Nucleic Acids Res.* 32: 4962–4968.
100 Benenson, Y. (2012). *Nat. Rev. Genet.* 13: 455–468.
101 Stojanovic, M.N., Stefanovic, D., and Rudchenko, S. (2014). *Acc. Chem. Res.* 47: 1845–1852.
102 Stojanovic, M.N. and Stefanovic, D. (2011). *J. Comput. Theor. Nanosci.* 8: 434–440.
103 Rinaudo, K., Bleris, L., Maddamsetti, R. et al. (2007). *Nat. Biotechnol.* 25: 795–801.
104 Benenson, Y. (2009). *Cur. Opinion. Biotechnol.* 20: 471–478.

3

Optimization of Enzyme-Based Logic Gates for Reducing Noise in the Signal Transduction Process

3.1 Introduction

The present chapter introduces, by illustrative model examples, concepts in noise reduction and control for scalability in biochemical computing. While the theoretical background for the noise reduction has been discussed elsewhere [1], this chapter is mostly concentrated on experimental realizations of noise-suppressing methods applied to different enzyme logic gates. The approach has been tested in experimental realizations for enzyme-reaction-based logic gates and networks [2–18]. However, the discussed theoretical ideas generally apply to a broad range of chemical and biomolecular computing systems, suggesting that, at present, typical networks of up to 10 binary gates can operate with the acceptable level of noise [18], similar to findings in the networking of neurons [19, 20]. For networks of more than order 10 binary steps, additional nonbinary network elements, as well as proper network design to utilize redundancy for digital error correction, will be needed for fault-tolerant operation [5, 17, 18]. The level of noise in the environments envisaged for applications of future chemical [21–24] and biomolecular [25–28] computing systems is quite high as compared to their electronic computer counterparts. Indeed, both the input/output signals and the "gate machinery" chemical concentrations can typically fluctuate several percent or more, on the scale normalized to the digital **0** to **1** range. Avoiding noise amplification, by careful design and parameter selection for gates and networks, is therefore quite important even for relatively small networks. The offered discussion does not address the origin/sources of stochastic and environmental noise in (bio)chemical reactions. Theoretical considerations and experimental realizations reviewed here apply to numerous reported chemical and biochemical computing systems. Indeed, chemical processes can be cast [29–33] in the language of computing operations, with signals represented by changes [21–24, 29–35] in structural, chemical, or physical properties, resulting due to physical [36, 37], chemical [38], or more than one type [39–41] of input. The output signals can be detected spectroscopically, electrically/electrochemically, or using other transduction tools [42]. Chemical computing can be done in the bulk, e.g., in solution [29–35] or at surfaces/interfaces [43–47], such as at electrodes or on Si-chips. Notably, additional noise in the logic gate operation can originate from interaction of biomolecules with signal transduction interfaces, being different for different transduction techniques.

Enzyme-Based Computing Systems, First Edition. Evgeny Katz.

3.2 Signal Transduction Function in the Enzyme-Based Logic Systems: Filters Producing Sigmoid Response Functions

While (bio)molecular realization of Boolean logic gates requires binary-defined input and output signals, all chemical systems, including those based on enzyme-catalyzed reactions, are not binary by their nature. Indeed, all reacting species can have variable concentrations, and the products generated have concentrations developed in time and dependent on the concentrations of the initial species. When specific low and high concentrations of the initial reactants (input signals) are selected to mimic the logic **0** and **1** values, in reality a range of concentrations will be applied. Variation of the input concentrations may have natural distribution (e.g., when physiologically produced inputs are applied) or could originate from technical irreproducibility (e.g., different sample preparations, pipetting, temperature variation, etc.). The input signal variability will be translated to the output signals functionally dependent on the concentrations of the initial reactants. In other words, some noise in the input and output signals is inevitable and should be considered for the system optimization.

A linear function of the produced output signal on the input signal obtained in a broad analyte concentration range (dynamic range) is always considered as a positive feature of analog sensors/biosensors, but it is an obvious disadvantage for digital systems where sharp transition between **0** and **1** outcome signals is needed. In analog sensors/biosensors, a high slope of the response function (signal vs. analyte concentration) is also a positive feature corresponding to a high sensitivity. However, this feature in digital systems with YES/NO (**1**, **0**) output, particularly in the areas near **0** and **1** inputs, means high analog noise produced by the digital system [16, 18] (Figure 3.1a). Many biocatalytic systems demonstrate signal

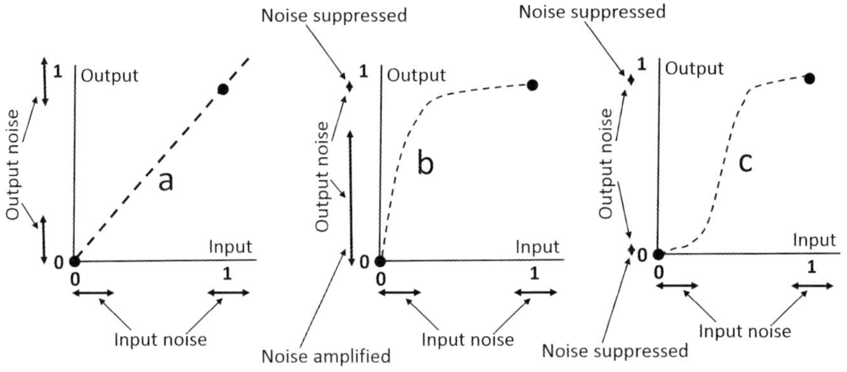

Figure 3.1 Different response functions of the Identity logic gate: (a) linear function, (b) convex function, and (c) sigmoid function. Notably, larger slope of the function results in the amplified noise in the output signal upon varying the input signal. The linear function (a) has the same noise produced near logic points **0** and **1**. The convex function (b) amplifies the noise near logic point **0** and suppresses noise near logic point **1**. The sigmoid function suppresses noise near both logic points **0** and **1**.

saturation at high concentrations of analyte species (e.g., enzyme substrates); thus the noise produced by the systems in the area of the signal saturation (possibly near digital **1** value) will be suppressed (Figure 3.1b). However, at low input concentrations the produced signal is increasing with the elevated concentrations (possibly with very sharp changes), thus significantly amplifying the analog noise near digital **0** value. This *convex shape* response function, which is very typical for biocatalytic processes, is suppressing noise only at high input concentrations, but it can significantly amplify analog noise at low input concentrations. The optimum transduction function (output signal vs. input signal) for binary operating systems should have sigmoidal shape with small slope near both **0** and **1** logic values corresponding to low and high concentrations, respectively (Figure 3.1c). This would allow analog noise suppression for both binary signals (**0** and **1**). Such *sigmoidal shape* suppressing analog noise is frequently produced in natural biochemical/physiological processes regulated through very sophisticated biological mechanisms [48] (Figure 3.2), which are very difficult or even impossible to reproduce in artificial biomolecular systems of limited complexity. However, similar results can be obtained for simple biocatalytic processes mimicking Boolean logic gates using very simple reaction steps, much easier than regulation mechanisms used by nature in biological processes. The simple chemical processes suppressing analog noise cannot compete with complex natural biological processes being the bases of life, but they can be effective enough to decrease noise in model logic gates converting *convex* response function to *sigmoidal* one. The change of the signal transduction function can be achieved by adding additional reaction steps named for brevity "filters." The following sections of the chapter overview different logic systems, including Identity gate, OR gate, AND gate, and XOR gate, with a sigmoidal dependence of the output signals on the input signal concentration. The theoretical part of the work,[1] where model systems have been considered and used to fit the experimental results, can be found elsewhere [2, 6, 8, 14, 49], while the following sections mostly overview the obtained experimental data.

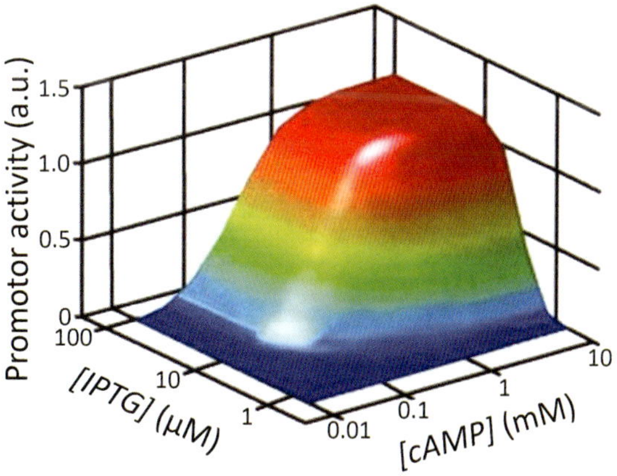

Figure 3.2 The map of a *cis*-regulatory input function in living *Escherichia coli* cells with parameters for AND-like function. Note that the sigmoid-like shape is produced through a complex biological regulatory mechanism. Isopropyl β-D-thiogalactoside (IPTG) and cyclic adenosine monophosphate (cAMP) are two inputs regulating cellular processes. Source: Adapted from Mayo et al. 2006 [48].

1 The theoretical work on modeling of the logic gates and fitting the experimental data, aiming at suppressing the analog noise and optimization of the function connecting output and input signals, was performed by Prof. Vladimir Privman (deceased in 2018).

3.2.1 Identity (YES) Logic Gate Optimization

Any chemical/biochemical reaction converting a primary substrate (reagent) to the final product can be considered as a binary operating Identity (YES) Boolean logic gate, if appropriate digitalization of the input signal (reagent/substrate concentration) is formulated. This means that the binary input signal **0** should be defined as a low (possibly zero) concentration of the reacting species, while the input signal **1** should be defined as a high concentration of these species, being significantly different from the low concentrations reserved for the definition of **0** input. When binary-defined input signals are applied, the output signal will be also digitized appearing at two levels, low and high, corresponding to **0** and **1** outputs generated by the reacting system. The exact concentrations of the produced species (output signal) are set by the (bio)chemical reaction process. While theoretically single concentration values should be used to define logic **0** and **1** values, in reality these values will be represented with a range of concentrations, thus introducing some fluctuations (noise) in the output signals. Figure 3.3A shows an example biocatalytic process [14], where the initial substrate H_2O_2 results in the oxidation of 2,2′-azino-*bis*(3-ethylbenzothiazoline-6-sulfonic acid), ABTS, catalyzed by horseradish peroxidase (HRP). The oxidized product ($ABTS_{ox}$) can be analyzed by measuring optical absorbance ($\lambda_{max} = 420$ nm), whose increase can be defined as the output signal. The optical absorbance corresponding to the biocatalytic production of $ABTS_{ox}$ increases with the initial concentration of the input signal (H_2O_2) reaching saturation controlled by the amount of added ABTS (assuming that the process is limited by ABTS and not by H_2O_2). This function (output vs. input signals) has a convex shape (Figure 3.3B) (black curve), demonstrating sharp increase at low input (H_2O_2) concentrations ended with a flat saturation region. The sharp increase of the output signal near the logic **0** value (at the low input concentrations) results in the high noise produced in the output signal around **0** input/output point. On the other hand, the function saturation achieved at the high input concentrations results in the low noise around **1** input/output point. Thus, in order to reduce the noise produced by the biocatalytic system, one should take care on the low input signal, decreasing the function slope at the **0** logic value. This can be achieved by adding one more reaction to the biocatalytic cascade named "filter" (Figure 3.3A). Ascorbic acid (Asc), operating as a reducing agent, returned back the oxidized ABTS form ($ABTS_{ox}$) to its initial reduced (noncolored) form, thus eliminating the output signal measured optically. The concentration of Asc can vary and should be optimized. Obviously, if the Asc concentration is higher than the H_2O_2 concentration, the $ABTS_{ox}$ will never be produced, thus inhibiting the output signal formation. However, if the H_2O_2 concentration exceeds the Asc concentration, the $ABTS_{ox}$ is produced after Asc is fully consumed. In other words, as long as Asc is present in the reaction system, the output signal formation is inhibited, but it sharply increases after Asc is finished in the filter reaction. This results in the change of the convex output function to the sigmoid function with low slopes at both binary input/output logic points (**0** and **1**) (Figure 3.3B). The extent of the flattening of the output function at low input concentrations depends on the Asc concentration. The best (optimized) output function should have the sharp increase in the middle concentration range

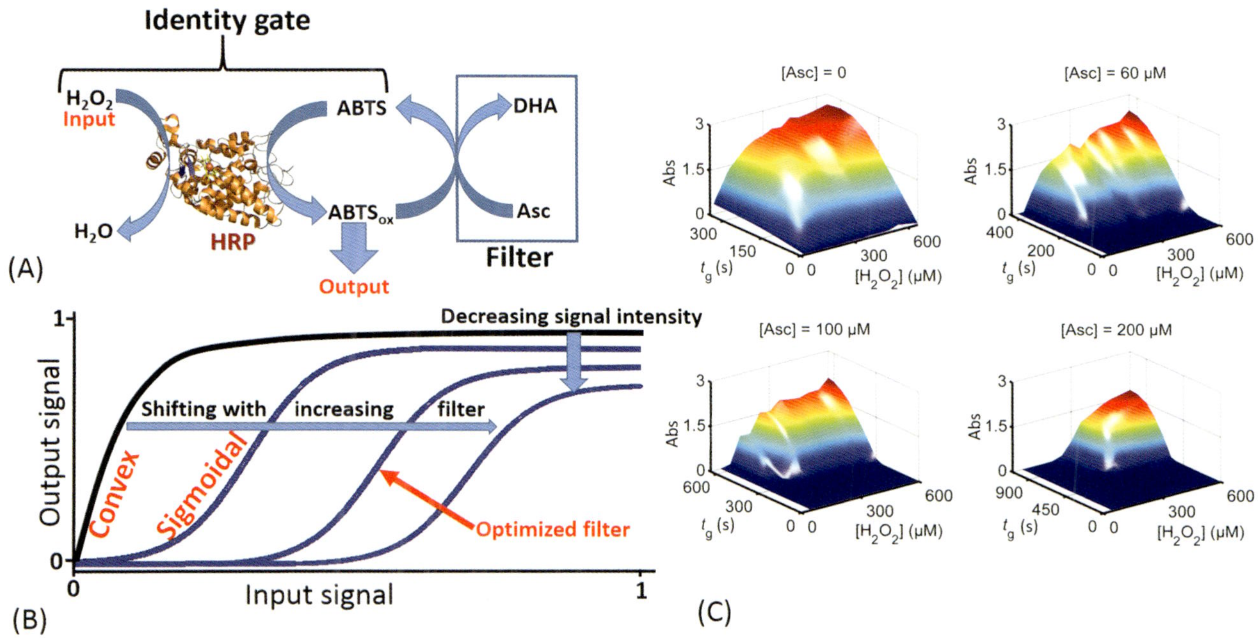

Figure 3.3 (A) The biocatalytic system mimicking the Identity (YES) Boolean logic gate. The input signal (H_2O_2) results in oxidation of ABTS in the reaction catalyzed by HRP. The oxidized product ($ABTS_{ox}$) is analyzed optically and represents the output signal. The reaction shown in the box represents the "filter" part of the system, where the $ABTS_{ox}$ product is returned back to the initial state (ABTS) being reduced by ascorbate (Asc); DHA is dehydroascorbic acid – the product of Asc oxidation. As long as Asc is present, the output signal formation is inhibited. (B) Mapping the logic gate response for varying the input signal concentration. A convex response function of the gate was obtained when the gate operates without the filter. Note the noise amplification near logic point **0** (for the small input concentrations). Adding the filter results in a sigmoid response with the inflection part shifted in the direction to the logic point **1** upon increasing the filter substrate (Asc) concentration. The optimized filter should have similar flattened areas near **0** and **1** logic points with the inflection located in the middle of the input concentration range. Note that the noise is suppressed by the low slope of the sigmoid response function near **0** and **1** points. (C) Experimental dependence of the optically measured output signal (corresponding to the $ABTS_{ox}$ concentration) on the initial concentration of the input signal (H_2O_2) and the reaction time. The plots show the transformation of the convex response function in the absence of the filter (Asc) to the sigmoid response function with the filter applied at different Asc concentrations. Note that the higher filter concentration results in longer flattened areas at the bottom part of the response function. Also, the intensity of the signal response (the maximum response) is decreasing with the filter concentration increase. Source: Privman et al. 2010 [14]. Adapted with permission of American Chemical Society.

between low and high input concentrations corresponding to the binary-defined **0** and **1** inputs. This would allow minimal noise at **0** and **1** inputs (see detailed theoretical explanations elsewhere [2, 6, 8, 14, 49]). The "price" paid for this gate optimization is the output signal intensity decrease. Indeed, the filter operation results in conversion of the reaction product back to the original reactant, thus finally decreasing the amount of the product generated. This decreases the optically measured output signal at the logic value **1** (Figure 3.3B). Overdoing the filter operation will result in significant output signal inhibition and in decrease of the flattened area at the logic value **1**, thus increasing noise at the high input concentrations. Figure 3.3C shows the input/output function (optically measured output signal vs. H_2O_2 concentration) and the kinetics of the output signal production for various concentrations of Asc (the filter substrate). It is easy to see that the convex function producing high noise at low H_2O_2 concentrations changes to the sigmoid function suppressing noise at the logic point **0** (low H_2O_2 concentrations) upon increasing the Asc concentration. The Asc concentration was optimized to have approximately the same flat areas at low and high concentrations of the input signal (H_2O_2), thus equally reducing noise at **0** and **1** logic points. It should be noted that the selection of Asc as the filter substrate is only an example and many other reactants converting the product back to the original substrate are possible. Particularly, different filter reactants should be used if the signal-producing reaction is different from the present example. Similar approach based on recycling of produced output species back to their initial state (or alternatively to another product) has been used to reduce noise at low input signal concentrations while mimicking other logic gates [7], e.g., AND [6, 9, 49] and OR [7] logic gates.

3.2.2 AND Logic Gate Optimization

Figure 3.4A shows a reaction biocatalyzed by horseradish peroxidase (HRP) where guaiacol (2-methoxyphenol) is oxidized by H_2O_2 to yield a colored product [50] analyzed by the optical absorbance changes. The system represents a simple realization of an AND logic gate with two reacting species operating as variable input signals [6] (Figure 3.4B). Chemically this system is almost the same as the one discussed before, which was used for realization of the Identity gate (except different species oxidized catalytically by HRP). However, logically it is different because two independent inputs, guaiacol (Input *A*) and H_2O_2 (Input *B*), are applied in four binary-defined combinations: **0,0**; **0,1**; **1,0**; **1,1**. The final optically analyzed product can be generated only in the presence of both inputs, thus producing the output signal **1** only at the input combination **1,1**, as expected for the Boolean AND gate. With the purpose of the noise analysis and gate optimization, both logic input signals (H_2O_2 and guaiacol) were applied at different concentrations ranging from the low concentration defined as logic **0** to the high concentration defined as logic **1**. When the response function (now represented by a surface in a 3D plot) was measured for different concentrations of H_2O_2 and guaiacol, the convex shape was found demonstrating a high slope of the signal increase, particularly at low concentrations of the reacting species (Figure 3.4C). This means that small fluctuations of the input concentrations

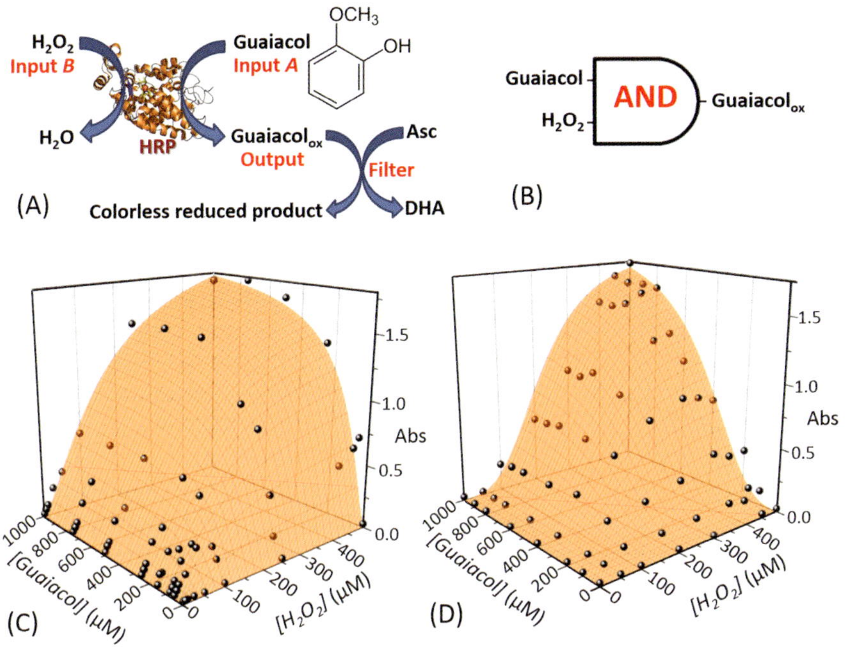

Figure 3.4 (A) A biocatalytic reaction mimicking an AND logic gate where the output **1** signal (oxidation of guaiacol observed as the increase of the optical absorbance) was obtained only in the presence of both input signals, guaiacol and H_2O_2. The operation of a "filter" system returns a fraction of the optically measured output signal back. (B) The AND logic gate shown schematically. (C) Response surface mapping of the output signal for different concentrations of the applied input signals when the biocatalytic system operated without the filter part. Note the convex shape of the response surface pattern with the sharp signal increase at low concentrations of the reacting species. (D) Response surface mapping of the output signal for different concentrations of the applied input signals including the filter part. Note that the system operation with the added filter changes the original convex response surface pattern to the sigmoid shape, thus decreasing the output signal at low concentrations of the reacting species and reducing the analog noise at the logic input **0**. Source: Bakshi et al. 2013 [6]. Adapted with permission of American Chemical Society.

near logic **0** point would result in large changes of the output signal, thus producing high analog noise. The convex response shape is typical for most of chemical processes, particularly in biocatalytic reactions, where the reaction rate is increasing with the elevated concentrations of the reacting species. On the other hand, the system does not produce much noise at higher concentrations of the reacting species (near logic value **1**), when the biocatalytic reaction reaches saturation, thus resulting in flattening of the response function. The noise at the logic value **0** can be decreased by decreasing the slope of the response function at low concentrations of the reacting species (taking into account both input signals, this will be at **0,0**; **0,1**; and **1,0** input combinations). In other words, the reaction should be inhibited at the low concentrations. This can be achieved by modifying the process with an additional step consuming the product of the reaction at its low concentration. Addition of ascorbic acid (Asc) to the system resulted in conversion of the reaction product (oxidized guaiacol) to another reduced colorless product (Figure 3.4A). Since Asc was added in limited (optimized) amount, after full consumption of Asc, the process inhibiting the output signal formation was stopped, and the output signal was increased, finally reaching the saturated value, thus converting the response function to the sigmoid shape (Figure 3.4D). The additional reaction step inhibiting the output signal formation at low concentrations of the input signals is considered as a filter function, similar to the system explained before for reducing analog noise in the Identity gate.

Another realization of the AND gate with the filter function [49] is shown in Figure 3.5A. The AND gate was realized with a two-step reaction cascade catalyzed by two enzymes, hexokinase (HK) and glucose 6-phosphate dehydrogenase (G6PDH). The first reaction step catalyzed by HK was activated with two substrates, adenosine 5'-triphosphate (ATP) (Input *A*) and glucose (Glc) (Input *B*), resulting in formation of glucose-6-phosphate (Glc6P) as an intermediate product in the reaction cascade. Then, Glc6P was used to reduce NAD^+ in the second reaction step catalyzed by G6PDH. This biocatalytic reaction resulted in the formation of NADH analyzed optically and defined as the final output signal. The output signal **1** (increased absorbance at $\lambda_{max} = 340$ nm corresponding to the produced NADH) was obtained only in the presence

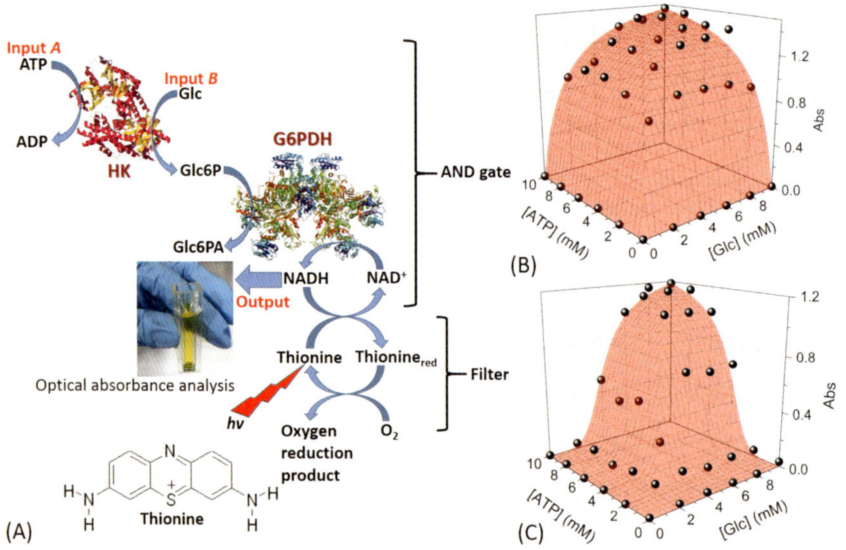

Figure 3.5 (A) The biocatalytic cascade for processing biomolecular input signals mimicking the AND logic gate followed by a photochemically activated filter process for converting a convex-shaped response function to a sigmoid response function (the system operation and all abbreviations are explained in the text). The convex response function in the absence of the filter (B) and the corresponding sigmoid function (C) obtained with the filter. The point symbols correspond to the experimentally measured outputs, and the surfaces were obtained with a theoretical model (see Ref. [49] for the theoretical aspects of this work). Source: Privman et al. 2013 [49]. Adapted with permission of American Chemical Society.

of both primary inputs (ATP and Glc) at the logic combination **1**,**1**, thus, mimicking the AND gate operation. As expected, mapping of the response function resulted at the convex shape surface (Figure 3.5B), with the sharp increase at low input concentrations, thus producing high noise near the logic points **0**,**0**; **0**,**1**; and **1**,**0**. The filter function aiming at the noise decrease at these points was realized with photochemical oxidation of NADH returning it to the original oxidized form NAD^+. The photochemical reaction, shown schematically in Figure 3.5A, resulted in the inhibition of the output signal formation at low input concentrations, thus yielding sigmoidal response function with small noise at the logic points **0**,**0**; **0**,**1**; and **1**,**0** (Figure 3.5C). Importantly, the optimization of the filter operation required optimization of the irradiation time and light intensity. Insufficient filter operation did not result in the formation of small slope areas at the bottom part of the response function (in other words did not decrease noise at low input concentrations). Overdoing the filter reaction inhibited the process too much, thus decreasing substantially the output signal at **1**,**1** logic point. Overall, the photochemical filter used in this system is only another option for reducing output signal at low input signals, while the concept of the filter operation is still the same as discussed before.

3.2.3 OR Logic Gate Optimization

The OR logic gate shown in Figure 3.6A is based on two biocatalytic reactions [8]. The first reaction catalyzed by esterase (Est) is activated with two substrates, ethyl butyrate (Input *A*) and methyl butyrate (Input *B*). Both of them are biocatalytically hydrolyzed to yield butyric acid, thus decreasing pH value of the reacting solution. The second reaction catalyzed by laccase results in oxidation of $[Fe(CN)_6]^{4-}$ with oxygen -producing $[Fe(CN)_6]^{3-}$ [51]. This process results in optical absorbance changes defined as the final output signal. The biocatalytic activity of laccase is pH dependent, being very low at pH values above 8 and sharply increasing at lower pH values [52, 53]. When a non-buffered reaction solution with initial pH 9.0 was used, the laccase reaction was inhibited, but it was sharply activated even with small pH decrease below pH 8. This resulted in significant noise at low input concentrations near the logic value **0**,**0** (Figure 3.6B). On the other hand, at high input concentrations corresponding to the logic value **1** for both inputs (**0**,**1**; **1**,**0**; and **1**,**1** input combinations), the laccase biocatalytic activity was rather stable demonstrating saturation (plateau), which originates from much weaker dependence of laccase activity at acidic pH values (meaning that the enzyme is already operating at the maximum activity) (Figure 3.6B). This plateau in the output function resulted in small noise at the logic points **0**,**1**; **1**,**0**; and **1**,**1** of the input signals. Addition of Tris-buffer suppressed pH changes for small amounts of butyric acid produced (Figure 3.6A) and resulted in sigmoid dependence of the output function on the input concentrations (Figure 3.6C), particularly suppressing noise near the logic point **0**,**0**, thus operating as the filter in the terms discussed above. It should be noted that this filter is effective for both input signals producing sigmoid function for both input concentrations.

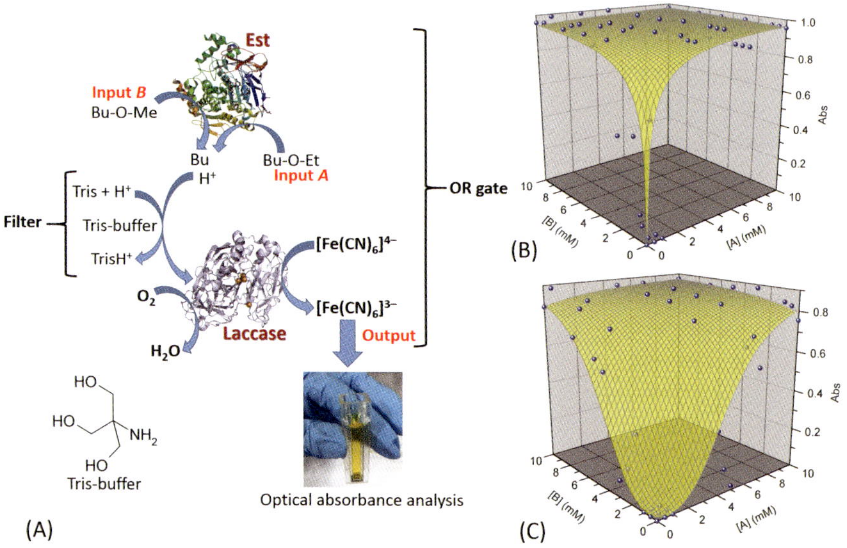

Figure 3.6 (A) The biocatalytic cascade mimicking an OR logic gate with the pH buffer operating as a filter converting the convex response function to the sigmoid response. The convex response function in the absence of the filter (B) and the corresponding sigmoid function obtained with the filter (C). The point symbols correspond to the experimentally measured outputs, and the surfaces were obtained with a theoretical model (see Ref. [8] for the theoretical aspects of this work). Note a very high slope (thus, high noise) in the response function near the logic point **0**, when the system operates without the filter. The filter decreases significantly the slope and noise near the logic point **0**. Source: Zavalov et al. 2012 [8]. Adapted with permission of American Chemical Society.

3.2.4 XOR Logic Gate Optimization

XOR logic gates based on enzyme-catalyzed reactions are difficult for realization and have been designed using different approaches [11, 54]. The system mimicking XOR gate discussed here is based on "filtering" (decomposing) input signals when they are both applied at the logic value **1** (means at **1**,**1** logic combination) [2]. In this realization the output signal measured optically was not artificially defined as an absolute value of the signal change, which was the case in other previously designed XOR systems [54]. In order to inhibit the output signal formation for the balanced **1**,**1** input combinations, the inputs were formulated as composition of two different substances in each input (Figure 3.7A). Input *A* was represented by ethanol (EtOH) with the added supplement of ATP (Input A'), and Input *B* was represented by glucose (Glc) with the added supplement of NAD^+ (Input B'). The reaction cascade proceeded in two steps shown schematically in boxes in Figure 3.7A. The first reaction step was aimed for inhibiting the final output signal formation in the second reaction step when both input signals ($A + A'$ and $B + B'$) are applied together. When the input signals are applied separately, the first reaction step is not activated. Let us consider the biocatalytic system operation with different input combinations.

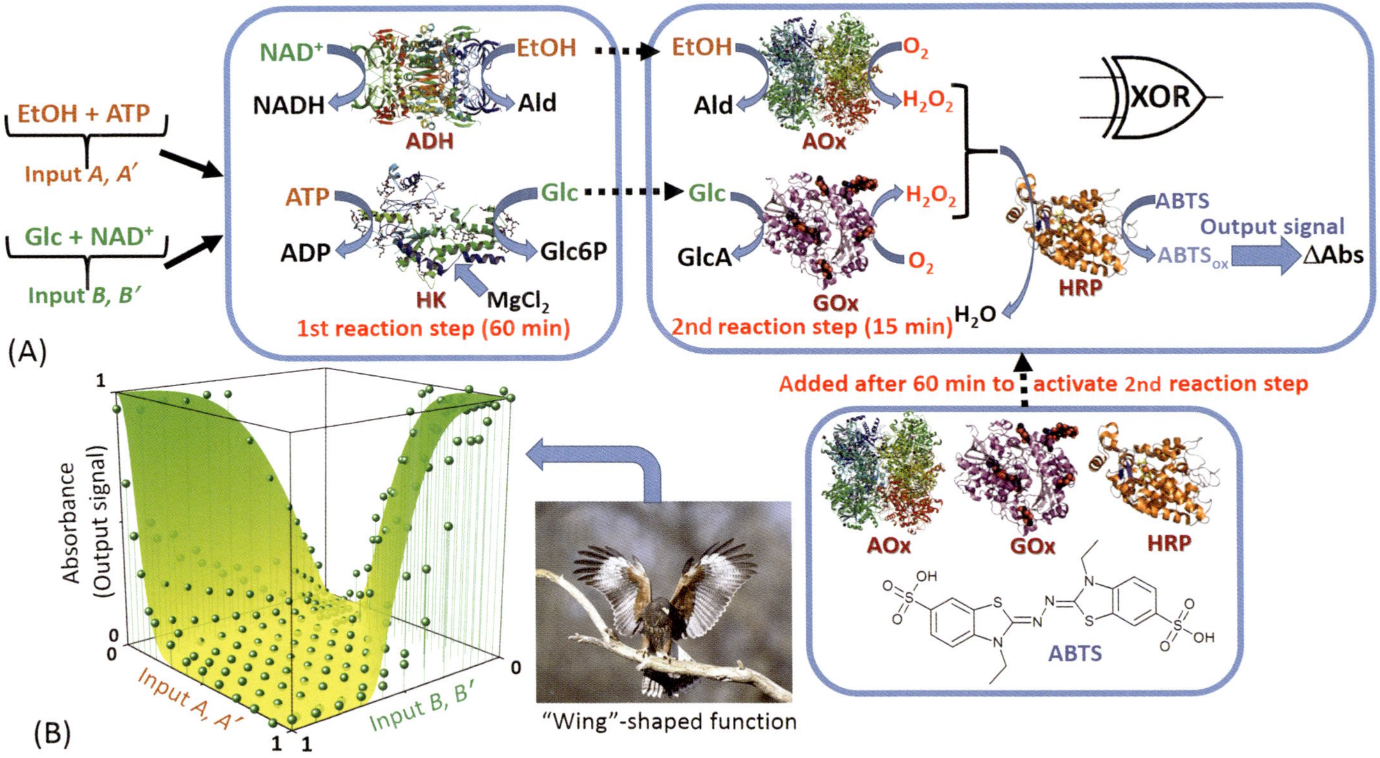

Figure 3.7 (A) Schematic of enzyme-catalyzed reactions for the realized biochemical XOR gate. The first stage of the reactions is carried out for 60 min. Then additional reactants, indicated in the scheme, are added, and 15 min later the system's response is measured as the change in the intensity of absorbance of $ABTS_{ox}$ at $\lambda_{max} = 415$ nm. The following abbreviations are used for the reaction products shown in the scheme: GlcA, gluconic acid; Glc6P, gluconate 6-phosphate acid; Ald, acetic aldehyde; $ABTS_{ox}$, ABTS oxidized product; ADP, adenosine 5′-diphosphate. All other chemicals (enzymes and their substrates) are specified with their abbreviations in the text. (B) Experimental realization of an optimized XOR gate (solid green dots) and the fitted model surface (see Ref. [2] for the theoretical model). Source: Filipov et al. 2017 [2]. Adapted with permission of John Wiley and Sons.

When both input signals (each composed of two chemicals as explained before) are not added to the gate "machinery" (**0**,**0** input combination), the reaction cascade is not activated. The whole set of reactions shown in the reaction scheme (Figure 3.7A) is staying mute and the final output signal defined as the optical absorbance increase is not produced (meaning output signal **0**). When chemicals corresponding to Input *A*,*A*′ (EtOH and ATP) are added, but chemicals corresponding to Input *B*,*B*′ (Glc and NAD^+) are not present (input combination **1**,**0**), the reactions shown in the first box (Figure 3.7A) are not activated. Indeed, the reaction catalyzed by alcohol dehydrogenase (ADH) is not active because NAD^+ is missing in the system, and the reaction catalyzed by hexokinase (HK) is mute because Glc is missing in the system. For this signal combination ATP is not used for any reaction, but EtOH is reacting at the second reaction step catalyzed by alcohol oxidase (AOx) being oxidized by oxygen and resulting in the production of H_2O_2. The generated H_2O_2 finally results in oxidation of ABTS catalyzed by horseradish peroxidase (HRP), thus yielding the optical absorbance increase corresponding to the output signal **1**. Similarly, when chemicals corresponding to Input *A*,*A*′ (EtOH and ATP) are not added, but chemicals corresponding to Input *B*,*B*′ (Glc and NAD^+) are present (input combination **0**,**1**), the reactions shown in the first box (Figure 3.7A) are not activated. In this case the reaction catalyzed by ADH is not active because of the absence of EtOH, and the reaction catalyzed by HK is not active because ATP is missing. For this input combination NAD^+ is not used in any reaction, but Glc is oxidized by oxygen in the reaction catalyzed by GOx during the second reaction step. This results in the H_2O_2 formation followed by the biocatalytic oxidation of ABTS, resulting in the optically measured output signal **1**. The most interesting process is activated by both input signals applied at the **1**,**1** logic combination. In this case both reactions catalyzed by ADH and HK are activated (note that all substrates/co-substrates required for both reactions are present). When all chemicals are applied at the experimentally optimized concentrations and the reaction time (60 minutes) is enough, EtOH and Glc are fully consumed, and the reactions catalyzed in the second step do not proceed. In this case H_2O_2 is not produced, thus the absorbance change is not present corresponding to the output signal **0**. Overall, the output signal production for different input combinations corresponds to the expected XOR logic function. In order to optimize the logic gate operation and to minimize noise, particularly close to the input signals **0**,**0**; **0**,**1**; **1**,**0**; and **1**,**1**, the system operation was studied with input concentrations varying between **0** and **1** binary logic values for both input signals [2]. In the previously discussed logic gates (Identity, OR, AND), the optimization resulted in transformation of the convex response function to the sigmoid function. The response surface corresponding to the XOR gate operation is much more complex (Figure 3.7B). The optimized (low noise) response function should have "flattened" bottom part between **0**,**0** and **1**,**1** logic points and "flattened" "wings" near **0**,**1** and **1**,**0** logic points. This was achieved by optimizing the input concentrations in pairs *A*,*A*′ and *B*,*B*′. While "flattening" the bottom part and the "wings" separately is a relatively easy task, "flattening" all of them requires very sophisticated optimization [2].

Overall, the discussed XOR gate represents a new type of biochemical gates, characterized by using pairs of reactants as inputs, which avoids pitfalls of most earlier XOR gate realizations in the biochemical computing framework. Most importantly, the output signals produced by the designed XOR gate appear as the same chemical, H_2O_2, which is then further transduced into $ABTS_{ox}$, allowing straightforward activation of downstream biomolecular logic gates and other processes concatenated to such XOR gate. Furthermore, the produced output signals can be used in chemical actuation processes, e.g., signal-controlled biomolecular release [55], which was not possible with previous biochemical/biocatalytic realizations of XOR gates that had their output signal defined as "absolute value" of a chemical concentration change. It should be noted that realizations of XOR gates have been attempted using switchable synthetic molecules [56], supramolecular "machines" [57], photochemically activated systems [58], and various biomolecular systems [59, 60]. In most of these systems, the problems have been similar to those addressed in the XOR design discussed here. Therefore, the present XOR gate optimization for the noise decrease is generally important in a context broader than enzyme-catalyzed reaction cascades.

3.3 Summary

The noise suppressing through the gate performance optimization, particularly using "filtering" of input/output signals, is even more important for logic networks with concatenated gates. Even small noise in each gate will be amplified when the signals go through several logic steps, finally degrading the binary logic process. Therefore, the gate optimization discussed in this chapter is particularly important for connecting gates in small networks. For larger networks, digital error correction [61, 62] by redundancy will also have to be implemented, and various network elements will have to be devised for filtering; signal splitting; signal balancing; gate-to-gate connectivity; memory; interfacing with external input, output, and control mechanisms; etc.

As new experiments on mapping out (bio)chemical gate functioning and network designs are reported, new features of noise and error control will be explored. Specifically, noise in the gate function itself, including spread of its values and imprecise mean values – not exactly at the expected reference output **0** or **1**, with deviations possibly also different for various inputs that should ideally yield the same logic output – will have to be considered and corrected, most likely by filtering. Indeed, we conclude by emphasizing that while in the longer term, network design and scaling up will be crucial, the shorter term challenges in (bio)chemical information processing have been to design and experimentally realize versatile and effective (bio)chemical filter processes and other nonbinary network elements that can be concatenated with various binary logic gates.

References

1 Privman, V. (2011). *Isr. J. Chem.* 51: 118–131.
2 Filipov, Y., Domanskyi, S., Wood, M.L. et al. (2017). *ChemPhysChem* 18: 2908–2915.

3 Verma, A., Fratto, B.E., Privman, V., and Katz, E. (2016). *Sensors* 16: 1042.
4 Privman, V., Domanskyi, S., Mailloux, S. et al. (2014). *J. Phys. Chem. B* 118: 12435–12443.
5 Privman, V., Zavalov, O., Halámková, L. et al. (2013). *J. Phys. Chem. B* 117: 14928–14939.
6 Bakshi, S., Zavalov, O., Halámek, J. et al. (2013). *J. Phys. Chem. B* 117: 9857–9865.
7 Zavalov, O., Bocharova, V., Halámek, J. et al. (2012). *Int. J. Unconventional Computing* 8: 347–365.
8 Zavalov, O., Bocharova, V., Privman, V., and Katz, E. (2012). *J. Phys. Chem. B* 116: 9683–9689.
9 Halámek, J., Zavalov, O., Halámková, L. et al. (2012). *J. Phys. Chem. B* 116: 4457–4464.
10 Halámek, J., Zhou, J., Halámková, L. et al. (2011). *Anal. Chem.* 83: 8383–8386.
11 Halámek, J., Bocharova, V., Arugula, M.A. et al. (2011). *J. Phys. Chem. B* 115: 9838–9845.
12 Pita, M., Privman, V., Arugula, M.A. et al. (2011). *PhysChemChemPhys* 13: 4507–4513.
13 Pedrosa, V., Melnikov, D., Pita, M. et al. (2010). *Int. J. Unconventional Computing* 6: 451–460.
14 Privman, V., Halámek, J., Arugula, M.A. et al. (2010). *J. Phys. Chem. B* 114: 14103–14109.
15 Privman, V., Pedrosa, V., Melnikov, D. et al. (2009). *Biosens. Bioelectron.* 25: 695–701.
16 Melnikov, D., Strack, G., Pita, M. et al. (2009). *J. Phys. Chem. B* 113: 10472–10479.
17 Privman, V., Arugula, M.A., Halámek, J. et al. (2009). *J. Phys. Chem. B* 113: 5301–5310.
18 Privman, V., Strack, G., Solenov, D. et al. (2008). *J. Phys. Chem. B* 112: 11777–11784.
19 Feinerman, O. and Moses, E. (2006). *J. Neurosci.* 26: 4526–4534.
20 Feinerman, O., Rotem, A., and Moses, E. (2008). *Nat. Phys.* 4: 967–973.
21 Szacilowski, K. (2012). *Infochemistry – Information Processing at the Nanoscale*. Chichester: Wiley.
22 de Silva, A.P. (2013). *Molecular Logic-Based Computation*. Cambridge: Royal Society of Chemistry.
23 Katz, E. (ed.) (2012). *Molecular and Supramolecular Information Processing – From Molecular Switches to Logic Systems*. Weinheim: Wiley-VCH.
24 Sienko, T., Adamatzky, A., Rambidi, N.G., and Conrad, M. (eds.) (2003). *Molecular Computing*. Cambridge, MA: MIT Press, USA.
25 Katz, E. (ed.) (2012). *Biomolecular Computing – From Logic Systems to Smart Sensors and Actuators*. Weinheim: Wiley-VCH.
26 Stojanovic, M.N., Stefanovic, D., and Rudchenko, S. (2014). *Acc. Chem. Res.* 47: 1845–1852.
27 Stojanovic, M.N. and Stefanovic, D. (2011). *J. Comput. Theor. Nanosci.* 8: 434–440.
28 Ezziane, Z. (2006). *Nanotechnology* 17: R27–R39.

29 Credi, A. (2007). *Angew. Chem. Int. Ed.* 46: 5472–5475.

30 Andréasson, J. and Pischel, U. (2015). *Chem. Soc. Rev.* 44: 1053–1069.

31 de Silva, A.P., Uchiyama, S., Vance, T.P., and Wannalerse, B. (2007). *Coord. Chem. Rev.* 251: 1623–1632.

32 de Silva, A.P. and Uchiyama, S. (2007). *Nature Nanotechnol.* 2: 399–410.

33 Szacilowski, K. (2008). *Chem. Rev.* 108: 3481–3548.

34 Bassani, D.M., Cucinotta, F., Bohne, C. et al. (2015). *Faraday Discuss.* 185: 399–411.

35 Daly, B., Ling, J., and de Silva, A.P. (2014). *Front. Chem. Sci. Eng.* 8: 240–251.

36 Andreasson, J., Pischel, U., Straight, S.D. et al. (2011). *J. Am. Chem. Soc.* 133: 11641–11648.

37 Vasilyev, S., Pita, M., and Katz, E. (2008). *Electroanalysis* 20: 22–29.

38 Ling, J., Daly, B., Silverson, V.A.D., and de Silva, A.P. (2015). *Chem. Commun.* 51: 8403–8409.

39 Liu, D., Liu, H.Y., and Hu, N.F. (2012). *J. Phys. Chem. B* 116: 1700–1708.

40 Saha, A., Manna, S., and Nandi, A.K. (2009). *Soft Matter* 5: 3992–3996.

41 Uchiyama, S., Kawai, N., de Silva, A.P., and Iwai, K. (2004). *J. Am. Chem. Soc.* 126: 3032–3033.

42 Katz, E. (2017). *ChemPhysChem* 18: 1688–1713.

43 Katz, E., Bocharova, V., and Privman, M. (2012). *J. Mater. Chem.* 22: 8171–8178.

44 Privman, M., Tam, T.K., Bocharova, V. et al. (2011). *ACS Appl. Mater. Interfaces* 3: 1620–1623.

45 Honarvarfard, E., Gamella, M., Poghossian, A. et al. (2017). *Appl. Mater. Today* 9: 266–270.

46 Molinnus, D., Poghossian, A., Keusgen, M. et al. (2017). *Electroanalysis* 29: 1840–1849.

47 Poghossian, A., Katz, E., and Schöning, M.J. (2015). *Chem. Commun.* 51: 6564–6567.

48 Mayo, A.E., Setty, Y., Shavit, S. et al. (2006). *PLoS Biol.* 4: e45.

49 Privman, V., Fratto, B.E., Zavalov, O. et al. (2013). *J. Phys. Chem. B* 117: 7559–7568.

50 Doerge, D.R., Divi, R.L., and Churchwell, M.I. (1997). *Anal Biochem.* 250: 10–17.

51 Morozova, O.V., Shumakovich, G.P., Gorbacheva, M.A. et al. (2007). *Biochemistry (Moscow)* 72: 1136–1150.

52 Holwerda, R.A. and Gray, H.B. (1975). *J. Am. Chem. Soc.* 97: 6036–6041.

53 Koudelka, G.B., Hansen, F.B., and Ettinger, M.J. (1985). *J. Biol. Chem.* 260: 15561–15565.

54 Privman, V., Zhou, J., Halámek, J., and Katz, E. (2010). *J. Phys. Chem. B* 114: 13601–13608.

55 Filipov, Y., Gamella, M., and Katz, E. (2018). *Electroanalysis* 30: 1281–1286.

56 de Silva, A.P. and McClenaghan, N.D. (2000). *J. Am. Chem. Soc.* 122: 3965–3966.

57 Credi, A., Balzani, V., Langford, S.J., and Stoddart, J.F. (1997). *J. Am. Chem. Soc.* 119: 2679–2681.

58 Szaciłowski, K., Macyk, W., and Stochel, G. (2006). *J. Am. Chem. Soc.* 128: 4550–4551.
59 Xia, F., Zuo, X., Yang, R. et al. (2010). *J. Am. Chem. Soc.* 132: 8557–8559.
60 Baron, R., Lioubashevski, O., Katz, E. et al. (2006). *Angew. Chem. Int. Ed.* 45: 1572–1576.
61 Baylis, J. (1998). *Error-Correcting Codes*. London: Chapman-Hall.
62 Blahut, R. (1983). *Theory and Practice of Error Control Codes*. Reading: Addison-Wesley.

4

Enzyme-Based Short Logic Networks Composed of Concatenated Logic Gates

4.1 Introduction: Problems in Assembling of Multistep Logic Networks

While realization of individual Boolean logic gates mimicked with enzyme-catalyzed reactions is straightforward and has been studied in detail [1] (see also Chapter 2), assembling multifunctional logic networks, which include several (potentially many) logic functions performed by many enzymes operating in a concerted way is not an easy task. In order to arrange well-operated biomolecular logic networks, several challenging problems have to be solved. These problems are different from those in electronic circuits, and they are more difficult to solve because of chemical nature of input/output signals passing through a reaction chain. Indeed, in electronics all signals are represented by voltage/current at all information-processing steps; thus signal transmission from step to step is easy. Opposite to that, in (bio)chemical, particularly enzyme-catalyzed reactions, a product in each reaction step is different from the substrate (reagent) prior to the reaction. Therefore, when biocatalytic reactions are integrated in a cascade, the input signal coming to the next reaction step is a product generated in the preceding reaction. This requires careful selection of the reaction steps and enzymes catalyzing these steps to allow each next reaction to be activated with a product generated before. When a multistep reaction proceeds in a homogeneous solution, all substrates/products at each reaction step should be different to avoid cross talking (shortcuts) between the reaction steps. In most cases, the enzymes operating together in biocatalytic cascades should be significantly different from each other (belonging to different classes) to avoid their cross talking. Very often logic gates perfectly operated separately cannot be easily applied together in a cascade performing a chain of logic operations because of possible interference between the reaction steps.

Another limitation for assembling logic networks based on enzyme reactions originates from analog noise amplification through the reaction chain. While individual logic gates can tolerate noise in the range of a few percent in input/output signal fluctuation, this noise is accumulated and amplified through the sequence of biocatalytic reaction steps, finally degrading the logic function. Therefore, the analog noise decrease, which is important for optimization of individual logic gates [2–6] (see also Chapter 3), becomes critically important

Enzyme-Based Computing Systems, First Edition. Evgeny Katz.

for multistep biocatalytic cascades operating as concatenated logic gates. The present chapter demonstrates a few examples of logic networks based on enzyme-catalyzed reactions, concentrating on the selection of reaction steps and decreasing noise propagation/amplification through the logic networks. Finally, an example of a logic network activated with a mixed combination of chemical and physical inputs is discussed.

4.2 Logic Network Composed of Concatenated Gates: An Example System

Figure 4.1A shows a biocatalytic cascade mimicking a logic network composed of three concatenated logic gates [7]: OR–AND–XOR (Figure 4.1B). The first reaction step catalyzed by acetylcholine esterase (AChE) was activated with two substrates, acetylcholine (Input *A*) and butyrylcholine (Input *B*), both resulting in the production of choline in the course of a hydrolytic reaction. Then, choline reacted with choline oxidase (ChOx) to yield betaine and to produce H_2O_2 upon reduction of O_2, which was a co-substrate for this reaction (Input *C*). The *in situ* produced H_2O_2 was used to oxidase NADH to NAD^+ in the reaction catalyzed by microperoxidase-11 (MP-11). Another reaction catalyzed by glucose dehydrogenase (GDH) resulted in the reversed process of NAD^+ reduction back to NADH. This reductive reaction was activated by glucose (Glc, Input *D*). All four inputs (*A*, *B*, *C*, and *D*) were defined as binary **0**/**1** logic values (absent/present in the optimized concentration) and applied in 16 possible combinations. The produced NADH was analyzed optically (measuring absorbance at $\lambda_{max} = 340$ nm) and the absolute value of the absorbance change ($|\Delta A|$) was defined as the final output signal. The reaction catalyzed by AChE was activated in the presence of either or both input signals (*A* and *B* applied in combinations **0**,**1**; **1**,**0**; and **1**,**1**), thus mimicking an OR logic gate. The next reaction step catalyzed by ChOx was activated only in the presence of both substrate and co-substrate (choline produced in the preceding reaction step and O_2 being an independent variable Input *C*), thus mimicking an AND logic gate connected to the first OR gate. Finally, NADH was produced in the reductive process catalyzed by GDH and activated with Glc, but it was also consumed in the reaction catalyzed by MP-11 in the presence of *in situ* produced H_2O_2. The optical changes (absorbance increase or decrease) were obtained when the NADH formation and consumption were not balanced. This reaction step was mimicking a XOR logic gate (note that the absorbance increase and decrease were considered as the same output signal **1**). Notably, to avoid cross talking between the biocatalytic reactions, all steps were catalyzed by enzymes corresponding to different classes: hydrolytic (AChE), oxidase (ChOx), peroxidase (MP-11), and dehydrogenase (GDH). The enzymatic cascade (Figure 4.1A) activated with different combinations of the binary-defined input signals (**0**, the input absence, and **1**, the input presence at the experimentally optimized concentration) demonstrated robust operation with clearly distinguished optically measured **0** and **1** output signals (Figure 4.2). However, longer cascades with more reaction steps mimicking larger logic networks are getting more affected

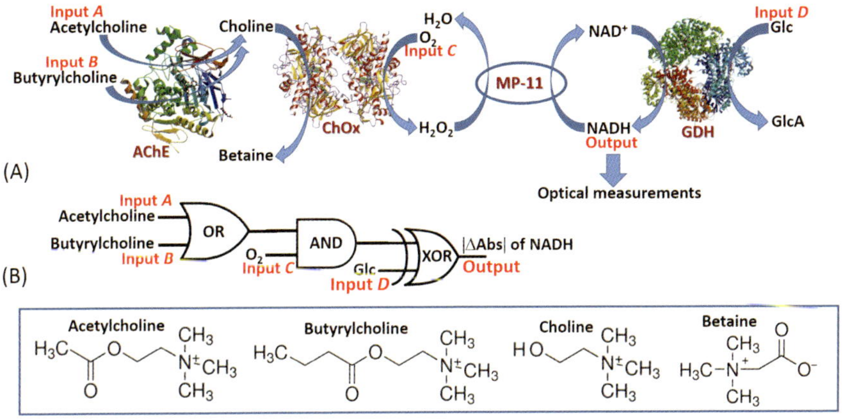

Figure 4.1 (A) Schematics of the biocatalytic cascade mimicking operation of concatenated logic gates based on four coupled biocatalysts. (B) Equivalent circuitry for the biocatalytic process.

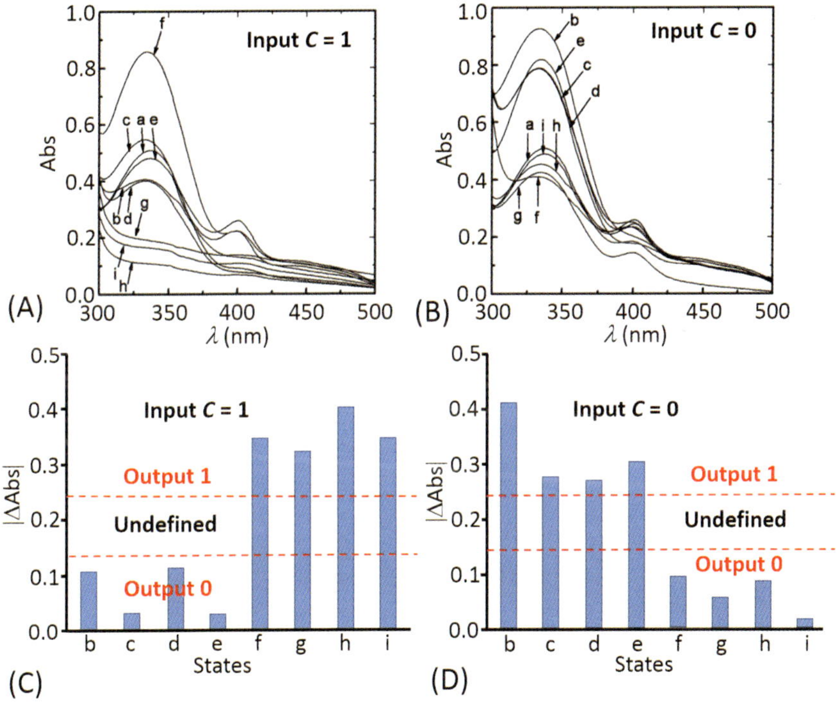

Figure 4.2 Operation and analysis of the concatenated biocatalytic gates in the presence of four inputs applied in 16 variable combinations. (A) Absorbance features of NADH in the concatenated four-enzyme system in the presence of O_2 (Input *C* = **1**; while other inputs are variable). Spectrum (a) corresponds to the initial state of the system prior to the input application. (C) Bar presentation of the output of the concatenated gates, derived from the absorbance modulus changes upon application of variable inputs *A*, *B*, *C*, and *D*: (b) **1,0,1,1**; (c) **0,1,1,1**; (d) **1,1,1,1**; (e) **0,0,1,0**; (f) **0,0,1,1**; (g) **1,0,1,0**; (h) **0,1,1,0**; and (i) **1,1,1,0**. (B) Absorbance features of NADH in the concatenated four-enzyme system in the absence of O_2 (Input *C* = **0**; while other inputs are variable). (D) Bar presentation of the output of the concatenated gates, derived from the absorbance modulus changes upon application of variable inputs *A*, *B*, *C*, and *D*: (b) **1,0,0,1**; (c) **0,1,0,1**; (d) **1,1,0,1**; (e) **0,0,0,1**; (f) **1,0,0,0**; (g) **0,1,0,0**; (h) **1,1,0,0**; and (i) **0,0,0,0**. The threshold absorbance values of 0.14 and 0.24 are marked for all bar presentations.

by the amplified noise. It has been theoretically predicted (based on a reaction model) that biocatalytic cascades are limited by approximately 10 reaction steps, after which the noise becomes too high and the difference between **0** and **1** outputs could be hardly observed [8]. It should be noted that the order of gates in the logic network shown in Figure 4.1A cannot be changed unless completely different biocatalytic reactions are used. The XOR gate mimicked with two biocatalytic reactions running in the opposite directions can be only placed at the end of a biocatalytic cascade. Indeed, the definition of its output as the absolute value of the absorbance change ($|\Delta A|$) is rather artificial. In reality, the used reactions produce different products (NADH and NAD^+ in the present example). These different chemical species react differently and they cannot be considered as the identical output if the reaction cascade should be continued after the XOR gate. In other words, the XOR gate defined in this way can be only placed at the end of a logic network where optical changes are measured without extending chemical reactions.

4.3 Logic Networks with Suppressed Noise in the Presence of Filter Systems

Sharp changes in the output signal upon variation of input signals, particularly near logic values **0** and **1**, result in the analog noise compromising binary operation of the logic gates. Since most of biochemical/biocatalytic systems demonstrate convex response function in the range of input concentrations between logic values **0** and **1**, the sharp response is mostly expected for low input concentrations near logic **0** point. If the concentration range of the used input signals is large enough, the response function near logic point **1**, corresponding to the high concentration, has saturation and, thus, flattened. This results in relatively small noise near logic point **1**. In order to suppress noise near logic point **0**, the use of "filter" function has been suggested [3–6] (see also Chapter 3). The chemical filter operating at low input concentrations consumes a part of the signal preventing its sharp increase, thus flattening the response function and converting the convex shape response to the sigmoid one. This approach has been studied experimentally and modeled theoretically for various OR/AND logic gates operating individually [3–6]. Based on the results obtained with the single gates, this approach has been used to decrease noise in a logic network composed of three concatenated AND gates [9] (Figure 4.3).

Figure 4.3A shows a reaction cascade catalyzed by three enzymes and activated by four input signals [9]. The first reaction catalyzed by maltose phosphorylase (MPh) was activated with maltose (Input *A*) and inorganic phosphate (Pi; Input *B*), then resulting in the formation of glucose (Glc) and glucose-1-phosphate (Glc1P). The next reaction step catalyzed by glucose oxidase (GOx) resulted in oxidation of glucose produced in the preceding reaction step by oxygen (Input *C*). Hydrogen peroxide (H_2O_2) produced upon O_2 reduction in this reaction was reacted in the next step catalyzed by horseradish peroxidase (HRP). This reaction resulted in oxidation of 3,3′,5,5′-tetramethylbenzidine (TMB)

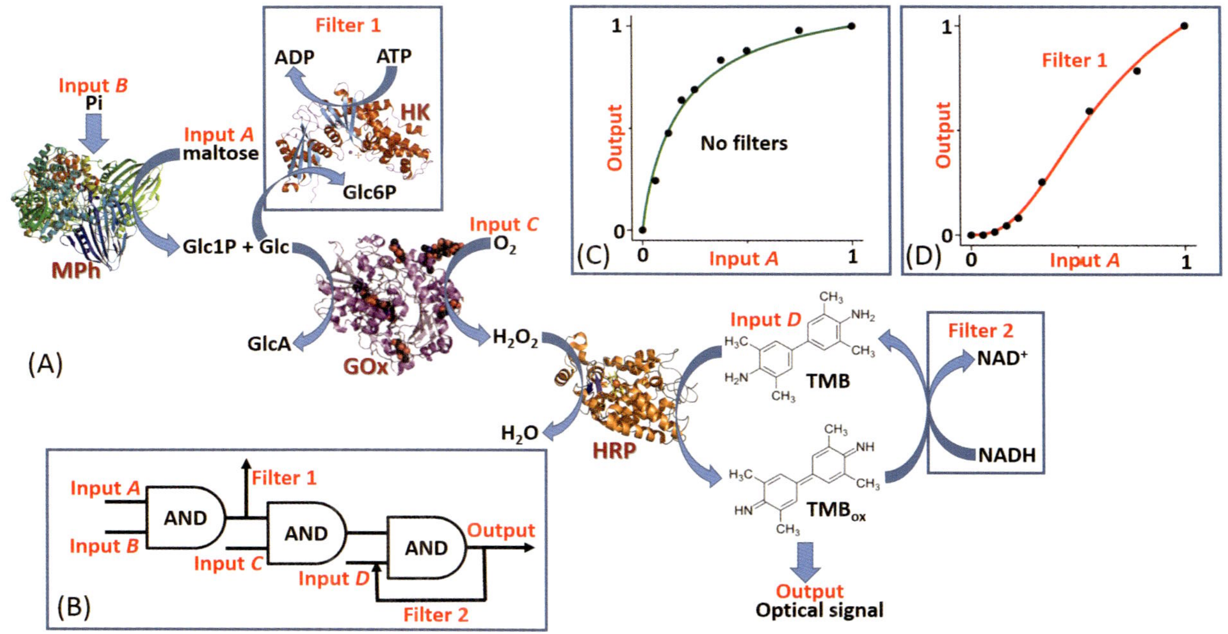

Figure 4.3 (A) Schematics of the biocatalytic cascade mimicking three concatenated AND gates with two filters added. The first filter based on the reaction catalyzed by hexokinase (HK) consumed Glc, thus inhibiting the second reaction step in the biocatalytic cascade. The second filter based on reduction of TMD_{ox} with NADH inhibited production of the final output signal. (B) The equivalent logic circuit composed of three concatenated AND gates with two filter units. (C) The convex response function in the absence of both filters. (D) The sigmoid response function obtained with the first filter (HK + ATP) operating with the optimized concentrations. Note that the operation of the second filter is not shown in this figure.

yielding blue-colored product (TMB_{ox}) analyzed optically ($\lambda_{max} = 655$ nm). The absorbance increase corresponding to formation of TMB_{ox} was defined as the final output signal produced by the biocatalytic cascade. Since each reaction step (Figure 4.3A) required two input signals, some of them externally applied (Inputs *A*, *B*, *C*, and *D*), and others produced *in situ* (Glc, H_2O_2), all reaction steps were considered as AND gates concatenated in the logic network (Figure 4.3B). In addition to the reaction steps processing the input signals and finally resulting in TMB_{ox} product, the system included two filters operating after the first and last reaction steps, Filter 1 and Filter 2, respectively (Figure 4.3A). Filter 1 consumed Glc, converting it to glucose-6-phosphate (Glc6P) in the reaction catalyzed by hexokinase (HK). This reaction proceeded as long as adenosine 5′-triphosphate (ATP) was present in the system. After the ATP was fully converted to adenosine 5′-diphosphate (ADP), the filter operation was stopped and glucose was proceeding to the next reaction step. It should be noted that the filter reaction and the GOx-catalyzed reaction were proceeding in the same solution at the same time; thus for the efficient filter operation, the activity of HK was much higher than the activity of GOx. Figure 4.3C shows convex response function measured for the final output signal (optical absorbance corresponding to TMB_{ox}) upon variation of Input *A* (maltose concentration) without any filter used. Note the high slop near logic point **0** resulting in the large noise at low concentration of maltose. When Filter 1 was activated and the activity of HK and concentration of ATP were optimized, the convex response function was changed to a sigmoid function with small increase near logic point **0** (Figure 4.3D). This resulted in decreasing noise near logic point **0**. It is quite obvious that Filter 1 was able to reduce noise only for fluctuation of Input *A* concentration, while all other inputs applied in the downstream part of the biocatalytic cascade were not affected by this filter. In order to reduce noise for Inputs *B*, *C*, and *D*, Filter 2 was applied. This resulted in sigmoid response functions for all input signals (not shown). The 3-AND gate concatenated circuit shown in Figure 4.3 was studied experimentally and modeled theoretically [9], and it represents an example system with the noise reduction by filtering signals processed through a multistep biocatalytic cascade. This work is only the very first step in assembling biomolecular logic networks with reduced noise. The study has to be continued to allow more sophisticated biocomputing systems with robust operation tolerating noise from the input signal fluctuation.

4.4 Logic Circuits Activated with Biomolecular Signals and Magnetic Field Applied

4.4.1 Biocatalytic Reactions Proceeding with Bulk Diffusion of Intermediate Substrates/Products and with Their Channeling

In biological systems, enzymes working cooperatively in biocatalytic pathways are organized in macromolecular complexes. The organized placement of enzymes in cells allows communication between catalytic sites without diffusion of biosynthetic intermediates into the bulk phase of the cell. The directed transport of substrates/products, called "*metabolic channeling,*" is highly important for life because it provides much more effective biocatalytic processes than can be achieved in chemical systems operating through random diffusion of intermediate species [10, 11]. Inspired by biological prototypes, many artificial biomolecular systems with organized placement of enzymes in synthetic nanoscale scaffolds have been designed [12, 13]. The major motivation for this research has been the enhancement of biocatalytic reactions [14], including those applied in biotechnological processes [15]. The important feature of these systems is *substrate channeling*, resulting in effective communication between enzymes similarly to biological processes. The common approach to the organized positioning of enzymes is based on the use of macromolecular scaffolds, frequently using DNA/RNA for their assembling [16–18]. In addition to static DNA scaffolds, dynamically operating DNA machines have been used to regulate enzyme cascades by external signals [19]. Despite the fact that the direct mimicking of biological structures requires exact positioning of cooperatively working enzymes at specific locations and with specific distances, much simpler approach based on random placement of enzymes in micro-containers [20] or on surfaces [21, 22] has been used and demonstrated some advantages comparing with solute state of the enzymes. In addition to highly challenging experimental work, the systems demonstrating substrate channeling have been studied theoretically, modeling the biocatalytic processes and comparing them with those proceeding through bulk solution diffusion [14, 23]. The important conclusion was that the substrate channeling is not always more effective than diffusional process [14] and the explanations for reaction facilitation due to substrate channeling is not always correct [24]. The substrate channeling results in the improved kinetics only under very specific conditions controlled by concentrations of reacting species and at specific time intervals of biocatalytic reactions [14, 25].

While the substrate channeling achieved by enzyme placement on nanoscale scaffolds has been demonstrated in many different systems [12, 13, 16–18], switchable systems with signal-controlled features [19], particularly switched reversibly between *substrate channeling* and *substrate diffusion*, have not been studied yet. Since dynamic changes in molecular (e.g., DNA) scaffolds are not easy to achieve [19], the switchable systems require different approach. The system discussed here aims at demonstrating magneto-controlled biocatalytic cascades including several reaction steps activated by a few enzymes communicating through *diffusional substrates* in the absence of magnetic field when the enzyme-functionalized magnetic nanoparticles are dispersed in a solution and through *substrate channeling* in the presence of magnetic field when the nanoparticles are aggregated and the enzymes are close to each other.

4.4.2 Magneto-Controlled Biocatalytic Cascade Switchable Between Substrate Diffusion and Substrate Channeling Modes of Operation

The studied biocatalytic system [26] included two kinds of magnetic nanoparticles (MNPs) functionalized with glucose oxidase (GOx) and horseradish peroxidase (HRP) (Figure 4.4). In the absence of the magnetic field, the MNPs were suspended in a buffer solution, and H_2O_2 produced in the

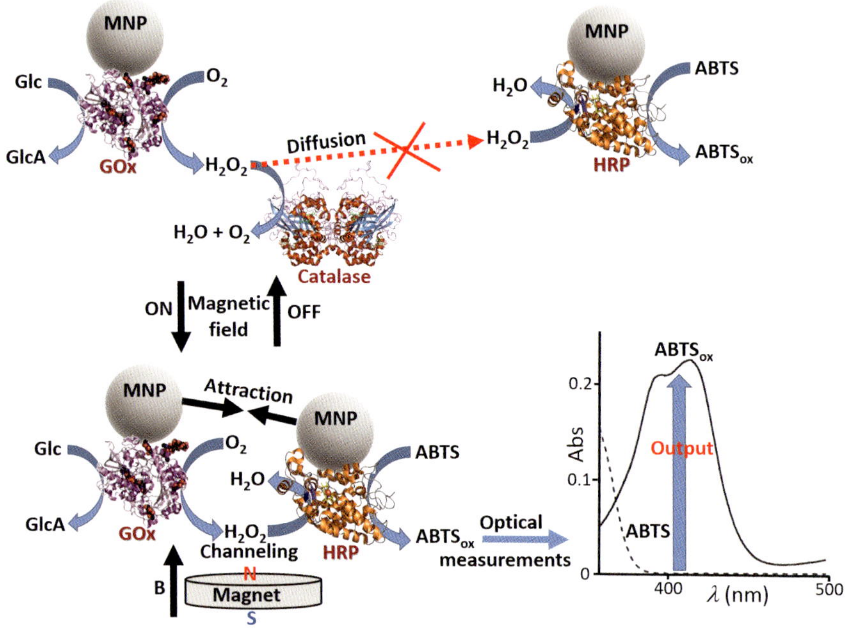

Figure 4.4 Schematics of the biocatalytic cascade realized in the presence of the GOx-functionalized MNPs and HRP-functionalized MNPs operating in the dispersed and aggregated states switchable by the external magnetic field. In the aggregated state produced in the presence of the magnetic field (bottom part of the scheme), the enzymes communicate via *substrate channeling* – the direct transfer of the H_2O_2 intermediate between enzymes in the cascade without the release of the intermediate into the bulk solution. In the dispersed state in the absence of the magnetic field (upper part of the scheme), the H_2O_2 intermediate is released to the bulk solution where it is intercepted and catalytically decomposed by soluble catalase, thus inhibiting its reaction with HRP. Note that the reaction proceeds to the HRP-catalyzed ABTS oxidation in the absence of catalase. The spectra shown correspond to the original ABTS optical absorbance and the absorbance change upon ABTS biocatalytic oxidation (formation of $ABTS_{ox}$). The following abbreviations are used: glucose (Glc), gluconic acid (GlcA), and magnetic field (B). Other abbreviations are explained in the text. Source: Filipov et al. 2018 [26]. Adapted with permission of John Wiley and Sons.

first reaction catalyzed by GOx was moving to the next reaction step catalyzed by HRP by diffusion through the bulk solution. Note that in this experiment catalase was not included in the solution; thus H_2O_2 was not intercepted and consumed on its way. Alternatively, in the presence of magnetic field, the MNPs were aggregated. It was expected that in this state the intermediate H_2O_2 produced by GOx can be delivered to the next enzyme, HRP, by the substrate channeling due to short distances separating the enzymes bound to the MNPs. Both systems, with and without MNPs aggregation, resulted in ABTS oxidation followed by the optical absorbance change (Figure 4.5A). Surprisingly, the time-dependent absorbance changes in both cases were not much different, thus reflecting similar kinetics of the ABTS oxidation through the diffusional pathway and the substrate channeling (it should be noted that at this point the "substrate channeling" can be only assumed, but it cannot be proved because

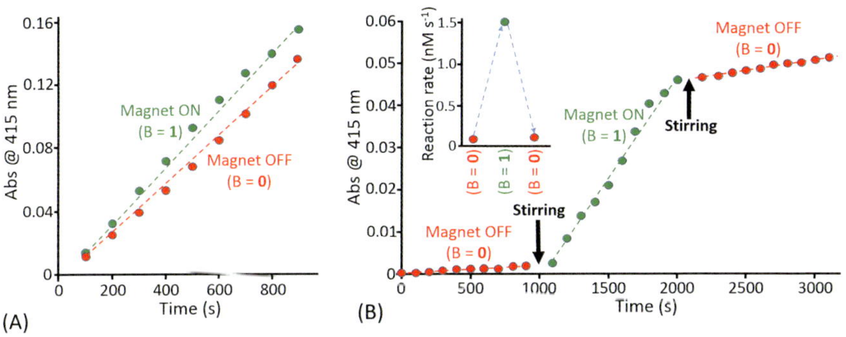

Figure 4.5 (A) Absorbance increase corresponding to the biocatalytic formation of $ABTS_{ox}$ in the presence of the GOx- and HRP-functionalized MNPs and in the presence and absence of the external magnetic field. The experiments were performed in the absence of catalase. (B) The same experiments performed in the presence of catalase in the bulk solution. The inset shows reversible change of the $ABTS_{ox}$ formation reaction rate upon switching ON–OFF the magnetic field. Source: Filipov et al. 2018 [26]. Adapted with permission of John Wiley and Sons.

of the kinetics similar to the diffusional reaction). The small difference in the kinetics of the diffusional reaction and the substrate channeling may originate from large substrate concentration and long duration of the reaction observed when the difference between two mechanisms becomes indistinguishable [14].

In order to demonstrate the difference between the reactions catalyzed by the dispersed and aggregated MNPs, the system composition was changed by adding catalase to the solution (Figure 4.4). This enzyme, operating in the solution, rapidly decomposed the diffusional intermediate product (H_2O_2) and inhibited the second reaction step, which is responsible for the absorbance change (Figure 4.5B) (magnet OFF). The function of this enzyme in the present system is similar to the "filter" operation studied in enzyme logic systems [3–6], where such reactions were applied for decreasing analog noise in binary signal processing [27, 28]. When the magnetic field was applied and the MNPs were aggregated, the second biocatalytic step in the cascade was reactivated and the corresponding absorbance increase was observed (Figure 4.5B) (magnet ON). This was explained by the substrate channeling, which was much less affected by the H_2O_2 decomposition catalyzed by catalase. Removing magnets and resuspending the MNPs again resulted in the process inhibition due to H_2O_2 decomposition during the diffusional path (Figure 4.5B) (magnet OFF again). Therefore, the changes in the reaction mechanism were reversible upon applying and removing the magnetic field, which resulted in the aggregation and disaggregation of the MNPs, respectively (Figure 4.5B, inset). The observed phenomenon is in agreement with the theoretical model [14], which predicts the improved kinetics for the substrate channeling only when the substrate concentration is low (being consumed by catalase in the present experiment). It should be noted that the rate of the ABTS oxidation (the last step in the biocatalytic cascade) in the aggregated system, where the substrate channeling is assumed, is still smaller than that without the "filter" (compare Figure 4.5A,B with the magnet ON). This can originate from partial decomposition of the intermediate substrate H_2O_2 by catalase even in the aggregated state of the MNPs.

4.4.3 Logic Signal Processing with the Switchable Biocatalytic System

The present system [26] is a particularly good example to illustrate logic processing of the input signals in multistep biocatalytic cascades. The reactions included in the biocatalytic cascade need the presence of two substrates: glucose (Glc) and ABTS. The second reaction step catalyzed by HRP can be inhibited in the presence of the filter enzyme catalase in the absence of magnetic field, but when magnetic field is applied, it cancels the effect of the filter. Overall, 4 input signals (Glc, ABTS, catalase, and magnetic field (Figure 4.6D)) can be applied and logically processed with the switchable biocatalytic cascade. To allow Boolean signal processing, the input signals should be defined in the binary (**0**,**1** or **NO**/**YES**) format. The logic value **0** for all input signals was defined as the absence of the reacting species (meaning their zero concentration) and the absence of the magnetic field. The logic **1** value of the input signals was represented with experimentally optimized concentrations of the reacting species and with the magnets located near the reacting volume. The input signals were applied in 16 different combinations of the **0** and **1** values outlined in the truth table (Figure 4.6B).

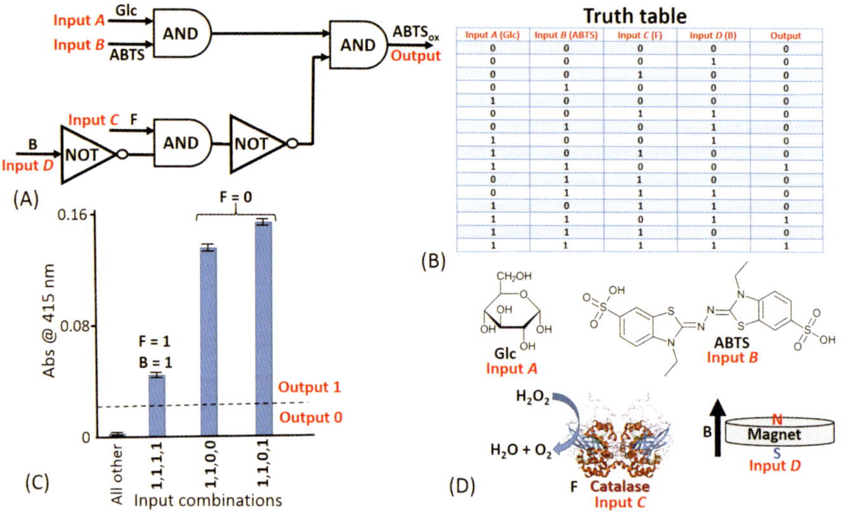

Input A (Glc)	Input B (ABTS)	Input C (F)	Input D (B)	Output
0	0	0	0	0
0	0	0	1	0
0	0	1	0	0
0	1	0	0	0
1	0	0	0	0
0	0	1	1	0
0	1	0	1	0
1	0	0	1	0
1	0	1	0	0
1	1	0	0	1
0	1	1	0	0
0	1	1	1	0
1	0	1	1	0
1	1	0	1	1
1	1	1	0	0
1	1	1	1	1

Figure 4.6 (A) Logic circuit corresponding to the biocatalytic cascade operating in the presence of the GOx- and HRP-functionalized MNPs activated with four variable input signals. (B) The truth table corresponding to the logic circuit shown in (A). (C) Optically measured output signals corresponding to the biocatalytic formation of $ABTS_{ox}$. The experiments were performed in 16 different combinations of the input signals (see the truth table). The dashed line separates the logic outputs **0** and **1**. (D) The input signals applied: Input *A* – glucose (Glc); Input *B* – ABTS; Input *C* – catalase; Input *D* – magnetic field. Logic **0** of the input signals corresponded to their physical absence (zero concentration of the reacting species and the absence of the magnets). Source: Filipov et al. 2014 [26]. Adapted with permission of John Wiley and Sons.

Figure 4.6A shows the equivalent logic circuit corresponding to the biocatalytic cascade shown in Figure 4.4. The logic circuit can be explained in the following way. To proceed to the end, the presence of both substrates is mandatory; thus Inputs *A* and *B* (Glc and ABTS, respectively) are processed through the AND logic gate, and the output **1** is produced only when both inputs are present (**1,1** combination for Inputs *A* and *B*). To produce $ABTS_{ox}$ at the end of the biocatalytic cascade, three options are possible (note that *A*, *B* should be always **1,1**): (a) magnetic field present, input $D = \mathbf{1}$; catalase present, input $C = \mathbf{1}$; (b) magnetic field present, input $D = \mathbf{1}$; catalase absent, input $C = \mathbf{0}$; and (c) magnetic field absent, input $D = \mathbf{0}$; catalase absent, input $C = \mathbf{0}$. Inputs *C* and *D* are logically processed through the cascade of NOT (Inverter)–AND–NOT logic circuit. Finally, this part of the circuit should deliver the input **1** to the final AND gate to get output **1** (ABTS oxidation). As the result of the input processing, output **1** (ABTS oxidation and absorbance increase) was obtained for the input combinations, **1,1,1,1**; **1,1,0,0**; and **1,1,0,1**, while all other input combinations did not show any meaningful absorbance increase (Figure 4.6C). Sum of products [29]

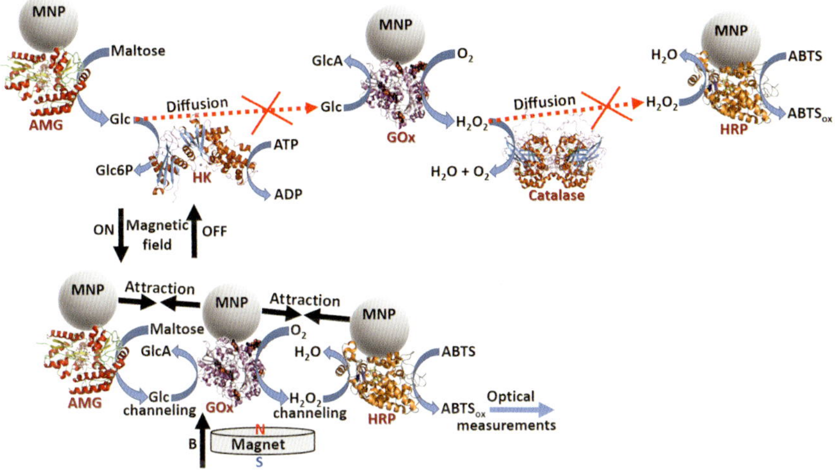

Figure 4.7 Schematics of the biocatalytic cascade realized in the presence of the AMG-functionalized MNPs, GOx-functionalized MNPs, and HRP-functionalized MNPs operating in the dispersed and aggregated states switchable by the external magnetic field. The biocatalytic cascade was activated with three input signals: maltose, O_2, and ABTS. In the aggregated state produced in the presence of the magnetic field (bottom part of the scheme), the enzymes communicate via *substrate channeling* – the direct transfer of the glucose and H_2O_2 intermediates between enzymes in the cascade without release of the intermediates into the bulk solution. In the dispersed state in the absence of the magnetic field (upper part of the scheme) the glucose intermediate and H_2O_2 intermediate (if it is produced) are released to the bulk solution where they are intercepted and catalytically decomposed by soluble hexokinase (HK)/ATP and catalase (both operating as "filters"), thus inhibiting their reactions in the cascade. The $ABTS_{ox}$ formation in the dispersed state is inhibited either in the presence of HK/ATP or catalase or both. The following abbreviations are used: amyloglucosidase (AMG), glucose-6-phosphate (Glc6P), hexokinase (HK), adenosine triphosphate (ATP), adenosine diphosphate (ADP), and gluconic acid (GlcA); other abbreviations are explained in the text. Source: Filipov et al. 2014 [26]. Adapted with permission of John Wiley and Sons.

(a short presentation of the logic operation equivalent to the truth table) of our logic gate system is $F = ABCD' + ABC'D + ABCD$.

While in past magnetic field applied as the logic signal together with a biomolecular input was used to mimic very simple single logic gates [30], the present system allowed its integration in a multi-gate logic circuit. Logic circuits of higher complexity are possible, when longer biocatalytic cascades with more biomolecular inputs are used [26] (Figure 4.7).

4.5 The Summary: Step Forward from Single Logic Gates to Complex Logic Circuits

Various Boolean logic gates mimicked with enzyme-catalyzed reactions (see Chapter 2) can be quite easily concatenated to short signal-processing networks, usually represented with three to four reaction steps. Furthermore, new demands for the biochemical gate networking will require a strategy to systematize the modeling, utilizing modular network analysis and detailed network optimization approaches, supplemented with single-gate optimization of key elements in the network (see Chapter 3). As the biocomputing networks become larger and more complex, new gates and non-Boolean network elements will have to be realized, characterized, and modeled. The latter non-Boolean elements should include filters functioning, for instance, by diverting some of the output or input(s). It has been argued [8] that this can induce sigmoid behavior reducing analog noise in the signal-processing systems. Additional networking elements addressing issues of signal splitting/balancing and signal amplification are needed for operation of the concatenated/branched logic networks (see Chapter 7). Ultimately, for larger networks the issues of network design for digital error correction will also come into play. Enzyme properties not presently explored for information processing might offer interesting avenues for optimization. For example, allosteric enzymes [31] frequently have the self-promoter (sigmoid response) property with respect to their substrate concentrations. Biochemical networks can offer robust error-free operation upon appropriate optimization of their components and interconnections [8, 32]. However, the limit of the biocomputing network complexity is set by the cross-reactivity of enzyme-catalyzed reactions. Only enzymes belonging to different biocatalytic classes (oxidases, dehydrogenases, peroxidises, hydrolases, etc.) could operate in a single "soup" without significant cross-reactivity. If chemical reasons require the use of cross-reacting enzymes in the system, they must be space separated to allow their operation. This might be achieved by the use of reaction–diffusion chemical processors separating the reacting species in time and space [33, 34]. Another approach would include compartmentalization of the reacting species using pattering on surfaces or in modular fluidic devices [35–38] (see Chapter 5). Microfluidic hardware required for digital processing of chemical signals is already being developed [39] and has been applied for some biomolecular, e.g., DNA computing, systems [40]. This technology can be easily adapted for the enzyme-based logic networks. Further development of this research area requires cooperative work of engineers, biochemists, and computer specialists. The ultimate goal of this work will be the design of a microfluidic lab-on-a-chip performing multienzyme-catalyzed cascades and operating similarly to an electronic chip by being able to integrate large networks for processing biochemical signals.

Another way to resolving the problem of the enzyme cross-reactivity in networks is by utilizing other than enzymatic biomolecular components with higher specificity. Combination of enzymes with other biomolecules, specifically with immune-recognition biomolecules (antigens–antibodies) [41], offers interesting avenues for new information-processing designs and applications. The use of immune species (antigens–antibodies), which offer a large variety of biorecognition combinations, could substantially increase the variability of biocomputing steps with minimum cross-reactivity. Some of the immune reactants could be labeled with enzymes, thus providing interconnections with the enzyme-based networks and resulting in integrated biocomputing systems based on biorecognition–biocatalysis processes [42]. Preliminary work has already demonstrated logic operations performed by antigen–antibody interactions coupled with enzyme logic gates [41]. The logically processed immune signals were utilized to control the operation of a biofuel cell [43] (see Chapter 13).

Finally, information-processing networks based on concatenated enzyme-based logic gates have been used for various applications including binary biosensing [44], signal-controlled bioelectrochemical systems [45] (e.g., biofuel cells [46]), signal-switchable materials [47], signal-triggered (bio)molecular release [22], information security systems [48], and various forensic applications [49]. All these applications are reviewed in the book.

References

1 Katz, E. (2019). *ChemPhysChem* 20: 9–22.
2 Filipov, Y., Domanskyi, S., Wood, M.L. et al. (2017). *ChemPhysChem* 18: 2908–2915.
3 Bakshi, S., Zavalov, O., Halámek, J. et al. (2013). *J. Phys. Chem. B* 117: 9857–9865.
4 Zavalov, O., Bocharova, V., Privman, V., and Katz, E. (2012). *J. Phys. Chem. B* 116: 9683–9689.
5 Privman, V., Halámek, J., Arugula, M.A. et al. (2010). *J. Phys. Chem. B* 114: 14103–14109.
6 Privman, V., Fratto, B.E., Zavalov, O. et al. (2013). *J. Phys. Chem. B* 117: 7559–7568.
7 Niazov, T., Baron, R., Katz, E. et al. (2006). *Proc. Natl. Acad. USA* 103: 17160–17163.
8 Privman, V., Strack, G., Solenov, D. et al. (2008). *J. Phys. Chem. B* 112: 11777–11784.
9 Privman, V., Zavalov, O., Halámková, L. et al. (2013). *J. Phys. Chem. B* 117: 14928–14939.
10 Winkel, B.S. (2004). *Annu. Rev. Plant. Biol.* 55: 85–107.

11 Mendes, P., Kell, D.B., and Welch, G.R. (1995). *Adv. Molec. Cell Biol.* 11: 1–19.
12 Idan, O. and Hess, H. (2013). *Curr. Opin. Biotechnol.* 24: 606–611.
13 Dueber, J.E., Wu, G.C., Malmirchegini, G.R. et al. (2009). *Nature Biotechnol.* 27: 753–759.
14 Idan, O. and Hess, H. (2013). *ACS Nano* 7: 8658–8665.
15 Zhang, Y.-H.P. (2011). *Biotechnol. Adv.* 29: 715–725.
16 Wilner, O.I., Weizmann, Y., Gill, R. et al. (2009). *Nature Nanotechnol.* 4: 249–254.
17 Fu, J., Liu, M., Liu, Y. et al. (2012). *J. Am. Chem. Soc.* 134: 5516–5519.
18 Delebecque, C.J., Lindner, A.B., Silver, P.A., and Aldaye, F.A. (2011). *Science* 333: 470–474.
19 Xin, L., Zhou, C., Yang, Z., and Liu, D. (2013). *Small* 9: 3088–3091.
20 Zhang, L., Shi, J., Jiang, Z. et al. (2011). *Green Chem.* 13: 300–306.
21 Mathesh, M., Liu, J., Barrow, C.J., and Yang, W. (2017). *Chem. Eur. J.* 23: 304–311.
22 Okhokhonin, A.V., Domanskyi, S., Filipov, Y. et al. (2018). *Electroanalysis* 30: 426–435.
23 Zhang, Y. and Hess, H. (2017). *ACS Catal.* 7: 6018–6027.
24 Zhang, Y., Tsitkov, S., and Hess, H. (2016). *Nature Commun.* 7: 13982.
25 Welch, G.R. and Easterby, J.S. (1994). *Trends Biochem Sci.* 19: 193–197.
26 Filipov, Y., Zakharchenko, A., Minko, S., and Katz, E. (2018). *ChemPhysChem* 19: 3035–3043.
27 Pedrosa, V., Melnikov, D., Pita, M. et al. (2010). *Int. J. Unconventional Computing* 6: 451–460.
28 Melnikov, D., Strack, G., Pita, M. et al. (2009). *J. Phys. Chem. B* 113: 10472–10479.
29 Daintith, J. and Wright, E. (2004). *A Dictionary of Computing*. Oxford University Press.
30 Vasilyev, S., Pita, M., and Katz, E. (2008). *Electroanalysis* 20: 22–29.
31 Ricard, J. and Cornish-Bowden, A. (1987). *Eur. J. Biochem.* 166: 255–272.
32 Privman, V., Arugula, M.A., Halámek, J. et al. (2009). *J. Phys. Chem. B* 113: 5301–5310.
33 Adamatzky, A., De Lacy Costello, B., and Asai, T. (2005). *Reaction-Diffusion Computers*. New York: Elsevier.
34 Adamatzky, A. and de Lacy Costello, B. (2003). *Phys. Lett. A* 309: 397–406.
35 Fratto, B.E. and Katz, E. (2017). *Int. J. Parallel, Emergent and Distributed Systems* 32: 139–156.
36 Fratto, B.E. and Katz, E. (2016). *ChemPhysChem* 17: 1046–1053.
37 Fratto, B.E. and Katz, E. (2015). *ChemPhysChem* 16: 1405–1415.
38 Fratto, B.E., Guz, N., and Katz, E. (2015). *Parallel Processing Lett.* 25 (01): 1540001.
39 Fair, R.B., Khlystov, A., Tailor, T.D. et al. (2007). *IEEE Design & Test of Computers* 24: 10–24.
40 Zhang, Y., Yu, H., Qin, J., and Lin, B. (2009). *Biomicrofluidics* 3: 044105.
41 Strack, G., Chinnapareddy, S., Volkov, D. et al. (2009). *J. Phys. Chem. B* 113: 12154–12159.

42 Halámek, J., Tam, T.K., Chinnapareddy, S. et al. (2010). *J. Phys. Chem. Lett.* 1: 973–977.
43 Tam, T.K., Strack, G., Pita, M., and Katz, E. (2009). *J. Amer. Chem. Soc.* 131: 11670–11671.
44 Halámek, J., Bocharova, V., Chinnapareddy, S. et al. (2010). *Mol. Biosyst.* 6: 2554–2560.
45 Katz, E., Bocharova, V., and Privman, M. (2012). *J. Mater. Chem.* 22: 8171–8178.
46 Tam, T.K., Pita, M., Ornatska, M., and Katz, E. (2009). *Bioelectrochemistry* 76: 4–9.
47 Katz, E. and Minko, S. (2015). *Chem. Commun.* 51: 3493–3500.
48 Strack, G. and Katz, E. (2012). *Int. J. Unconventional Computing* 8: 419–432.
49 Bakshi, S., Halámková, L., Halámek, J., and Katz, E. (2014). *Analyst* 139: 559–563.

5

Sophisticated Reversible Logic Systems

5.1 Introduction

The present chapter summarizes the advances achieved in experimental realization of reversible logic gates [1] (Feynman gate [Controlled NOT; CNOT] with 2-input/2-output channels; Double Feynman gate [DFG], Toffoli gate, Peres gate, and Fredkin gate all with 3-input/3-output channels) using enzyme-catalyzed reactions performed in modular flow systems. The chemical inputs and outputs can be potentially extended to include additional information processing steps, thus allowing for further increase of system complexity. The studied systems were composed of simple single enzyme-functionalized cells connected in specific networks with a modular design and are presented in the chapter in the order of increasing of their complexity.

5.1.1 Reversible Logic Gates and Their Features

Molecular [2–7] and biomolecular [8–16] systems mimicking operation of reversible logic gates (e.g., Toffoli [8, 14, 15], Fredkin [8–10, 13], Peres [14, 15], Feynman [3, 4, 11], and Double Feynman [14–16] gates) are of particular interest, since they provide unique output patterns for each combination of input signals. The reversible logic gates have been pioneered and researched by a group of scientists (Figure 5.1), who formulated their conceptual features and possible applications. It should be noted that the term "reversibility" in the definition of logic operations has the meaning different from commonly used in chemistry. In information processing the "logic reversibility" means the possibility to recover initial information from the processed information. This is not possible for most of trivial Boolean gates such as AND, OR, etc. Indeed, the logic output **0** can be generated by AND gate for different input combinations: **0**,**0**; **0**,**1**; and **1**,**0** – thus the original pattern of the input signals cannot be recovered when the output signal **0** is generated. Special property of logic reversibility can be achieved when the number of channels for the output and input signals is the same, for example, 2-input/2-output, 3-input/3-output. In such case each combination of the input signals can produce a unique pattern of the output signals that allows recovering the initial inputs. Depending on the signal processing function, which is unique to the specific reversible logic gate, the output signals will change differently when various combinations of the inputs are applied. Using this change, the

Enzyme-Based Computing Systems, First Edition. Evgeny Katz.

Figure 5.1 Scientists greatly contributed to the formulation of reversible logic gates, after whom the gates were named: Richard Feynman, Tommaso Toffoli, Edward Fredkin, and Asher Peres. Source: The photos were reproduced from the Wikipedia public domain.

unknown input signals can always be recovered from the measured output signals if the logic function is known. This inherent property of reversible logic gates is important, particularly for biosensing applications of logic gates [17, 18] when all the information in regards to the inputs is needed. For example, medical analysis of biomarkers signaling on injuries [18] can be performed using AND logic gates combining together two biomarkers of low specificity. The decision about the medical conditions, e.g., specific kind of injury, would be based on the simultaneous presence of two biomarkers appearing at concentrations above specific thresholds. While this is enough for the medical conclusion about the presence of the injury conditions, the negative result (logic output **0**) has no clear meaning because it can originate from various combinations of biomarkers, when one or both of them appear below the thresholds. Knowing exactly the original combination of the biomarkers would be the great advantage for biomedical applications. This could be achievable with the use of reversible logic gates.

5.1.2 Logic Reversibility vs. Physical Reversibility

While justifying the importance of reversible logic gates, another inherent physical property that is frequently emphasized is the ability to save energy. During irreversible logic computations, each bit of information lost generates $kT \ln 2$ joules of heat energy based on thermodynamic consideration [19]. This release of heat results in serious problems for computer engineering, particularly with densely packed nano-sized elements. Theoretically [20, 21] and experimentally [22], in some electronic realizations, reversible information processing should not be accompanied by an entropy increase [23], thus classifying it as an energy saving process. This is considered to be a very important feature of reversible information processing and will be needed in future electronic computers to allow fast computing in nanoscale elements without generation of heat. However, this is not applicable to chemical systems where energy dissipation is inevitable. Therefore, the molecular/biomolecular realizations of reversible logic gates keep only *logic* reversibility, while the energy savings are illusory goals. This rather trivial and obvious conclusion should not be overlooked when conceiving and/or implementing reversible logic gates in (bio)molecular systems [8, 14].

5.1.3 Integration of Reversible Logic Gates into Biomolecular Computing Systems

In order to be useful in a practical sense, reversible logic gates should be integrated in complex information processing networks, which include multistep Boolean and non-Boolean (e.g., memory) operations. Unfortunately, some of the experimentally realized systems were all photonic [7], where the input/output signals were optical signals. This does not allow for easy integration with molecular networks, thus resulting in stand-alone operations of the gates. Other reversible logic gates used chemical input signals (e.g., metal ions) [2–5], which cannot be easily produced by preceding chemical reactions. This makes integration of the designed reversible gates in complex logic networks very difficult. Although these systems illustrate novel functions, it is certainly not enough for any practical application. Only a few recent experimental examples have demonstrated reversible logic operations with input/output signals that are represented by biomolecules using DNA [8, 10] and enzyme [11–16] reactions, thus allowing for the extension of information processing and integration of reversible logic gates into complex biomolecular networks that can be designed to mimic biological systems. The most sophisticated reversible logic systems have integrated DNA- and enzyme-based logic networks [24].

5.1.4 Spatial Separation of Enzyme Logic Operation: The Use of Flow Devices

Recently designed logic networks based on enzyme-biocatalyzed multistep reactions [25–28] are the easiest for the practical realization of systems where logic operations are particularly useful in biomedical sensing [17, 18] as well as diagnostic applications [29, 30]. Despite the fact that complex multi-input/multistep information processing systems have been successfully realized with enzymatic cascades [25, 26], the requirement of several independently read output signals for realization of reversible logic gates is not a simple task. Cross talking between enzymatic pathways and chemically produced output signals limit the complexity of the enzyme-based logic systems when they are realized in a homogeneous system. In order to assemble more complex systems for realization of reversible logic gates, "clocking" (temporal control) and spatial separation (compartmentalization) [5] of various steps (e.g., in flow devices) are needed [14, 31, 32].

Figure 5.2A shows a single flow cell that was used for assembling flow devices mimicking the operation of reversible logic gates. Before the flow cells were assembled in circuitries to organize biocatalytic cascades composed of several interconnected biocatalytic cells, each cell was modified with one immobilized enzyme (Figure 5.2B). For realizing different logic gates, Feynman gate (Controlled NOT; CNOT), Double Feynman gate (DFG), Toffoli gate, Peres gate, and Fredkin gate, the physical layout of the flow circuitries varied depending

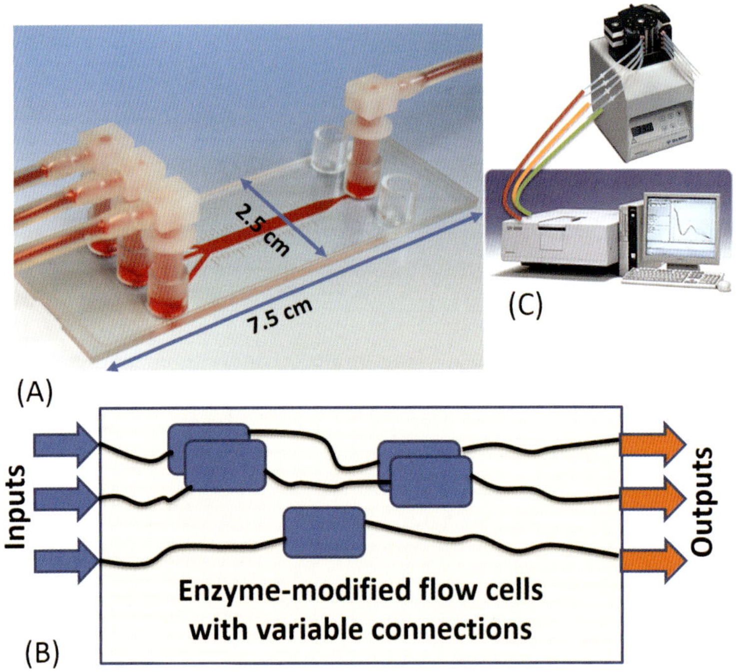

Figure 5.2 (A) A single flow cell used as a component of the system (the red color in the cell does not correspond to the real experimental solutions). (B) Schematic of the system with many interconnected enzyme-modified cells assembled in a circuitry with variable connections to mimic different logic gates (the scheme does not correspond to any specific logic network; for specific circuitries see Figures 5.5, 5.9, 5.13, 5.17, and 5.21). (C) Spectrophotometer connected to the flow system with help of a multichannel peristaltic pump. Source: Fratto and Katz 2015 [15]. Adapted with permission of John Wiley and Sons.

on the needs of the logic operations. The flow systems that were composed of several interconnected cells had three input and three output channels (except CNOT gate, which included 2-input/2-output channels). The input channels were connected to test tubes filled with stock solutions of the input signal chemicals, while the output channels were connected to the flow cuvette in a spectrophotometer for optical reading of the output signals (Figure 5.2C). The flow of reacting species moving from one enzyme-modified cell to another and finally to the spectrophotometer was maintained with a peristaltic pump.

5.2 Feynman Gate: Controlled NOT (CNOT) Gate

Here we illustrate the realized Feynman gate (Controlled NOT (CNOT) gate) [11], which is the only nontrivial reversible gate with two input and two output signals [33]. The first experimental realization of a CNOT gate was accomplished in 1995 [34]. Since its realization the CNOT gate has been implemented in many physical systems. This implementation demonstrates great importance for future quantum computing [35]; however, research aiming at its integration with

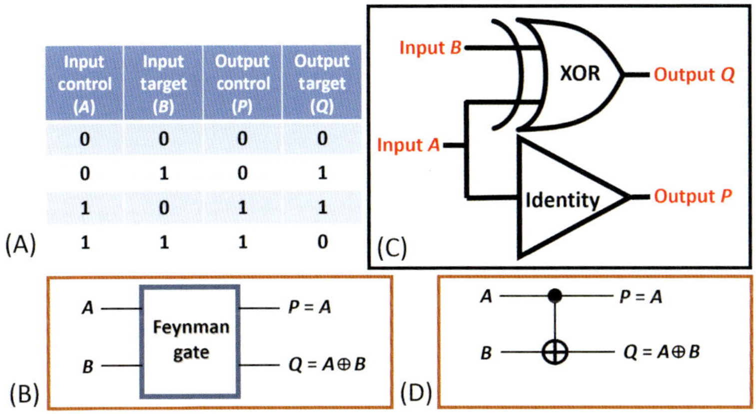

Figure 5.3 The truth table (A), block diagram (B), logic circuitry (C) and equivalent electronic circuitry (D) for Feynman (CNOT) gate. Standard Boolean notations are used in the schemes [36].

biomolecular logic networks is limited by very few publications [11]. Figure 5.3A shows the truth table of the CNOT gate. The CNOT gate flips the second Input *B* (the target input) and directs it to the second Output *Q* (the target output) if and only if the first Input *A* (the control input) is **1**, while the first Input *A* is always copied to Output *P* (the control output). In other words, Output *P* represents an Identity (ID) gate for Input *A*, while Output *Q* is processing Inputs *A* and *B* according to an XOR logic operation. The logic scheme illustrating CNOT operation is shown in Figure 5.3B–D, where Input *A* is directed to an Identity gate and XOR gate is activated by both Inputs *A* and *B*. In present experimental realization Input *A* was defined as a Tris-buffer solution (0.1 M, pH 7.1) containing *p*-nitrophenyl phosphate (PNPP) and pyruvate (Pyr) with concentrations 10 and 1 mM, respectively, to represent logic **1** value. Logic **0** value for Input *A* was defined as the absence of PNPP and Pyr in the solution. Input *B* was defined as a Tris-buffer solution (0.1 M, pH 7.1) containing glucose-6-phosphate (Glc6P), 6 mM, for logic **1** value. The absence of Glc6P was used to encode logic **0** for Input *B*. Note that the concentrations of the reacting species were optimized experimentally as will be described later. PNPP representing one of the reacting species in Input *A* was converted to *p*-nitrophenol (PNP) in the reaction biocatalyzed by alkaline phosphatase (AP), resulting in the optical absorbance increase (λ_{max} = 420 nm) (Figure 5.4). This biocatalytic reaction represented the ID gate (Figure 5.3C), and the produced absolute values of absorbance changes were used as Output *P*, which was defined as logic **1** or **0** when $|\Delta Abs| > 0.1$ or $|\Delta Abs| < 0.1$, respectively. Input *A* was also directed to the biochemical system performing an XOR logic operation (Figure 5.3C).

The XOR gate in the CNOT circuitry was defined with two oppositely directed biocatalytic redox reactions [37] activated by pyruvate (Pyr; second reacting species in Input *A*) and glucose-6-phosphate (Glc6P; Input *B*) (Figure 5.4). Input *B* was defined as logic **0** and **1** for absence of Glc6P and its presence with the concentration of 6 mM, respectively. The reaction media (all input solutions)

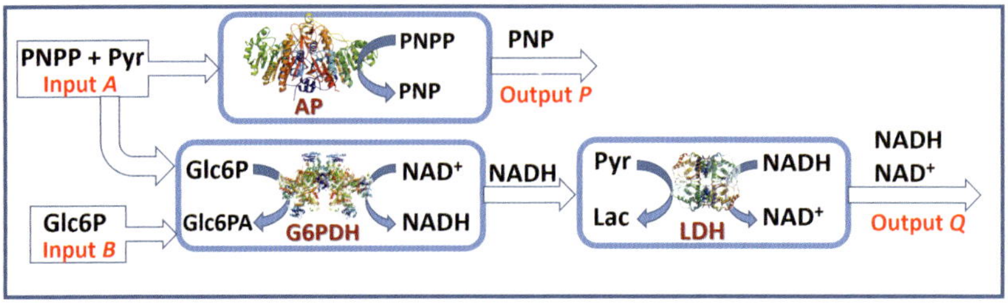

Figure 5.4 The biocatalytic cascade mimicking the CNOT gate operation (Glc6PA, gluconate-6-phosphate acid; Lac, lactate; all other abbreviations and reactions are explained in the text).

also included NADH (0.4 mM) and NAD^+ (10 mM) cofactors as a part of the gate "machinery" (the non-variable part of the biocatalytic system). NADH was oxidized by Pyr in the reaction biocatalyzed by lactate dehydrogenase (LDH) producing NAD^+ and decreasing the NADH absorbance at $\lambda_{max} = 340$ nm. NAD^+ was reduced by Glc6P in the reaction biocatalyzed by glucose-6-phosphate dehydrogenase (G6PDH) producing NADH and increasing its absorbance. Output Q produced by the XOR gate (Figure 5.3C) was measured as absolute values of the absorbance changes at $\lambda_{max} = 340$ nm and defined as logic **1** and **0** for $|\Delta Abs| > 0.2$ or $|\Delta Abs| < 0.2$.

The Identity and XOR gates operating in parallel were realized in a flow system schematically outlined in Figure 5.4. The biocatalytic reactions proceeding in the flow cells are shown schematically in each box representing a single cell (Figure 5.4). Figure 5.5 shows a photo of the experimental setup for realization of the Feynman gate (CNOT gate). The solutions containing biochemicals representing Inputs A and B with the variable binary logic values **0** and **1**, as well as the constant composition of the "machinery" represented by mixed $NADH/NAD^+$, were pumped through flow devices containing immobilized enzymes biocatalyzing chemical transformations mimicking logic operations.

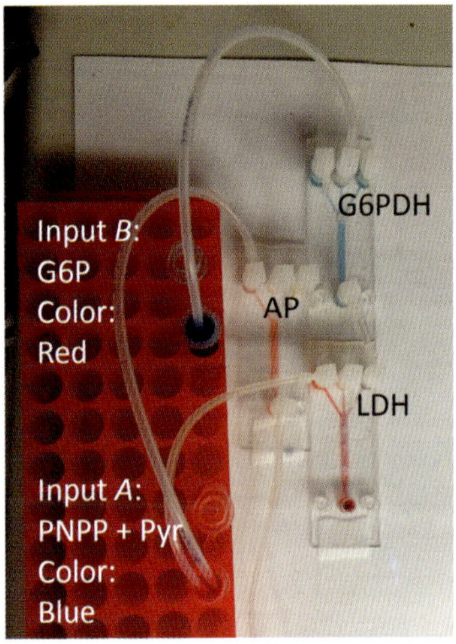

Figure 5.5 The experimental setup for realization of the Feynman (CNOT) gate. Different colored dyes are used in this image to illustrate the experimental realization including the mixing of channels where it is applicable. The colors used in the photo do not correspond to the real reacting solutions. G6P is used here instead of Glc6P abbreviation used in the text.

The flow design of the biochemical device allowed "clocking" (temporal control) and spatial separation of the reaction steps. While the Identity gate realized in a single flow unit performed a simple biocatalytic transformation of PNPP to PNP yielding an optically readable signal, the XOR gate was composed of two flow-through units: one modified with G6PDH that was reducing NAD^+ when Glc6P was available and another modified with LDH that was oxidizing NADH in the presence of Pyr (Figure 5.4). Therefore, the optical absorbance corresponding to the NADH concentration was decreasing in the presence of Pyr and absence of Glc6P (input combination **1,0**) and increasing in the absence of Pyr and presence of Glc6P (input combination **0,1**). Since the NADH concentration and the corresponding absorbance were changed in different directions (decreasing and increasing), the Output Q value **1** was defined as the absolute value of the absorbance change. The flow cells modified with G6PDH and LDH were optimized (balanced) in such a way that simultaneous presence of Pyr and Glc6P (input combination **1,1**) resulted in no changes in the concentration of NADH, meaning logic value **0** for Output Q. Balancing of the biocatalytic reaction rates was achieved by optimization of input concentrations for Pyr and Glc6P for the specific activity of the enzymes immobilized in the flow units. Obviously, the absence of Pyr and Glc6P (input combination **0,0**) did not result in any reaction and preserved the NADH absorbance unchanged, resulting in logic **0** for Output Q. These biocatalytic processes performed in the flow system allowed operation of the XOR part of the CNOT gate. It should be noted that the absorbance measurements for the solutions reacted in the flow device were performed vs. the "machinery" solution containing $NADH/NAD^+$ applied to the reference channel of the spectrophotometer, thus reflecting absorbance difference in the reacting solutions rather than their full absorbance.

Figure 5.6 shows the experimental data obtained upon application of the input signals in different logic combinations. Figure 5.6A shows the absorbance changes observed in the Output P (from the Identity gate). In the absence of PNPP, only minor absorbance changes were observed, meaning an Output P logic **0** value. Each time application of PNPP (regardless presence or absence of any other species) resulted in the absorbance increase at λ_{max} = 420 nm

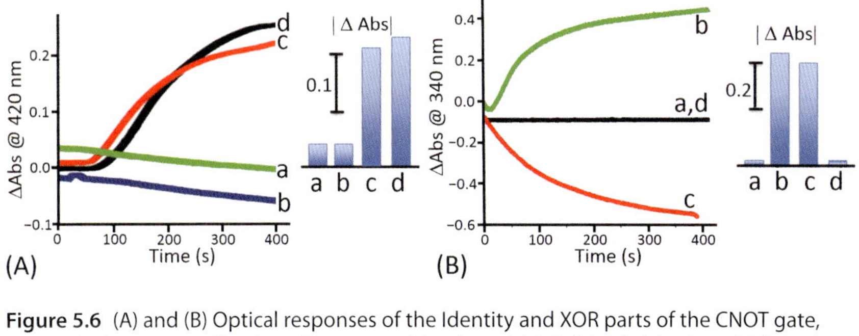

Figure 5.6 (A) and (B) Optical responses of the Identity and XOR parts of the CNOT gate, respectively, upon application of input combinations: (a) **0,0**; (b) **0,1**; (c) **1,0**; and (d) **1,1**. Insets show the optical outputs obtained after 400 seconds of the flow system operation. Source: Moseley et al. 2014 [11]. Adapted with permission of Royal Society of Chemistry.

reflecting the formation of PNP and resulting in the output logic **1**; thus, Input *A* was directly copied to Output *P*. Figure 5.6B shows the XOR gate performance where only unbalanced input signals **0**,**1** and **1**,**0** resulted in the absorbance changes (output **1**), while the absence of the reacting Pyr and Glc6P (inputs **0**,**0**) or their balanced application (inputs **1**,**1**) resulted in no absorbance changes (output **0**).

5.3 Double Feynman Gate (DFG) Operation

Figure 5.7A shows the truth table of DFG. DFG operates with three input and three output signals, where Input *A* is copied to Output *P*, while two other output signals *Q* and *R* are the results of two XOR logic functions performed on Inputs *A*, *B* and *A*, *C*, respectively (Figure 5.7B). In other words, DFG can be represented as the Identity (ID) gate and two XOR gates operating in parallel (Figure 5.7C,D). Therefore, when realized in a biochemical system [15], the DFG is a very convenient example of a parallel computing system. This convenience is illustrated by the simplistic design of the gate as a flow-through device with parallel channels.

The ID and two XOR gates operating in parallel were realized in a flow system outlined in Figure 5.8. The solutions containing biochemicals, pyruvate (Pyr), glucose-6-phosphate (Glc6P), and glucose (Glc) representing Inputs *A*, *B*, and *C*, respectively, with the variable binary logic values **0** and **1**, as well as the constant composition of the "machinery" represented by mixed NADH/NAD$^+$ were pumped through the flow devices containing immobilized enzymes biocatalyzing

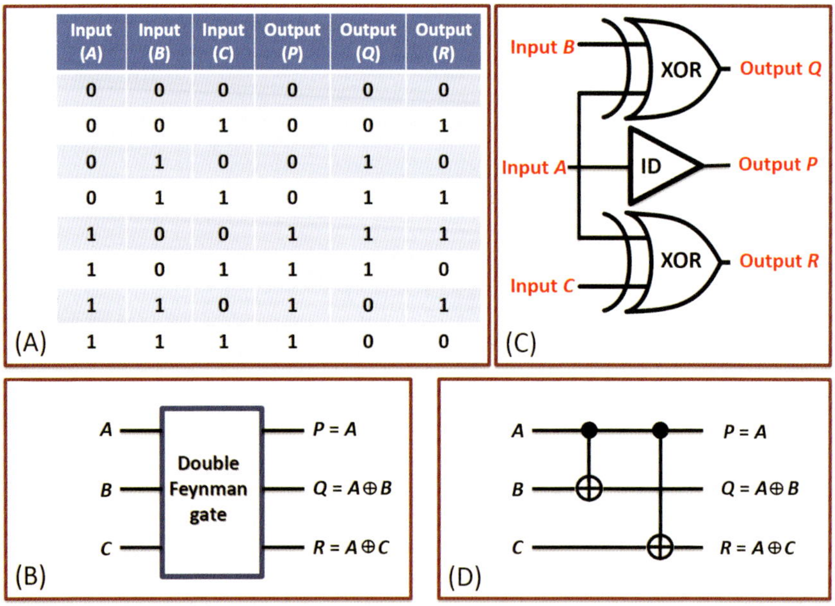

Input (*A*)	Input (*B*)	Input (*C*)	Output (*P*)	Output (*Q*)	Output (*R*)
0	0	0	0	0	0
0	0	1	0	0	1
0	1	0	0	1	0
0	1	1	0	1	1
1	0	0	1	1	1
1	0	1	1	1	0
1	1	0	1	0	1
1	1	1	1	0	0

Figure 5.7 The truth table (A), block diagram (B), logic circuitry (C), and equivalent electronic circuitry (D) for Double Feynman gate (DFG). Standard Boolean notations are used in the schemes [36]. Source: Adapted from Fratto and Katz 2015 [15].

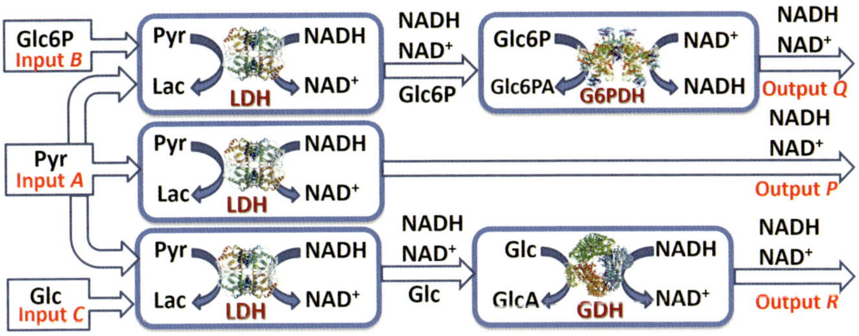

Figure 5.8 Experimental realization of the biocatalytic Double Feynman gate (DFG) in the flow device (GlcA, gluconic acid; Lac, lactate; Glc6PA, gluconate-6-phosphate acid; all other abbreviations and processes are explained in the text). Source: Adapted from Fratto and Katz 2015 [15].

chemical transformations mimicking logic operations. The flow design of the biochemical device allowed "clocking" (temporal control) and spatial separation of the reaction steps. While the ID gate realized in a single flow unit performed a simple biocatalytic transformation (NADH oxidation) yielding an optically readable signal, the XOR gates were composed of two flow-through units connected in serial: one modified with the enzyme oxidizing NADH to yield NAD^+ and another with the enzyme reducing NAD^+ resulting in the formation of NADH, thus decreasing and increasing optical absorbance corresponding to NADH concentration. The input signals were applied to the 3-channel flow system, where the biocatalytic reactions were performed in parallel. The biocatalytic reactions proceeding in the flow cells are shown schematically in each box representing a single cell (Figure 5.8).

Input *A* represented by Pyr was applied to the flow cell functionalized with LDH (Figure 5.8). Pyr, if it is present in Input *A* (logic **1** value), resulted in oxidation of NADH in the reaction biocatalyzed by LDH, thus resulting in the optical absorbance decrease (λ_{max} = 340 nm). In the absence of Pyr (logic **0** value), the NADH oxidation was not possible and the absorbance was not changed. This biocatalytic reaction represented the ID gate (Figure 5.7C), and the produced absolute values of absorbance changes were used as Output *P*, which was defined as logic **1** or **0** when $|\Delta Abs| > 0.2$ or $|\Delta Abs| < 0.2$, respectively. Input *A* (Pyr) was also directed to biochemical systems performing XOR logic operations (Figure 5.7C). The first XOR channel was fed with Inputs *A* and *B* (Pyr and Glc6P, respectively). This channel was composed of two flow cells functionalized with LDH and G6PDH operating in sequence (Figure 5.8). The XOR gate was designed to operate with two oppositely directed biocatalytic redox reactions [37] activated by Pyr and Glc6P. In the presence of Pyr, NADH was oxidized in the reaction biocatalyzed by LDH in the first flow cell, thus decreasing its absorbance at 340 nm. On the other hand, in the presence of Glc6P, NAD^+ was reduced in the reaction biocatalyzed by G6PDH in the second flow cell, thus increasing absorbance at 340 nm corresponding to the formation of NADH. Output *Q* in this channel was measured as absolute values of the absorbance changes at λ_{max} = 340 nm and defined as logic **1** and **0** for $|\Delta Abs| > 0.2$ or

|ΔAbs| < 0.2, respectively. The second XOR channel was activated with Inputs *A* and *C* (Pyr and Glc, respectively). This channel was composed of two flow cells functionalized with LDH and GDH operating in sequence (Figure 5.8). This XOR gate was also designed to operate with two opposite biocatalytic reactions. In the presence of Pyr, NADH was oxidized in the reaction biocatalyzed by LDH in the first flow cell, thus decreasing its absorbance at 340 nm. In the presence of Glc, NAD^+ was reduced in the reaction biocatalyzed by GDH in the second flow cell, thus increasing absorbance at 340 nm corresponding to the formation of NADH. Output *R* in this channel was measured as absolute values of the absorbance changes at $\lambda_{max} = 340$ nm and defined as logic **1** and **0** for |ΔAbs| > 0.2 or |ΔAbs| < 0.2, respectively. Figure 5.9 shows the experimental setup for realization of the DFG.

The flow cells modified with enzymes oxidizing NADH and reducing NAD^+ were optimized (balanced) in such a way that simultaneous presence of the oxidizing input (Pyr) and reducing input (Glc6P or Glc) (input combination **1**,**1**) resulted in negligible overall changes in the concentration of NADH, meaning logic value **0** for Outputs *Q* and *R*. Balancing of the biocatalytic reaction rates was achieved by optimization of input concentrations for Pyr and Glc6P in one channel and Pyr and Glc in another channel for the specific activity of the enzymes immobilized in the flow units. Obviously, the absence of Pyr and Glc6P or Glc (input combinations **0**,**0**) did not result in any reaction and preserved the NADH absorbance unchanged, resulting in logic **0** for Outputs *Q* and *R*. The unbalanced input signals (**0**,**1** or **1**,**0**) applied to the XOR channels resulted in the NADH absorbance changes considered as the output signal **1**. These biocatalytic processes performed in the flow system allowed operation of the XOR parts of DFG. It should be noted that the absorbance measurements for the solutions reacted in the flow device were performed vs. the "machinery" solution containing a constant amount of NADH/NAD^+ applied to the reference channel of the spectrophotometer, thus reflecting the absorbance difference in the reacting solutions rather than their full absorbance.

Figure 5.10 shows the experimental data obtained upon application of the input signals in eight different logic combinations. Figure 5.10A,B show the

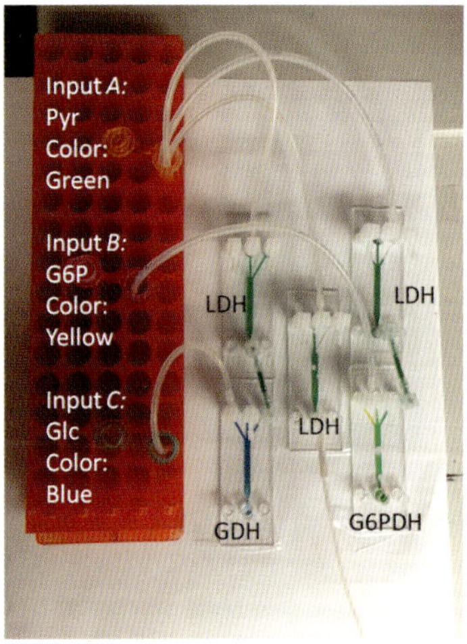

Figure 5.9 Experimental realization of DFG (photo of the flow cell circuitry). Different colored dyes are used in this image to illustrate the experimental realization including the mixing of channels where it is applicable. The colors used in the photo do not correspond to the real reacting solutions. G6P is used here instead of Glc6P abbreviation used in the text. Source: Adapted from Fratto and Katz 2015 [15].

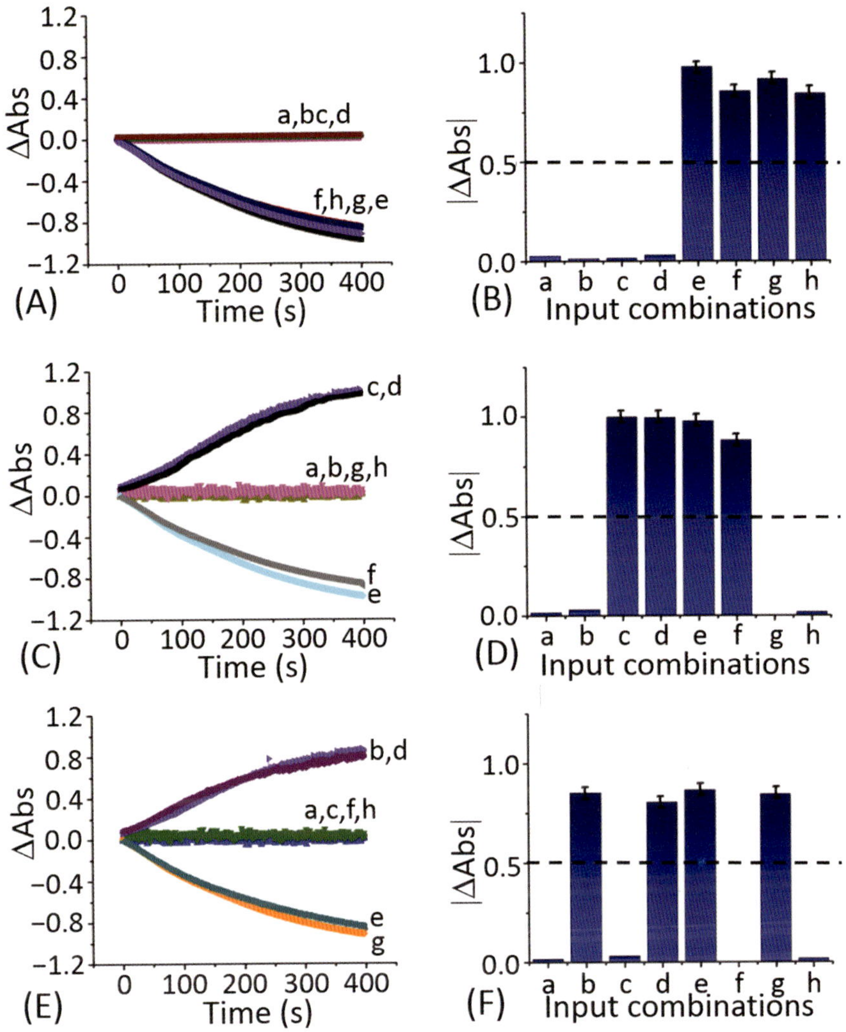

Figure 5.10 Optical responses of the Double Feynman gate (DFG) system to various combinations of the input signals: (a) **0,0,0**; (b) **0,0,1**; (c) **0,1,0**; (d) **0,1,1**; (e) **1,0,0**; (f) **1,0,1**; (g) **1,1,0**; (h) **1,1,1**. (Note that the logic values are shown for the input signals in the following order: *A*, *B*, *C*.) A and B panels show *P* output corresponding to the Identity operation copying *A* input signal. C–D panels show *Q* output corresponding to the XOR operation on *A* and *B* inputs. E–F panels show *R* output corresponding to the XOR operation on *A* and *C* inputs. A, C, and E plots show kinetics of the signal generation. B, D, and F bar charts show the signals obtained after 400 seconds of the flow system operation. All output signals were read at λ_{max} 340 nm. The dashed lines are thresholds separating **0** and **1** output values. The data shown in the bar charts are average of three independent experiments. Source: Adapted from Fratto and Katz 2015 [15].

absorbance changes observed in Output *P* (from the ID gate). In the absence of Pyr no absorbance changes were observed, meaning a logic **0** value for Output *P*. Each application of Pyr (regardless presence or absence of any other species) resulted in the decrease of absorbance at λ_{max} = 340 nm, thus reflecting the oxidation of NADH and resulting in the output logic **1**. Figure 5.10C,D and Figure 5.10E,F show the XOR gate performance where only unbalanced *A*, *B* and *A*, *C* input signals (**0**,**1** and **1**,**0**) resulted in the change of absorbance (output **1**), while the absence of the reacting species (inputs **0**,**0**) or their balanced application (inputs **1**,**1**) resulted in no change of absorbance (output **0**). Overall, the obtained experimental data show the output signal pattern (Figure 5.10), corresponding to the truth table (Figure 5.7A), characteristic of DFG.

5.4 Toffoli Gate Operation

Figure 5.11A shows the truth table of Toffoli gate. Toffoli gate operates with three input and three output signals, where Inputs *A* and *B* are directly copied to Outputs *P* and *Q*, thus representing two ID gates operating in parallel. In addition to their role in the ID gates, Inputs *A* and *B* are directed to an AND gate to generate an intermediate output that is then directed to an XOR gate together with Input *C* to yield Output *R* (Figure 5.11B–D). In other words, Toffoli gate can be represented as two ID gates operating in parallel with the AND–XOR concatenated gates (Figure 5.11C).

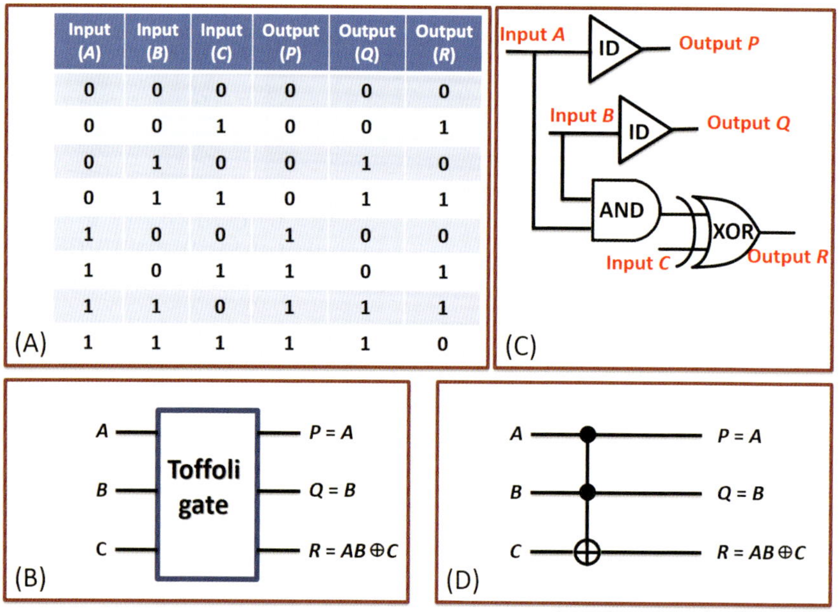
(A)

Input (*A*)	Input (*B*)	Input (*C*)	Output (*P*)	Output (*Q*)	Output (*R*)
0	0	0	0	0	0
0	0	1	0	0	1
0	1	0	0	1	0
0	1	1	0	1	1
1	0	0	1	0	0
1	0	1	1	0	1
1	1	0	1	1	1
1	1	1	1	1	0

Figure 5.11 The truth table (A), block diagram (B), logic circuitry (C), and equivalent electronic circuitry (D) for Toffoli gate. Standard Boolean notations are used in the schemes [36]. Source: Adapted from Fratto and Katz 2015 [15].

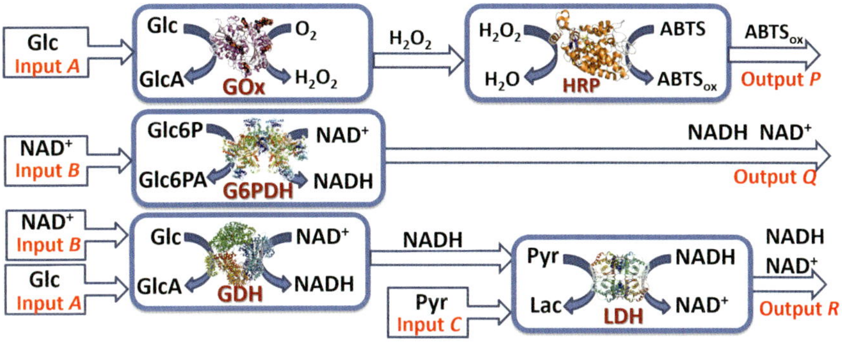

Figure 5.12 Experimental realization of the biocatalytic Toffoli gate in the flow device (GlcA, gluconic acid; Lac, lactate; G6PA, gluconate-6-phosphate acid; all other abbreviations and processes are explained in the text). Source: Adapted from Fratto and Katz 2015 [15].

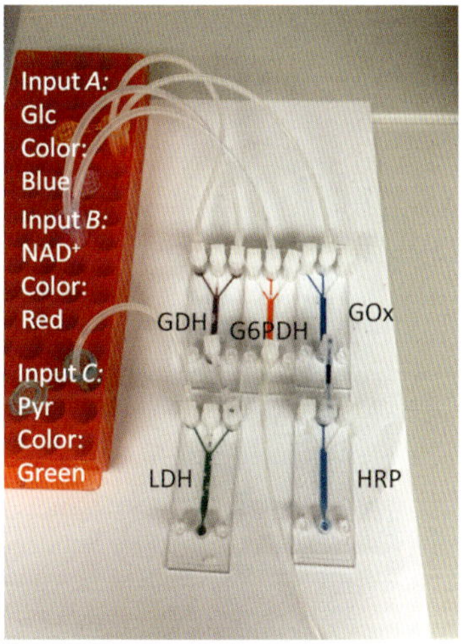

Figure 5.13 Experimental realization of Toffoli gate (photo of the flow cell circuitry). Different colored dyes are used in this image to illustrate the experimental realization including the mixing of channels where it is applicable. The colors used in the photo do not correspond to the real reacting solutions. Source: Adapted from Fratto and Katz 2015 [15].

Two ID gates and the AND–XOR concatenated gates operating in parallel were realized [15] in a flow system outlined in Figure 5.12. Figure 5.13 shows a photo of the experimental setup for realization of the Toffoli gate. The solutions containing biochemicals, Glc, NAD^+, and Pyr, representing Inputs *A*, *B*, and *C*, respectively, with the variable binary logic values **0** and **1**, as well as the constant composition of the "machinery" that is represented by a mixture of NADH/Glc6P/ABTS (note that O_2 was always present in the solution in equilibrium with air) were pumped through flow cells containing immobilized enzymes biocatalyzing chemical transformations mimicking logic operations. The first ID gate activated by Input *A* was composed of two flow cells operating in sequence: the first cell modified with GOx produced H_2O_2 in the presence of glucose (Glc) (if Input *A* was applied at logic value **1**). Then for convenient optical detection, the produced H_2O_2 was reacted with ABTS in the second flow cell modified with HRP to yield colored oxidized ABTS ($ABTS_{ox}$), which represented Output *P*. Obviously, in the absence of glucose (logic value **0** for

Input *A*), H_2O_2 was not produced and ABTS was not oxidized, thus preserving the optical absorbance in this channel without changes. Output *P* in this channel was measured as the absorbance changes at $\lambda_{max} = 415$ nm characteristic of $ABTS_{ox}$ and defined as logic **1** and **0** for $\Delta Abs > 0.2$ or $\Delta Abs < 0.2$, respectively. The second ID gate activated with Input *B* was represented by a single flow cell modified with G6PDH. In the presence of NAD^+ (logic value **1** for Input *B*), this cell produced NADH (note that the reducing species Glc6P were always present in the "machinery" solution), which was considered as Output *Q*. In the absence of NAD^+ (logic value **0** for Input *B*), NADH was not produced, thus keeping the initial absorbance without changes. It should be noted that NADH was also present in the "machinery" solution; thus the NADH produced in the flow cells was added to the background amount of NADH. Output *Q* in this channel was measured as absolute values of the absorbance changes at $\lambda_{max} = 340$ nm characteristic of NADH and defined as logic **1** and **0** for $|\Delta Abs| > 0.2$ or $|\Delta Abs| < 0.2$, respectively. Overall, Inputs *A* and *B* were directly copied to Outputs *P* and *Q*.

The AND gate was activated with a combination of Inputs *A* and *B* (Glc and NAD^+, respectively). In the presence of both reacting species (inputs **1**,**1**), GDH catalytically reduced NAD^+ to NADH, which was moving with the flow to the next cell modified with LDH. If either or both reacting species were absent (inputs **0**,**1**; **1**,**0**; **0**,**0**), NADH was not produced, thus demonstrating the AND logic features. The cell functionalized with LDH was additionally fed with Input *C* (Pyr). If both reacting species (NADH and Pyr) were present (**1**,**1**,**1** combination for Inputs *A*, *B*, *C*), the reaction biocatalyzed by LDH resulted in oxidation of NADH, thus bringing its concentration down to the original level and keeping the optical absorbance with no changes (Output *R* **0**). It should be noted that the Pyr concentration was carefully optimized to keep the balance and to compensate the NADH production in the first reacting cell. The same result was achieved if no reactions were activated in both connected flow cells (**0**,**0**,**0** combination for Inputs *A*, *B*, *C*). When NADH was produced in the first cell but Pyr was not present (**1**,**1**,**0** combination for Inputs *A*, *B*, *C*), Output *R* demonstrated increasing absorbance corresponding to the produced NADH (Output *R* **1**). When NADH was not produced in the first cell, but Pyr was present (**0**,**0**,**1**; **0**,**1**,**1**; **1**,**0**,**1** combinations for Inputs *A*, *B*, *C*), the reaction in the second cell resulted in the consumption of NADH present in the background solution, thus resulting in the absorbance decrease. Output *R* in this channel was measured as absolute values of the absorbance changes at $\lambda_{max} = 340$ nm characteristic of NADH and defined as logic **1** and **0** for $|\Delta Abs| > 0.2$ or $|\Delta Abs| < 0.2$, respectively.

Figure 5.14 shows the experimental data obtained upon application of the input signals in eight different logic combinations. Figure 5.14A,B show the absorbance changes observed in Output *P* (from the first ID gate). In the absence of Glc, no absorbance changes were observed, meaning logic **0** value for Output *P*. Each application of Glc (regardless presence or absence of any other species) resulted in the increase of absorbance at $\lambda_{max} = 415$ nm, reflecting the oxidation of ABTS and resulting in the output logic **1**. Figure 5.14C,D show the absorbance changes observed in Output *Q* (from the second ID gate). In the absence of NAD^+, no absorbance changes were observed, meaning logic **0** value for Output *Q*. When

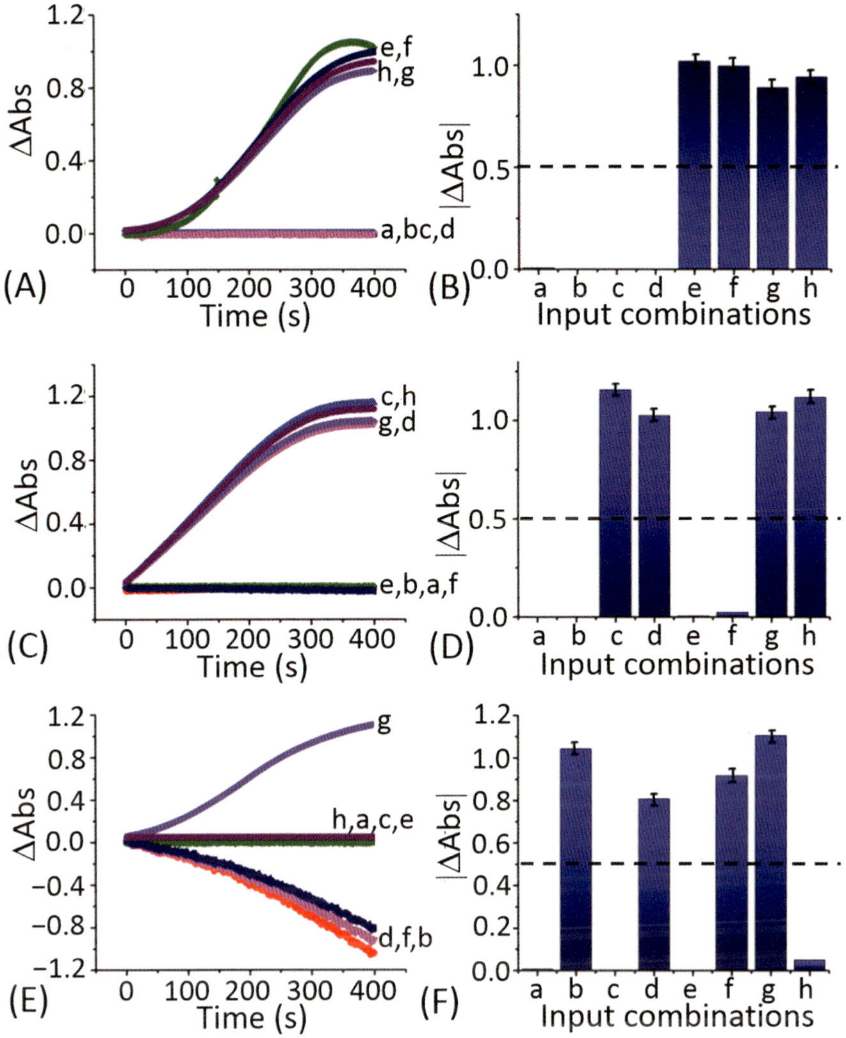

Figure 5.14 Optical responses of the Toffoli gate system to various combinations of the input signals: (a) **0,0,0**; (b) **0,0,1**; (c) **0,1,0**; (d) **0,1,1**; (e) **1,0,0**; (f) **1,0,1**; (g) **1,1,0**; (h) **1,1,1**. (Note that the logic values are shown for the input signals in the following order: *A*, *B*, *C*.) A and B panels show *P* output corresponding to the Identity operation copying *A* input signal. C–D panels show *Q* output corresponding to the Identity operation copying *B* input signal. E–F panels show *R* output corresponding to the AND operation on *A* and *B* inputs followed by the XOR operation on the output from the AND gate and *C* input. A, C, and E plots show kinetics of the signal generation. B, D, and F bar charts show the signals obtained after 400 seconds of the flow system operation. Output *P* was read at λ_{max} 415 nm and Outputs *Q* and *R* were read at λ_{max} 340 nm. The dashed lines are thresholds separating **0** and **1** output values. The data shown in the bar charts are average of three independent experiments. Source: Adapted from Fratto and Katz 2015 [15].

NAD^+ was applied (regardless presence or absence of any other species), the absorbance at $\lambda_{max} = 340$ nm was increased corresponding to the production of NADH and resulting in the output logic **1**. Figure 5.14E,F show the AND–XOR gate performance where only unbalanced *A*, *B*, and *C* input signals (**0,0,1**; **0,1,1**; **1,0,1**; **1,1,0**) resulted in the change of absorbance (output **1**), while the absence of the reacting species (inputs **0,0,0**) or their balanced application (inputs **0,1,0**; **1,0,0**; **1,1,1**) resulted in no change of absorbance (output **0**). Overall, the obtained experimental data show the output signal pattern (Figure 5.14), corresponding to the truth table (Figure 5.11A), characteristic of Toffoli gate.

5.5 Peres Gate Operation

Figure 5.15A shows the truth table of Peres gate. Peres gate operates with three input and three output signals, where Input *A* is directly copied to Output *P*, while two other output signals *Q* and *R* are the results of complex logic operations. Inputs *A* and *B* are processed through an XOR gate to yield Output *Q*. Inputs *A* and *B* are also processed through an AND gate, and the resulting intermediate goes to another XOR gate together with Input *C* to generate Output *R* at the end of the concatenated logic operations (Figure 5.15B). In other words, Peres gate can be represented as the ID gate operating in parallel with an XOR gate and an AND–XOR circuit (Figure 5.15C,D).

The ID, XOR, and AND–XOR gates operating in parallel were realized [15] in a flow system outlined in Figure 5.16. Figure 5.17 shows a photo of the

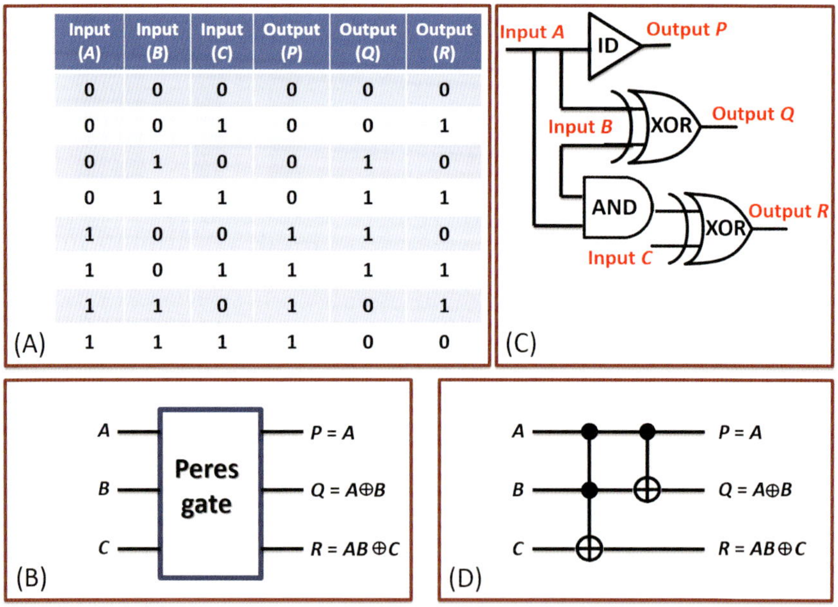

Input (*A*)	Input (*B*)	Input (*C*)	Output (*P*)	Output (*Q*)	Output (*R*)
0	0	0	0	0	0
0	0	1	0	0	1
0	1	0	0	1	0
0	1	1	0	1	1
1	0	0	1	1	0
1	0	1	1	1	1
1	1	0	1	0	1
1	1	1	1	0	0

Figure 5.15 The truth table (A), block diagram (B), logic circuitry (C), and equivalent electronic circuitry (D) for Peres gate. Standard Boolean notations are used in the schemes [36]. Source: Adapted from Fratto and Katz 2015 [15].

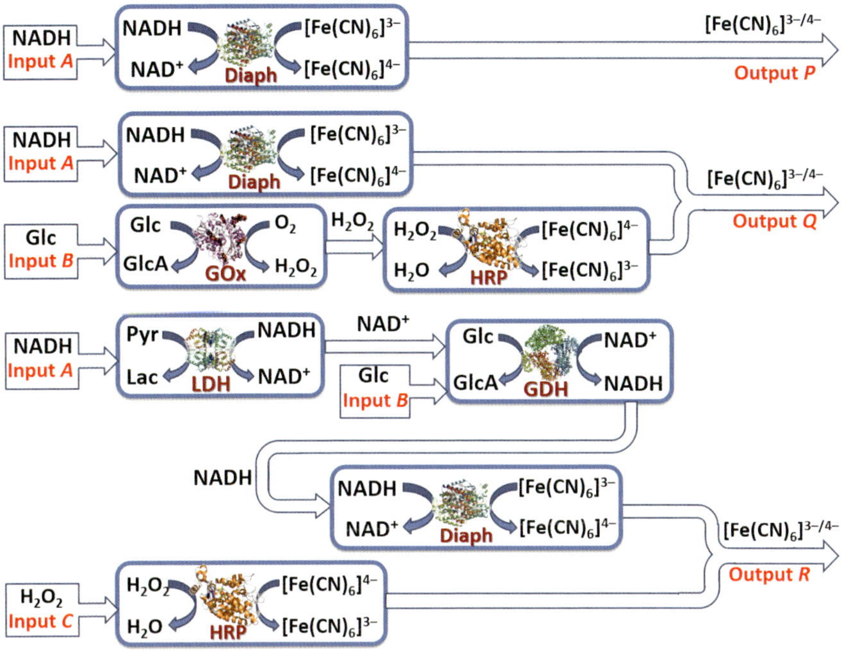

Figure 5.16 Experimental realization of the biocatalytic Peres gate in the flow device (GlcA, gluconic acid; Lac, lactate; all other abbreviations and processes are explained in the text). Source: Adapted from Fratto and Katz 2015 [15].

Figure 5.17 Experimental realization of Peres gate (photo of the flow cell circuitry). Different colored dyes are used in this image to illustrate the experimental realization including the mixing of channels where it is applicable. Source: Fratto and Katz 2015 [15]. Adapted with permission of John Wiley and Sons.

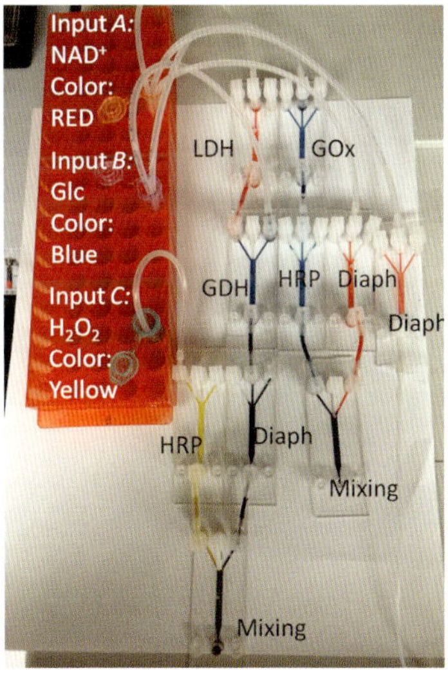

experimental setup for realization of the Peres gate. The solutions containing biochemicals, NADH, Glc, and H_2O_2, representing Inputs *A*, *B*, and *C*, respectively, with the variable binary logic values **0** and **1**, as well as the constant composition of the "machinery" represented by mixed $K_3[Fe(CN)_6]$, $K_4[Fe(CN)_6]$ and Pyr (note that O_2 was always present in the solution in equilibrium with air) were pumped through flow cells containing immobilized enzymes biocatalyzing chemical transformations mimicking logic operations. The ID gate activated by Input *A* (NADH) was represented with a flow cell functionalized with Diaph, which catalyzed NADH oxidation by $[Fe(CN)_6]^{3-}$. This process resulted in the production of reduced $[Fe(CN)_6]^{4-}$ and absorbance decrease at $\lambda_{max} = 420$ nm characteristic of $[Fe(CN)_6]^{3-}$ when Input *A* was applied at logic **1** value. In the absence of NADH (logic **0**), the reaction did not proceed, and the absorbance was not changed regardless of logic values of other inputs.

The XOR gate activated with Inputs *A* and *B* was realized in the following way. Input *A* (NADH) was applied to a flow cell functionalized with Diaph and operating in the way similar to described above. This biocatalytic pathway resulted in the consumption of $[Fe(CN)_6]^{3-}$. Input *B* (Glc) was applied to another flow cell functionalized with GOx. The reaction in this cell resulted in biocatalytic oxidation of glucose and concomitant production of H_2O_2 (note that O_2 was present in the solution). The *in situ* produced H_2O_2 was directed to the next flow cell modified with HRP where $[Fe(CN)_6]^{4-}$ was oxidized by H_2O_2 to yield $[Fe(CN)_6]^{3-}$. This reaction chain was activated only in the presence of Glc (Input *B* in logic **1** value). The solutions flowing out of the Diaph-modified cell fed with Input *A* and out of the GOx–HRP-modified cells fed with Input *B* were mixed after the biochemical reactions to yield finally Output *Q*. The overall result showed increasing or decreasing absorbance at $\lambda_{max} = 420$ nm corresponding to the increasing or decreasing concentration of $[Fe(CN)_6]^{3-}$ when Inputs *A* and *B* were applied with the combinations **0,1** or **1,0**, respectively. When Inputs *A* and *B* were applied at logic values **0,0** or **1,1**, the overall absorbance was not changed because the final concentration of $[Fe(CN)_6]^{3-}$ coming from two parallel channels was not changed due to the absent (**0,0** inputs) or balanced (**1,1** inputs) reactions. Output *Q* in this channel was measured as the absolute values of the absorbance changes at $\lambda_{max} = 420$ nm, characteristic of $[Fe(CN)_6]^{3-}$, and defined as logic **1** and **0** for $|\Delta Abs| > 0.1$ or $|\Delta Abs| < 0.1$, respectively.

The most sophisticated function was realized for Output *R* (Figure 5.16). First Input *A* (NADH) was processed in the cell modified with LDH where NADH was biocatalytically oxidized with Pyr (note that Pyr was always present in the background solution) to yield NAD^+, which was then directed to the next cell modified with GDH. In that cell, NAD^+ was reduced back to NADH in the presence of Input *B* (Glc). The two-step biocatalytic cascade was terminated with the production of NADH if both reaction steps were activated (combination **1,1** for Inputs *A* and *B*). When NADH input was absent in the first cell (Input *A* **0**), the biocatalytic cascade was not activated at all, regardless of the absence or presence of Glc in the second cell (input combinations **0,0** and **0,1** for Inputs *A* and *B*). In the presence of NADH in the first cell modified with LDH and in the absence of Glc in the second cell modified with GDH (combination **1,0** for Inputs *A* and *B*), the overall result was production of NAD^+ in the first cell, which was not returned back to NADH in the second cell. Thus, the final output in the form of the NADH production after two consecutive cells was only in the presence of both Inputs *A* and *B* in **1,1** combination, thus demonstrating the features of the AND function. The produced NADH was applied to the next flow cell modified with Diaph where NADH was used to reduce $[Fe(CN)_6]^{3-}$ yielding $[Fe(CN)_6]^{4-}$. Another reaction operating in parallel was performed in the cell modified with HRP where Input *C* (H_2O_2) oxidized $[Fe(CN)_6]^{4-}$ to yield $[Fe(CN)_6]^{3-}$. The solution produced in this cell was mixed with the solution coming out of the cascade of reactions realized in the cells modified with LDH–GDH–Diaph. The solution coming out of HRP-modified cell increases the $[Fe(CN)_6]^{3-}$ concentration if Input *C* is applied at logic value **1** (meaning presence of H_2O_2). The solution coming out of LDH–GDH–Diaph cells decreases the $[Fe(CN)_6]^{3-}$ concentration if Inputs *A* and *B* are applied at **1,1** combination. The overall result is the unchanged $[Fe(CN)_6]^{3-}$ concentration in case of **0,0,0**; **0,1,0**; **1,0,0**; and **1,1,1** combinations for Inputs *A*, *B*, and *C* because of the absence of the corresponding reactions or due to their balancing. All other input combinations (**0,0,1**; **0,1,1**; **1,0,1**; **1,1,0**) resulted in the $[Fe(CN)_6]^{3-}$ concentration change, thus resulting in the corresponding absorbance change. These results correspond to the logic operation of the concatenated AND–XOR gates as needed for the part of Peres gate. Output *R* in this channel was measured as absolute values of the absorbance changes at $\lambda_{max} = 420$ nm characteristic of $[Fe(CN)_6]^{3-}$ and defined as logic **1** and **0** for $|\Delta Abs| > 0.1$ or $|\Delta Abs| < 0.1$, respectively. It should be noted that the absorbance measurements for the solutions reacted in the flow device were performed vs. the "machinery" solution that contained a constant concentration of $[Fe(CN)_6]^{3-}/[Fe(CN)_6]^{4-}$ that was applied to the reference channel of the spectrophotometer, thus reflecting absorbance difference in the reacting solutions rather than their full absorbance.

Figure 5.18 shows the experimental data obtained upon application of the input signals in eight different logic combinations. Figure 5.18A,B show the absorbance changes observed in Output *P* (from the ID gate). In the absence of NADH, no absorbance changes were observed, meaning logic **0** value for Output *P*. Each application of NADH (regardless presence or absence of any other species) resulted in the decrease of absorbance at $\lambda_{max} = 420$ nm, reflecting the reduction of $[Fe(CN)_6]^{3-}$ and resulting in the output logic **1**. Figure 5.18C,D show the absorbance changes observed in Output *Q* (from the XOR gate). In the absence of NADH and Glc, no reactions proceeded and no absorbance changes were observed. The same overall result was observed for the balanced reactions in the presence of both NADH and Glc. Only unbalanced reactions (meaning the presence of one of the reacting species, NADH or Glc) resulted in the concentration changes for $[Fe(CN)_6]^{3-}$ and corresponding absorbance changes at $\lambda_{max} = 420$ nm. This channel represented the XOR gate operation activated by Inputs *A* and *B*. Figure 5.18E,F show the absorbance changes in Output *R* (from AND–XOR concatenated gates). This channel shows the unchanged absorbance at $\lambda_{max} = 420$ nm (meaning the unchanged concentration of $[Fe(CN)_6]^{3-}$) for input combinations **0,0,0** (when no reactions proceed) and **0,1,0**; **1,0,0**; **1,1,1** (when the reactions are balanced), thus resulting in logic **0** for Output *R*. All other input combinations resulted in the unbalanced reactions and the corresponding optical changes, thus resulting in the logic value **1** for Output

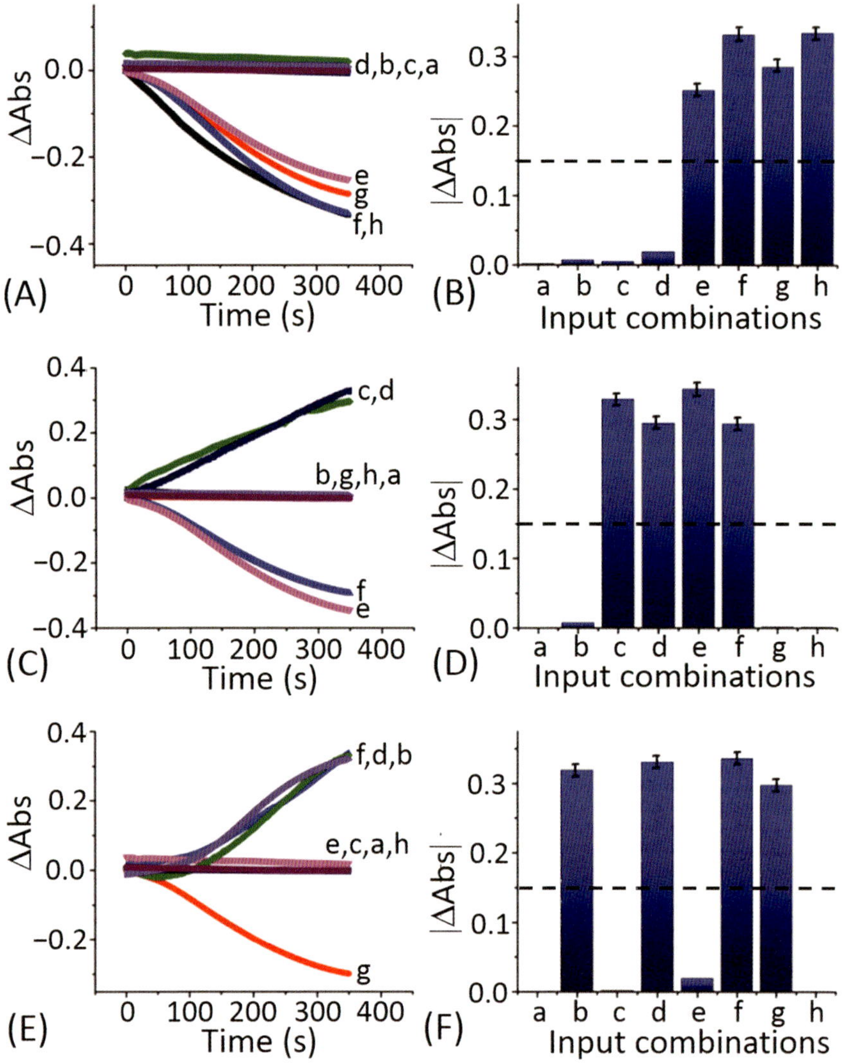

Figure 5.18 Optical responses of the Peres gate system to various combinations of the input signals: (a) **0,0,0**; (b) **0,0,1**; (c) **0,1,0**; (d) **0,1,1**; (e) **1,0,0**; (f) **1,0,1**; (g) **1,1,0**; (h) **1,1,1**. (Note that the logic values are shown for the input signals in the following order: *A*, *B*, *C*.) A and B panels show *P* output corresponding to the Identity operation copying *A* input signal. C–D panels show *Q* output corresponding to the XOR operation on *A* and *B* inputs. E–F panels show *R* output corresponding to the AND operation on *A* and *B* inputs followed by the XOR operation on the output from the AND gate and *C* input. A, C, and E plots show kinetics of the signal generation. B, D, and F bar charts show the signals obtained after 350 seconds of the flow system operation. All output signals were read at λ_{max} 420 nm. The dashed lines are thresholds separating **0** and **1** output values. The data shown in the bar charts are average of three independent experiments. Source: Fratto and Katz 2015 [15]. Adapted with permission of John Wiley and Sons.

R. Overall, the obtained experimental data show the output signal pattern (Figure 5.18), corresponding to the truth table (Figure 5.15A), characteristic of Peres gate.

5.6 Gates Redirecting Output Signals

While some of the reversible logic gates (specifically, Toffoli, Peres, and Feynman gates) can be realized as a complex combination of simple Boolean logic operations [15], such as Identity, AND, XOR (see Figures 5.3C, 5.7C, 5.11C, and 5.15C), other reversible logic systems have much higher complexity and features that cannot be easily mimicked with concatenated standard Boolean gates, thus requiring a different approach. This is specifically the situation with realization of the Fredkin gate [21]. In the Fredkin gate two Data input signals are directed to two output channels or switched from one channel to another depending on the logic value of the Control input signal. Thus, the Fredkin gate represents a Controlled-Swap gate (CSWAP) with the swapping (exchange) of the output channels depending on the Control channel. While the Fredkin gate is switching two Data channels, a simple version of this gate is switching only one Data channel to one of the two Output channels depending on the Control input signal. This Controlled-Switch gate is discussed in the following section prior to the Fredkin gate because both of them have been realized using a similar approach.

5.6.1 Controlled-Switch Gate

The Controlled-Switch gate changes processing of Data Input *B* depending on the logic value of Control Input *A*. Figure 5.19A–C show the truth table, block diagram, and equivalent electronic circuitry, respectively, for the Controlled-Switch gate. Note that the system operation cannot be formulated as combination of Boolean logic gates, at least in their trivial definition. In the biomolecular realization, Data Input *B* was defined as the absence or presence of NAD^+ (**0** and **1** logic values, respectively) in the solution pumped through the flow cells (Figure 5.20). On the other hand, the Control Input *A* had a very special definition different from most of other enzyme-based logic gates. The logic values **1** and **0** were defined as the presence of glucose (Glc) and lactate (Lac), respectively (note that it was not the absence or presence of one substance, but the presence of different substances to define **0** and **1** inputs). Control Input *A* is always directly copied to Output *P*, thus performing Identity gate (ID) operation. When Control Input *A* has **0** logic value (in the present system defined as the presence of Lac), Data Input *B* is copied to Output *Q*. On the contrary, when Control Input *A* has **1** logic value (in the present system defined as the presence of Glc), Data Input *B* is redirected to Output *R*. It should be noted that Output *R* shows **0** logic value when the signal is directed to Output *Q*, while Output *Q* shows **0** logic value when the signal is directed to Output *R*. In other words (in the experimental chemical realization), the non-active output channel shows only the background solution (meaning **0** output), while the active output channel is copying the logic value of Data Input

(A)

Input (*A*)	Input (*B*)	Output (*P*)	Output (*Q*)	Output (*R*)
0	0	0	0	0
1	0	1	0	0
0	1	0	1	0
1	1	1	0	1

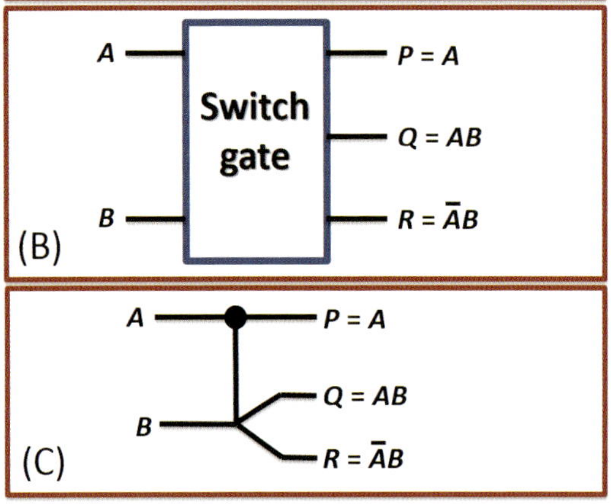

Figure 5.19 The truth table (A), block diagram (B), and equivalent electronic circuitry (C) for the Switch gate. Standard Boolean notations are used in the schemes [36]. Source: Fratto and Katz 2016 [13]. Adapted with permission of John Wiley and Sons.

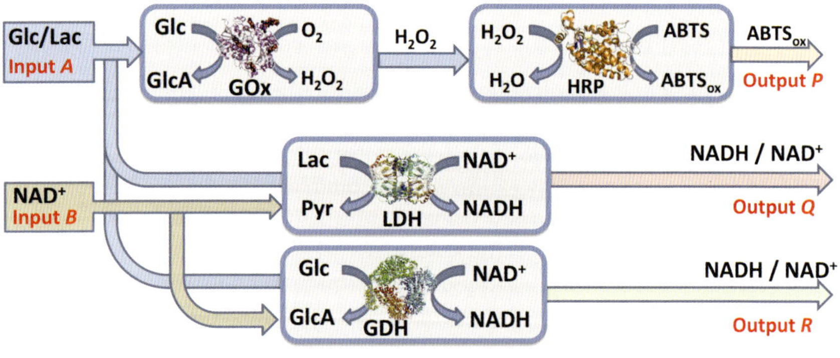

Figure 5.20 Experimental realization of the biocatalytic Switch gate in the flow device (GlcA, gluconic acid; Pyr, pyruvate; all other abbreviations and processes are explained in the text). Source: Fratto and Katz 2016 [13]. Adapted with permission of John Wiley and Sons.

B, which could be **0** or **1**. This logic function is easy to follow by looking at the truth table (Figure 5.19A). Indeed, Control Input *A* is directly copied to Output *P* showing **0** output for **0** input and **1** output for **1** input. Data Input *B* is copied to Output *Q* (**0** output for **0** input; **1** output for **1** input) when Control Input *A* has value **0**. Alternatively, Data Input *B* is copied to Output *R* (**0** output for **0** input; **1** output for **1** input) when Control Input *A* has value **1**. Whenever Output *Q* is active (showing **0** or **1**), Output *R* shows only **0**; similarly, when Output *R* is active (showing **0** or **1**), Output *Q* shows only **0**.

Figure 5.20 shows schematically the experimental chemical realization of this logic behavior in the flow cell system (the photo of the experimental setup is shown in Figure 5.21). Control Input *A* represented by the presence of Lac (**0** input) or Glc (**1** input) is passing through the flow cell functionalized with GOx. There is no any chemical reaction in this cell when the input is Lac. Also, this input is passing the next cell functionalized with HRP without any chemical reaction, thus producing no changes in the solution optical absorbance considered as **0** logic output. On the contrary, when Input *A* is represented by Glc, the reaction in the first cell functionalized with GOx results in the formation of H_2O_2, which results in the oxidation of ABTS in the second cell functionalized with HRP; both reactions are standard biocatalytic processes for these enzymes. Note that ABTS and O_2 are included in the "machinery" composition of the background solution to allow these reactions, when Glc is present. The produced $ABTS_{ox}$ shows the increased absorbance at $\lambda_{max} = 415$ nm, which is considered as output **1**. In addition to this pathway, Controlled Input *A* is directed to flow into two cells organized in parallel, one functionalized with LDH and another functionalized with GDH. Data Input *B* is also applied to these two cells. Splitting chemical signals for Inputs *A* and *B* in two different pathways was achieved with the use of Y-tubes to enable a fan-out operation (Figure 5.21). The routing of the solution into two pathways resulted in dilution of the dissolved chemicals, thus requiring additional experimental optimization of their concentrations. When Control Input *A* has **0** value (containing Lac), it can activate the cell functionalized with LDH producing NADH, only if NAD^+ is present; when NAD^+ is absent no reaction proceeds (meaning values **1** or **0** of Input *B*, respectively). If NADH

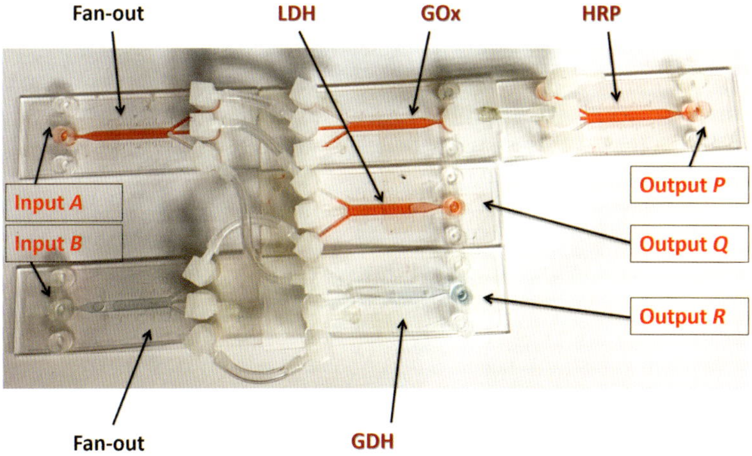

Figure 5.21 Experimental realization of the Switch gate (photo of the flow cell circuitry). Different colored dyes are used in this image to illustrate the experimental realization including the mixing of channels where it is applicable. Note that the colors are used for the illustration only and do not correspond to the real view of the system upon its operation. The abbreviated enzyme names were explained in the paper. The fan-out channels served for mixing and distributing flows in the system. Source: Fratto and Katz 2016 [13]. Adapted with permission of John Wiley and Sons.

is produced, the optical absorbance at λ_{max} = 340 nm was increasing, being considered as output **1**; otherwise the channel shows no absorbance change (output **0**). Regardless of the Input *B* value, when Output *R* is not active and shows no optical changes, it is considered as output **0**, if Control Input *A* is **0**. This is easy to understand, since the presence of Lac cannot activate any reaction in the cell functionalized with GDH. When Control Input *A* has value **1** (the presence of Glc), it activates the cell functionalized with GDH, thus producing NADH in the channel *R*, if NAD^+ is available. In this case the pathway functionalized with LDH stays mute. This process and the corresponding input/output signals can be explained exactly the same way as above.

Figure 5.22 shows the experimental data obtained upon application of the input signals in four different logic combinations. Figure 5.22A,B show the absorbance changes observed in Output *P* (from the ID gate corresponding to the Control input). In the presence of Lac (Input *A* is **0**), no absorbance changes were observed, meaning a logic **0** value for Output *P*. Application of Glc (Input *A* is **1**) resulted in the increase of absorbance at λ_{max} = 415 nm, thus reflecting the oxidation of ABTS and resulting in the output logic **1**. Figure 5.22C,D and 3E,F show output signals in Data channels *Q* and *R*, respectively. Output *Q* shows the absorbance increase corresponding to the NADH formation (λ_{max} = 340 nm) (output **1**) only when Control Input *A* is **0** and Data Input *B* is **1**. Output *R* shows the similar absorbance increase (output **1**) only when Control Input *A* is **1** and Data Input *B* is **1**. Overall, the obtained experimental data show the output signal pattern (Figure 5.22), corresponding to the truth table (Figure 5.19A), characteristic of the Controlled-Switch gate.

5.6.2 Fredkin (Controlled-Swap) Gate

After discussion on the Controlled-Switch gate, the Fredkin gate, operating similarly, is easy to understand. The Fredkin gate operates as a Controlled-Swap gate, where the logic value of the Control input (Input *A*) determines whether the two Data Inputs, Inputs *B* and *C*, are switched between the corresponding output channels or remain unchanged (Figure 5.23). In this gate Control Input *A* is directly copied to Output *P*, while Data Inputs *B* and *C* are copied to Outputs *Q* and *R* if Input *A* has logic value **0**. Alternatively, when Input *A* has the logic value **1**, Input *B* is redirected to Output *R*, and Input *C* is redirected to Output *Q*, thus switching them from the original pathways. Importantly, the logic value read at Output *P* indicates whether Inputs *B* and *C* are directly transposed to Output *Q* and Output *R* or are switched. Figure 5.23A shows the truth table corresponding to the operation of the Fredkin gate, while Figure 5.23B,C show the block diagram and equivalent electronic circuitry, respectively. In order to realize the switchable performance of the Fredkin gate [13], Control Input *A* was defined as the presence of Lac (corresponding to logic **0** value) or presence of Glc (corresponding to logic value **1**) (note that it was not simply the presence and absence of a single component as in many other logic gates discussed above). Data Inputs *B* and *C* were defined as the absence and presence of NAD^+ for logic values **0** and **1**, respectively. All three inputs were applied in eight possible combinations (Figure 5.23A).

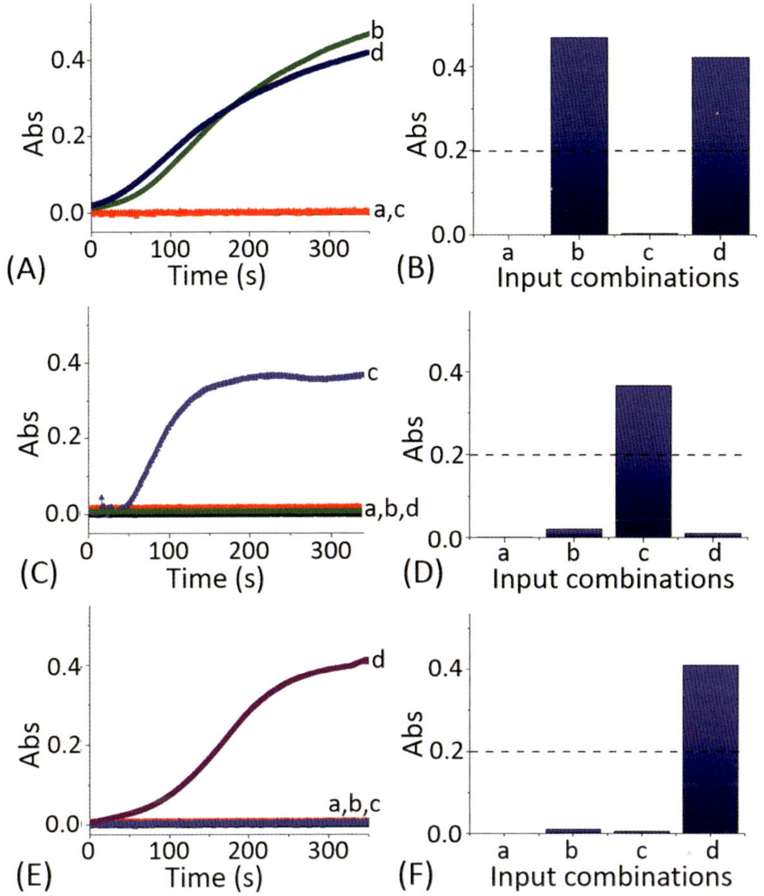

Figure 5.22 Optical responses of the Switch gate system to various combinations of the input signals: (a) **0**,**0**; (b) **1**,**0**; (c) **0**,**1**; (d) **1**,**1**. (Note that the logic values are shown for the input signals in the following order: *A*, *B*.) A and B panels show *P* output corresponding to the Identity operation copying Input *A*. C–D panels show *Q* output, and E–F panels show *R* output corresponding to various combinations of *A* and *B* inputs. A, C, and E plots show kinetics of the signal generation. B, D, and F bar charts show the signals obtained after 350 seconds of the flow system operation. Output signal *P* was read at λ_{max} 415 nm, whereas output signals *Q* and *R* were measured at λ_{max} 340 nm. Absorbance change, ΔA, of 0.2 corresponded to the threshold separating logic **0** and **1** values of the output signals. The data shown in the bar charts are average of three independent experiments. Source: Fratto and Katz 2016 [13]. Adapted with permission of John Wiley and Sons.

It should be noted that the electronic realization of the Fredkin gate is rather complicated; for example, it can be assembled in a very complicated manner as a combination of two Switch gates and two Inverse Switch gates [38]. On the contrary, the chemical realization in the flow cell device is relatively straightforward and simple, following the approach described above for the Controlled-Switch gate (Figure 5.24) (the photo of the experimental setup is shown in Figure 5.25). In the Fredkin gate, Control Input *A*, when represented

Input (A)	Input (B)	Input (C)	Output (P)	Output (Q)	Output (R)
0	0	0	0	0	0
0	0	1	0	0	1
0	1	0	0	1	0
0	1	1	0	1	1
1	0	0	1	0	0
1	0	1	1	1	0
1	1	0	1	0	1
1	1	1	1	1	1

(A)

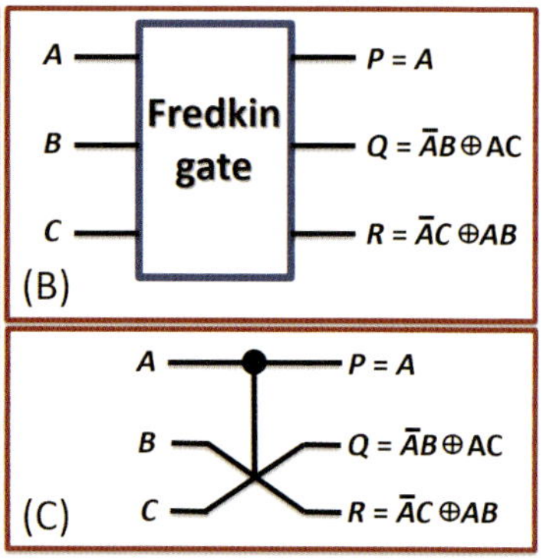

Figure 5.23 The truth table (A), block diagram (B) and equivalent electronic circuitry (C) for the Fredkin gate. Standard Boolean notations are used in the schemes [36]. Source: Fratto and Katz 2016 [13]. Adapted with permission of John Wiley and Sons.

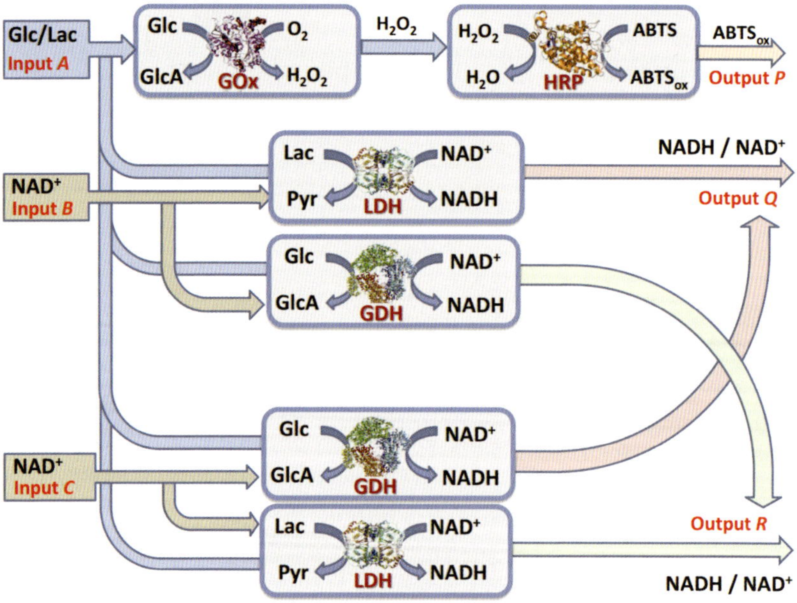

Figure 5.24 Experimental realization of the biocatalytic Fredkin gate in the flow device (GlcA, gluconic acid; Pyr, pyruvate; all other abbreviations and processes are explained in the text). Source: Fratto and Katz 2016 [13]. Adapted with permission of John Wiley and Sons.

by Lac (logic value **0**), resulted in no reaction in the cells functionalized with GOx and HRP, thus producing no optical changes in the output channel and resulting in Output *P* being **0**. When Input *A* is represented by Glc (logic value **1**), the biocatalytic reactions in the cells functionalized with GOx and HRP resulted in the formation of H_2O_2 and finally $ABTS_{ox}$, thus increasing the optical absorbance at $\lambda_{max} = 415\,nm$, which was considered as output **1**. This process mimicked the ID gate copying Input *A* to Output *P*. Control Input *A* was also

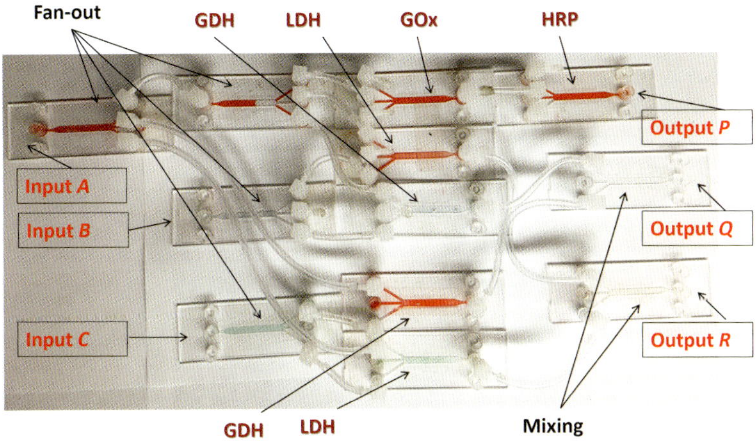

Figure 5.25 Experimental realization of the Fredkin gate (photo of the flow cell circuitry). Different colored dyes are used in this image to illustrate the experimental realization including the mixing of channels where it is applicable. Note that the colors are used for the illustration only and do not correspond to the real view of the system upon its operation. The abbreviated enzyme names were explained in the paper. The fan-out channels served for mixing and distributing flows in the system. Source: Fratto and Katz 2016 [13]. Adapted with permission of John Wiley and Sons.

split and directed to the flow cells functionalized with LDH and GDH operating in parallel. This was similar to the construction of the Controlled-Switch gate with only one difference – two pairs of the LDH and GDH cells operating in parallel, each of them fed with NAD^+ (logic value **1**) or background solution with no NAD^+ added (logic value **0**), have been assembled in the flow system (should be compared with the Controlled-Switch gate having only one pair of the LDH and GDH cells; see Section 5.6.1). Both Data Inputs *B* and *C* were represented with similar solutions but applied to different pairs of LDH and GDH cells through the different tubing. This was possible due to the modular design of the flow system and cannot be realized in a single solution due to the inability to distinguish the difference in logic inputs if the inputs are represented by the same chemical signal. The separation of pathways allows the logically different inputs to be represented by the same chemical substance (NAD^+ in the present case), since the routing of the system ensures the channels will remain separate. Data Inputs *B* and *C* were both denoted as logic value **1** in the presence of NAD^+ and as logic value **0** by the background solution. The two Data Inputs were each split into two pathways and passed through the cells functionalized with LDH and GDH enzymes (Figure 5.24). The LDH and GDH enzymes were mute (giving no reactions) if NAD^+ was absent (meaning logic **0** for Inputs *B* and *C*) regardless of the logic value of Input *A*. When NAD^+ was present (logic value **1** for Inputs *B* and *C*), the LDH or GDH cells were activated depending on the presence of Lac or Glc in Input *A*. The change of Input *A* from **0** (presence of Lac) to **1** (presence of Glc) resulted in the inhibition of LDH reaction and activation of GDH reaction, thus redirecting Data Inputs *B* and *C* from one output channel to the other output channels (Figure 5.24). Each output channel (*Q* and *R*) was connected with the tubing to LDH and GDH pathways corresponding to Inputs *B* and *C* channels, thus allowing Outputs *Q* and *R* to be activated by Inputs *B* and *C*, respectively, if Input *A* is **0**. Alternatively, Outputs *Q* and *R* responded to Inputs *C* and *B*, respectively (note the switch in their order), if Input *A* is **1**, thus following the signal pattern shown in the truth table (Figure 5.23A),

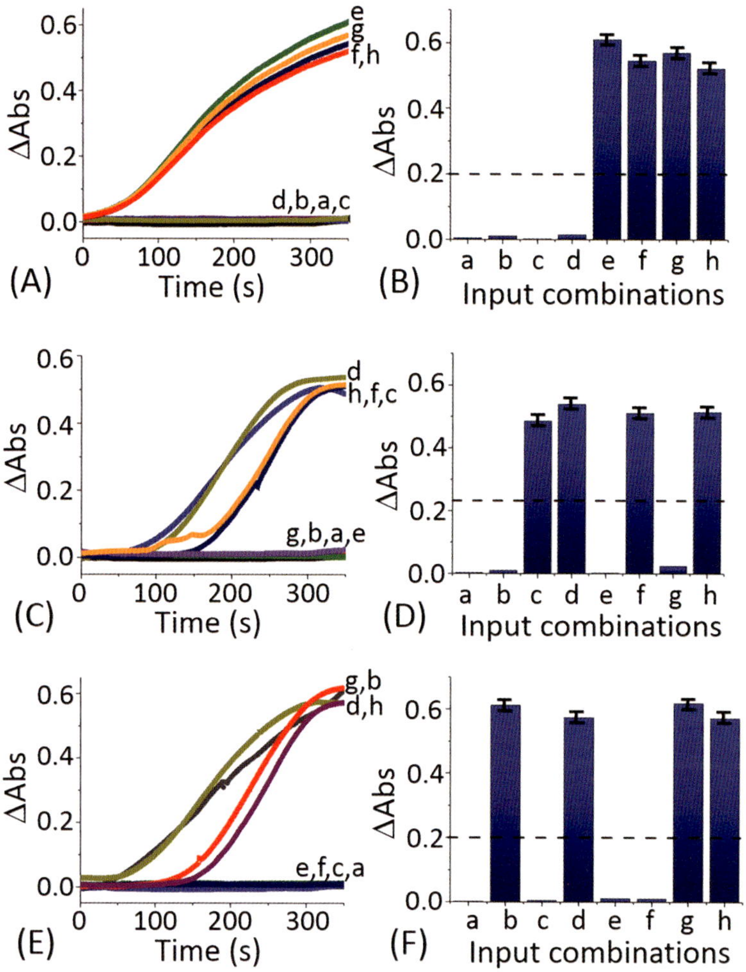

Figure 5.26 Optical responses of the Fredkin gate system to various combinations of the input signals: (a) **0,0,0**; (b) **0,0,1**; (c) **0,1,0**; (d) **0,1,1**; (e) **1,0,0**; (f) **1,0,1**; (g) **1,1,0**; (h) **1,1,1**. (Note that the logic values are shown for the input signals in the following order: *A*, *B*, *C*.) A and B panels show *P* output corresponding to the Identity operation copying *A* input signal. C–D panels show *Q* output and E–F panels show *R* output corresponding to various combinations of *A*, *B*, and *C* inputs. A, C, and E plots show kinetics of the signal generation. B, D, and F bar charts show the signals obtained after 350 seconds of the flow system operation. Output signal *P* was read at λ_{max} 415 nm, whereas output signals *Q* and *R* were measured at λ_{max} 340 nm. Absorbance change, ΔA, of 0.2 corresponded to the threshold separating logic **0** and **1** values of the output signals. The data shown in the bar charts are average of three independent experiments. Source: Fratto and Katz 2016 [13]. Adapted with permission of John Wiley and Sons.

corresponding to the features of the Fredkin gate. The experimental results illustrating the performance of the Fredkin gate are shown in Figure 5.26. Finally, it should be noted that the present design of the Fredkin gate, in the biochemical realization, is based on the use of a single Controlled-Switch gate, which is much simpler than its electronic counterpart, which requires a complex network structure. Overall, both gates, Controlled-Switch and Controlled-Swap, are important for routing signals in complex information processing systems. While at the present stage these logic gates are shown to operate as stand-alone units, the real use of them will be when they are included into complex information processing networks. Since the Data channel outputs are represented with the NAD^+/NADH cofactors, these output signals can be extended to downstream logic operations relatively easy.

5.7 Advantages and Disadvantages of the Developed Approach

5.7.1 Advantages

Biochemical systems of high complexity composed of many operating enzymes biocatalyzing several reactions cannot be realized in a single solution without compartmentalization of the reacting species similarly to the approach used by nature in biological cell. Therefore, the approach applied in the present study, where each biocatalytic reaction is running in a separate volume in individual reacting cells and communicating via flow moving from one cell to another, is a good solution for increasing the system complexity. While biocatalytic systems organized in a single solution were limited by maximum three to four reaction steps [28, 39] and required extremely complicated optimization [40], the reactions separated in individual cell are much simpler for performance and optimization, thus allowing the increased complexity of the systems. Most important, the modular design with individual flow cells, each modified with one enzyme, allows easy combination of them in various networks for performing different logic operations (AND, XOR, etc.) being parts of the complex information processes, as it is demonstrated in the realized reversible logic gates.

If the signal processing system needs to have several output signals and when it is organized in one solution, the output signals must have different optical properties (when the output signal is read by optical measurements) being represented by different chemical species. For example, one of the recently reported systems with several inputs and three outputs used individually readable absorbance changes corresponding to NADH and $ABTS_{ox}$ (λ_{max} 340 and 420 nm, respectively) and a fluorescence output produced by luciferase/luciferin system in the presence of ATP (λ_{max} 552 nm) [41]. If output signals are read by any other means, for example, using redox species analyzed electrochemically, the problem still persists since the produced species should be chemically different to demonstrate different redox potentials that are individually readable by cyclic voltammetry or by any other electrochemical technique. This requirement puts serious limitations on the use of biochemical reactions. The present approach does not have this limitation because the output signals are read from separated channels, thus allowing the outputs to be represented by the same chemical species; for example, NADH can be the output from all logic processes proceeding in parallel in different channels.

Another advantage of the designed biomolecular systems can be found in comparison with the reversible logic systems based on complex synthetic macromolecules. All-photonic gates activated by light signals and producing light emission as output cannot be easily extended to other chemical information processing steps [7]. On the other hand, the output signals represented by chemical species (e.g., NADH) could be easily connected to the extending information processing steps or even used for chemical actuation processes (e.g., stimulation of drug release) [42, 43].

5.7.2 Disadvantages

In electronic realizations of reversible logic gates, the input and output signals are represented by electrical potential/current changes. Therefore, the end of one logic gate can be easily connected to the beginning of another. This common property of electronic systems allows assembling reversible logic gates in complex networks for sophisticated computation. In the present chemical realization, the output signals are different by their nature from the input signals. Thus, connecting the logic elements in high hierarchy systems is difficult or even impossible, at least for some of the presently developed systems. This certainly limits the complexity of possible systems. In other words, the chemical extensions of the present logic systems are possible, but connecting the same components to each other is certainly not as trivial as in electronic system.

It should be noted that the developed systems allow *logic reversibility*, meaning the possibility to recover the whole set of input signals by analyzing the pattern of the output signals. The biocatalytic reactions cannot run in the opposite directions; thus the physical reversibility is not possible in these systems. Also, the biochemical systems do not have energy saving properties expected theoretically for electronic realization of the reversible computation.

Obviously, the time scale of the chemical systems operation is in minutes, thus being incomparably longer than that of electronic systems [44]. This makes the use of chemical systems very problematic for realistic computational applications. Even with these drawbacks, the new systems are very interesting for their possible integration with biosensors for logic processing biosensor inputs. However, this drawback is not as problematic as it may seem, since it could be at least partially resolved if the systems are scaled down to microfluidic devices operating on a shorter time scale [45].

Overall, it should be noted that the disadvantages briefly discussed above are characteristic of molecular/biomolecular computing systems in general, rather than being specific for the present approach based on the use of flow devices. On the other hand, the advantages of the developed approach clearly demonstrated the possibility of increasing complexity of the information processing systems, which is not achievable in a homogeneous system without compartmentalization of reaction steps.

5.8 Conclusions and Perspectives

The obtained results demonstrated for the first time information processing in reversible logic gates, such as Feynman gate (CNOT), Double Feynman gate (DFG), Toffoli gate, Peres gate, and Fredkin gate, performed in enzyme-based biocatalytic systems. The first results for CNOT, DFG, Toffoli, and Peres gates realized in biocatalytic flow systems [11, 13, 15] are promising for integrating the reversible gates into complex biomolecular logic networks. The logic processing of biomolecular signals in the reversible mode will be particularly beneficial for biosensing applications that need each combination of the output signals to correspond with a unique pattern of the input signals, thus allowing restoration of the original input values. The application of logically reversible gates for the analysis of biomedically important biomarker signaling for various physiological dysfunctions, similarly to previously reported injury diagnostics [29, 30], is feasible. Technological realization of the information processing systems in flow devices allows for "clocking" (temporal control) as well as spatial separation of the various steps of multistage biochemical processes, thus providing novel options for their sophistication and functional flexibility. The developed approach allows designing other reversible logic gates of higher complexity. Still technological realization of reversible logic systems with flow networks requires a lot of additional work. Increasing complexity of the interconnected flow networks may result in difficulties, which require fundamental research on biochemical reactions and transport in complex structured systems [46, 47]. Practical use of the systems discussed in this chapter will certainly require transition from the macroscale flow devices used in the present preliminary study to microfluidic lab-on-a-chip devices, allowing their miniaturization and potentially faster response operation.

References

1 Garipelly, R., Madhu Kiran, P., and Santhosh Kumar, A. (2013). *Int. J. Emerging Technol. Adv. Eng.* 3: 417–423.
2 Pérez-Inestrosa, E., Montenegro, J.-M., Collado, D. et al. (2007). *J. Phys. Chem. C* 111: 6904–6909.
3 Cervera, J. and Mafé, S. (2010). *ChemPhysChem* 11: 1654–1658.
4 Remón, P., Ferreira, R., Montenegro, J.-M. et al. (2009). *ChemPhysChem* 10: 2004–2007.
5 Remón, P., Hammarson, M., Li, S. et al. (2011). *Chem. Eur. J.* 17: 6492–6500.
6 Semeraro, M. and Credi, A. (2010). *J. Phys. Chem. C* 114: 3209–3214.
7 Andréasson, J., Pischel, U., Straight, S.D. et al. (2011). *J. Am. Chem. Soc.* 133: 11641–11648.
8 Orbach, R., Remacle, F., Levine, R.D., and Willner, I. (2012). *Proc. Natl. Acad. USA* 109: 21228–21233.
9 Roy, S. and Prasad, M. (2010). *Optical Eng.* 49: 065201.
10 Klein, J.P., Leete, T.H., and Rubin, H. (1999). *Biosystems* 52: 15–23.

11 Moseley, F., Halámek, J., Kramer, F. et al. (2014). *Analyst* 139: 1839–1842.
12 Fratto, B.E. and Katz, E. (2017). *Int. J. Parallel, Emergent and Distributed Systems* 32: 139–156.
13 Fratto, B.E. and Katz, E. (2016). *ChemPhysChem* 17: 1046–1053.
14 Katz, E. and Fratto, B.E. (2017, Volume 2, Chapter 2). *Advances in Unconventional Computing* (ed. A. Adamatzky), 29–59. Springer.
15 Fratto, B.E. and Katz, E. (2015). *ChemPhysChem* 16: 1405–1415.
16 Fratto, B.E., Guz, N., and Katz, E. (2015). *Parallel Processing Letters* 25 (01): 1540001.
17 Katz, E., Wang, J., Privman, M., and Halámek, J. (2012). *Anal. Chem.* 84: 5463–5469.
18 Wang, J. and Katz, E. (2011). *Israel J. Chem.* 51: 141–150.
19 Landauer, R. (1961). *IBM J. Res. Develop.* 5: 261–269.
20 Toffoli, T. (1982). *Int. J. Theor. Phys.* 21: 165–175.
21 Fredkin, E. and Toffoli, T. (1982). *Int. J. Theor. Phys.* 21: 219–253.
22 Takeuchi, N., Yamanashi, Y., and Yoshikawa, N. (2014). *Scientific Rep.* 4: 6354.
23 Bennett, C.H. (1973). *IBM J. Res. Develop.* 17: 525–532.
24 Guz, N., Fedotova, T.A., Fratto, B.E. et al. (2016). *ChemPhysChem* 17: 2247–2255.
25 Katz, E. and Privman, V. (2010). *Chem. Soc. Rev.* 39: 1835–1857.
26 Katz, E. (2015). *Curr. Opin. Biotechnol.* 34: 202–208.
27 Zhou, J., Arugula, M.A., Halámek, J. et al. (2009). *J. Phys. Chem. B* 113: 16065–16070.
28 Privman, V., Arugula, M.A., Halámek, J. et al. (2009). *J. Phys. Chem. B* 113: 5301–5310.
29 Halámková, L., Halámek, J., Bocharova, V. et al. (2012). *Analyst* 137: 1768–1770.
30 Halámek, J., Windmiller, J.R., Zhou, J. et al. (2010). *Analyst* 135: 2249–2259.
31 Toepke, M.W., Abhyankar, V.V., and Beebe, D.J. (2007). *Lab on Chip* 7: 1449–1453.
32 Scida, K., Li, B.L., Ellington, A.D., and Crooks, R.M. (2013). *Anal. Chem.* 85: 9713–9720.
33 O'Brien, J.L., Pryde, G.J., White, A.G. et al. (2003). *Nature* 426: 264–267.
34 Monroe, C., Meekhof, D.M., King, B.E. et al. (1995). *Phys. Rev. Lett.* 75: 4714–4717.
35 Siomau, M. and Fritzsche, S. (2010). *Eur. Phys. J. D* 60: 417–421.
36 Roth, C.H. Jr. and Kinney, L.L. (2010). *Fundamentals of Logic Design*. Cengage Learning.
37 Privman, V., Zhou, J., Halámek, J., and Katz, E. (2000). *J. Phys. Chem. B* 114: 13601–13608.
38 Lee, J., Huang, X., and Zhu, Q.-S. (2010). *Int. J. Digital Content Technology and its Applications* 4: 153–158.
39 Privman, V., Zavalov, O., Halámková, L. et al. (2013). *J. Phys. Chem. B* 117: 14928–14939.
40 Halámek, J., Bocharova, V., Chinnapareddy, S. et al. (2010). *Molec. Biosyst.* 6: 2554–2560.

41 Guz, N., Halámek, J., Rusling, J.F., and Katz, E. (2014). *Anal. Bioanal. Chem.* 406: 3365–3370.
42 Mailloux, S., Guz, N., Zakharchenko, A. et al. (2014). *J. Phys. Chem. B* 118: 6775–6784.
43 Mailloux, S., Halámek, J., and Katz, E. (2014). *Analyst* 139: 982–986.
44 De Vos, A. (2010). *Reversible Computing: Fundamentals, Quantum Computing, and Applications*. Weinheim: Wiley-VCH.
45 Zhao, Y. and Chakrabarty, K. (2010). *IEEE Trans. Biomed. Circuits Syst.* 4: 250–262.
46 Konkoli, Z. (2005). *Phys. Rev. E* 72: 011917.
47 Verma, A., Fratto, B.E., Privman, V., and Katz, E. (2016). *Sensors (MDPI)* 16: 1042.

6

Transduction of Signals Generated by Enzyme Logic Gates

While other book chapters are concentrated on formulation, assembling, operation, and applications of various enzyme-based Boolean logic gates, the present chapter is overviewing different methods (optical, electrochemical, etc.) used for transduction of biomolecular output signals generated by enzyme systems [1] (Figure 6.1). This chapter is a little bit technical and more related to biosensing aspects of the systems, rather than their computing features. Still all signal transduction methods are exemplified with specific logic operations, thus giving clear explanations on the use of biosensing methods for the analysis of logic operations. It should be clearly understood that the described methods are much more general than the given examples and can be applied to other logic gates performing different logic operations. Also, the same or similar methods can be applied to read out signals produced upon more complex computing operations performed, for example, by memory systems (Chapter 8) or arithmetic functions (Chapter 9).

6.1 Optical Analysis of Output Signals Generated by Enzyme-Based Logic Systems

Optical analysis of chemical products in enzyme-biocatalyzed reactions is frequently used in standard enzyme assay tests, particularly for the assay of enzymatic reactions biocatalyzed by NAD^+/NADH-dependent dehydrogenases (e.g., lactate dehydrogenase [LDH]), oxidases (e.g., glucose oxidase [GOx]) or peroxidases (e.g., horseradish peroxidase [HRP]) [2]. The LDH-biocatalyzed production or oxidation of NADH (note that the reaction is reversible) can be easily followed by measuring optical absorbance at $\lambda_{max} = 340$ nm characteristic of NADH [3]. The same optical analysis can be performed for the reaction biocatalyzed by any other NAD^+/NADH-dependent enzyme. The GOx-biocatalyzed glucose oxidation can be analyzed through the analysis of the concomitant product H_2O_2 [4]. This reaction is usually coupled to the HRP-biocatalyzed formation of a colored product in the presence of H_2O_2 [5]. The chromogenic substrates used in the HRP reaction are usually 2,2′-azino-*bis*(3-ethylbenzothiazoline-6-sulfonic acid) (ABTS), resulting in the oxidized colored product ($ABTS_{ox}$) absorbing light at $\lambda_{max} = 420$ nm or 3,3′,5,5′-tetramethylbenzidine (TMB) that is converted to the

Enzyme-Based Computing Systems, First Edition. Evgeny Katz.

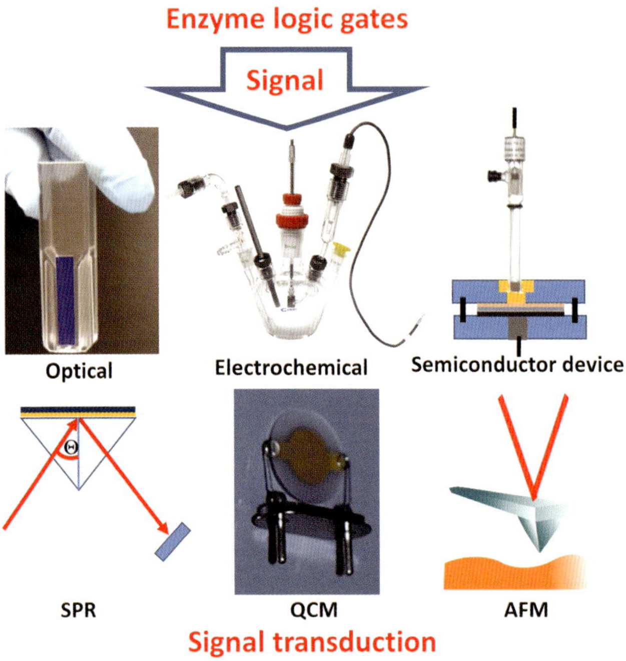

Figure 6.1 Transduction of chemical output signals generated by enzyme-based logic gates using different methods: optical measurements, electrochemical analysis, semiconductor devices used, surface plasmon resonance (SPR) measurements, quartz crystal microbalance (QCM) measurements, and atomic force microscopy (AFM) analysis.

oxidized product absorbing light at $\lambda_{max} = 650$ nm. The optical analysis of the ABTS or TMB oxidation products can be used to follow any biocatalytic reaction, resulting in the formation of H_2O_2 when it is coupled to the process biocatalyzed by HRP. Optical analysis of ATP can be performed using bioluminescence produced by luciferase/luciferin biocatalytic system [6]. Light emission produced by this system is a simple way to analyze ATP, which can be coupled to any other biocatalytic reaction producing or consuming ATP utilized in enzyme logic gates. In the sections below the optical absorbance and bioluminescence assays coupled to biocatalytic reactions mimicking various Boolean logic gates are discussed. Other optical methods, for example, surface plasmon resonance (SPR), can be used to follow biocatalytic reactions and analyze output signals produced by various enzyme logic gates.

6.1.1 Optical Absorbance Measurements for Transduction of Output Signals Produced by Enzyme-Based Logic Gates

Various enzyme-catalyzed reactions have been used to mimic basic Boolean logic operations with the output signals measured as optical absorbance changes due to formation or consumption of colored biochemicals, including AND [7–11], OR [11–13], NAND [14], NOR [11, 14], XOR [12, 15, 16], INHIB [11, 12], Identity [11], and Inverter [11] gates. The individual logic gates with the optical absorbance readout have been assembled to form logic networks of different structures and various complexity [17, 18]. Even more complex multienzyme/multi-logic gate branched network with switchable logic operation and optical readout has been applied for analysis of injury biomarkers [19]. This system represented the most sophisticated logic network still operating in a one-pot solution. Since many enzymes operated together performing different logic functions, the optimization of the reacting conditions was extremely complicated. This work was extended to reversible logic gates composed of several Boolean operations organized in the form of complex networks: Feynman (Controlled NOT [CNOT]) gate [20], Double Feynman gate [21], Toffoli gate [21], Peres gate [21], and Fredkin (Controlled-Swap) gate [22]. These logic gates included many enzyme-catalyzed reactions organized in flow systems with modular architecture allowing for the optical analysis of multiple products in different channels. It should be noted that systems of such high complexity cannot be assembled in a homogeneous solution because of "cross talking" between different logic gates. Thus, spatial separation of the reacting processes and time difference between them ("clocking") are mandatory for their realization in flow devices composed of separate reacting cells and channels. Also, the multichannel design of the logic systems allowed the use of the same chemical species (e.g., NADH) for different logic outputs analyzed separately in different channels. In this section we will discuss example systems of limited complexity based on the enzyme-catalyzed reactions mimicking OR, AND, and XOR logic gates with the output signals measured as optical absorbance changes. The interested readers can find more complex systems based on the same optical absorbance readout method in many published papers [17–19] as well as in other chapters of the book (see particularly Chapters 2, 4, 5, and 9).

The OR gate was realized using two parallel biocatalytic reactions activated with two different input signals and producing the same chemical species in both reactions considered as the output signal (Figure 6.2A). Two enzymes, glucose dehydrogenase (GDH) and alcohol dehydrogenase (ADH), and the NAD^+ cofactor represented the non-variable part of the system ("machinery"), which was the same for all combinations of the applied inputs. Glucose (Glc) and ethanol (EtOH) were defined as input signals, Input *A* and Input *B*, respectively, and they were applied to the gate "machinery" in four different combinations: **0**,**0**; **0**,**1**; **1**,**0**; and **1**,**1**, where logic value **0** corresponded to the absence of the input chemical (meaning its zero concentration), while the input concentration corresponding to logic value **1** was optimized experimentally to produce conveniently high output signals. Input *A* (Glc) was converted to the oxidized product, gluconic acid (GlcA), reducing in the concomitant reaction NAD^+ to yield NADH. This reaction was catalyzed by GDH. In another reaction catalyzed by ADH, EtOH was oxidized to acetaldehyde (Ald), and NAD^+ was reduced to NADH. Overall, NADH, considered as the final output product, was produced in the presence of either input or both of them (input combinations **0**,**1**; **1**,**0**; and **1**,**1**). The only input combination resulting in no production of NADH was **0**,**0** (in other words the absence of both substrates for the biocatalytic reactions). Figure 6.2B shows the truth table corresponding to the OR gate, and Figure 6.2C shows the experimental results for the enzyme-based system mimicking OR gate function. Note that the

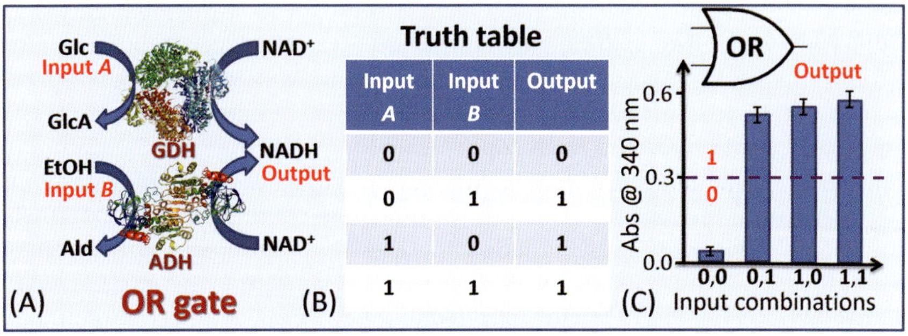

Figure 6.2 Optical absorbance measurements used for the analysis of output signals produced by an enzyme OR logic gate. (A) The OR logic gate based on enzyme catalytic reactions – schematics. (B) Truth table of a Boolean OR gate. (C) Optical absorbance corresponding to the biocatalytically produced NADH measured for different input combinations. The dashed line shows the threshold value separating logic **1** and logic **0** output signals. Source: Katz 2017 [1]. Adapted with permission of John Wiley and Sons.

biocatalytically produced NADH was analyzed by following optical absorbance at $\lambda = 340$ nm characteristic of NADH. The low and high absorbance values separated by a threshold were considered as **0** and **1** output signals.

The AND gate was realized using two consecutive biocatalytic reactions activated with two input signals represented by Glc and ABTS, Input *A* and Input *B*, respectively (Figure 6.3A). The "machinery" part of the logic gate

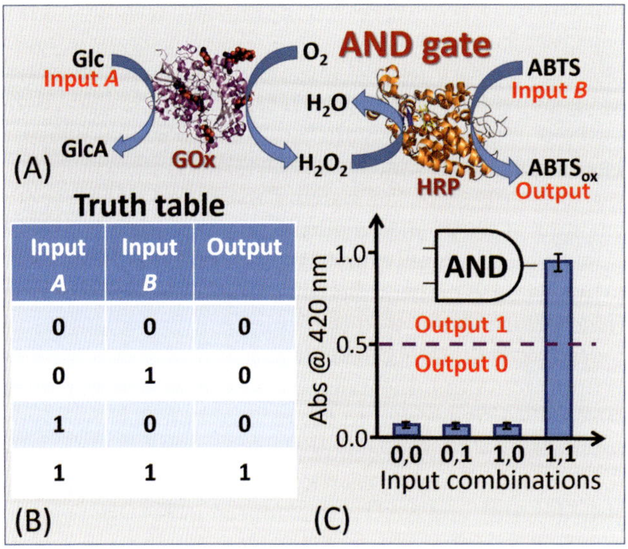

Figure 6.3 Optical absorbance measurements used for the analysis of output signals produced by an enzyme AND logic gate. (A) The AND logic gate based on enzyme catalytic reactions – schematics. (B) Truth table of a Boolean AND gate. (C) Optical absorbance corresponding to the biocatalytically produced $ABTS_{ox}$ measured for different input combinations. The dashed line shows the threshold value separating logic **1** and logic **0** output signals. Source: Katz 2017 [1]. Adapted with permission of John Wiley and Sons.

included glucose oxidase (GOx) and horseradish peroxidase (HRP); oxygen dissolved in the solution was also present in the system. Glc oxidation and O_2 reduction yielding H_2O_2 was catalyzed by GOx. The *in situ* produced H_2O_2 reacted with HRP, resulting in the biocatalyzed oxidation of ABTS yielding the colored $ABTS_{ox}$ product. The two-step biocatalytic process yielded the final $ABTS_{ox}$ product, considered as the output signal, only in the presence of both reacting input species (**1**,**1** input combination). If any or both of the input chemicals were missing (input combinations **0**,**0**; **0**,**1**; and **1**,**0**), the biocatalytic reaction was not proceeding till the very end and $ABTS_{ox}$ was not produced. The biocatalytic process was mimicking the AND logic gate, where the optical absorbance at $\lambda = 420$ nm characteristic of $ABTS_{ox}$ was considered as the output signal. Figure 6.3B shows the truth table characteristic of the AND logic operation, while Figure 6.3C shows the experimental realization of the AND logic gate with the output signals measured optically.

The XOR gate was realized using two biocatalytic reactions driven in the opposite directions by two input signals represented by Glc and pyruvate (Pyr), Input *A* and Input *B*, respectively. The "machinery" part of the logic gate included GDH and lactate dehydrogenase (LDH) (Figure 6.4A). NAD^+ and NADH cofactors (oxidized and reduced forms, respectively) were added to the system in equal concentrations as a part of the "machinery." In the presence of Glc, the reaction catalyzed by GDH resulted in the reduction of NAD^+, thus increasing the NADH concentration and the corresponding absorbance at $\lambda = 340$ nm. On the other hand, in the presence of Pyr, the reaction catalyzed by LDH resulted in the

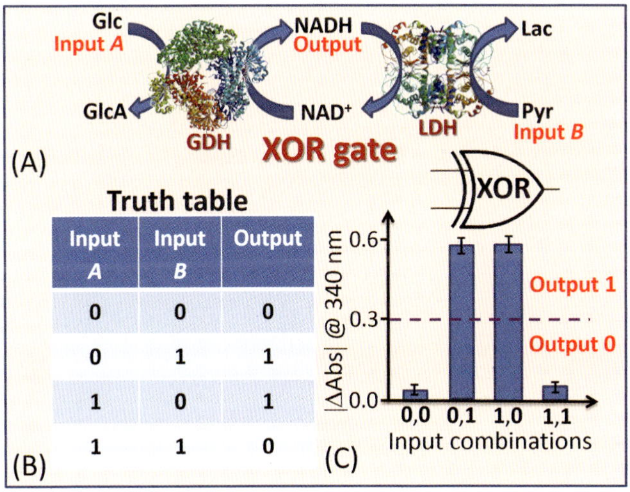

Input *A*	Input *B*	Output
0	0	0
0	1	1
1	0	1
1	1	0

Figure 6.4 Optical absorbance measurements used for the analysis of output signals produced by an enzyme XOR logic gate. (A) The XOR logic gate based on enzyme catalytic reactions – schematics. (B) Truth table of a Boolean XOR gate. (C) Absolute value of the optical absorbance changes corresponding to the biocatalytically produced/consumed NADH measured for different input combinations. The dashed line shows the threshold value separating logic **1** and logic **0** output signals. Source: Katz 2017 [1]. Adapted with permission of John Wiley and Sons.

oxidation of NADH, thus decreasing its absorbance. Thus, the optical absorbance was changed, increased or decreased, in the presence of Glc or Pyr (input signal combinations **0**,**1**; **1**,**0**), respectively. The biocatalytic reactions were optimized in such a way that both of them running simultaneously (inputs **1**,**1**) compensated each other, thus resulting in no absorbance changes. Obviously, applying no chemical inputs (inputs **0**,**0**), the biocatalytic reactions were not activated, and the absorbance was not changed. In order to fit the XOR gate logic function, the output signal was defined as the absolute value of the absorbance change, $|\Delta Abs|$, corresponding to either increase or decrease of the NADH concentration. Figure 6.4B shows the truth table of the XOR gate, and Figure 6.4C demonstrates results for its experimental realization. Note that the exemplified XOR gate produces the output signal **0** for the balanced input signals (**0**,**0** and **1**,**1**), while the unbalanced inputs (**0**,**1** and **1**,**0**) result in the output **1** measured as the absolute value of the absorbance change.

Three logic gates discussed above are the most typical examples of the enzyme-based systems used for mimicking Boolean logic operations. Many other gates with different logic [11] and more complex signal processing functions [20–22] have been experimentally realized using optical absorbance measurements as the transduction technique to follow the output signals.

A different approach to the optical analysis of output signals produced by enzyme logic gates was realized using shift of the absorbance band rather than change of the absorbance intensity [23]. The optical analysis was based on the dependence of plasmon energy in Au nanostructures controlled by the electrical charge formed on their surfaces. Magnetic core ($CoFe_2O_4$)–Au-shell nanoparticles (NPs; 18 ± 3 nm) [24] were functionalized with a thiolated monolayer bearing carboxylic groups (Figure 6.5A). The charge produced on the surface of the NPs was controlled by the dissociation state of the surface-bound carboxylic groups. At the initial neutral pH, the carboxylic groups were dissociated, thus producing a negative charge on the surface. When pH was decreased below pK_a of the carboxylic groups (ca. 5.2 ± 0.1; note that the pK_a value of the surface-confined carboxylic groups is different from the pK_a of a corresponding acid in a solution) [25], the carboxylic groups were protonated, and the surface charge became neutral. This pH-dependent change of the surface charge resulted in the change of the surface-localized plasmon energy, thus shifting λ_{max} of the NPs absorbance band (the neutral charge resulted in the shift of the λ_{max} to longer wavelength – red shift of the absorbance band). The pH change was produced by the enzyme-catalyzed reactions mimicking OR and AND logic gates [23]. The enzymes were covalently bound to SiO_2 particles (ca. 74 μm diameter) used as a platform for the biocatalysts. Figure 6.5A shows schematically the biocatalytic process mimicking the AND logic gate (the OR gate was mimicked in a similar way but using different enzymes catalyzing different reactions). The biocatalytic process was activated with two inputs, sucrose and O_2, Inputs *A* and *B*, respectively. Sucrose was cut to fructose (Frc) and glucose (Glc) in the reaction catalyzed by invertase (Inv). Then glucose was oxidized by O_2 in the reaction catalyzed by GOx, yielding gluconic acid (GlcA), thus, decreasing

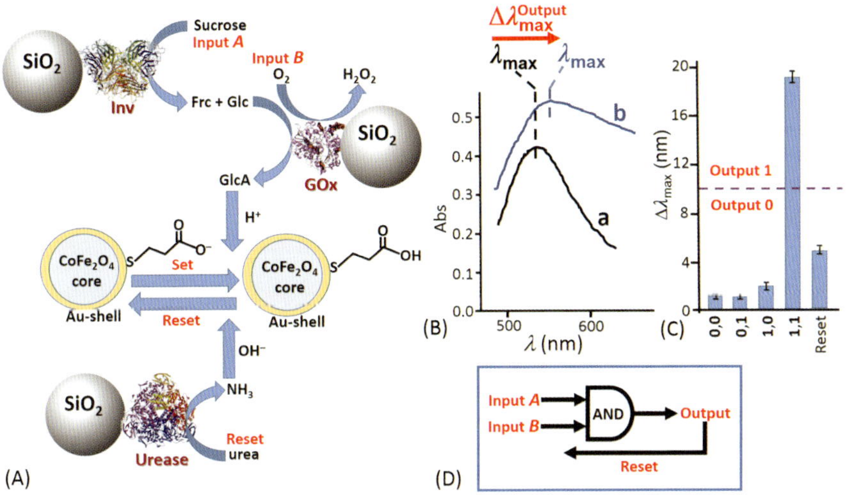

Figure 6.5 Optical absorbance measurements used for the analysis of output signals produced by an enzyme logic gate. (A) The AND-Reset logic system based on the biocatalyzed reactions and integrated with the magnetic core/Au-shell nanoparticles functionalized with a thiolated monolayer containing carboxylic groups. (B) Absorbance spectra corresponding to the LSPR of the NPs measured at different pH values: (a) pH 7.0, (b) pH ca. 4. (C) Logic output defined as the λ_{max} shift generated by the system upon application of various combinations of the input signals and the reset signal. The dashed line shows the threshold value separating logic **1** and logic **0** output signals. (D) Logic scheme of the AND-Reset system. Source: Pita et al. 2008 [23]. Adapted with permission of John Wiley and Sons.

the pH value. The two-step biocatalytic reaction proceeded to the end only in the presence of both input signals (logic combination **1**,**1**), thus mimicking the AND logic gate. The *in situ* produced low pH resulted in protonation of the surface-confined carboxylic groups and resulted in the absorbance wavelength shift originating from the change of the localized surface plasmon resonance (LSPR) (Figure 6.5B). The large shift of the λ_{max} (ca. 20 nm) was observed only for the **1**,**1** combination of the input signals, while λ_{max} was preserved almost unchanged for all other input combinations (**0**,**0**; **0**,**1**; **1**,**0**), as expected for the AND gate (Figure 6.5C). The logic gate was reset to the initial pH and to almost initial λ_{max} by the reaction catalyzed by urease converting urea to NH_3 (Figure 6.5A,C). The whole biocatalytic system represented the two-input AND gate with the Reset function (Figure 6.5D). Different components of the logic system played different roles in the signal processing and output readout. The SiO_2 particles suspended in the solution operated as a convenient platform for the enzyme immobilization, the superparamagnetic core ($CoFe_2O_4$), allowed easy manipulation of the signal-reading NPs in the presence of an external magnetic field, while the Au-shell was the reporting unit transducing pH changes to the optical signals. Other logic systems based on the enzyme-catalyzed pH changes are discussed in Sections 6.2.3–6.2.6.

6.1.2 Bioluminescence Measurements for Transduction of Output Signals Produced by Enzyme-Based Logic Gates

While optical absorbance measurements are the most frequently used method for analyzing output signals generated by enzyme logic systems, the method is limited to the processes yielding or consuming chemicals absorbing light in the visible (including near-UV) range of wavelength. The use of short-UV or IR spectra is not convenient for multicomponent complex biochemical systems (note that proteins/enzymes absorb light around 280 nm, thus restricting optical measurements of the output signals in this UV area). Therefore, analysis of biomolecules that are not colored requires different techniques. The most typical example is bioluminescent analysis of adenosine 5′-triphosphate (ATP) through a biocatalytic reaction in the presence of luciferase/luciferin system [6].

Figure 6.6A shows a two-step biocatalytic reaction in the presence of two enzymes, pyruvate kinase (PK) and luciferase (Luc), activated by three inputs, phosphoenol pyruvate (PEP), adenosine 5′-diphosphate (ADP), and luciferin (Lucif), Inputs *A*, *B*, and *C*, respectively. The first reaction catalyzed by PK results in consumption of PEP and concomitant conversion of ADP to ATP. Then, the second reaction catalyzed by Luc results in bioluminescence when ATP is produced through the first reaction and luciferin is present in the system. The two-step reaction activated with three input signals resembles a logic network composed of two concatenated AND gates (Figure 6.6B). Indeed, both biocatalytic reactions proceed till the end only in the presence of all reacting species (input combination **1,1,1**), while the absence of any of the input chemicals should inhibit the bioluminescence. Figure 6.6C, curve b, shows

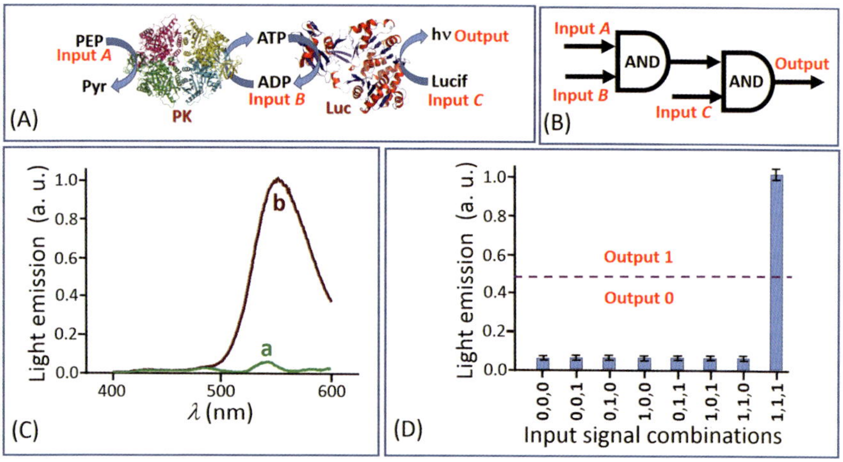

Figure 6.6 Bioluminescence measurements used for the analysis of output signals produced by an enzyme logic gate. (A) Two concatenated AND logic gates based on enzyme catalytic reactions – schematics. (B) Logic scheme of the concatenated AND gates. (C) Bioluminescence spectra measured for the logic system when it produces output **0** (a) and **1** (b). (D) Light emission by the bioluminescent system measured for different combinations of the input signals. The dashed line shows the threshold value separating logic **1** and logic **0** output signals. Source: Guz et al. 2014 [26]. Adapted with permission of John Wiley and Sons.

the light emission spectrum (bioluminescence) defined as the output signal **1** and observed for the input combination **1,1,1**. All other input signal combinations (**0,0**; **0,1**; **1,0**) resulted in the output signal **0** observed as low intensity background luminescence (Figure 6.6C, curve a). The bar chart (Figure 6.6D) shows all experimentally measured output signals for eight combinations of the inputs. The present system exemplifies application of bioluminescence as the method for observing output signals generated in the presence of ATP, which is a very common component of various enzyme-based logic systems of various complexity [26]. Another system mimicking a branched logic network was based on bioluminescence produced upon oxidation of luminol catalyzed by horseradish peroxidase (HRP) in the presence of enzymatically produced H_2O_2 [27]. Since H_2O_2 is a common product of many enzyme reactions (represented by different oxidases, e.g., glucose oxidase, lactate oxidase, etc.), the bioluminescence produced by the H_2O_2–HRP–luminol system is a very convenient method of the output signal transduction, being particularly useful when the absorbance measurements are difficult, for example, in heterogeneous systems.

6.1.3 Surface Plasmon Resonance (SPR) Measurements for Transduction of Output Signals Produced by Enzyme-Based Logic Gates

Redox reactions taking place in polymer thin films deposited on a gold surface result in changes in surface reflectance and can be detected by surface plasmon resonance (SPR) measurements [28, 29]. These surface redox processes can be coupled to biocatalytic reactions [30, 31] and used to analyze output signals produced by enzyme-based logic gates. Figure 6.7A shows schematically the reaction catalyzed by LDH and activated with two input chemicals: lactate (Lac) and NAD^+, Input *A* and Input *B*, respectively. The reaction resulted in the reduction of NAD^+ to NADH with the concomitant oxidation of Lac to pyruvate (Pyr). The biocatalytic system operated in a solution in contact with a chemically modified Au surface (SPR plate). Prussian blue (PB), an inorganic, insoluble, three-dimensional polymer, was deposited electrochemically on the Au surface and used to transduce the biocatalytic reaction output (NADH) to the surface reflectance changes measured by SPR [31]. The processes was started when PB was in the oxidized state, $Fe_4^{III}[Fe^{II}(CN)_6]_3$, produced by applying 0.3 V (vs. Ag/AgCl reference electrode [RE]) on the modified electrode. This state of PB demonstrated a relatively high reflectance measured by SPR method (Figure 6.7B, curve a). In the presence of the biocatalytically produced NADH, PB was chemically reduced to $K_4Fe_4^{II}[Fe^{II}(CN)_6]_3$ state with a smaller reflectance (Figure 6.7B, curve b); thus the reflectance value responded to the absence or presence of NADH in the solution. Before any reaction the modified surface demonstrated the high reflectance corresponding to the initial oxidized state of PB, but the reflectance was decreased after the biocatalytic production of NADH due to formation of the reduced state of PB on the surface. Input *A* and Input *B* signals (Lac and NAD^+) were applied in four different combinations, but the reaction resulted in the NADH formation only in the presence of both inputs (**1,1** combination). Therefore, the original high reflectance was preserved when

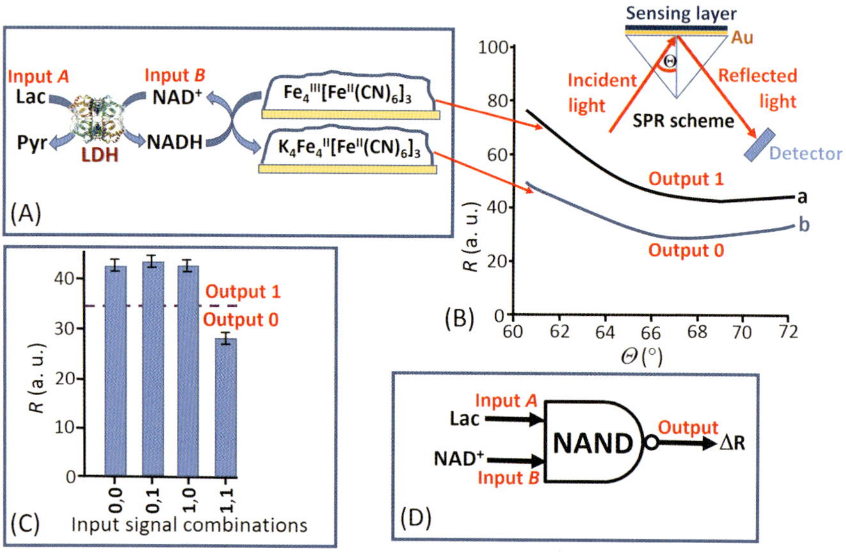

Figure 6.7 SPR measurements used for the analysis of the output signals produced by an enzyme logic gate. (A) The NAND logic gate based on enzyme catalytic reactions coupled to redox transformations of Prussian blue thin film on an SPR sensing surface. (B) Reflectance spectra measured by SPR and corresponding to the output **1** (a) and **0** (b). (C) Reflectance measured by SPR at $\Theta = 68°$ for different combinations of the input signals. The dashed line shows the threshold value separating logic **1** and logic **0** output signals. (D) NAND gate scheme. Source: Katz 2017 [1]. Adapted with permission of John Wiley and Sons.

the inputs were applied in **0**,**0**; **0**,**1**; and **1**,**0** combinations, and it was decreased in the case of **1**,**1** combination (Figure 6.7C). The high reflectance was defined as the output signal **1**, while the smaller reflectance was considered as the output **0**. The function demonstrated by the system resembles NAND logic gate with the SPR transduction of the output signal (Figure 6.7D). The SPR transduction of the output signal can be beneficial if the direct optical measurements of NADH are difficult, for example, in the presence of other colored components in systems absorbing light similar to the NADH spectrum.

6.2 Electrochemical Analysis of Output Signals Generated by Enzyme-Based Logic Systems

Electrochemical methods can be conveniently applied to the analysis of output signals produced by enzyme logic systems when redox-active species (e.g., H_2O_2 or NADH) are produced or consumed as the result of the logic operations [32–34]. The electroanalytical methods used in such applications could be similar to those used in electrochemical biosensors. Various electroanalytical methods, including chronoamperometry [33, 34] and potentiometry [35–37], can be employed for detecting the output signals. The potentiometry methods can be extended to the use of ion-selective and pH-sensing electrodes [38–40], thus extending the output signal measurements to the analysis of various species, which are not necessary redox active. Coupling of enzyme-catalyzed reactions with switchable stimuli-responsive materials deposited on electrode surfaces allows for indirect electrochemical analysis of the output signals, which are not redox active and not directly detectable by electrochemical means [38–46]. For example, pH variation in the course of enzyme-catalyzed reactions mimicking logic operations has been used to activate electrodes functionalized with pH-switchable polymer-based thin films. The pH changes produced by the biocatalytic system resulted in switching of a polymer-brush thin film between a nonpermeable shrunk state and a permeable swollen state, thus allowing electrochemical analysis of the interfacial state using cyclic voltammetry or impedance spectroscopy [39, 40]. Based on the same concept, but using a pH-switchable nanoporous membrane deposited on an electrode surface, enzyme logic operations were transduced to electrical outputs by impedance spectroscopy [47]. Notably, in these systems the electrochemical analysis was used to follow the interfacial properties of the modified electrode rather than direct analysis of the chemical species produced by the enzyme reactions. Signal-switchable materials can be used not only as thin films on sensing electrodes but also as switchable components of a bulk liquid phase, for example, in the form of a microemulsion switchable between oil-in-water (O/W) and water-in-oil (W/O) states [48]. This system can be switched between the states with a low and high electrical resistance upon getting signals produced by an enzyme logic gate. The change in the system state can be followed using simple DC electrical conductivity measurements. In addition to electrochemical measurements on conducting electrodes, semiconducting devices (usually in the form of field-effect transistors [FET]) have been successfully used for electronic transduction of chemical output signals produced by enzyme logic systems [49–54].

6.2.1 Chronoamperometric Transduction of Chemical Output Signals Produced by Enzyme-Based Logic Systems

Figure 6.8 exemplifies the use of chronoamperometric measurements for analysis of output signals produced by an enzyme logic system [34]. The biocatalytic cascade was initiated by glutamate (Glu) and glucose dehydrogenase (LDH), Input *A* and Input *B*, respectively. Note that in this hybrid system one of the inputs was represented by a substrate for an enzymatic reaction (Glu), while the second input was an enzyme (LDH) (Figure 6.8A). Two other enzymes, glutamate oxidase (GluOx) and alanine transaminase (ALT), as well as required substrates, alanine (Ala), NADH, and O_2, were non-variable parts of the logic gate ("machinery"). Both inputs used in the logic system were represented with biomarkers of traumatic brain injury (TBI), thus aiming at the logic processing of the biomarker concentration changes and TBI diagnosis. In this biomedical/bioanalytical application, the logic values **0** and **1** of the input signals were defined as their physiologically normal and elevated pathophysiological concentrations, respectively (note the input signal definition different from most of the examples discussed in the book). The multistep biocatalytic reaction (Figure 6.8A) used in the logic analysis of the input signals (biomarkers) was proceeding at any combination of the inputs because the logic **0** was not physically zero concentration of the reacting species. However, the rate of the process was strongly dependent on the logic combination of the input signals reflecting difference in their concentrations. The

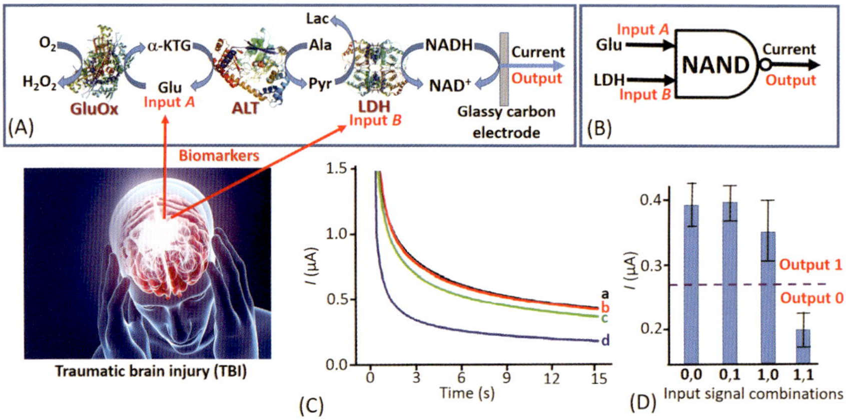

Figure 6.8 Electrochemical transduction of the enzyme generated output signals – chronoamperometric measurements. (A) The NAND logic gate based on enzyme catalytic reactions with the input signals corresponding to traumatic brain injury (TBI) biomarkers. NADH participating in the biocatalytic process is analyzed by chronoamperometry (at −0.1 V vs. Ag/AgCl) on a glassy carbon electrode. (B) NAND gate scheme. (C) Chronoamperometric curves measured for different combinations of the input signals: (a) **0**,**0**; (b) **0**,**1**; (c) **1**,**0**; and (d) **1**,**1**. (D) Current measured by chronoamperometry after 15 seconds from the beginning of the potential step for different combinations of the input signals. The dashed line shows the threshold value separating logic **1** and logic **0** output signals. Source: Zhou et al. 2011 [34]. Adapted with permission of John Wiley and Sons.

highest rate of the biocatalytic reaction was achieved at **1**,**1** combination of the inputs, while all other combinations (**0**,**0**; **0**,**1**; and **1**,**0**) resulted in low rates of the biocatalytic reactions. The output signal was defined as the current on a sensing glassy carbon electrode corresponding to electrochemical oxidation of NADH. Since the analyzed species was consumed (not produced) by the biocatalytic reaction, the output signal was smaller for the higher concentrations of the input species, thus representing the NAND gate (Figure 6.8B). The biocatalytic reaction was allowed for six minutes after adding the input signals to the gate "machinery," and then the chronoamperometric analysis of the residual NADH was performed. Figure 6.8C shows the experimental chronoamperometric current changes obtained at different logic combinations of the inputs, and Figure 6.8D shows the current values measured after 15 seconds of the chronoamperometric measurements. The low output signal (logic value **0**) obtained at **1**,**1** combination of the inputs was signaling on the TBI conditions, thus allowing the biomedical diagnosis based on the combined consideration of both TBI biomarkers. Despite the fact that the logic analysis of the biomarkers was performed in a model system, the developed approach keeps promise for future rapid diagnostics of TBI in the binary "YES/NO" format, particularly useful for point-of-care applications.

6.2.2 Potentiometric Transduction of Chemical Output Signals Produced by Enzyme-Based Logic Systems

Electrochemical analysis of NADH participating in biocatalytic reactions can be performed by different electrochemical methods, including potentiometric

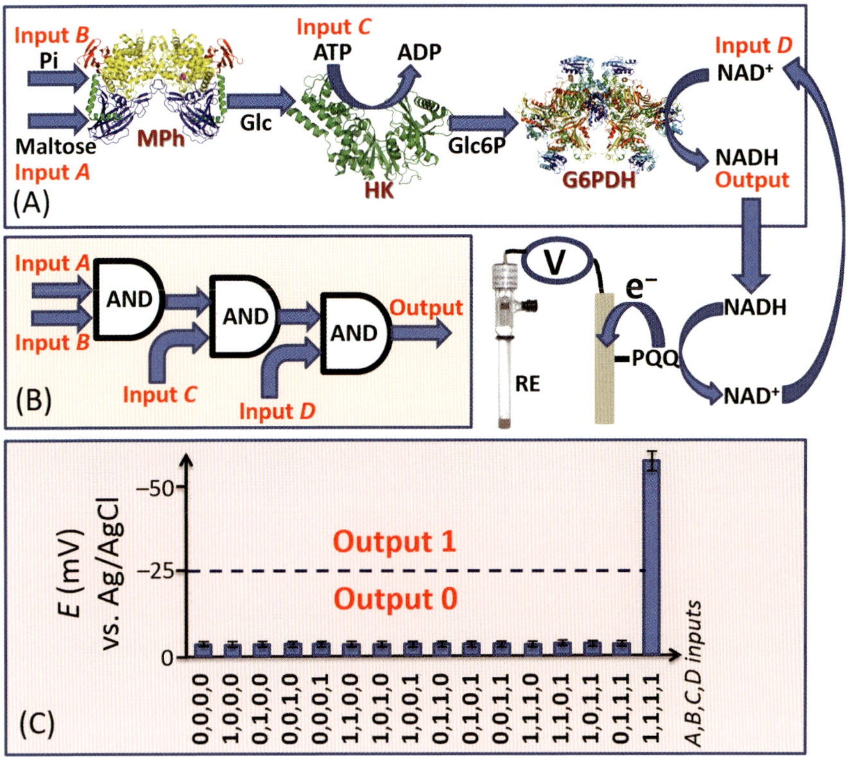

Figure 6.9 Electrochemical transduction of the enzyme generated output signals – potentiometric measurements. (A) The biocatalytic cascade activated with four input signals and ended with the reduction of NAD^+ to NADH. The produced NADH was analyzed potentiometrically on an electrode modified with PQQ catalyzing its oxidation. The negative potential produced on the PQQ-sensing electrode in the presence of NADH was measured vs. a reference electrode. (B) The logic scheme composed of three concatenated AND logic gates corresponding to the biocatalytic cascade shown in (A). (C) Potentials produced on the PQQ-sensing electrode after applying four input signals in 16 different combinations. The dashed line shows the threshold value separating logic **1** and logic **0** output signals. Source: Mailloux et al. 2015 [36]. Adapted with permission of John Wiley and Sons.

measurements (Figure 6.9). An example system [36] represented by a biocatalytic cascade activated with four input signals (Figure 6.9A) was mimicking three concatenated AND logic gates (Figure 6.9B). The biocatalytic process proceeded to the very end only if all input signals appeared at the logic values **1** (**1**,**1**,**1**,**1** combination), thus resulting in the formation of NADH as the final product. The sensing electrode was modified with pyrroloquinoline quinone (PQQ), which is a known catalyst for the NADH oxidation [55]. The electrocatalytic oxidation of NADH at the PQQ-modified electrode can proceed spontaneously (this was used previously in biofuel cells based on NADH oxidation [56, 57]), resulting in the formation of a negative potential measured vs. a reference electrode (RE). Figure 6.9C shows the potential produced on the PQQ-modified sensing electrode at different combinations of the variable logic inputs (16 variants), where only one input combination (**1,1,1,1**) resulted in the high negative potential corresponding to the NADH production in the biocatalytic cascade, as expected for the system mimicking three concatenated AND gates. Similar potentiometric measurements have been used for transduction of NADH formation to the electronic output in other logic systems of high complexity [35].

6.2.3 pH Measurements as a Tool for Transduction of Chemical Output Signals Produced by Enzyme-Based Logic Systems

Many enzyme-catalyzed processes produce acids or bases upon redox or hydrolytic reactions. The most typical examples are formation of gluconic acid (GlcA) upon oxidation of glucose catalyzed by GOx or GDH and production of organic acids from esters in reactions catalyzed by esterase. These reactions, if they proceed in a solution with low buffer capacity or without buffer at all, result in bulk pH changes, which can be used to follow the enzymatic reaction and can serve as the final output signal. Enzyme-based logic gates (e.g., OR, AND) [23, 38, 40, 47, 58–62] and logic networks of different complexity [39, 42, 59, 63] have been realized using decreasing pH (ΔpH) as the output signal measured with a regular pH-sensitive electrode and a standard pH meter. On the other hand, hydrolytic decomposition of urea catalyzed by urease produces ammonia, thus increasing pH value. This reaction can be used to reset the initial neutral pH value after it was decreased by the biocatalytic reactions producing acids [47].

Figure 6.10A shows a reaction cascade catalyzed by four enzymes operating as a "machinery," alcohol dehydrogenase (ADH), glucose dehydrogenase (GDH), amyloglucosidase (AMG), and invertase (Inv), activated with four input signals represented by NADH, acetaldehyde (Ald), maltose, and sucrose, Inputs *A*, *B*, *C*, and *D*, respectively [59]. The final product of the biocatalytic reactions is gluconic acid (GlcA), which results in acidification of the reaction solution when the reactions proceed to the very end. The pH decrease (ΔpH) in the bulk solution was considered as the output signal measured with a standard pH-sensing electrode. The reaction cascade included two pathways. One was represented by two parallel reactions catalyzed by AMG and Inv and triggered with maltose and sucrose, respectively. This part of the reaction cascade mimics an OR logic gate and results in the production of glucose. The second pathway was composed of two consecutive reactions biocatalyzed by ADH and GDH and activated with Ald and NAD^+, respectively. The first reaction step activated with Ald and NADH represents an AND logic gate. The final reaction step converting glucose to GlcA was catalyzed by GDH only when both intermediate products, NAD^+ and glucose, were produced, thus also representing an AND gate connected to the AND–OR gates operating in parallel (Figure 6.10B). The final output (ΔpH) was large (meaning logic value **1**) only when both intermediate products, glucose and NAD^+, were produced through the OR and AND gates. This was achieved at the input logic combinations **1,1,1,0**; **1,1,0,1**; and **1,1,1,1**; otherwise the output signal was **0** (meaning small pH changes for all other input combinations) (Figure 6.10C). While pH changes can be measured directly in the bulk solution, they can be further used to change interfacial properties of signal-responsive materials associated with electrode surfaces or emulsions and then analyzed using different

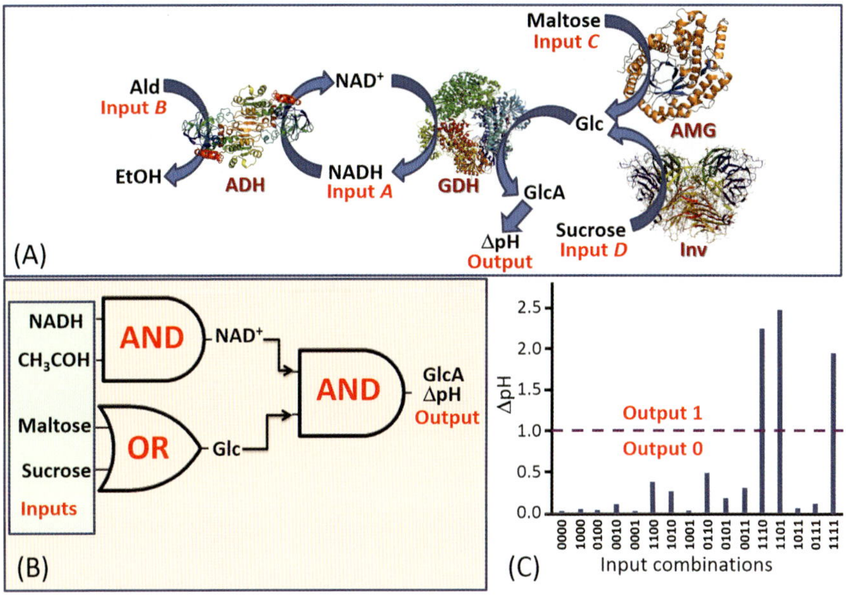

Figure 6.10 Electrochemical transduction of the enzyme generated output signals – pH measurements. (A) The biocatalytic cascade activated with four input signals and ended with oxidation of glucose to gluconic acid. The pH change was defined as the output signal measured with a standard pH-sensitive electrode. (B) The equivalent logic network corresponding to the biocatalytic cascade shown in (A). (C) Bulk pH changes obtained with different combinations of the logic inputs. The dashed line shows the threshold value separating logic **1** and logic **0** output signals. Source: Tam et al. 2009 [59]. Adapted with permission of John Wiley and Sons.

electrochemical methods (not measuring pH changes directly). These systems are overviewed in Sections 6.2.4 and 6.2.5 of the chapter.

6.2.4 Indirect Electrochemical Analysis of Output Signals Generated by Enzyme-Based Logic Systems Using Electrodes Functionalized with pH-Switchable Polymers

Stimuli-responsive materials, exemplified by organic polymers [64–66] and more specifically by polyelectrolyte brushes attached to a modified electrode surface [67–71], can be used to analyze electrochemically changes in the environment, particularly variation of the pH value. Indeed, protonation/deprotonation of the polyelectrolyte brushes results in their restructuring between swollen hydrophilic state permeable for ionic species of the opposite charge and collapsed hydrophobic state nonpermeable for ionic species, thus inhibiting electrochemical processes at the electrode surface. Various polyelectrolyte brushes have been produced on electrodes and demonstrated pH-switchable electrode activity [67–71]. These pH-switchable modified electrodes have been functionally coupled to enzyme-based logic systems producing pH changes *in situ* as output signals of logic operations [39, 40, 42, 63]. While in the majority of the reported systems enzyme reactions have been performed in solutions, in some structurally advanced assemblies, the enzymes were physically bound to modified electrode surfaces producing local pH changes at interfaces, thus resulting in switching of the electrode activity without bulk pH changes [72].

Two important factors should be taken into account when pH-switchable polyelectrolyte brushes are used as stimuli-responsive materials in connection with pH changes produced by enzyme logic systems. (i) The range of the pH changes produced by the biocatalyzed reactions should include the pK_a value of the polyelectrolyte. In other words, the pH change should be enough to change the protonation state of the polymer brushes. (ii) The direction of the pH change should correspond to the specific application. For example, decreasing pH upon production of acids (e.g., gluconic acid in the biocatalytic oxidation of glucose) can result in protonation of polyelectrolyte brushes, thus converting them to the positively

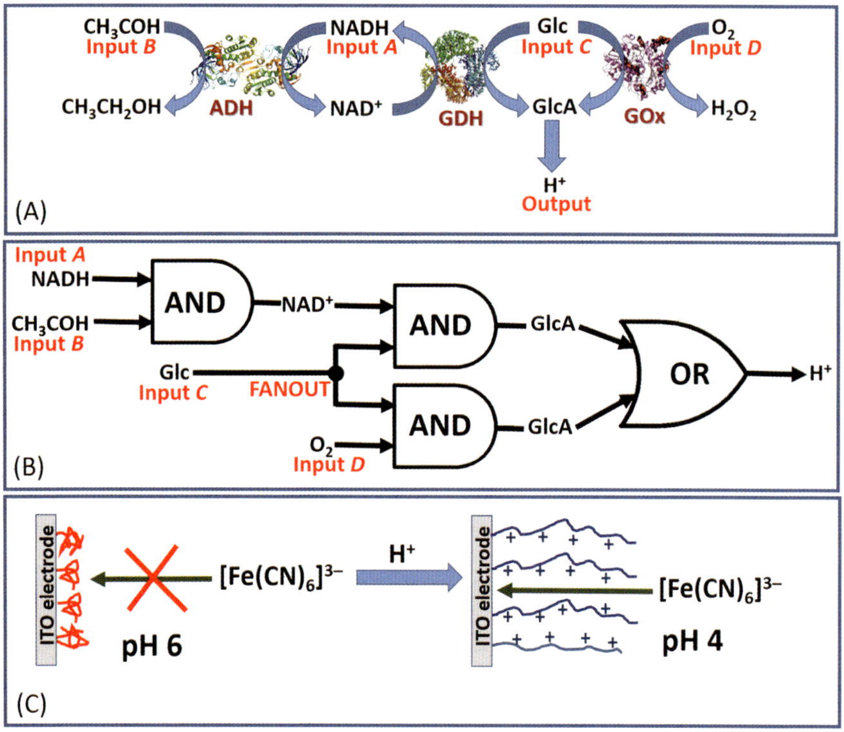

Figure 6.11 (A) The biocatalytic cascade activated with four input signals and ended with oxidation of glucose to gluconic acid yielding acidic pH value. (B) The equivalent logic network corresponding to the biocatalytic cascade shown in (A). (C) pH-switchable electrode surface modified with a polymeric brush. The hydrophobic interface (left) produced at pH 6 is not permeable for the solution redox species, while the protonated hydrophilic interface (right) produced at pH 4 is permeable and active for the redox process of the negatively charged redox probe. Source: Katz 2017 [1]. Adapted with permission of John Wiley and Sons.

charged state permeable for anionic redox species, thus activating the modified electrode in the presence of the anionic redox probe. The opposite pH change (possibly in the reset process) can inhibit the electrochemical reactions on the modified electrode surface.

A biocatalytic cascade, outlined schematically in Figure 6.11A, results in pH decrease due to production of GlcA as the final product of a multistep enzyme-catalyzed reaction [39]. Three enzymes, alcohol dehydrogenase (ADH), glucose dehydrogenase (GDH), and glucose oxidase (GOx), activated with four input signals, NADH, acetaldehyde, glucose, and oxygen, Inputs *A*, *B*, *C*, and *D*, respectively, operate in concert performing logic operations on the inputs applied in different combinations. Figure 6.11B explains the biocatalytic reactions in terms of logic operations, in the form of a network composed of four concatenated/branched logic gates. The biocatalytic cascade finally resulting in the pH change was controlled by a pattern of the applied inputs. The pH decrease generated *in situ* by the biochemical reactions was coupled with restructuring of a pH-responsive poly(4-vinyl pyridine)-(P4VP)-brush-functionalized electrode. The reorganization of the pH-sensitive P4VP-brush polymer from its neutral and hydrophobic state to the protonated and swollen state resulted in the switch of the electrode interface from the OFF state when electrochemical reactions are inhibited to the ON state when the electrode is electrochemically active (Figure 6.11C). The ON–OFF switch of the electrode activity was analyzed following an electrochemical reaction of a soluble diffusional redox probe, $Fe(CN)_6]^{3-/4-}$ (Figure 6.12). This allowed to analyze the logic output generated by the enzyme-catalyzed reactions by electrochemical means. Cyclic voltammograms (Figure 6.12A) and Faradaic impedance spectra (Figure 6.12C) measured on the P4VP-modified electrode demonstrated its switchable features controlled by the pH value, changed by the enzyme reactions depending on the combinations of the chemical inputs applied to the logic system. Since the number of the variable inputs was 4, the whole set of the input signal combinations included 16 variants. The input combinations resulting in the production of GlcA and thus decreasing the pH value were considered to generate the logic output **1**. Otherwise, in case of no pH changes because of no production of GlcA, the system generated the output signal **0**. The produced pattern of the output signals was determined by the Boolean logic encoded in the logic circuitry (Figure 6.12B,D). The use of the switchable P4VP-modified electrode controlled by the pH changes allows amplification of the output signals. Indeed, the concentrations of the chemical inputs applied for activating the biocatalytic reactions and concentrations of the biocatalytically produced outputs might be much smaller than the concentration of the redox probe (in the present example $[Fe(CN)_6]^{3-/4-}$) operating with the switchable electrode. Therefore, the response corresponding to the redox probe measured by electrochemical means in the form of changes in the cyclic voltammograms or Faradaic impedance spectra might be much higher compared with the electrochemical responses measured directly for the biochemical species participating in the enzyme-catalyzed reactions.

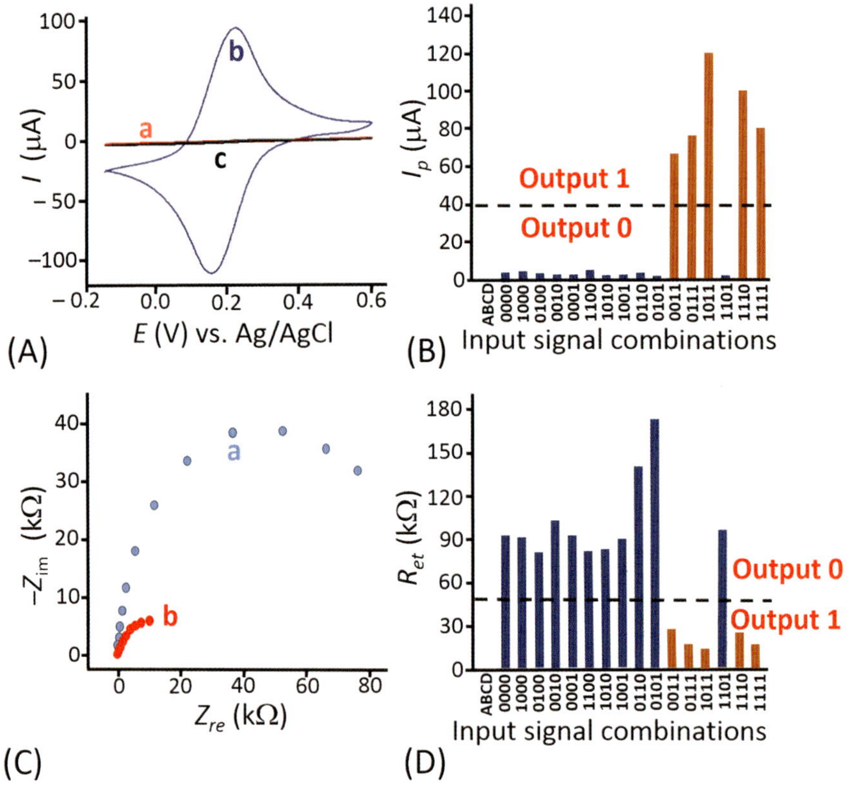

Figure 6.12 Experimental results corresponding to the system shown schematically in Figure 6.11. (A) Cyclic voltammograms obtained for the indium tin oxide (ITO) electrode modified with the P4VP-polymer brush in (a) the initial OFF state, pH ca. 6.7; (b) the ON state enabled by the input combination **1**,**1**,**1**,**0**, recorded at pH ca. 4.3; and (c) *in situ* reset to the OFF state, pH ca. 8.8. Potential scan rate was 100 mV s^{-1}. (B) Anodic peak currents, I_p, for the 16 possible input combinations. The data were obtained from cyclic voltammograms recorded under the same conditions as in panel (A). (C) Faradaic impedance spectra (in the form of Nyquist plots) for (a) the initial OFF state, pH ca. 6.5; (b) the ON state enabled with the input combination **1**,**1**,**1**,**1**, recorded at pH ca. 4.0. (D) Electron transfer resistance, R_{et}, for the 16 possible input combinations. The data were derived from impedance spectra obtained under the same conditions as in panel (C). The dashed lines in (B) and (D) show the threshold values separating logic **1** and logic **0** output signals. Source: Privman et al. 2009 [39]. Adapted with permission of John Wiley and Sons.

6.2.5 Conductivity Measurements as a Tool for Transduction of Chemical Output Signals Produced by Enzyme-Based Logic Systems

Signal-responsive materials (specifically, polyelectrolyte brushes) can be associated not only with electrode surfaces discussed in Section 6.2.4, but can also be integrated with various micro/nano-heterogeneous systems bringing to them signal-switchable features [48, 73]. For example, silica nanoparticles (SiO_2-NPs; 200 nm diameter) were functionalized with polyelectrolyte brushes to allow their switchable hydrophobicity–hydrophilicity features controlled by external pH changes generated by enzymatic reactions [48]. Figure 6.13A shows reactions resulting in the pH decrease due to formation of acetic acid (Ac) or gluconic

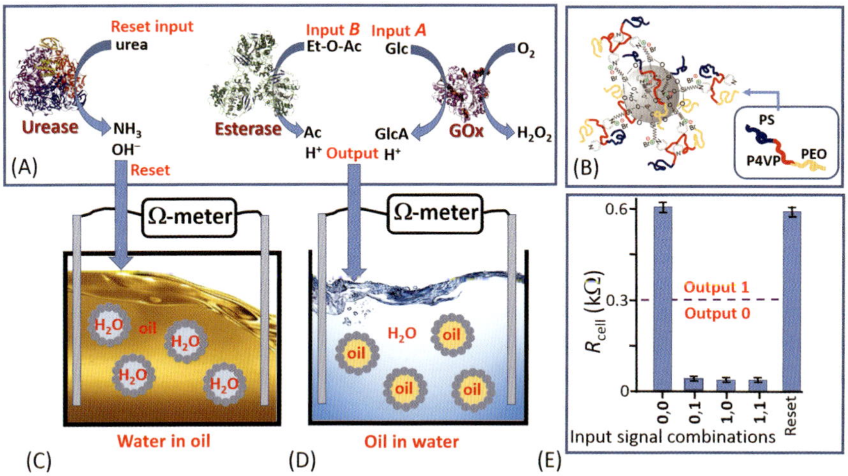

Figure 6.13 The use of DC conductivity measurements for transducing output signals generated by an enzyme logic gate. (A) The biocatalytic cascade mimicking OR gate and the biocatalytic reset system – schematics. (B) SiO_2-NPs functionalized with the block copolymer, PS-b-P4VP-b-PEO, pH-switchable brushes. (C) The W/O emulsion demonstrating the high ohmic resistance for the DC. This emulsion was in the beginning of the experiments when pH of the aqueous phase was ca. 6. Also, this emulsion was produced upon applying the Reset input. (D) The O/W emulsion demonstrating the low ohmic resistance for the DC. This emulsion was produced when the pH of the aqueous phase was decreased to pH ca. 4 by the production of acids in the enzymatic reactions. (E) The bar chart showing the output signals in the form of the ohmic DC resistance measured in the cell after applying the input signals in different combinations. The dashed line shows the threshold value separating logic **1** and logic **0** output signals. Source: Motornov et al. 2009 [48]. Adapted with permission of John Wiley and Sons.

acid (GlcA) catalyzed by esterase and GOx, respectively [48]. This part of the enzymatic process represents the OR logic gate activated by glucose (Glc) and ethyl acetate (Et-O-Ac), Inputs *A* and *B*, respectively. The acidic pH produced enzymatically *in situ* was reset to the initial neutral pH value by producing NH_3 in the reaction catalyzed with urease and activated with the Reset input of urea. The pH-signal-responsive system included SiO_2-NPs functionalized with a block copolymer, PS-b-P4VP-b-PEO, composed of polystyrene (PS), poly(4-vinyl pyridine) (P4VP), and poly(ethylene oxide) (PEO) (Figure 6.13B). The switchable amphiphilic SiO_2-NPs have been used as emulsifiers of Pickering emulsions [74–78].

In Pickering emulsions, nanoparticles with amphiphilic properties are located at the water/oil interface stabilizing different kinds of the emulsions, water-in-oil (W/O) or oil-in-water (O/W), depending on the wetting properties of the nanoparticles. The nanoparticles with more hydrophilic properties stabilize O/W emulsions, while nanoparticles with more hydrophobic features preferably stabilize W/O emulsions. When the wetting properties of the nanoparticles are changed upon protonation/deprotonation of their acid/base groups, the emulsion can be converted from O/W to W/O or *vice versa*. As an example, the nanoparticles functionalized with the PS-b-P4VP-b-PEO grafted to the SiO_2-NPs surface stabilize W/O emulsions at pH ca. 6 and O/W emulsions at

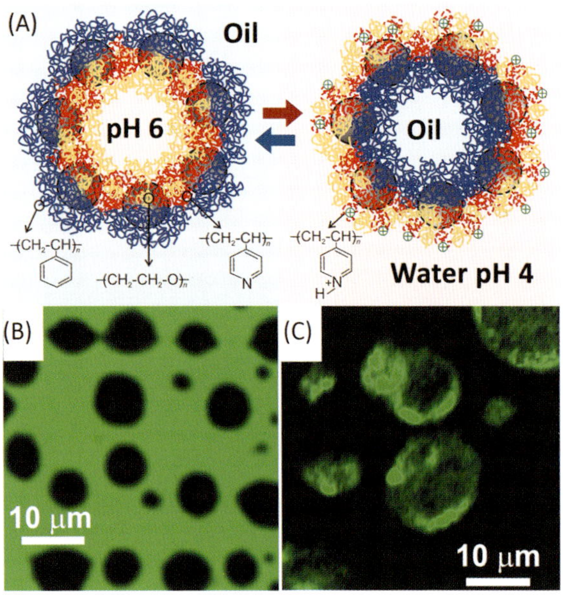

Figure 6.14 (A) Water-in-oil (W/O) and oil-in-water (O/W) emulsions obtained at different pH values due to different protonation states of the pyridine units in the polymer brush bound to the SiO_2-NPs. (B) and (C) Microscopic images of the W/O and O/W emulsions, respectively, with the oil (toluene) phase colored by adding the oil-soluble 1,1,4, 4-tetraphenyl-1,3-butadien dye (note the green color associated with the oil phase). Source: Motornov et al. 2009 [48]. Adapted with permission of John Wiley and Sons.

pH ca. 4 due to non-protonated (neutral) and protonated (positively charged) states of P4VP at the corresponding pH values (Figure 6.14A).

As the result, the functionalized SiO_2-NPs perform a "command" function to switch between inverse (W/O) and direct (O/W) emulsions upon receiving signals from enzyme-catalyzed reactions. The change in the emulsion structure was read out by DC conductivity measurements demonstrating dramatic changes in the conductivity upon the emulsion inversion (Figure 6.13C,D). After producing the O/W emulsion with a low ohmic resistance upon application of **0,1**; **1,0**; **1,1** input signals (note that the enzyme system was mimicking the OR gate), the emulsion was converted back to the W/O state with a high resistance by applying the Reset input bringing pH back to pH ca. 6. Figure 6.13E shows the resistance measured between two electrodes immersed in the emulsion after applying different input combinations to the enzyme logic system mimicking OR gate. Note that the final output plotted in Figure 6.13E as the cell resistance (R_{cell}) corresponds to the NOR logic gate since the resistance is decreasing when the hydrogen ion concentration is increased. Importantly, the switchable features of the SiO_2-NPs interface were transduced to the changes of the bulk conductivity of the liquid system [48]. In addition to the resistance measurements, the reversible transition between the W/O and O/W emulsions was visualized by adding an oil-soluble dye coloring the oil phase (Figure 6.14B,C).

6.2.6 Transduction of Chemical Output Signals Produced by Enzyme-Based Logic Systems Using Semiconductor Devices

Integration of enzyme-based logic gates with field-effect devices based on an electrolyte–insulator–semiconductor (EIS) system [79–81] is the most attractive and promising approach for transduction of biomolecular logic signals into electrical output signals. Indeed, the EIS devices are electrochemical analogs of electronic elements used in conventional electronic logic gates and computing systems. Previous research, not always related to the logic gates, has demonstrated the use of EIS devices for detection of pH changes [82] and for analysis of enzymatic reactions [83, 84] and charged macromolecules (DNA, proteins, polyelectrolytes) [85–87]. Thus, the use of the EIS electronic devices is straightforward for transduction of various output signals generated by enzyme logic systems.

It should be noted that the experimental work on the enzyme logic systems associated with the EIS devices was mostly limited by relatively simple AND, OR Boolean logic gates producing pH changes readable by the EIS device [50–53]. However, more sophisticated logic networks, e.g., reversible Controlled-NOT (CNOT) logic gate, interfaced with EIS devices are also possible [88]. An example system mimicking OR logic gate realized by two parallel reactions catalyzed by esterase and GOx in a solution is shown schematically in Figure 6.15A. Both reactions resulted in acidification of the background solution due to formation of either butyric acid or gluconic acid or both of them if the reactions are activated by the corresponding input signals of glucose (Glc) and ethyl butyrate, Inputs *A* and *B*, respectively [53]. Application of the input signals in three combinations (**0**,**1**; **1**,**0**; and **1**,**1**) resulted in pH decrease due to activation of either or both biocatalytic reactions. The obtained result was considered as the output signal **1**. Obviously, the absence of both inputs (meaning **0**,**0** logic combination) did not result in any reaction and did not produce any pH change, thus generating the output signal **0**. The reaction solution containing the enzymes and inputs applied in various combinations was analyzed by the pH-sensitive EIS device with the interface functionalized with Au nanoparticles coated with a thiolated monolayer containing carboxylic groups (Figure 6.15B). When the EIS sensing interface was exposed to the initial neutral background solution (pH 7), the carboxylic groups associated with the nanoparticles were dissociated and negatively charged. Their negative charge was preserved in case of the output signal **0** (meaning no pH change) generated by the enzyme logic system in response to **0**,**0** input combination. However, in the presence of the acids produced by the enzyme reactions (output signal **1**), the carboxylic groups were protonated, and the surface charge became neutral (Figure 6.15B).

Impedance spectroscopy, particularly analyzing the interfacial capacitance, was applied to follow the charge variation on the surface of the EIS device (Figure 6.15C). Mott–Schottky plots [89, 90] derived from the impedance spectra allowed analysis of the flat band voltage (V_{FB}) changed as the result of the biocatalytic reaction (Figure 6.15D). The interfacial capacitance and V_{FB} derived from the impedance measurements were considered as the electronic output signals converted by the EIS device from the pH variation produced by the enzyme logic gate. The changes in the EIS interfacial capacitance (Figure 6.15C) or shift of the Mott–Schottky function (Figure 6.15D) were observed when the input signals (Glc and ethyl butyrate) were applied in three different combinations (**0**,**1**; **1**,**0**; **1**,**1**), but not in the **0**,**0** combination, thus demonstrating the Boolean OR gate features. The present example is really simple in its realization, but the EIS devices can be applied to much more sophisticated logic

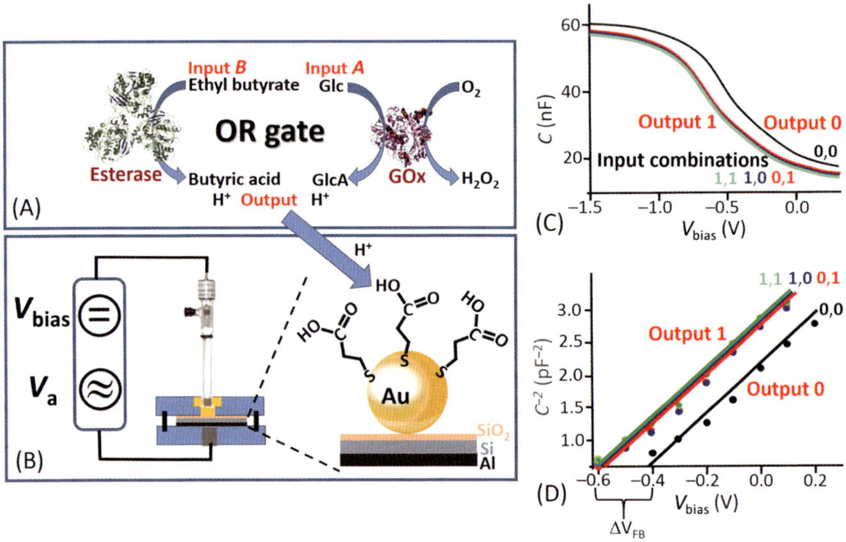

Figure 6.15 Electronic transduction of the enzyme generated output signals – the use of a semiconductor device. (A) The enzyme-based OR logic gate producing pH changes as the output signal realized with a biocatalytic reaction. (B) Electronic scheme of the semiconductor signal-transducing device. The transducer surface is functionalized with Au-NPs coated with a thiolated monolayer containing carboxylic groups. (C) Capacitance vs. bias voltage applied to the gate of the EIS device measured after application of the input signals in various combinations. (D) Mott–Schottky plots derived from the impedance spectra obtained for the signal-transducing Si-chip after application of the input signals in various combinations. Note the flat band voltage shift (ΔV_{FB}) corresponding to the output signal **1**. Source: Krämer et al. 2009 [53]. Adapted with permission of John Wiley and Sons.

systems [88]. The transduction of the biochemical signals, e.g., pH changes, to the electronic format with the use of the semiconductor (EIS) device is conceptually important due to the demonstrated interfacing of the biomolecular logic and truly electronic systems. The full advantage of this approach can be achieved when solid-state electronics is integrated with a "soft matter" biological logic system without soluble species involved in the signal processing steps [88, 91].

6.3 Macro/Micro/Nano-mechanical Transduction of Chemical Output Signals Produced by Enzyme-Based Logic Systems

Enzyme-catalyzed reactions, particularly when the produced chemicals communicate with signal-responsive materials can result in mechanical movements of objects at macro-, micro-, and nanoscales [47, 73, 92, 93]. These processes could be considered as prototypes of futuristic "smart" nanomachines and nanorobotic systems [94, 95], where the mechanical operations are controlled by logically processed biomolecular/biological signals. In the example systems overviewed in the sections below, the mechanical transformations are considered as the final output signals produced after processing the input signal combinations through enzyme biocatalytic cascades mimicking Boolean logic gates. Depending on the dimensions of the mechanically moving species, the technical tools used to follow the transformations are ranging from simple photos for macro-objects [93] to atomic force microscopy (AFM) for nano-objects [47, 73].

6.3.1 Mechanical Bending of a Cantilever Used for Transduction of Chemical Output Signals Produced by Enzyme-Based Logic Systems

Various materials, often represented by conducting electrodes, can be functionalized with conductive polymers [96, 97], particularly using polypyrrole [98], polythiophene [99], and polyaniline [100], deposited on surfaces to allow control over interfacial properties dependent on the redox state of the polymer film. Counter-ion exchange between the polymer matrices and external solutions triggered by redox reactions in the conducting polymers results in shrinking–swelling of the polymer films allowing their application as artificial "muscles" [101]. Deposition of the conducting polymers on flexible supports (e.g., polymeric strips [102–106] or micro-cantilevers [107, 108]) allowed for their mechanical bending upon reduction–oxidation of the polymer films, being usually induced by electrochemical means. In this case the flexible support was represented by a conducting material chemically modified with the redox polymer. Unfortunately, little attention has been given to chemically induced mechanical actuation triggered by redox transformations of the conducting polymers upon their reactions with reducing/oxidizing species applied in a solution [93, 109]. The chemical, or better biochemical, activation of mechanical processes can be important in the future implantable biomedical devices operating upon physiological commands, thus representing an important research direction, particularly when the mechanical operations are controlled by logically processed multiple biomolecular signals.

Figure 6.16A shows a reaction cascade catalyzed by three enzymes, glucose-6-phosphate dehydrogenase (G6PDH), diaphorase (Diaph), and laccase, and activated with three signals, glucose-6-phosphate (Glc6P), NAD^+, and O_2, Inputs *A*, *B*, and *C*, respectively [93]. The first enzymatic reaction catalyzed by G6PDH resulted in the production of NADH reduced species when both chemical Inputs *A* and *B* appeared, thus representing the AND logic gate. The produced NADH resulted in the formation of $[Fe(CN)_6]^{4-}$ reduced species through the Diaph catalyzed process. On the other hand, in the presence of O_2 (Input *C*), the reaction catalyzed by laccase resulted in $[Fe(CN)_6]^{3-}$ oxidized species. The process started in a solution containing both $[Fe(CN)_6]^{4-}$ and $[Fe(CN)_6]^{3-}$ redox species. The deviation of their concentrations from the initial values to the dominated reduced or oxidized species was considered as the logic **1** output (meaning that the increase of the reduced or oxidized species is essentially the same logic output). Therefore, the process catalyzed by Diaph and laccase represents an XOR logic gate where the unbalanced operation (input signals **0**,**1** and **1**,**0**) produces output **1** and the balanced operation (input signals **0**,**0** and **1**,**1**) results in the output **0**. Indeed, the balanced biocatalytic reactions

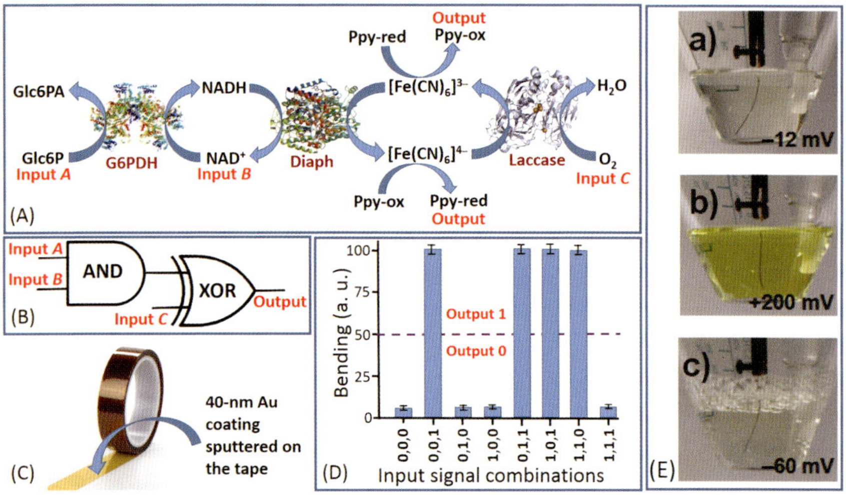

Figure 6.16 Mechanical transduction of the output signals produced by an enzyme logic gate. (A) The biocatalytic cascade activated with three input signals controlling the oxidation state of $[Fe(CN)_6]^{3-}$/$[Fe(CN)_6]^{4-}$ redox species. (B) The equivalent logic network corresponding to the biocatalytic cascade shown in (A). (C) Flexible Au-coated tape used as a conducting support for the Ppy film. (D) The output signals in the form of tape bending after applying the input signals in different combinations. (E) The photos showing the bending tape when the Ppy film was in its reduced (a), oxidized (b), and again reduced (c) states. The potentials measured (vs. Ag/AgCl electrode) on the conducting tape are shown in the photos. Source: Strack et al. 2010 [93]. Adapted with permission of John Wiley and Sons.

(upon appropriate optimization) resulted in no changes in the $[Fe(CN)_6]^{4-}$ and $[Fe(CN)_6]^{3-}$ concentrations, while the unbalanced reactions catalyzed by either Diaph or laccase produced more $[Fe(CN)_6]^{4-}$ or more $[Fe(CN)_6]^{3-}$, respectively. Figure 6.16B shows the equivalent logic network composed of the concatenated AND–XOR logic gates corresponding to the biocatalytic reactions. A flexible polymeric support (polyimide tape) coated with a 40-nm Au conducting film (Figure 6.16C) was modified with a polypyrrole (Ppy) film (ca. 8.6 μm) deposited electrochemically. When the Ppy-modified strip was contacted with the solution containing the $[Fe(CN)_6]^{4-}$/$[Fe(CN)_6]^{3-}$ redox species and the enzyme logic system, the redox state of Ppy was controlled by the redox state of the species dominated in the solution. In other words, the presence of $[Fe(CN)_6]^{4-}$ resulted in the reduced state of Ppy, while $[Fe(CN)_6]^{3-}$ yielded the oxidized state of Ppy. The Ppy-modified strip demonstrated bending upon variation of the polymer redox state, originating from the redox-induced volume changes of the polymer film, thus representing a mechanical actuator controlled by the logically processed biochemical signals. The macroscopic bending of the Ppy strip was photographed using a digital camera (Figure 6.16E). The photos shown in images a–b–c correspond to the reduced–oxidized–reduced states of PPy, and the potentials measured on the Ppy-modified strip electrodes are given in the images. The mechanical bending of the PPy-modified strip was defined as the final output signal produced by the enzyme logic system, and it is shown in

Figure 6.16D as the function of the input signal combinations processed by the enzymatic cascade mimicking the AND–XOR concatenated network.

6.3.2 Quartz Crystal Microbalance (QCM) Transduction of Chemical Output Signals Produced by Enzyme-Based Logic Systems

Enzyme-catalyzed reactions can produce insoluble in water products precipitating on a solid support modified with the working enzymes. These biocatalytic reactions have been used in various biosensors, mostly based on formation of biomolecular complexes due to biorecognition processes (immune-recognition or DNA complementarity), while the enzymes producing the precipitating products have been used as the labels in the biomolecular complexes [110–114]. Accumulation of the insoluble product on the sensing interface was analyzed using impedance spectroscopy or quartz crystal microbalance (QCM) measurements, thus providing the biosensing signals. Similar biocatalytic systems mimicking logic operations and producing insoluble precipitates on solid supports can be used for microgravimetric transduction of the chemical outputs. The QCM measurements are based on change in vibration frequency of the piezoelectric device (Figure 6.17A) upon deposition of the insoluble material on its surface [115].

Figure 6.17B shows schematically the biocatalytic cascade operated on a QCM Au surface and mimicking AND logic gate. Since the insoluble product should be precipitated on the QCM surface for the microgravimetric analysis, the enzymes catalyzing the reactions have to be immobilized on the sensing surface. The Au surface of the QCM sensor was modified with a self-assembled thiolated monolayer bearing active ester groups, which were used for covalent binding of horseradish peroxidase (HRP) through formation of amide bonds [113]. In the next modification step, the second enzyme, glucose oxidase (GOx), was immobilized on the HRP layer by bridging the enzymes with glutaric dialdehyde. The bi-enzyme layers biocatalytically operated in the presence of two chemical inputs, glucose (Glc) and 4-chloro-1-naphthol (CN), Inputs *A* and *B*, respectively. The glucose oxidation catalyzed by GOx resulted in the concomitant production of H_2O_2, and the second reaction catalyzed by HRP resulted in oxidation of CN to insoluble oxidized product (CN-ox) precipitated on the QCM surface. The CN-ox formation was only possible in the presence of Glc and CN, in other words when Inputs *A* and *B* were applied at **1**,**1** combination. When one or both inputs were missing (input combinations **0**,**0**; **0**,**1**; and **1**,**0**), the biocatalytic reactions did not proceed till the very end, and the insoluble product was not formed on the QCM surface. Therefore, the biocatalytic cascade mimicked the AND logic gate. Figure 6.17C shows time-dependent frequency changes measured on the QCM when the input signals were applied in different combinations. Only **1**,**1** input combination resulted in the frequency change (corresponding to the output signal **1**) due to deposition of the insoluble reaction product, while all other input combinations were not affecting the QCM frequency (meaning output signal **0**). Figure 6.17D summarizes the results in the form of the bar chart showing frequency changes measured after 30 minutes from the time when the inputs were applied in different combinations. The obtained results demonstrated microgravimetric transduction of the output signals generated by

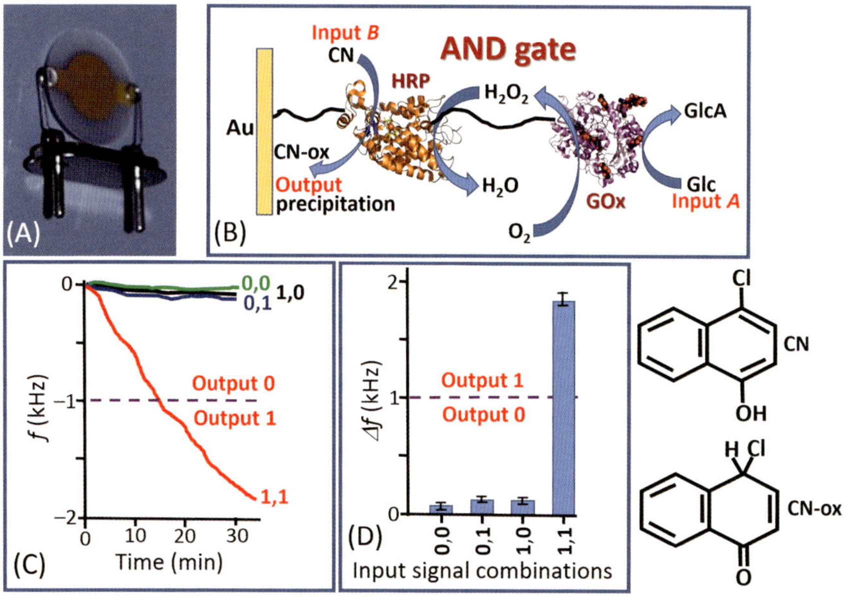

Figure 6.17 Microgravimetric transduction of the output signals produced by an enzyme logic gate. (A) A QCM device used for transducing the enzyme logic gate output signals. (B) The biocatalytic cascade mimicking AND gate and operating on the surface of the QCM device. The enzyme reactions result in the formation of an insoluble product precipitating on the QCM surface. (C) Frequency variation measured with the QCM device upon application of the input signals in various combinations. (D) The bar chart showing the frequency changes measured with the QCM device after 30 minutes from application of the input signals in different combinations. The structures of 4-chloro-1-naphthol (CN; Input *B*) and its oxidized product (CN-ox) are shown. Source: Katz 2017 [1]. Adapted with permission of John Wiley and Sons.

the enzyme logic gate, where the final signal was presented with the mechanical oscillation of the quartz crystal.

6.3.3 Atomic Force Microscopy (AFM) Transduction of Chemical Output Signals Produced by Enzyme-Based Logic Systems

Nanostructured objects functionalized with signal-responsive materials can respond to signals produced by enzyme logic gate. The responding nano-objects can be represented by nanoporous membranes with variable nanoporosity [47] and by nanoparticles with aggregation/disaggregation features [73] dependent on chemical signals produced by enzymatic reactions. Nanoparticle assembly–disassembly can be triggered by chemical signals processed through biomolecular logic gates representing a biocomputing nanoplatform for therapeutics and diagnostics [116]. Biomolecular logic gates have been used to arrange dynamically gold nanoparticles on DNA origami [117]. Assembling of gold nanoparticles has been demonstrated using a pH-responsive DNA nanomachine logically processing biomolecular input signals [118].

Figure 6.18A shows the biocatalytic cascade mimicking an OR logic gate [47]. The gate "machinery" was composed of two enzymes, invertase (Inv) and glucose oxidase (GOx), and oxygen in the solution. The biocatalytic reactions were initiated by two biomolecular substrates, sucrose and glucose, Inputs *A* and *B*, respectively, applied in four different combinations. Sucrose was cut into fructose and glucose in the reaction catalyzed by Inv. Glucose produced *in situ* from sucrose or added as an independent input was oxidized in the reaction catalyzed by GOx yielding gluconic acid, thus decreasing the solution pH value. The acidic solution was produced in the presence of either or both inputs, sucrose or/and glucose, as expected for the OR logic gate. The produced acidic pH was reset to the initial neutral value by a reaction converting urea (Reset input) to NH_3 catalyzed by urease. Overall, the biocatalytic process represented the OR gate with the Reset function (Figure 6.18B). The solution with the pH controlled by the external input signals was in the immediate contact with a pH-switchable nanoporous membrane (Figure 6.18C), composed of alginate hydrogel [119] comprising D-mannuronic acid and L-guluronic acid residues cross-linked with Ca^{2+} cations (Figure 6.18D). At the initial neutral pH, the alginate polymer contains deprotonated (negatively charged) carboxylic groups, thus being hydrophilic and swollen. The swollen polymer matrix expends in its volume and closes completely the nanopores in the membrane body. When the solution pH is decreased below pK_a of the alginate polymer ($pH < 4$), the carboxylic groups are getting protonated resulting in their neutral charge, thus yielding the hydrophobic shrunk polymer. This results in the opening of the nanopores (380 ± 116 nm measured by AFM at the half-depth of the pores). The reset of the solution pH to the initial neutral value returns the nanopores to their closed state (Figure 6.18C). The operation of the enzyme logic system was based on the pH variation due to the biocatalytic reactions. Obviously, the pH variation can be measured directly with a pH-sensitive electrode or indirectly following the membrane permeability controlled by the pH value. The latter was measured by observing molecular diffusion through the pores and by measuring the membrane impedance when the membrane was deposited on an electrode surface. However, for the present section of the review, it is important to emphasize that the membrane state was analyzed with atomic force microscopy (AFM) (Figure 6.18C). The membrane state with the open nanopores (output signal **1**) was achieved when the input signals were applied at **0**,**1**; **1**,**0**; and **1**,**1** combinations; otherwise the closed-pore state (output signal **0**) was obtained with the input combination **0**,**0** (Figure 6.18C). The membrane was reset to the initial state by the Reset signal after the output signal **1** was produced.

A similar biocatalytic cascade (Figure 6.19A), but using enzymes as the input signals (Inv and GOx, Inputs *A* and *B*, respectively), was used to variate the solution pH and then control aggregation / dissociation of SiO_2-NPs (200 nm) functionalized with poly(2-vinyl pyridine) (P2VP) polyelectrolyte [73]. Despite the fact that the biocatalytic cascade was the same as in the previous example, the new definition of the input signals resulted in the different logic function. Indeed, the final output signal (GlcA producing acidic pH value) was obtained only when both biocatalytic reactions proceeded in the presence of both input signals (both enzymes present). At the initial neutral pH value, the P2VP polymer was deprotonated and neutral. This resulted in the aggregation of the neutral NPs, thus yielding their aggregates with a large size. When the pH value was decreased

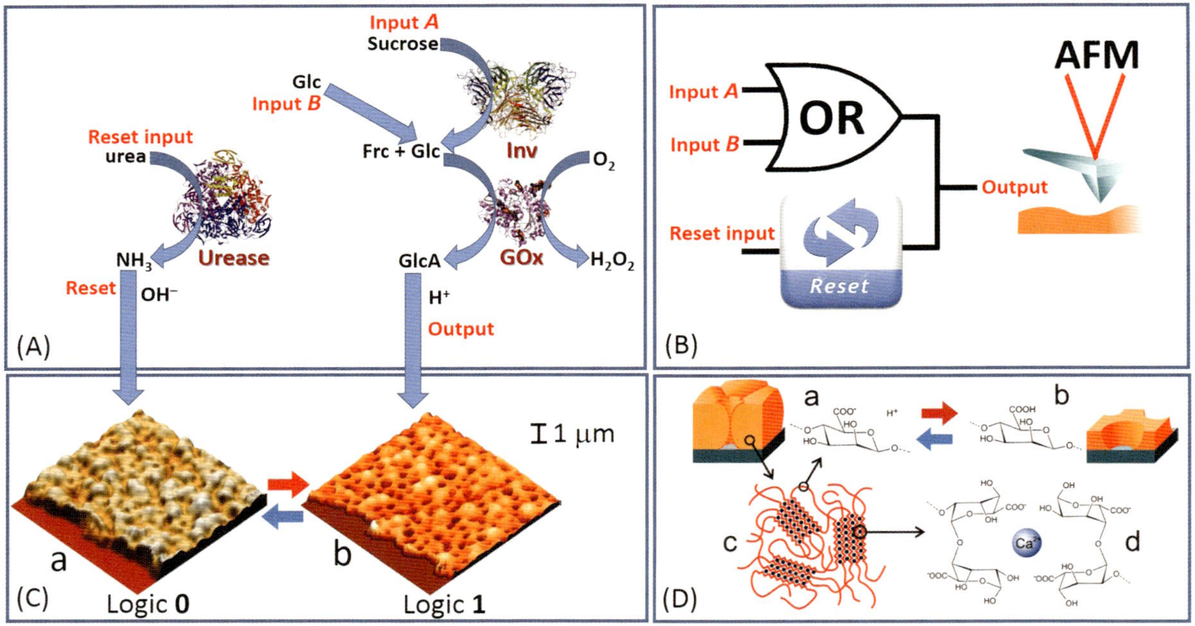

Figure 6.18 AFM transduction of the output signal produced by an enzyme logic gate – switchable porosity in a membrane. (A) The biocatalytic cascade mimicking an OR logic gate and a Reset function. (B) The schematics of the OR-Reset logic device with the output transduced by AFM. (C) The nanoporous membrane visualized with AFM in the closed-pore state at neutral pH (a) and open-pore state at the acidic pH (b). (D) Schematics showing the swollen alginate polymer resulting in the closed-pore state and shrunk polymer producing the open-pore state of the membrane. The structure of the Ca^{2+}-cross-linked alginate is shown. Source: Tokarev et al. 2009 [47]. Adapted with permission of John Wiley and Sons.

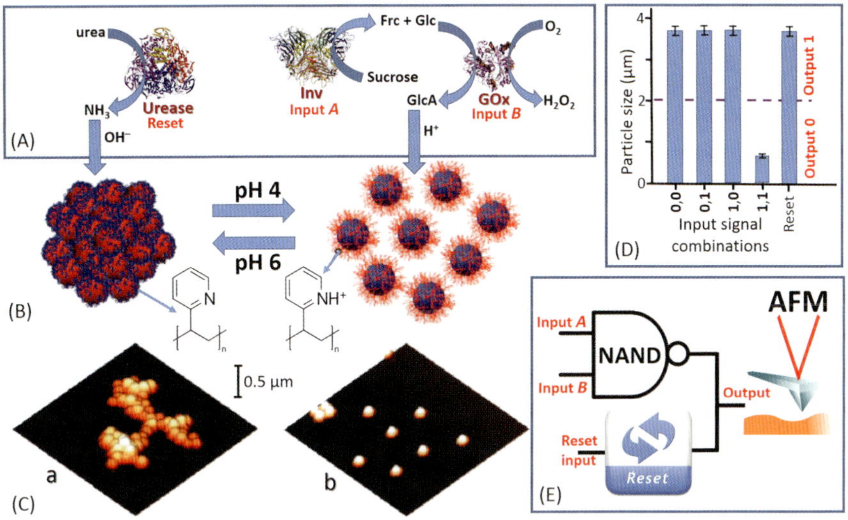

Figure 6.19 AFM transduction of the output signal produced by an enzyme logic gate – switchable aggregation/disaggregation of nanoparticles (NPs). (A) The biocatalytic cascade mimicking an NAND logic gate and a Reset function. (B) Reversible aggregation/disaggregation of the NPs controlled by pH values. The NPs were functionalized with P2VP-polymer brushes changing their charge at different pH values due to the pyridine group protonation (pH 4) and deprotonation (pH 6). (C) The AFM images of the aggregated (a) and disaggregated (b) nanoparticles after their deposition on a surface. (D) The bar chart showing the size of the NPs in their aggregated and disaggregated states obtained upon application of the input signals in different combinations and after applying the reset signal. The dashed line shows the threshold value separating logic **1** and logic **0** output signals. (E) Schematics of the NAND-Reset logic device with the output transduced by AFM. Source: Motornov et al. 2008 [73]. Adapted with permission of John Wiley and Sons.

($pH < 4$) due to acidification of the solution with the *in situ* produced GlcA, the pyridine units of P2VP polymer were protonated bringing positive charge to the NPs. The positively charged NPs were dissociated from their aggregates revealing their small size (Figure 6.19B). The size of the nano-species in their aggregated and dissociated states was measured *ex situ* observing the structures deposited from the suspensions on Si wafers using AFM (Figure 6.19C). The large size aggregates were defined as the output signal **1**, while the small size disaggregated NPs were considered as the output signal **0**. The output **0** was obtained only when both enzyme inputs were present to complete the biocatalytic cascade (input signal combination **1**,**1**), while all other input combinations (**0**,**0**; **0**,**1**; **1**,**0**) resulted in no formation of acidic pH, thus keeping the NPs in their aggregated state (meaning the output signal **1**). This logic operation corresponds to the NAND gate (Figure 6.19D). The reset to the initial neutral pH returning the dissociated NPs to the aggregated state was achieved with the biocatalytic reaction producing NH_3 (Figure 6.19A). Overall, the studied system resembles the NAND-Reset function, which was followed by AFM measurements (Figure 6.19E).

The systems exemplified above are also interesting because they demonstrated the nano-mechanical actuation logically controlled by the pH signals produced *in situ* through biocatalytic reactions. Also, it should be noted that simple reformulation of the input signal definitions can result in different logic operations (OR/NAND in the given examples).

6.4 Conclusions and Perspectives

The majority of research papers, review articles, and books overviewing progress in biomolecular [120–125] (including enzyme-based [126, 127]) and molecular [128–139] logic gates concentrate on their composition and performance while giving little attention to the methods of the output signal transduction. The present chapter aims at highlighting the underrepresented aspects of biomolecular computing, specifically collecting examples of the enzyme logic systems connected to different physical methods of the output signal transduction. While the most frequently used method is based on optical absorbance measurements after performing biocatalytic reactions in solutions, other methods, especially based on electrochemical techniques, are highly important. Electrochemical interfaces, particularly functionalized with signal-responsive materials, operating with the enzyme-based logic systems demonstrated significant advantages over the optical readout of the chemical output signals. It should be noted that the optical tools applied to the analysis of the output signals allow only the signal readout, while electrochemical interfaces can operate as actuators triggered by the biomolecular computing systems. Indeed, electrochemical systems can release molecular/biomolecular species in response to the signals processed by the enzyme logic gates, thus activating downstream reactions and processes [35, 36], for example, releasing DNA species activating next logic steps in the DNA computing process. Also, the electrochemical signal transducers can switch ON/OFF various bioelectrochemical devices (e.g., biofuel cells) [140] in response to the logically processed biomolecular signals.

Electrochemical interfaces [141] releasing DNA species (short artificial oligonucleotides) in response to output signals generated by enzyme logic systems resulted in the unique integration of enzyme-based and DNA-based logic elements in highly sophisticated hybrid computing systems [35, 36], benefiting from the features of the enzyme and DNA components. These systems allowed for the logically reversible processing of the initial input signals through the enzyme and DNA reactions, being particularly beneficial for biosensing applications when the restoration of the initial signal pattern from the final output signals is important.

Simple single-output producing logic gates (e.g., AND, OR, NAND, XOR, etc.) can easily operate as a "one-pot" system, being assembled in a solution or on a transducer surface. However, when more sophisticated logic systems composed of concatenated/branched logic elements and producing several output signals are aimed, the use of a "one-pot" composition becomes difficult because of cross talking of the logic elements and overlapping of the output signals. This makes the readout of the output signals particularly difficult when optical absorbance is used as the signal transduction method. One of the approaches used to solve the problem and to allow the increasing complexity of the logic systems is the spatial separation of the logic elements and organizing the output signals through physically different channels, for example, in multichannel flow devices [21, 22, 142–144]. Application of different transduction methods to follow the output signals is highly important for increasing complexity of the biomolecular logic systems, particularly when the system generates several signals measured separately. An example system composed of several concatenated logic gates based on a biocatalytic reaction cascade was used to illustrate the advantages of reading the output signals through different transduction mechanisms, including optical absorbance measurements and bioluminescence [26]. Application of AFM for reading output signals generated by biomolecular logic systems allows extremely high sensitivity, up to detection of single molecules, because the detection process can be localized on a very small surface area [145–147].

While the present chapter is concentrated specifically on the readout methods for the output signals generated by the enzyme logic systems, other biomolecular computing assemblies, particularly including DNA-based logic systems [124, 125, 148–152], can be combined with various detection methods discussed in this chapter. Particularly, catalytic reactions activated by DNAzymes in the presence of various combinations of inputs can be analyzed by the methods developed for the analysis of the biocatalytic logic systems based on the enzyme reactions [35, 36]. Also, many other transducing methods can be envisaged for various biomolecular logic systems, including, for example, electroluminescence as the signal detection tool [153].

Overall, the biomolecular computing systems, and particularly enzyme-based logic gates and networks, can contribute not only to the design of future "biocomputers," which are not possible at the present level of technology, but also on a short term to the development of novel binary-operating biosensors with various transduction tools of the logically processed bio-signals. It should be noted that the biomolecular information processing systems demonstrated promising results while operating in a biological environment [154] dealing with real biological samples [155–157], thus giving rise to the expectations of their real practical applications, particularly in therapeutics and diagnostics [116] and in forensic investigations [157].

References

1 Katz, E. (2017). *ChemPhysChem* 18: 1688–1713.

2 Bergmeyer, H.-U. (ed.) (1974). *Methods of Enzymatic Analysis*. Weinheim: Verlag Chemie GmbH.

3 Passonneau, J.V. and Lowry, O.H. (1993). *Enzymatic Analysis: A Practical Guide*, Chapter 1, 3–22. Totowa, NJ: Humana Press.

4 Wilson, R. and Turner, A.P.F. (1992). *Biosens. Bioelectron.* 7: 165–185.

5 Dunford, H.B. (1991). Horseradish peroxidase: structure and kinetic properties. In: *Peroxidases in Chemistry and Biology*, vol. 2 (ed. J. Everse, K.E. Everse and M.B. Grisham), 1–24. Boca Raton, FL: CRC Press.

6 Fraga, H. (2008). *Photochem. Photobiol. Sci.* 7: 146–158.

7 Bakshi, S., Zavalov, O., Halámek, J. et al. (2013). *J. Phys. Chem. B* 117: 9857–9865.
8 Privman, V., Fratto, B.E., Zavalov, O. et al. (2013). *J. Phys. Chem. B* 117: 7559–7568.
9 Halámek, J., Zavalov, O., Halámková, L. et al. (2012). *J. Phys. Chem. B* 116: 4457–4464.
10 Baron, R., Lioubashevski, O., Katz, E. et al. (2006). *Org. Biomol. Chem.* 4: 989–991.
11 Baron, R., Lioubashevski, O., Katz, E. et al. (2006). *J. Phys. Chem. A* 110: 8548–8553.
12 Strack, G., Pita, M., Ornatska, M., and Katz, E. (2008). *ChemBioChem* 9: 1260–1266.
13 Zavalov, O., Bocharova, V., Privman, V., and Katz, E. (2012). *J. Phys. Chem. B* 116: 9683–9689.
14 Zhou, J., Arugula, M.A., Halámek, J. et al. (2009). *J. Phys. Chem. B* 113: 16065–16070.
15 Halámek, J., Bocharova, V., Arugula, M.A. et al. (2011). *J. Phys. Chem. B* 115: 9838–9845.
16 Privman, V., Zhou, J., Halámek, J., and Katz, E. (2010). *J. Phys. Chem. B* 114: 13601–13608.
17 Niazov, T., Baron, R., Katz, E. et al. (2006). *Proc. Natl. Acad. USA.* 103: 17160–17163.
18 Privman, V., Arugula, M.A., Halámek, J. et al. (2009). *J. Phys. Chem. B* 113: 5301–5310.
19 Halámek, J., Bocharova, V., Chinnapareddy, S. et al. (2010). *Mol. Biosyst.* 6: 2554–2560.
20 Moseley, F., Halámek, J., Kramer, F. et al. (2014). *Analyst* 139: 1839–1842.
21 Fratto, B.E. and Katz, E. (2015). *ChemPhysChem* 16: 1405–1415.
22 Fratto, B.E. and Katz, E. (2016). *ChemPhysChem* 17: 1046–1053.
23 Pita, M., Krämer, M., Zhou, J. et al. (2008). *ACS Nano* 2: 2160–2166.
24 Pita, M., Abad, J.M., Vaz-Dominguez, C. et al. (2008). *J. Colloid Interface Sci.* 321: 484–492.
25 Zhao, J., Luo, L., Yang, X. et al. (1999). *Electroanalysis* 11: 1108–1111.
26 Guz, N., Halámek, J., Rusling, J.F., and Katz, E. (2014). *Anal. Bioanal. Chem.* 406: 3365–3370.
27 Tam, T.K., Pita, M., and Katz, E. (2009). *Sens. Actuat. B* 140: 1–4.
28 Wang, Y. and Knoll, W. (2006). *Anal. Chim. Acta* 558: 150–157.
29 Chegel, V.I., Raitman, O.A., Lioubashevski, O. et al. (2002). *Adv. Mater.* 14: 1549–1553.
30 Raitman, O.A., Katz, E., Bückmann, A.F., and Willner, I. (2002). *J. Am. Chem. Soc.* 124: 6487–6496.
31 Raitman, O.A., Katz, E., Willner, I. et al. (2001). *Angew. Chem. Int. Ed.* 40: 3649–3652.
32 Windmiller, J.R., Santhosh, P., Katz, E., and Wang, J. (2011). *Sens. Actuat. B* 155: 206–213.
33 Chuang, M.-C., Windmiller, J.R., Santhosh, P. et al. (2011). *Chem. Commun.* 47: 3087–3089.

34 Zhou, N., Windmiller, J.R., Valdés Ramírez, G. et al. (2011). *Anal. Chim. Acta* 703: 94–100.

35 Guz, N., Fedotova, T.A., Fratto, B.E. et al. (2016). *ChemPhysChem* 17: 2247–2255.

36 Mailloux, S., Gerasimova, Y.V., Guz, N. et al. (2015). *Angew. Chem. Int. Ed.* 54: 6562–6566.

37 Mailloux, S., Guz, N., Zakharchenko, A. et al. (2014). *J. Phys. Chem. B* 118: 6775–6784.

38 Zhou, J., Tam, T.K., Pita, M. et al. (2009). *ACS Appl. Mater. Interfaces* 1: 144–149.

39 Privman, M., Tam, T.K., Pita, M., and Katz, E. (2009). *J. Am. Chem. Soc.* 131: 1314–1321.

40 Privman, M., Tam, T.K., Bocharova, V. et al. (2011). *ACS Appl. Mater. Interfaces* 3: 1620–1623.

41 Wang, J. and Katz, E. (2010). *Anal. Bioanal. Chem.* 398: 1591–1603.

42 Katz, E., Bocharova, V., and Privman, M. (2012). *J. Mater. Chem.* 22: 8171–8178.

43 Katz, E. and Minko, S. (2015). *Chem. Commun.* 51: 3493–3500.

44 Katz, E., Minko, S., Halámek, J. et al. (2013). *Anal. Bioanal. Chem.* 405: 3659–3672.

45 Yu, X., Lian, W.J., Zhang, J.N., and Liu, H.Y. (2016). *Biosens. Bioelectron.* 80: 631–639.

46 Tam, T.K., Zhou, J., Pita, M. et al. (2008). *J. Am. Chem. Soc.* 130: 10888–10889.

47 Tokarev, I., Gopishetty, V., Zhou, J. et al. (2009). *ACS Appl. Mater. Interfaces* 1: 532–536.

48 Motornov, M., Zhou, J., Pita, M. et al. (2009). *Small* 5: 817–820.

49 Katz, E., Poghossian, A., and Schöning, M.J. (2017). *Anal. Bioanal. Chem.* 409: 81–94.

50 Poghossian, A., Katz, E., and Schöning, M.J. (2015). *Chem. Commun.* 51: 6564–6567.

51 Poghossian, A., Malzahn, K., Abouzar, M.H. et al. (2011). *Electrochim. Acta* 56: 9661–9665.

52 Poghossian, A., Krämer, M., Abouzar, M.H. et al. (2009). *Procedia Chemistry* 1: 682–685.

53 Krämer, M., Pita, M., Zhou, J. et al. (2009). *J. Phys. Chem. C* 113: 2573–2579.

54 Molinnus, D., Bäcker, M., Iken, H. et al. (2015). *Phys. Status Solidi A* 212: 1382–1388.

55 Katz, E., Lötzbeyer, T., Schlereth, D.D. et al. (1994). *J. Electroanal. Chem.* 373: 189–200.

56 Willner, I., Arad, G., and Katz, E. (1998). *Bioelectrochem. Bioenerg.* 44: 209–214.

57 Koushanpour, A., Gamella, M., Guz, N., and Katz, E. (2017). *Electroanalysis* 29: 950–954.

58 Pita, M., Minko, S., and Katz, E. (2009). *J. Mater. Sci.: Materials in Medicine* 20: 457–462.

59 Tam, T.K., Pita, M., Ornatska, M., and Katz, E. (2009). *Bioelectrochemistry* 76: 4–9.
60 Wang, X., Zhou, J., Tam, T.K. et al. (2009). *Bioelectrochemistry* 77: 69–73.
61 Bychkova, V., Shvarev, A., Zhou, J. et al. (2010). *Chem. Commun.* 46: 94–96.
62 Pita, M., Privman, V., Arugula, M.A. et al. (2011). *PhysChemChemPhys* 13: 4507–4513.
63 Pita, M., Privman, M., and Katz, E. (2012). *Top. Catal.* 55: 1201–1216.
64 Minko, S. (2006). *J. Macromol. Sci. C – Polymer Rev.* 46: 397–420.
65 Luzinov, I., Minko, S., and Tsukruk, V.V. (2004). *Prog. Polym. Sci.* 29: 635–698.
66 Stuart, M.A.C., Huck, W.T.S., Genzer, J. et al. (2010). *Nature Mater.* 9: 101–113.
67 Tam, T.K., Pita, M., Motornov, M. et al. (2010). *Electroanalysis* 22: 35–40.
68 Tam, T.K., Pita, M., Motornov, M. et al. (2010). *Adv. Mater.* 22: 1863–1866.
69 Tam, T.K., Pita, M., Trotsenko, O. et al. (2010). *Langmuir* 26: 4506–4513.
70 Motornov, M., Tam, T.K., Pita, M. et al. (2009). *Nanotechnology* 434006.
71 Tam, T.K., Ornatska, M., Pita, M. et al. (2008). *J. Phys. Chem. C* 112: 8438–8445.
72 Bocharova, V., Tam, T.K., Halámek, J. et al. (2010). *Chem. Commun.* 46: 2088–2090.
73 Motornov, M., Zhou, J., Pita, M. et al. (2008). *Nano Lett.* 8: 2993–2997.
74 Fujii, S., Armes, S.P., Binks, B.P., and Murakami, R. (2006). *Langmuir* 22: 6818–6825.
75 Fujii, S., Read, E.S., Binks, B.P., and Armes, S.P. (2005). *Adv. Mater.* 17: 1014–1017.
76 Binks, B.P., Murakami, R., Armes, S.P., and Fujii, S. (2005). *Angew. Chem. Int. Ed.* 44: 4795–4798.
77 Duan, H.W., Wang, D.Y., Sobal, N.S. et al. (2005). *Nano Lett.* 5: 949–952.
78 Saleh, N., Phenrat, T., Sirk, K. et al. (2005). *Nano Lett.* 5: 2489–2494.
79 Konorov, P.P., Yafyasov, A.M., and Bogevolnov, V.B. (2006). *Field Effect in Semiconductor-Electrolyte Interfaces*. Princeton, NJ: Princeton University Press.
80 Shinwari, M.W., Deen, M.J., and Landheer, D. (2007). *Microelectronics Reliability* 47: 2025–2057.
81 Vlasov, Y.G., Tarantov, Y.A., and Bobrov, P.V. (2003). *Anal. Bioanal. Chem.* 376: 788–796.
82 Poghossian, A.S. (1992). *Sens. Actuat. B* 7: 367–370.
83 Siqueira, J.R., Abouzar, M.H., Bäcker, M. et al. (2009). *Phys. Status Solidi A* 206: 462–467.
84 Weng, W.H., Wang, C.W., Pang, S.T., and Pan, T.M. (2016). *J. Electrochem. Soc.* 163: B445–B452.
85 Abouzar, M.H., Poghossian, A., Cherstvy, A.G. et al. (2012). *Phys. Status Solidi A* 209: 925–934.
86 Poghossian, A., Weil, M., Cherstvy, A.G., and Schöning, M.J. (2013). *Anal. Bioanal. Chem.* 405: 6425–6436.
87 Poghossian, A., Bäcker, M., Mayer, D., and Schöning, M.J. (2015). *Nanoscale* 7: 1023–1031.

88 Honarvarfard, E., Gamella, M., Poghossian, A. et al. (2017). *Appl. Mater. Today* 9: 266–270.
89 Albery, W.J., O'Shea, G.J., and Smith, A.L. (1996). *J. Chem. Soc., Faraday Trans.* 92: 4083–4085.
90 Lasia, A. (2014, Chapter 10). *Electrochemical Impedance Spectroscopy and its Applications*, 251–255. New York: Springer.
91 Koushanpour, A., Gamella, M., Guo, Z. et al. (2017). *J. Phys. Chem. B* 121: 11465–11471.
92 Munkhjargal, M., Matsuura, Y., Hatayama, K. et al. (2013). *Sens. Actuat. B* 188: 831–836.
93 Strack, G., Bocharova, V., Arugula, M.A. et al. (2010). *J. Phys. Chem. Lett.* 1: 839–843.
94 Wang, J. (2013). *Nanomachines: Fundamentals and Applications*. Weinheim: Wiley-VCH.
95 Mavroidis, C. and Ferreira, A. (eds.) (2013). *Nanorobotics – Current Approaches and Techniques*. New York: Springer.
96 Inzelt, G. (2008). *Conducting Polymers*. Berlin, Heidelberg: Springer.
97 Skotheim, T.A., Elsenbaumer, R.L., and Reynolds, J.R. (eds.) (1998). *Handbook of Conducting Polymers*. New York: Marcel Dekker, Inc.
98 Wang, L.-X., Li, X.-G., and Yang, Y.-L. (2001). *React. Func. Polym.* 47: 125–139.
99 Schopf, G. and Koßmehl, G. (1997). *Polythiophenes: Electrically Conductive Polymers*. Berlin: Springer.
100 Kang, E.T., Neoh, K.G., and Tan, K.L. (1998). *Prog. Polym. Sci.* 23: 277–324.
101 Baughman, R.H. (1996). *Synth. Metals* 78: 339–353.
102 Otero, T.F. and Sansiñena, J.M. (1997). *Bioelectrochem. Bioenerg.* 42: 117–122.
103 Otero, T.F. and Sansiñena, J.M. (1995). *Bioelectrochem. Bioenerg.* 38: 411–414.
104 Otero, T.F. and Sansiñena, J.M. (1998). *Adv. Mater.* 10: 491–494.
105 Takashima, W., Kaneko, M., Kaneto, K., and MacDiarmid, A.G. (1995). *Synth. Metals* 71: 2265–2266.
106 Qibing, P. and Inganläs, O. (1992). *Adv. Mater.* 4: 277–278.
107 Smela, E., Inganas, O., and Lundstrom, I. (1995). *Science* 268: 1735–1738.
108 Lahav, M., Durkan, C., Gabai, R. et al. (2001). *Angew. Chem. Int. Ed.* 40: 4095–4097.
109 Küttel, C., Stemmer, A., and Wei, X. (2009). *Sens. Actuat. B* 141: 478–484.
110 Alfonta, L., Bardea, A., Khersonsky, O. et al. (2001). *Biosens. Bioelectron.* 16: 675–687.
111 Bardea, A., Katz, E., and Willner, I. (2000). *Electroanalysis* 12: 1097–1106.
112 Alfonta, L., Katz, E., and Willner, I. (2000). *Anal. Chem.* 72: 927–935.
113 Patolsky, F., Zayats, M., Katz, E., and Willner, I. (1999). *Anal. Chem.* 71: 3171–3180.
114 Patolsky, F., Katz, E., Bardea, A., and Willner, I. (1999). *Langmuir* 15: 3703–3706.
115 Johannsmann, D. (2015). *The Quartz Crystal Microbalance in Soft Matter Research: Fundamentals and Modeling*. Heidelberg: Springer.

116 Evans, A.C., Thadani, N.N., and Suh, J. (2016). *J. Controlled Release* 240: 387–393.
117 Yang, J., Song, Z.C., Liu, S. et al. (2016). *ACS Appl. Mater. Interfaces* 8: 22451–22456.
118 Yao, D.B., Li, H., Guo, Y.J. et al. (2016). *Chem. Commun.* 52: 7556–7559.
119 Rehm, H.A. (ed.) (2009). *Alginates: Biology and Applications*. Heidelberg: Springer.
120 Katz, E. (ed.) (2012). *Biomolecular Computing – From Logic Systems to Smart Sensors and Actuators*. Weinheim: Willey-VCH.
121 Fan, D.Q., Zhu, J.B., Liu, Y.Q. et al. (2016). *Nanoscale* 8: 3834–3840.
122 Benenson, Y. (2012). *Nature Rev. Genet.* 13: 455–468.
123 Qian, L. and Winfree, E. (2011). *J. Royal Soc. Interface* 8: 1281–1297.
124 Stojanovic, M.N., Stefanovic, D., and Rudchenko, S. (2014). *Acc. Chem. Res.* 47: 1845–1852.
125 Stojanovic, M.N. and Stefanovic, D. (2011). *J. Comput. Theor. Nanosci.* 8: 434–440.
126 Katz, E. and Privman, V. (2010). *Chem. Soc. Rev.* 39: 1835–1857.
127 Katz, E. (2015). *Curr. Opin. Biotechnol.* 34: 202–208.
128 Szacilowski, K. (2012). *Infochemistry – Information Processing at the Nanoscale*. Chichester: Wiley.
129 de Silva, A.P. (2013). *Molecular Logic-Based Computation*. Cambridge: Royal Society of Chemistry.
130 Katz, E. (ed.) (2012). *Molecular and Supramolecular Information Processing – From Molecular Switches to Logic Systems*. Weinheim: Willey-VCH.
131 Sienko, T., Adamatzky, A., Rambidi, N.G., and Conrad, M. (eds.) (2003). *Molecular Computing*. Cambridge, MA: MIT Press.
132 de Silva, A.P. (2008). *Nature* 454: 417–418.
133 de Silva, A.P. (2007). *Nature Nanotechnol.* 2: 399–410.
134 de Silva, A.P., Uchiyama, S., Vance, T.P., and Wannalerse, B. (2007). *Coord. Chem. Rev.* 251: 1623–1632.
135 Credi, A. (2007). *Angew. Chem. Int. Ed.* 46: 5472–5475.
136 Pischel, U., Andreasson, J., Gust, D., and Pais, V.F. (2013). *ChemPhysChem* 14: 28–46.
137 Pischel, U. (2010). *Angew. Chem. Int. Ed.* 49: 1356–1358.
138 Szacilowski, K. (2008). *Chem. Rev.* 108: 3481–3548.
139 Andreasson, J. and Pischel, U. (2015). *Chem. Soc. Rev.* 44: 1053–1069.
140 Katz, E. and Pita, M. (2009). *Chem. Eur. J.* 15: 12554–12564.
141 Jia, Y.M., Duan, R.X., Hong, F. et al. (2013). *Soft Matter* 9: 6571–6577.
142 Fratto, B.E., Lewer, J.M., and Katz, E. (2016). *ChemPhysChem* 17: 2210–2217.
143 Fratto, B.E., Roby, L.J., Guz, N., and Katz, E. (2014). *Chem. Commun.* 50: 12043–12046.
144 Katz, E. and Fratto, B.E. (2016). *Advances in Unconventional Computing*, vol. 2, Chapter 2 (ed. A. Adamatzky), 29–59. Springer.
145 Volkov, D., Strack, G., Halámek, J. et al. (2010). *Nanotechnology* 21: 145503.
146 Strack, G., Chinnapareddy, S., Volkov, D. et al. (2009). *J. Phys. Chem. B* 113: 12154–12159.
147 Pita, M., Cui, L., Gaikwad, R.M. et al. (2008). *Nanotechnology* 19: 375502.

148 Ezziane, Z. (2006). *Nanotechnology* 17: R27–R39.

149 Fan, D., Wang, E., and Dong, S. (2017). *ACS Appl. Mater. Interfaces* 9: 1322–1330.

150 Fan, D., Wang, K., Zhu, J. et al. (2015). *Chem. Sci.* 6: 1973–1978.

151 Fan, D., Wang, E., and Dong, S. (2017). *Chemical Science* 8: 1888–1895.

152 Fan, D., Zhu, X., Dong, S., and Wang, E. (2017). *ChemPhysChem* 18: 1767–1772.

153 Liu, S., Li, M., Yu, X. et al. (2015). *Chem. Commun.* 51: 13185–13188.

154 Kahan, M., Gil, B., Adar, R., and Shapiro, E. (2008). *Physica D* 237: 1165–1172.

155 Halámková, L., Halámek, J., Bocharova, V. et al. (2012). *Analyst* 137: 1768–1770.

156 Zhou, J., Halámek, J., Bocharova, V. et al. (2011). *Talanta* 83: 955–959.

157 Katz, E. and Halámek, J. (2014). *Ann. Forensic Res. Analysis* 1: 1002.

7
Circuit Elements Based on Enzyme Systems

Rapid progress in the assembling of the enzyme-based logic gates in complex multicomponent networks requires developments of additional components mimicking electronic counterparts, such as FANOUT gate and feedback circuitries [1], amplifier [2], signal convertor [2], flip-flop memory unit [3], multiplexer and demultiplexer [4, 5], etc., which are necessary for the effective operation of biochemical computing networks and devices performing at least basic arithmetic operations. The present chapter overviews some of these networking circuit elements based on enzyme-catalyzed reactions.

7.1 Enzyme-Based Multiplexer and Demultiplexer

The important network components needed for branching biocomputing pathways are multiplexer and demultiplexer, which are needed for connecting branched biocatalytic pathways and redirecting signals processed in these reactions. With understanding of the importance of these networking elements, several example systems performing multiplexing and demultiplexing functions have been designed using non-biochemical (synthetic) molecules [6–15]. It should be noted that these molecular multiplexer/demultiplexer systems were mostly based on optical (usually fluorescent) output signals, which are not compatible with biochemical pathways and cannot be used to switch signals between different branches of biochemical networks. While several examples of the multiplexer/demultiplexer biomolecular design, particularly based on DNA and DNAzyme systems [16–22], were recently published, the enzyme-based systems performing these networking functions are represented by very few recently developed systems overviewed below [4, 5].

7.1.1 General Definition of the Multiplexer and Demultiplexer Functions

In electronics a multiplexer is a device that performs multiplexing – selection of one of many analog or digital input signals – and forwards the selected input into a single output channel. When it is combined with digitally processed inputs,

Enzyme-Based Computing Systems, First Edition. Evgeny Katz.

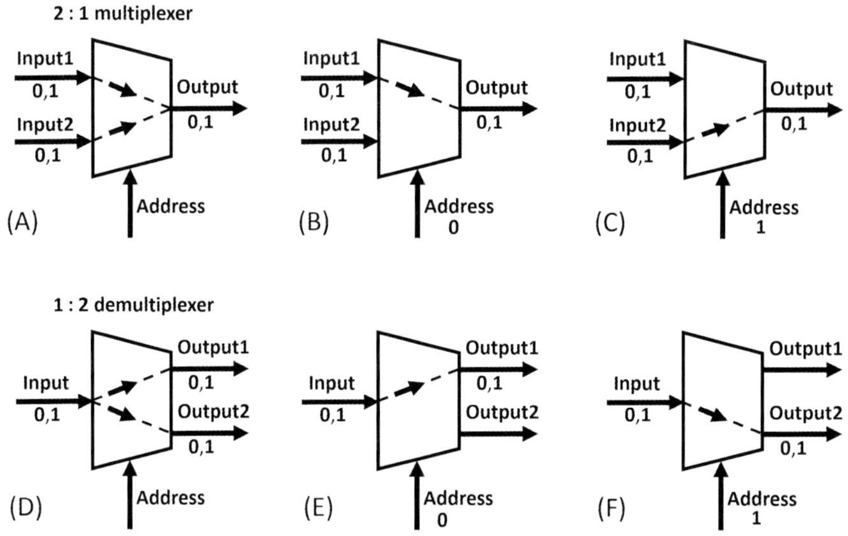

Figure 7.1 (A) Schematics of the 2-to-1 digital multiplexer. (B) Operation of the multiplexer when Address input is **0**: Input1 is directed to the Output channel regardless its value, while Input2 is not in use. (C) Operation of the multiplexer when Address input is **1**: Input2 is directed to the Output channel regardless its value, while Input1 is not in use. Overall, the Address signal controls which data input is directed to the output. (D) Schematics of the 1-to-2 digital demultiplexer. (E) Operation of the demultiplexer when Address input is **0**: Input is directed to the Output1 channel regardless its value, while Output2 is not in use. (F) Operation of the demultiplexer when Address input is **1**: Input is directed to the Output2 channel regardless its value, while Output1 is not in use. Overall, the Address signal controls which output channel is accepting to the input signal.

the signals are represented by logic **0** and **1** values. In the simplest example of a 2-to-1 multiplexer, the device should select one of two digital data input signals (Input1; Input2) and direct it to the single output channel (Output). An additional input signal (addressing input, Address) should switch between the two input channels, selecting one of them to direct the signal to the output channel (Figure 7.1A). When the addressing signal Address appears as **0**, the input data signal from Input1 channel is directed to the Output channel regardless of its logic value (**0** or **1**) (Figure 7.1B). If the addressing channel Address has value **1**, the data input signal is directed from Input2 to Output also regardless of its value (**0** or **1**) (Figure 7.1C). In general, the number of the input channels might be more than two (n-to-1 multiplexer); in this case the number of the addressing channels should be also increased.

The demultiplexer performs the opposite operation. It directs the signal from a single Input channel to one of the output channels (Figure 7.1D) (in case of 1-to-2 demultiplexer to the Output1 or Output2 channels; in general the number of output channels might be more than two). The selection of the output channel is controlled by the Address signal. In case of a binary 1 : 2 demultiplexer, the **0** and **1** values in the Input are moving to Output1 when Address is **0** (Figure 7.1E); on the other hand the **0** and **1** values from the Input are copied to Output2 when Address is **1** (Figure 7.1F).

The examples discussed below represent the realizations of a 2-to-1 binary multiplexer and 1-to-2 binary demultiplexer based on reactions catalyzed by enzymes.

7.1.2 2-to-1 Digital Multiplexer Based on the Enzyme-Catalyzed Reactions

In order to mimic the 2-to-1 digital multiplexer function, the designed system [4] included two enzymes, glucose oxidase (GOx) and laccase, used as the data input signals (Input1 and Input2, respectively) activating two biocatalytic input channels (Figure 7.2A). The presence of these enzymes in optimized concentrations was considered as logic **1** input for the data input channels, while their absence was used as logic **0** input. It should be noted that the optimized concentrations of the enzymes were selected based on the experimental convenience for generating similar output signals in convenient time. In the present work the "optimized" concentrations imply neither noise reduction [23–26] nor physiological levels for biomedical applications [27, 28]. The "machinery" of the biochemical system included horseradish peroxidase (HRP), glucose, $K_4[Fe(CN)_6]$, and oxygen (in equilibrium with air) dissolved in phosphate buffer solution at the initial pH 7.5. The biocatalytic activities of the both input enzymes have optimums at very different pH values. GOx reveals the highest activity at pH ca. 7, while laccase has the best activity at pH ca. 4 [29]. Thus, different pH values can be applied to separate the enzyme activities and to perform the biocatalytic reactions selectively. However, as laccase shows significant activity at the GOx optimum pH and *vice versa*, the applied pH values were 8.6 (above the optimum value) to activate GOx and pH 2.0 (below the optimum value) to activate laccase to be sure that only the appropriate enzyme is working. It was demonstrated in preliminary experiments that GOx activates the biocatalytic cascade finally resulting in the oxidation of ferrocyanide at pH 8.6, while laccase was unable to oxidize ferrocyanide at this pH value. On the other hand, at pH 2.0

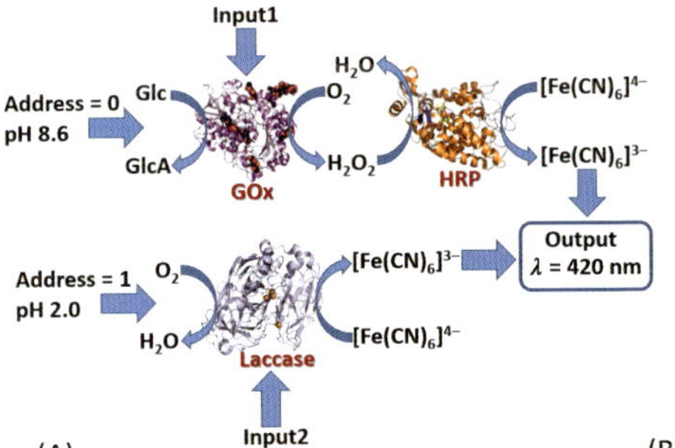

(A)

(B)

Data input		Address input	Data output
Input1	Input2		
GOx	Laccase	pH 8.6 (0) pH 2.0 (1)	λ = 420 nm
0	0	0	0
1	0	0	1
0	1	0	0
1	1	0	1
0	0	1	0
1	0	1	0
0	1	1	1
1	1	1	1

Figure 7.2 (A) The biocatalytic system mimicking the 2-to-1 digital multiplexer, where glucose oxidase (GOx) and laccase are the data input signals (Input1 and Input2, respectively) and the pH change is the addressing signal (Address). The same Output, $[Fe(CN)_6]^{3-}$, is produced in the both reacting pathways and then detected optically by measuring absorbance increase at $\lambda = 420$ nm. (B) The truth table of the 2-to-1 digital multiplexer.

only laccase was catalytically active to oxidize ferrocyanide, while GOx was almost inactive for the reaction (Figure 7.2A).

Keeping this result in mind, a set of experiments to mimic the 2-to-1 digital multiplexer using the biocatalytic reactions shown in Figure 7.2A was performed. Various pH values were applied as addressing inputs to select between the biocatalytic channels: pH = 8.6 was defined as the logic **0** input for Address, while pH = 2.0 was considered as the logic input **1**. The reacting solution was adjusted to the selected pH values by the titration with H_2SO_4 or NaOH. The final output signal of the biochemical multiplexer was defined as the biocatalytic formation of ferricyanide analyzed by the optical absorbance measurements at $\lambda = 420$ nm. Figure 7.3A shows the absorbance spectra obtained at pH 8.6 (Address = **0**). The absorbance peak, $\lambda_{max} = 420$ nm, characteristic of $[Fe(CN)_6]^{3-}$ was obtained when GOx was added regardless of the presence or absence of laccase (input combinations **1**,**1** and **1**,**0**, respectively, where the first digit corresponds to Input1 (GOx) and the second to Input2 (laccase)). On the other hand, the absorbance was not changed when GOx was absent regardless of the presence or absence of laccase (input combinations **0**,**1** and **0**,**0**, respectively). This experiment demonstrated that Address = **0** selects the channel Input1 to generate the output signal, while the channel Input2 is mute (Figure 7.3A, inset). Figure 7.3B shows the absorbance spectra obtained at pH 2.0 (Address = **1**). In this case the high absorbance reflecting the $[Fe(CN)_6]^{3-}$ formation was obtained only when laccase was present (input signals **0**,**1** and **1**,**1**), while in the absence of laccase (input signals **1**,**0** and **0**,**0**), the absorbance was not changed. This experiment

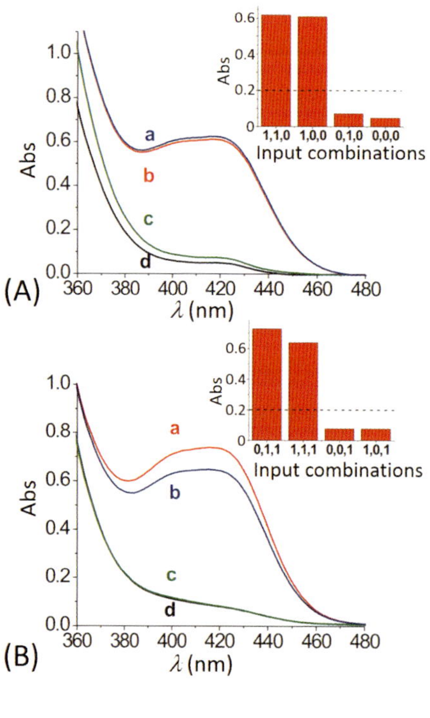

Figure 7.3 Spectra generated by the biocatalytic 2-to-1 multiplexer upon different combinations of the data and address signals. (A) Address = **0**: (a) Input1 = **1**, Input2 = **1**; (b) Input1 = **1**, Input2 = **0**; (c) Input1 = **0**, Input2 = **1**; (d) Input1 = **0**, Input2 = **0**. Inset: Bar chart of the output signals (Output) generated by different combinations of the Input1, Input2, and Address signals. (B) Address = **1**: (a) Input1 = **0**, Input2 = **1**; (b) Input1 = **1**, Input2 = **1**; (c) Input1 = **1**, Input2 = **0**; (d) Input1 = **0**, Input2 = **0**. Inset: Bar chart of the output signals (Output) generated by different combinations of the Input1, Input2, and Address signals. The dashed line is the threshold separating the logic **0** and **1** output values. Source: Arugula et al. 2010 [4]. Adapted with permission of American Chemical Society.

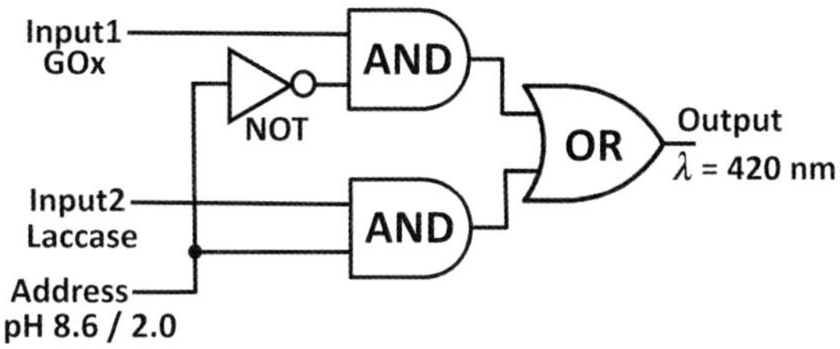

Figure 7.4 The electronic equivalent circuitry of the 2-to-1 digital multiplexer based on the enzyme-catalyzed reactions.

demonstrated that Address = **1** selects the channel Input2 to generate the output signal, while the channel Input1 is inactive (Figure 7.3B, inset).

The demonstrated pH-controlled switching between two biocatalytic channels mimics the 2-to-1 digital multiplexer (Figure 7.2B), and it can be represented by the electronic equivalent circuitry shown in Figure 7.4. In this scheme two AND logic gates operate in parallel, while the Address input has the opposite effect on them being inverted for one of the AND gates. The concerted operation of the gates results in the production of the same output signal being however activated by two different data inputs. This equivalent circuit allows the multiplexer operation similarly to the electronic counterpart [30]. It should be noted that application of pH changes as the addressing input would be very convenient in many biocomputing networks, and many examples of enzyme logic gates [31–35] and their networks [1] producing pH changes were already designed. Also, application of enzymes as input signals in the logic systems has been studied [36].

7.1.3 1-to-2 Digital Demultiplexer Based on the Enzyme-Catalyzed Reactions

The opposite process of forwarding one data input signal (Input) into two different output channels (Output1 and Output2) can be performed by a 1-to-2 digital demultiplexer. In this case the additional addressing input (Address) selects either output channel will be used to take the data input signal (Figure 7.1D–F). In order to mimic this electronic device, a biochemical system with the "machinery" part composed of three enzymes was designed: glucose oxidase (GOx), glucose dehydrogenase (GDH), and horseradish peroxidase (HRP). In addition to the enzymes, the system included 2,2′-azino-*bis*(3-ethylbenzothiazoline-6-sulfonic acid) (ABTS) and oxygen (in equilibrium with air) dissolved in phosphate buffer solution, pH 7.5. Glucose (Glc) was defined as logic data signal **1** in the only Input channel, while the absence of glucose was defined as logic input **0**. The addressing signal was the addition of NAD^+ for the logic value **1**, and the absence of NAD^+ was defined as logic input **0**. The output signals were detected optically at two different wavelengths: $\lambda = 420$ nm (Output1) corresponding to the oxidized ABTS ($ABTS_{ox}$) and $\lambda = 340$ nm (Output2) corresponding to the reduced NAD^+ (NADH) (Figure 7.5A). When the data Input channel had logic value **0** (the absence of glucose), none of the reactions was activated, and the both output channels had the logic value **0** regardless of the Address logic value. However, in the presence of glucose (Input = **1**), the biocatalytic reactions were activated, and depending on the presence (Address = **1**) or absence (Address = **0**) of NAD^+, different catalytic processes resulted in the different optical changes,

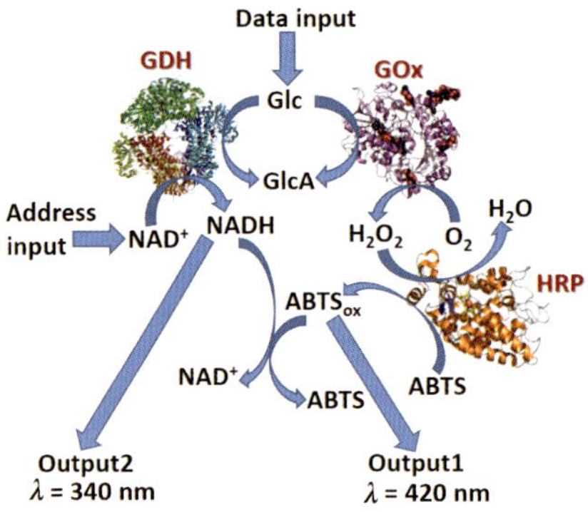

(A)

Data input	Address input	Data output	
		Output1	Output2
Glucose	NAD+	λ = 420 nm	λ = 340 nm
0	0	0	0
1	0	1	0
0	1	0	0
1	1	0	1

(B)

Figure 7.5 (A) The biocatalytic system mimicking the 1-to-2 digital demultiplexer, where glucose is the data input signal (Input) and NAD^+ is the addressing signal (Address). Two different output channels (Output1 and Output2) are represented by $ABTS_{ox}$ and NADH, which production is triggered by the same data input and selected by the addressing input. (B) The truth table of the 1-to-2 digital demultiplexer.

thus activating Output1 or Output2 channels. Specifically, in the absence of NAD^+, glucose was oxidized by GOx yielding H_2O_2 and finally resulting in the oxidation of ABTS biocatalyzed by HRP (Figure 7.5A). This resulted in the increased absorbance at $\lambda = 420$ nm (Output1). However, in the presence of NAD^+, the glucose oxidation was mostly performed by GDH (present in 100-fold higher activity of GDH compared to GOx), resulting in the formation of NADH and the increase of the absorbance at $\lambda = 340$ nm (Output2). The biocatalytically generated NADH reduced $ABTS_{ox}$, which was produced through the reaction catalyzed by GOx, thus eliminating the optical changes at $\lambda = 420$ nm and inhibiting the Output1.

The experimental spectra obtained for different combinations of the Input and Address logic values are shown in Figure 7.6 (note that the initial absorbance of ABTS was subtracted from the obtained spectra, thus resulting in the differential spectra). In the absence of glucose (Input = **0**), the absorbance was practically unchanged at both analyzed wavelengths (Figure 7.6, curves b and c for Address = **0** and **1**, respectively). In the presence of glucose (Input = **1**) and in the absence of NAD^+ (Address = **0**), the absorbance was increased at $\lambda = 420$ nm (Output1 = **1**) due to the formation of $ABTS_{ox}$ (Figure 7.6, curve d). Note that the absorbance was also decreased at 340 nm in the differential spectrum reflecting conversion of ABTS to $ABTS_{ox}$. This change was opposite to the increasing absorbance in case when Output2 channel was activated. In the presence of glucose (Input = **1**) and in the presence of NAD^+ (Address = **1**), the absorbance was increased at $\lambda = 340$ nm (Output2 = **1**) due to the formation of NADH (Figure 7.6, curve a). In this case the absorbance at $\lambda = 420$ nm was not changed at all.

The demonstrated NAD^+-controlled switching between two output channels mimics the 1-to-2 digital multiplexer (Figure 7.5B), and it can be represented by the electronic equivalent circuitry shown in Figure 7.7. In this scheme two AND

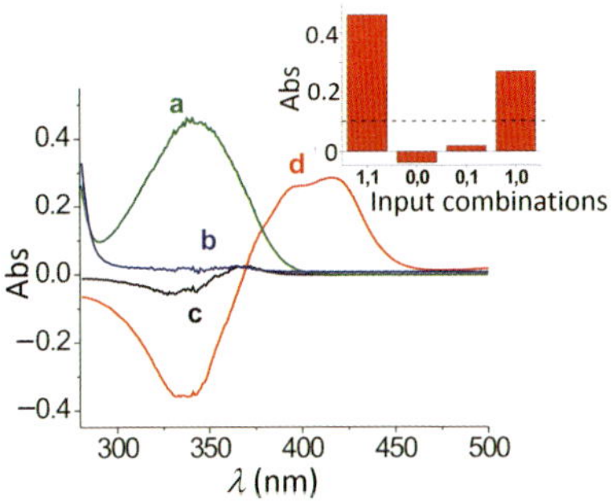

Figure 7.6 Spectra generated by the biocatalytic 1-to-2 digital demultiplexer upon application of different combinations of the data and address signals: (a) Input = **1**, Address = **1**; (b) Input = **0**, Address = **0**; (c) Input = **0**, Address = **1**; (d) Input = **1**, Address = **0**. Inset: Bar chart of the output signals (Output1 and Output2) generated by different combinations of the Input and Address signals. Note that Output1 and Output2 were measured at $\lambda = 420$ nm and $\lambda = 340$ nm, respectively. The dashed line is the threshold separating the logic **0** and **1** output values. Source: Arugula et al. 2010 [4]. Adapted with permission of American Chemical Society.

Figure 7.7 The electronic equivalent circuitry of the 1-to-2 digital demultiplexer based on the enzyme-catalyzed reactions.

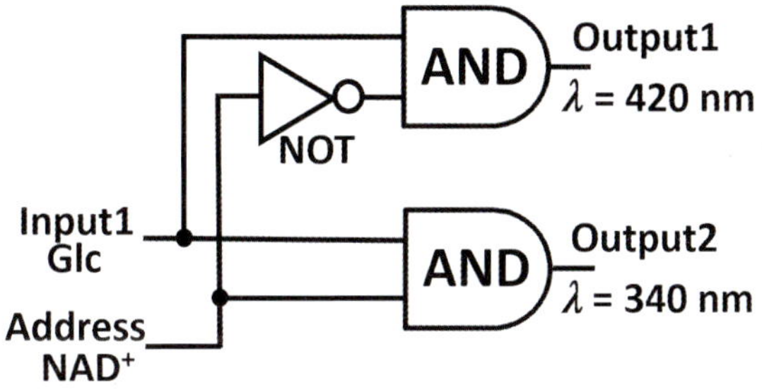

logic gates operate in parallel producing; however, different output signals read optically at different wavelengths. The selection of the output channel depends on the logic value of the addressing input being inverted at one of the gates. This equivalent circuit allows the demultiplexer operation similarly to the electronic counterpart [30]. The use of glucose as the data input and NAD^+ as the addressing input is very convenient since the both biochemicals are typical components of many developed enzyme logic systems [1, 31–35].

It should be noted that the designed biochemical demultiplexer was activated (Data Input and Address) by the biomolecular signals represented by the enzyme substrate and cofactor (glucose and NAD^+), which can be produced *in situ* through different biocatalytic reactions. Also the products (Output1 and Output2) were redox species ($ABTS_{ox}$ and NADH), which can be consumed by downstream chemical processes. This allows integration of the demultiplexer into complex biocatalytic cascades with the reaction steps before and after the demultiplexer. Unfortunately, the multiplexer described above was based on the signals (Input1, Input2) represented by the enzymes (GOx and laccase) that cannot be produced *in situ* as the result of any biochemical reaction; thus the designed multiplexer cannot be integrated with any preceding reaction.

7.1.4 1-to-2 Digital Demultiplexer Interfaced with an Electrochemical Actuator

The next generation of the enzyme-based digital demultiplexer has been realized in a flow device significantly simplifying the system design and operation [5]. More importantly, the designed system allowed extension of the demultiplexer to downstream electrochemical processes, finally resulting in signal-controlled biomolecular release.

In order to mimic this electronic device, a biochemical system was designed with the "machinery" part composed of three enzymes, hexokinase (HK), glucose dehydrogenase (GDH), and glucose-6-phosphate dehydrogenase (G6PDH), immobilized in three different flow cells connected in a complex flow device (Figure 7.8) (see Chapter 5 for similar flow devices used for mimicking reversible logic gates). The experimental setup is shown in Figures 7.9A and 7.10 and its operation is explained below. It should be noted that the biocatalytic reaction cascade performing the demultiplexer function operated in a much simpler way compared to the electronic counterpart. The first flow cell functionalized with HK was activated with the Input and Address signals represented by glucose (Glc) and adenosine triphosphate(ATP), respectively. Both signals were defined as logic **0** value for the absence of Glc and ATP (meaning their zero concentrations), while their presence was defined as logic value **1** for the Input and Address signals. The Glc and ATP concentrations were experimentally optimized for the specific enzyme activity in the flow cells. This optimization was paramount due to the different amounts of immobilized enzymes, thus resulting in their different catalytic activities. The different activity of the immobilized enzymes was compensated by the careful experimental selection of the substrate-input concentrations. This optimization was aimed at the production of output signals with comparable intensities upon application of different combinations of the input signals. The output signals produced by the biocatalytic cascades in both output channels were defined as the optical absorbance (Abs) at $\lambda = 340$ nm corresponding to the formation of NADH, which is the final product of the reaction cascade (Figure 7.11). Output1 and Output2 were considered as logic value **0** when Abs < 0.6 and **1** when Abs > 0.6 (Figure 7.11C,D).

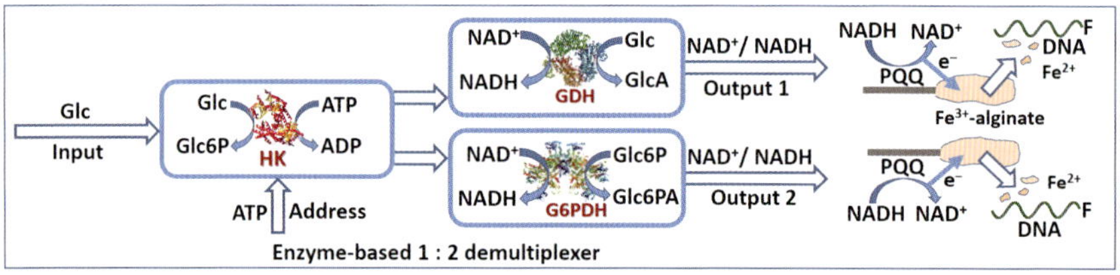

Figure 7.8 Biocatalytic realization of the 1 : 2 demultiplexer in the flow device, where each cell is functionalized with a different enzyme and performs different reactions. The far right part of the scheme shows the operation of the electrochemical actuators associated with two output channels and releasing DNA molecules in response to the NADH output signals produced by the demultiplexer. Source: Fratto et al. 2017 [5]. Adapted with permission of John Wiley and Sons.

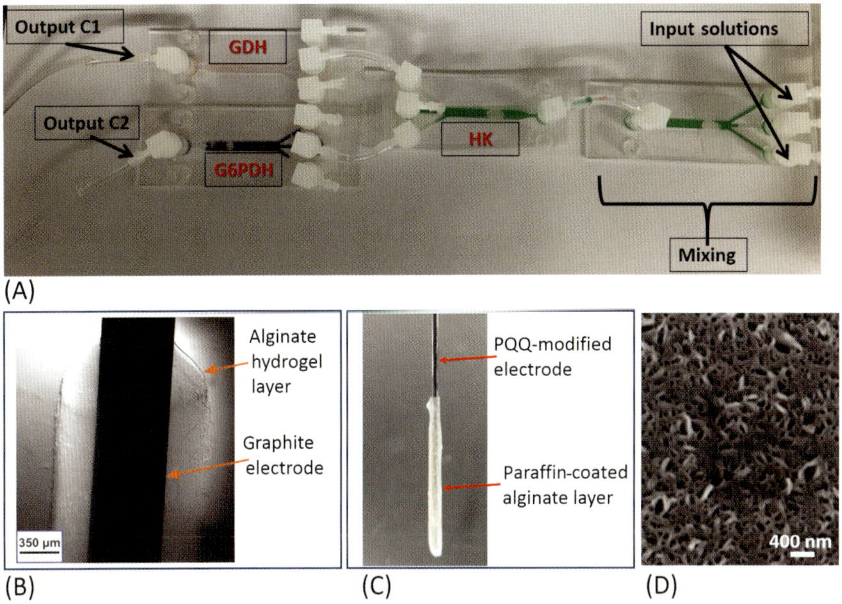

Figure 7.9 (A) The experimental realization of the 1 : 2 demultiplexer (photo of the flow cell circuitry). Different colored dyes are used in this image to illustrate the experimental realization including the mixing of channels where it is applicable. Note that the colors are used for the illustration only and do not correspond to the real view of the system upon its operation. The mixing cell served for mixing and distributing flow in the system. The release chambers are omitted from the system shown in the photo. (B) The alginate film produced at the graphite electrode surface by electrochemical deposition. Note that it was deposited only at a part of the electrode surface. (C) The "Janus" electrode after protecting the alginate film with paraffin wax and modification of the second part with PQQ. (D) An SEM image of the paraffin film with the porosity. Source: Fratto et al. 2017 [5]. Adapted with permission of John Wiley and Sons.

When the Input appeared at its logic value **0** (meaning the absence of Glc), no biocatalytic reactions were activated because all of the enzymes were missing the required substrates, thus producing the output signals with the value **0** in both channels, regardless of the logic value of the Address (Figure 7.11A,B, spectra a and c, and Figure 7.11C,D, bars a and c). When the Input was applied at the logic value **1** (corresponding to the presence of Glc), this value was translated to different output channels depending on the logic value **0** or **1** in the Address channel. Indeed, when the Address was applied at the logic value **0** (corresponding to the absence of ATP), the first biocatalytic reaction in the flow cell modified with HK did not proceed, and Glc was moved unchanged to the next flow cells organized in parallel (Figures 7.8 and 7.9A). Then Glc was biocatalytically oxidized in the flow cell modified with GDH with the concomitant reduction of NAD^+ yielding NADH (note that NAD^+ was always present as a part of the "machinery" solution). This resulted in the formation of NADH product in the Output1 channel, meaning output **1** (Figure 7.11A, spectrum b, and Figure 7.11C, bar b). On the other hand, Glc was passing through the flow cell modified with G6PDH without any reaction, thus producing no NADH in

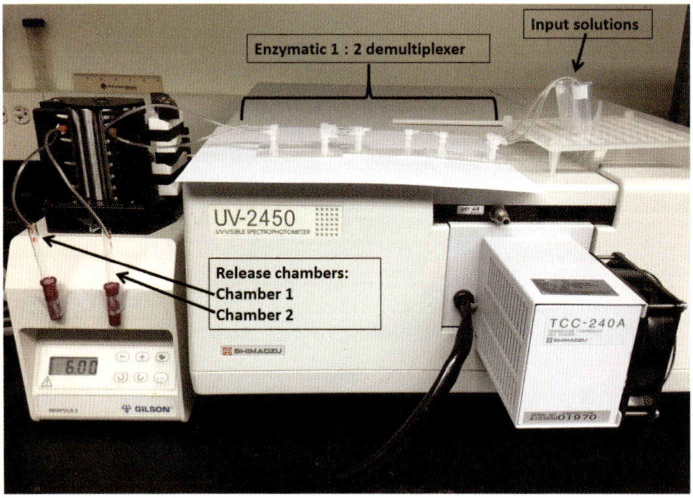

Figure 7.10 The experimental realization of the 1 : 2 demultiplexer (whole experimental setup). This photo serves to illustrate the placement of the input wells, enzymatic 1 : 2 demultiplexer, peristaltic pump, and release chambers. Sample solutions were taken from the release chambers and analyzed with the fluorometer. Note that the absorbance measurements for the Output1 and Output2 channels were performed with the different connections of the tubing – instead of connecting them to the release chambers (as it is show in the present photo), they were used for collecting samples for the optical absorbance measurements with the spectrophotometer, thus analyzing the NADH concentration in both output channels. Source: Fratto et al. 2017 [5]. Adapted with permission John Wiley and Sons.

the Output2 channel, corresponding to output **0** (Figure 7.11B, spectrum b, and Figure 7.11D, bar b). When ATP appeared in the Address channel (logic value **1**), Glc was phosphorylated in the reaction biocatalyzed by HK. The product of this reaction, glucose-6-phosphate (Glc6P), was moved with the liquid flow to the next flow cells instead of Glc. In this case the flow cell modified with GDH was mute because it was missing the reaction substrate, but the flow cell modified with G6PDH was activated and produced gluconate-6-phosphate acid, while NAD^+ was reduced to NADH. Therefore, Output1 was **0** and Output2 was **1** (in reverse to the situation discussed above when the Address was **0**) (Figure 7.11A,B, spectra d, and Figure 7.11C,D, bars d). Overall, the optically measured Output1 and Output2 logic values followed the expected dependence on the Input and Address logic values (Figure 7.5B). It should be noted that the complete consumption of Glc in the reaction biocatalyzed by HK is mandatory for the process in the presence of ATP. The ATP concentration and the flow rate have been optimized to ensure almost 100% conversion of Glc to Glc6P (actually it was never exactly 100%, thus resulting in minor responses in the output channels that supposed to be mute).

The important feature of the designed demultiplexer is production of NADH as the output signals in the Output1 or Output2 channels. As it has been already shown [37–41], the biocatalytically produced NADH is able to generate a negative potential on an electrode, while being electrocatalytically oxidized by pyrroloquinoline quinone (PQQ) immobilized on an electrode surface. PQQ is

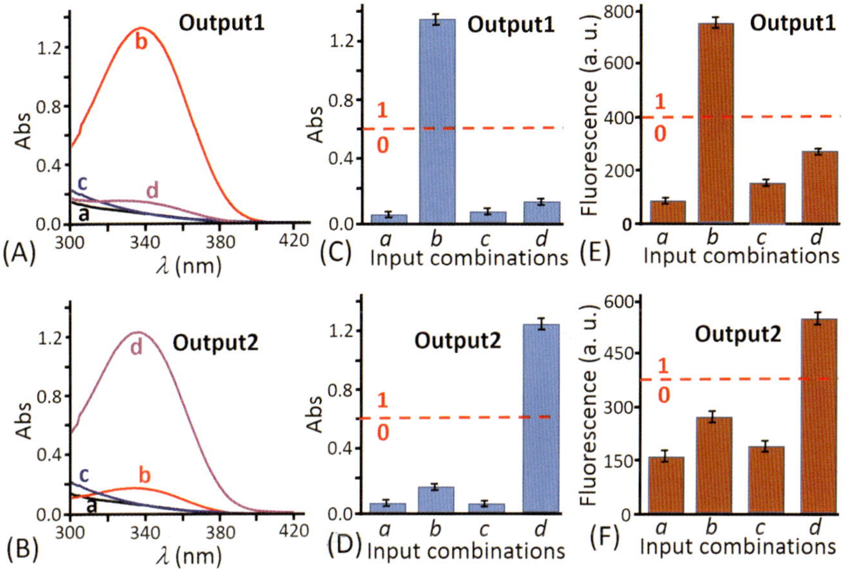

Figure 7.11 (A) Absorbance spectra measured as the result of the biocatalytic cascade in the Output1 channel for different logic value combinations of Input and Address. (B) The same as (A) but for the Output2 channel. (C) The bar chart showing the output signals in the Output1 channel extracted from the absorbance spectra at $\lambda = 340$ nm for different logic value combinations of Input and Address. (D) The same as (C) but for the Output2 channel. (E) The bar chart showing the output signals in the Output1 channel corresponding to the electrochemically stimulated DNA release measured as the fluorescence intensity (note that DNA was labeled with a fluorescent dye). (F) The same as (E) but for the Output2 channel. Dashed lines in (C–F) correspond to the thresholds separating logic values **0** and **1** of the output signals. The following notations are used for the input combinations: (a) **0**,**0**; (b) **1**,**0**; (c) **0**,**1**; (d) **1**,**1** where the fist number corresponds to the Input and second for the Address values. Source: Fratto et al. 2017 [5]. Adapted with permission of John Wiley and Sons.

a well-known electrocatalyst for the NADH oxidation [42], particularly when it is used in biofuel cells [43]. It should be noted that the NADH oxidation electrocatalyzed by PQQ proceeds without application of an external potential and the process proceeds spontaneously, resulting in the formation of a negative potential of ca. −60 mV (vs. Ag/AgCl reference) on the electrode [37–41]. This negative potential was used in other studies to dissolve electrochemically an Fe^{3+}-cross-linked alginate hydrogel thin film on a connected electrode, resulting in the release of biomolecules entrapped in the alginate gel [37–41] (see Chapter 12 for the related systems). In the present demultiplexer device, the electrochemical system responding to the NADH signals was simplified by deposition of both components: PQQ electrocatalyst and alginate hydrogel on two different domains of the same electrode. The PQQ monolayer was covalently bound through a carbodiimide procedure to amino groups of polyethylene imine (PEI) [44] that was adsorbed on a graphite electrode surface. Note that a pencil graphite rod was used as the electrode support [45, 46] and the PQQ monolayer occupied a half of the length of this electrode. Figure 7.12 shows a cyclic voltammogram obtained for the PQQ-modified electrode, indicating

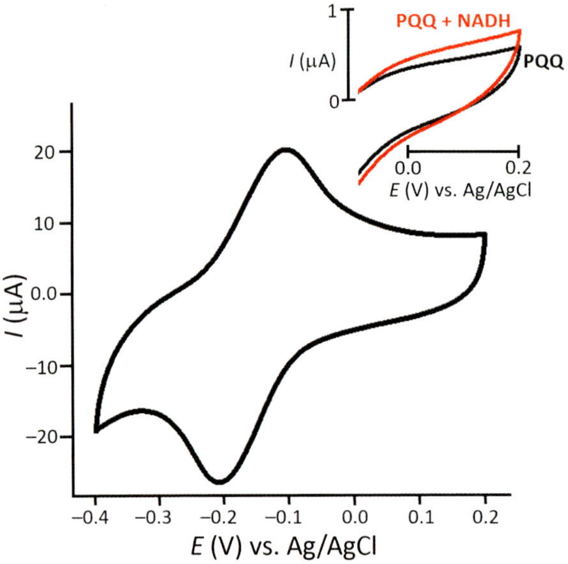

Figure 7.12 The cyclic voltammogram of the PQQ-modified graphite electrode recorded in 0.1 M Tris-buffer, pH 7.4, in the absence of oxygen with the potential scan rate, 50 mV s^{-1}. Inset: Fragments of cyclic voltammograms of the PQQ-modified electrode in the absence and presence of NADH (0.2 mM) obtained at the potential scan rate, 1 mV s^{-1}. The difference in the anodic currents observed at positive potentials corresponds to the electrocatalytic oxidation of NADH. Source: Fratto et al. 2017 [5]. Adapted with permission of John Wiley and Sons.

the $E_{1/2}$ potential of −0.15 V, being consistent with the known redox potential of PQQ [47]. The surface concentration of the immobilized PQQ (vs. geometrical area of the electrode surface, without taking into account the electrode roughness) was ca. 1×10^{-9} mole cm^{-2}, which is reasonable for a monolayer coverage [47], assuming the electrode roughness factor of ca. 10 [45]. Cyclic voltammograms obtained on the PQQ-modified electrode in the absence and presence of NADH (Figure 7.12, inset) show the increase of the anodic current when NADH is present corresponding to the NADH electrocatalytic oxidation. The electrocatalytic current is rather small because of the low concentration of NADH (ca. 0.2 mM) produced *in situ* by the biocatalytic cascade (compared with much higher NADH concentrations used in other studies [42]). Note that the cyclic voltammograms were obtained in a 3-electrode configuration including the PQQ-modified working electrode, a bare counter electrode, and a reference electrode, while the solution containing NADH (logic output **1**) was applied to the electrochemical cell.

The second half of the graphite electrode was covered with the electrochemically deposited Fe^{3+}-cross-linked alginate hydrogel (Figure 7.9B), containing entrapped DNA that was labeled with a fluorescent dye: 6-FAM-5′-TGC AGA CGT TGA AGG ATC CTC-3′ (FAM attached to the DNA is a fluorescein derivative). After formation of the alginate film on the electrode surface, it was coated with a film of paraffin for protection (Figure 7.9C). When the biocatalytically produced NADH reacted with the PQQ-modified electrode, it generated a negative potential resulting in the reduction of the cross-linking Fe^{3+} cations to the Fe^{2+} state, which is not capable of cross-linking alginate [39]. This resulted in the alginate hydrogel dissolution and the release of the entrapped DNA. The protecting paraffin film did not prevent the DNA release from the dissolved alginate hydrogel due to the porosity of paraffin [48] despite its hydrophobic nature (Figure 7.9D). In this experiment the bifunctional PQQ alginate electrode was not connected to any other electrodes and was not polarized by a potentiostat. Two identical bifunctional ("Janus") electrodes were placed in capillaries connected to the Output1 or Output2 channels. When one of the output channels contained NADH (output logic value **1**), the PQQ part of the "Janus" electrode was activated, and the second part received the negative potential and reducing current, resulting in the dissolution of the alginate film and the release of DNA. Figure 7.10 shows the connection of the flow device to the release chambers containing the modified PQQ alginate electrodes. Notably, the selection of the channel where this process occurred was controlled by the logic value of the Address signal. The released DNA was labeled with a fluorescent dye (FAM), thus allowing the release process to be followed by measuring a fluorescence signal in the solution (Figure 7.11E,F). While the fluorescence signals for the released DNA followed the signal pattern of the NADH produced, the ratio of the signals **1**/**0** was smaller for the released DNA compared with the NADH production (compare Figure 7.11C–E and Figure 7.11D–F). This effect was explained by partial dissolution of alginate in the presence of NAD^+ and ATP. Indeed, both molecules (NAD^+ and ATP) have phosphate groups capable of producing complexes with Fe^{3+} cations, thus removing the cross-linking iron cations from the alginate hydrogel and resulting in its partial dissolution. This process increased the background signal for the DNA release corresponding to the logic value **0** of the output signal.

Overall, the performance of the biomolecular 1 : 2 demultiplexer was extended to incorporate the electrochemical actuation function represented by the electrochemically stimulated DNA release controlled by the Input and Address binary signals. Upon further investigation, an increase in the system's complexity is easily possible. Indeed, the input signals, Glc and ATP, can be biochemically produced through various logic gates and circuits, while the released DNA can be included in different logic operations based on various DNA or DNAzyme systems [37, 38]. Importantly, the modular design of the flow system should not be overlooked as it allows an easy method for the connections to the input and output channels, their extensions, and variations. This form of a modular system is best illustrated by its comparison to a chemical "lego" system. Therefore, the important feature of the present 1 : 2 demultiplexer realization is its extendability and easy integration in complex biomolecular computing and actuating systems. The present realization of the 1 : 2 demultiplexer allows other bioelectronic devices operating in the same way. For example, one-bit half-adder and half-subtractor have been realized in a very similar biocatalytic system [49].

7.2 Biomolecular Signal Amplifier Based on Enzyme-Catalyzed Reactions

When concatenated logic gates operating in a complex network are aimed, the issue of the signal amplification between the gates should be addressed. Many natural biochemical pathways, particularly responsible for signal transduction, include amplification steps sometimes reaching 1000-fold increasing signals [50]. The electronic counterparts in integrated schemes also include amplification steps for the processed signals [30]. However, the electronic systems always operate with electrical signals, while biochemical systems use various chemicals through all biocatalytic steps. Different methods should be applied for the amplification of these chemical signals depending on their nature. In order to illustrate this approach, applied specifically to enzyme logic gates, a biocatalytic system amplifying NADH output signal produced by the gates was designed (note that NADH cofactor is a very common output generated by enzyme logic systems) [2]. Alcohol dehydrogenase (ADH) was selected as the biocatalyst producing NADH in the presence of NAD^+ and ethanol [51]. The tricky part of the process was activation of this biocatalytic reaction by the NADH signal, in other words aiming the amplified production of NADH activated by the NADH signal (Figure 7.13). To achieve this goal, disulfiram (DS) was selected to cause a reversible inhibition of ADH and further reactivation of the enzyme upon biocatalytic removal of the inhibitor initiated by NADH. It is known that disulfiram inhibits ADH only when it is in the disulfide state and the enzyme can be reactivated when the disulfide bond in the inhibitor is reduced [52]. Glutathione reductase (GR) can be used to reduce disulfide bonds in the oxidized state of glutathione (GSSG) using NADH as the reductive agent [53, 54]; however, GR is not efficient for the direct reduction of disulfiram [55]. Thus, GR was applied to reduce oxidized glutathione (GSSG) in the presence of NADH; then the

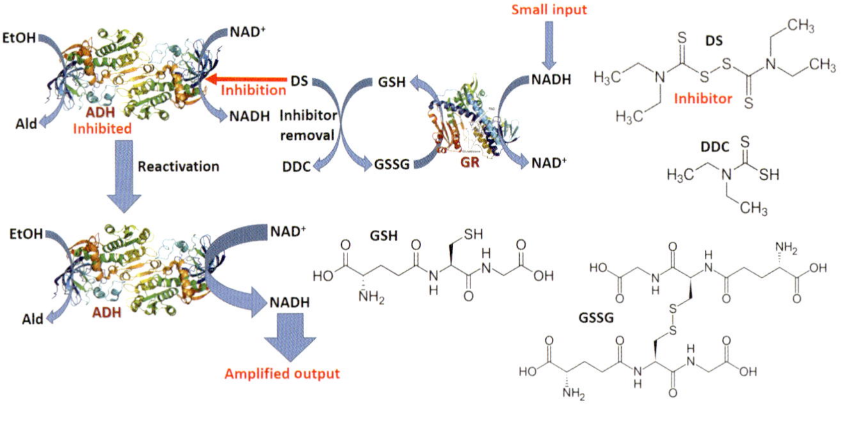

Figure 7.13 The self-promoted biocatalytic system for the amplification of the NADH signal. The inhibited ADH enzyme operates with a low reaction rate, while inhibitor removal increases the rate of the biocatalytic reaction. Formulas at the right show structures of some reacting species used in the system.

biocatalytically reduced glutathione (GSH) induced the reduction of disulfiram to yield diethyldithiocarbamate (DDC) due to the thiol–disulfide exchange [56]. This resulted in the removal of the inhibitor, reactivation of ADH, and production of NADH, thus amplifying the initial NADH input (Figure 7.13). The following experiment was performed to demonstrate the amplification process. First, ADH was inhibited by the addition of the optimized concentration of disulfiram upon overnight incubation. The optimum disulfiram concentration implied significant inhibition of ADH, which still could be reactivated by the thiol–disulfide exchange with the reduced glutathione produced *in situ* by the GR reaction. A lower disulfiram concentration did not result in substantial inhibition of ADH, while a higher concentration did not allow the enzyme reactivation. The inhibited ADH was tested for the production of NADH in the presence of NAD^+ and ethanol. Despite its inhibition, the enzyme ADH demonstrated some residual activity producing NADH (Figure 7.14, curve a). The addition of NADH input (50 μM) to the solution in the absence of the disulfide reducing system (GR and GSSG) did not affect the rate of the biocatalytic production of NADH (Figure 7.14, curve b). The same experiment performed in the presence of GR, GSSG, and NADH (50 μM) resulted in the enhanced production of NADH, witnessing the inhibitor reductive removal and the ADH reactivation (Figure 7.14, curve c). The rate of the NADH production was increased by twofold, thus demonstrating the NADH signal amplification. While the achieved twofold amplification of the NADH signal is not very impressive, the present experiment exemplifies the general approach to the biocatalytic signal amplification by removing an inhibitor and reactivating a biocatalytic cascade [2].

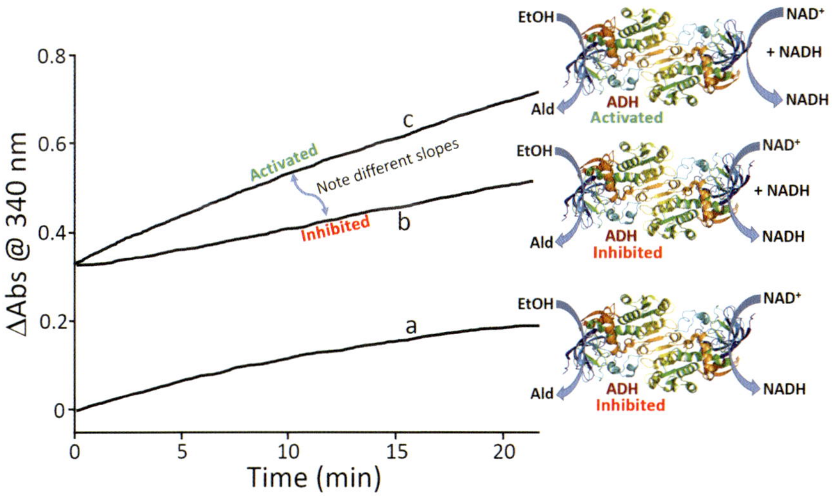

Figure 7.14 Absorbance changes (λ_{max} = 340 nm) corresponding to the NADH production in the signal amplification system: (a) with the inhibited ADH enzyme, (b) upon addition of the NADH signal without activation system, and (c) in the presence of the activation system and NADH signal. Note ca. twofold increase in the rate of the NADH production after removing the inhibitor in the system.

7.3 Biomolecular Signal Converter Based on Enzyme-Catalyzed Reactions

All chemical [57–60] and biochemical [61–63] networks composed of concatenated logic gates (particularly including enzyme-based networks) [64–66] reported until now do not fulfill Boolean properties of associativity, distributivity, and commutativity [67]. Indeed, each chemical step consumes chemicals and produces other chemicals different from the original inputs. Individual chemical reactions can mimic Boolean logic operations and even their sequences, but they cannot be exchanged in their order because of the chemical difference between the input and output signals at each step. For example, two AND gates cannot be interchanged in the order of their operation: AND 1 → AND 2 cannot operate in the opposite order AND 2 → AND 1 because the molecular output generated by AND 2 is different from the input required by AND 1 and it cannot be consumed by AND 1 (Figure 7.15). This difference might be used in some specific applications, for example, to realize the IMPLICATION logic operation, when the final result depends not only on the logic values but also on the order of the input signals [65, 68]. However, to mimic the electronic networks, meeting the requirements of Boolean properties, chemical/biochemical logic gates should be exchangeable regardless of the chemical nature of the input/output signals. To achieve this goal, at least partially, novel methods for converting the output signals to the initial input signals should be developed, thus potentially allowing the sequential operation of the logic gates with the same inputs at each individual logic step. To illustrate the concept, biochemical systems transforming the final NADH output signal generated by the enzyme logic gates to some of the initial input signals consumed by the gates, maltose and phosphate, have been designed [2].

A two-step conversion method of the NADH signal was designed (Figure 7.16). First, the NADH signal was used to activate a biocatalytic system composed of glucose dehydrogenase (GDH) and gluconic acid (GlcA), able to generate glucose (Glc). Using the reversibility of the reaction biocatalyzed by GDH, gluconic acid

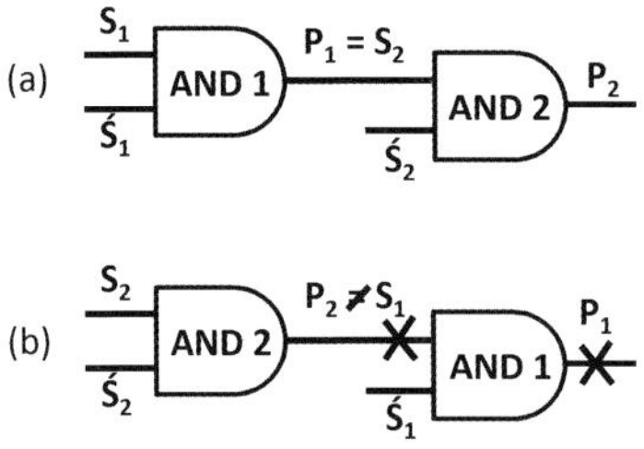

Figure 7.15 The scheme illustrating that the enzyme logic gates (e.g., AND gates in the present example) cannot be exchanged in their order. Thus, the biomolecular logic systems do not have commutativity property characteristic of Boolean logic. (a) AND 1 logic gate can be followed by another AND 2 logic gate if the product generated by the first gate (P_1) is the same as the substrate required for the operation of the second gate ($P_1 = S_2$). (b) The opposite order of the gates cannot be realized since the product generated by AND 2 (P_2) is not the substrate needed for AND 1 (S_1). $\acute{S}_1$ and $\acute{S}_2$ are co-substrates (additional inputs) consumed by the enzyme-based logic gates.

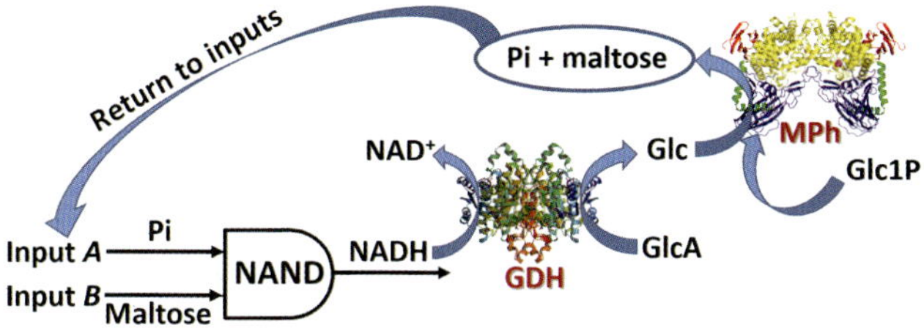

Figure 7.16 The two-step biocatalytic system converting the final output signal of NADH to the initial input signals of maltose and phosphate.

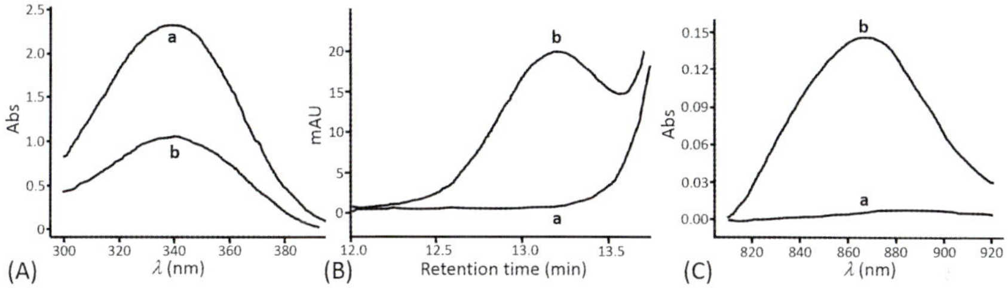

Figure 7.17 Conversion of the final output signal of NADH produced by the NAND gate to the intermediate signal of glucose and to the initial signals of maltose and phosphate. (A) Absorbance spectra of NADH in the converter system: (a) before and (b) after 10 minutes of the reaction. (B) HPLC analysis of maltose in the converter system: (a) before and (b) after the reaction for two hours (mAU = instrumental milli-arbitrary units). (C) Spectrophotometric analysis of phosphate using Splittgerber's reagent: (a) before and (b) after five minutes of the reaction. Source: Zhou et al. 2009 [2]. Adapted with permission of American Chemical Society.

was reduced to glucose in the presence of NADH. The reaction was followed by the decrease of the NADH optical absorbance (Figure 7.17A). Then another reversible reaction biocatalyzed by maltose phosphorylase (MPh) resulted in the conversion of glucose-1-phosphate and *in situ* generated glucose to maltose and phosphate, where maltose formation was analyzed by high-performance liquid chromatography (HPLC) (Figure 7.17B) and the phosphate production was proved by a colorimetric test [69] (Figure 7.17C). The designed system allowed the digital conversion (in terms of YES/NO; **1**/**0**) of the final output signal NADH to the initial signals maltose and phosphate consumed by the logic gate, thus potentially allowing assembling of the same gates in their various combinations and fulfilling the Boolean network properties. The signal-converting chemical system illustrated here is very far from being a universal converter, but at least it demonstrates this feature and highlights the importance of such systems for future biocomputing devices.

7.4 Utilization of a Fluidic Infrastructure for the Realization of Enzyme-Based Boolean Logic Circuits

Even with the distinctive advantages that enzyme-based biocomputing can bring to the field of unconventional molecular computing, the major drawback of such systems may still be directly related to their original realization in homogeneous solutions [62, 63]. When developing such complex systems in homogeneous solutions, the enzymes have the ability to show cross-reactivity, giving them the potential to convolute the system. An additional problem is the inherent limitations that are placed on the system in terms of complexity. For example, if a homogeneous system is being measured optically and is designed to have three input signals and three output signals, such a system would require the measurement of the output signals to be carefully considered. The simplest method would be to design a system where the output signals are three species that absorb at different wavelengths. If the system was designed to have the output signals absorbing at the same wavelength, then a threshold-based system would need to be implemented. Although a threshold-based system would be possible, the user is no longer able to distinguish which chemical output is related to a specific logic output like a system that is composed of three different output chemicals. The problem could be resolved by using different techniques for detecting the output signals (see Chapter 6). The ability to monitor different components with different measuring techniques has inherent advantages. On the contrary, the added machinery and methods that are required to sense multiple output species can hinder the goal of creating simple, streamline systems that are able to perform complex logic operations. For this reason, the transition from homogeneous solution-based systems to systems that utilize a fluidic infrastructure is a new and beneficial avenue for the continued development of biocomputing systems [70] (see also Chapter 5).

The use of a fluidic infrastructure as a base for a biocomputing system has many underwritten advantages. These intrinsic advantages are rarely present or in some cases can impose unwanted complexity to biocomputing systems that are realized in homogeneous solutions [71]. The first advantage of a fluidic infrastructure is the ability to physically separate both the different input channels and their respective logic operations [72]. The triviality of this advantage may in some ways downplay its inherent importance. The second advantage of a fluidic infrastructure is the modularity of the system. In the application of a fluidic infrastructure, the systems have the ability to compartmentalize the necessary enzymes in their respective fluidic devices [72]. With this compartmentalization, specific enzymes are able to be restricted to defined reacting chambers, thus giving the system the ability to be reconfigured and reset without changing the overarching physical setup [73]. The third advantage of a fluidic infrastructure is the ability for the system to be optimized via the rate that the solution is moved through the system. In such applications, the flow of solution through the system is able to control the production of the signal by directly influencing the time that the input species are in contact with the reacting chambers that house the compartmentalized enzymes [74]. This means that the optimization of the system is no longer relegated to the concentration of the input species, the units of enzyme present, or the buildup of the product from an ongoing reaction. The continuous flow of solution not only decreases the buildup of product in the reacting chamber, but it also limits of the consumption of the input species [74]. This allows for the catalytic consumption of the input signal to be influenced by the flow rate of the signal, thus giving the ability to further tune the system based on the flow rate. In addition to the flow rate, the timing of the signal may also be used to control the output signal [74]. For such an application, a successful mapping of the fluidic properties would be pivotal. It may be surmised that upon completion of this mapping, a characterized fluidic skeleton could have multiple applications that all depend on the different enzymes that are needed for specific processes [75]. The logic-based inputs may be paired with the properties of the fluidic skeleton to illicit distinct responses from the system. Signal processing in flow systems, particularly composed of several reacting units performing different biocatalytic reactions, requires additional optimization [74]. The problems associated with the reacting flow could be substantially different from the enzyme-catalyzed reactions proceeding in one-pot chemical device. With the implementation of properly designed cells for noise reduction, it should be possible to advance the use of flow systems to multistep consecutive connectivity in order to increase the complexity of the digital signal processing, as well as the inherent networking capabilities.

7.5 Other Circuit Elements Required for the Networking of Enzyme Logic Systems and General Conclusions

As soon as the biocomputing systems start to increase in their complexity, moving from stand-alone Boolean logic gates to their networks and sophisticated functional systems, the networking problems resolving connectivity issues between individual units become important. While individual Boolean logic gates can be optimized for their performance, their cooperative operation might be far from optimum and require different conditions. The problems might be similar to optimization of electronic circuitries, but they might be also specific for the chemical realizations. For example, enzymes included in the complex multicomponent systems may require different concentrations of substrates/cofactors or different pH values for their optimized performance. Some of the optimization problems could be resolved by compromising between the optimal conditions required for different enzymes if they are not too far from each other. On the other hand, special networking elements could be designed and added to the systems to allow their operation. One of the most important additional components suppressing noise produced by individual components is a "filter" unit consuming a part of the produced signal and converting the output function from a convex concentration-dependent shape typical for biocatalytic reactions to sigmoidal shape more characteristic of electronic counterparts (see the detailed discussion in Chapter 3) [76–86]. While stand-alone operating logic gates could tolerate to some extent noise produced by the reacting system (the noise can be related to uncontrolled fluctuation of input signal concentrations), the systems composed of many (even a few) interconnected gates can produce substantial noise amplified by the signal transfer from one gate to another, thus degrading the final output to a meaningless random answer. In such case the filtering of signals produced by the logic gates could be highly important and even mandatory for the proper operation of logic systems.

Some other networking components could originate from mimicking biological way of information processing. Feed-forward operation and the corresponding biomolecular systems representing the "hardware" could be important for improving dynamics of time-dependent operation of the biocatalytic logic systems. While general approach to the enzyme-based systems performing feed-forward function has been extensively discussed [87, 88] (see also Chapter 15), the experimental realization of these systems is awaiting for further work, which seems to be not easy but important for future biocomputing systems.

Overall, the connectivity of individual logic elements into integrated computing circuitries is a challenging goal requiring new elements performing novel functions. Some of these functions can be inspired by biological processes, but they should be realized with much simpler molecular systems compared with biological systems. On the other hand, other ideas could be borrowed from electronic computing systems. However, even in this case the molecular systems are not necessary based on the same structural realization, but still performing similar functions.

References

1 Privman, M., Tam, T.K., Pita, M., and Katz, E. (2009). *J. Am. Chem. Soc.* 131: 1314–1321.
2 Zhou, J., Arugula, M.A., Halámek, J. et al. (2009). *J. Phys. Chem. B* 113: 16065–16070.
3 Pita, M., Strack, G., MacVittie, K. et al. (2009). *J. Phys. Chem. B* 113: 16071–16076.
4 Arugula, M.A., Bocharova, V., Pita, M. et al. (2010). *J. Phys. Chem. B* 114: 5222–5226.
5 Fratto, B.E., Guz, N., Fallon, T.T., and Katz, E. (2017). *ChemPhysChem* 18: 1721–1725.
6 Andréasson, J., Straight, S.D., Bandyopadhyay, S. et al. (2007). *J. Phys. Chem. C* 111: 14274–14278.
7 Amelia, M., Baroncini, M., and Credi, A. (2008). *Angew. Chem. Int. Ed.* 47: 6240–6243.
8 Perez-Inestrosa, E., Montenegro, J.M., Collado, D., and Suau, R. (2008). *Chem. Commun.* 1085–1087.
9 Andréasson, J., Straight, S.D., Bandyopadhyay, S. et al. (2007). *Angew. Chem. Int. Ed.* 46: 958–961.
10 Singh, P., Singh, H., Bhargava, G., and Kumar, S. (2017). *Sens. Actuat. B* 245: 1004–1014.
11 He, Y. and Yu, H.L. (2015). *Anal. Bioanal. Chem.* 407: 6741–6746.
12 Kaur, N. and Kumar, S. (2014). *J. Chem. Sci.* 126: 49–54.
13 Chen, K. and Schmittel, M. (2016). *Anal. Bioanal. Chem.* 408: 7077–7083.
14 Erbas-Cakmak, S., Bozdemir, O.A., Cakmak, Y., and Akkaya, E.U. (2013). *Chem. Sci.* 4: 858–862.
15 Lewandowska, K. and Szacilowski, K. (2011). *Austr. J. Chem.* 64: 1409–1413.
16 Ran, X., Wang, Z.Z., Ju, E.G. et al. (2018). *Nanotechnology* 29: 065501.

17 Orbach, R., Remacle, F., Levine, R.D., and Willner, I. (2014). *Chem. Sci.* 5: 1074–1081.
18 Pan, L.Q., Wang, Z.Y., Li, Y.F. et al. (2017). *Nanoscale* 9: 18223–18228.
19 Gao, R.R., Yao, T.M., Lv, X.Y. et al. (2017). *Chem. Sci.* 8: 4211–4222.
20 Wang, K., He, M.Q., Wang, J. et al. (2016). *Sci. Rep.* 6: 34810.
21 Sarker, A., Babu, H.M.H., and Rashid, S.M.M. (2015). *IET Nanobiotechnol.* 9: 226–238.
22 Wu, C.T., Wang, K., Fan, D.Q. et al. (2015). *Chem. Commun.* 51: 15940–15943.
23 Privman, V., Pedrosa, V., Melnikov, D. et al. (2009). *Biosens. Bioelectron.* 25: 695–701.
24 Melnikov, D., Strack, G., Pita, M. et al. (2009). *J. Phys. Chem. B* 113: 10472–10479.
25 Privman, V., Arugula, M.A., Halámek, J. et al. (2009). *J. Phys. Chem. B* 113: 5301–5310.
26 Privman, V., Strack, G., Solenov, D. et al. (2008). *J. Phys. Chem. B* 112: 11777–11784.
27 Manesh, K.M., Halámek, J., Pita, M. et al. (2009). *Biosens. Bioelectron.* 24: 3569–3574.
28 Pita, M., Zhou, J., Manesh, K.M. et al. (2009). *Sens. Actuat. B* 139: 631–636.
29 BRENDA - The comprehensive enzyme information system. http://www.brenda-enzymes.org (accessed 27 February 2019).
30 Bostock, G. (1988). *Programmable Logic Devices. Technology and Applications.* New York, NY: McGraw-Hill.
31 Zhou, J., Tam, T.K., Pita, M. et al. (2009). *ACS Appl. Mater. Interfaces* 1: 144–149.
32 Krämer, M., Pita, M., Zhou, J. et al. (2009). *J. Phys. Chem. B* 113: 2573–2579.
33 Tokarev, I., Gopishetty, V., Zhou, J. et al. (2009). *ACS Appl. Mater. Interfaces* 1: 532–536.
34 Motornov, M., Zhou, J., Pita, M. et al. (2009). *Small* 5: 817–820.
35 Motornov, M., Zhou, J., Pita, M. et al. (2008). *Nano Lett.* 8: 2993–2997.
36 Strack, G., Pita, M., Ornatska, M., and Katz, E. (2008). *ChemBioChem* 9: 1260–1266.
37 Guz, N., Fedotova, T.A., Fratto, B.E. et al. (2016). *ChemPhysChem* 17: 2247–2255.
38 Mailloux, S., Gerasimova, Y.V., Guz, N. et al. (2015). *Angew. Chem. Int. Ed.* 54: 6562–6566.
39 Katz, E., Pingarrón, J.M., Mailloux, S. et al. (2015). *J. Phys. Chem. Lett.* 6: 1340–1347.
40 Gamella, M., Guz, N., Mailloux, S. et al. (2014). *Electroanalysis* 26: 2552–2557.
41 Mailloux, S., Guz, N., Zakharchenko, A. et al. (2014). *J. Phys. Chem. B* 118: 6775–6784.
42 Katz, E., Lötzbeyer, T., Schlereth, D.D. et al. (1994). *J. Electroanal. Chem.* 373: 189–200.
43 Willner, I., Arad, G., and Katz, E. (1998). *Bioelectrochem. Bioenerg.* 44: 209–214.

44 Moore, A.N.J., Katz, E., and Willner, I. (1996). *Electroanalysis* 8: 1092–1094.
45 Kariuki, J.K. (2012). *J. Electrochem. Soc.* 159: H747–H751.
46 Bond, A.M., Mahon, P.J., Schiewe, J., and Beckett, V.V. (1997). *Anal. Chim. Acta* 345: 67–74.
47 Katz, E., Schlereth, D.D., and Schmidt, H.-L. (1994). *J. Electroanal. Chem.* 367: 59–70.
48 Pechook, S. and Pokroy, B. (2013). *Soft Matter* 9: 5710–5715.
49 Fratto, B.E., Lewer, J.M., and Katz, E. (2016). *ChemPhysChem* 17: 2210–2217.
50 Wilson, K. and Walker, J. (eds.) (2005). *Principles and Techniques of Biochemistry and Molecular Biology*. New York: Cambridge University Press.
51 De Smidt, O., Du Preez, J.C., and Albertyn, J. (2008). *FEMS Yeast Res.* 8: 967–978.
52 Carper, W.R., Dorey, R.C., and Beber, J.H. (1987). *Clin. Chem.* 33: 1906–1908.
53 Rietveld, P., Arscott, L.D., Berry, A. et al. (1994). *Biochemistry* 33: 13888–13895.
54 Chung, Y.C. and Hurlbert, R.E. (1975). *J. Bacteriol.* 123: 203–211.
55 Nagendra, S.N., Shetty, K.T., Subhash, M.N., and Guru, S.C. (1991). *Life Sci.* 49: 23–28.
56 Gilbert, H.F. (1995). *Methods Enzymol.* 251: 8–28.
57 Szacilowski, K. (2012). *Infochemistry – Information Processing at the Nanoscale*. Chichester: Wiley.
58 de Silva, A.P. (2013). *Molecular Logic-Based Computation*. Cambridge: Royal Society of Chemistry.
59 Sienko, T., Adamatzky, A., Conrad, M., and Rambidi, N.G. (eds.) (2003). *Molecular Computing*. Cambridge, MA: MIT Press.
60 Katz, E. (ed.) (2012). *Molecular and Supramolecular Information Processing – From Molecular Switches to Logic Systems*. Weinheim: Wiley-VCH.
61 Katz, E. (ed.) (2012). *Biomolecular Computing – From Logic Systems to Smart Sensors and Actuators*. Weinheim: Wiley-VCH.
62 Katz, E. and Privman, V. (2010). *Chem. Soc. Rev.* 39: 1835–1857.
63 Katz, E. (2015). *Curr. Opin. Biotechnol.* 34: 202–208.
64 Katz, E. (2019). *ChemPhysChem* 20: 9–22.
65 Strack, G., Ornatska, M., Pita, M., and Katz, E. (2008). *J. Am. Chem. Soc.* 130: 4234–4235.
66 Niazov, T., Baron, R., Katz, E. et al. (2006). *Proc. Natl. Acad. USA.* 103: 17160–17163.
67 Whitesitt, J.E. (1995). *Boolean Algebra and Its Applications*. New York: Dover Publications.
68 Margulies, D., Felder, C.E., Melman, G., and Shanzer, A. (2007). *J. Am. Chem. Soc.* 129: 347–354.
69 Parsons, T.R., Maita, Y., and Lalli, C.M. (1984). *A Manual of Chemical and Biological Methods for Seawater Analysis*. New York: Pergamon Press.
70 Fratto, B.E. and Katz, E. (2017). *Int. J. Parallel, Emergent Distributed Systems* 32: 139–156.
71 Toepke, M.W., Abhyankar, V.V., and Beebe, D.J. (2007). *Lab Chip* 7: 1449–1453.
72 Fratto, B.E. and Katz, E. (2015). *ChemPhysChem* 16: 1405–1415.

73 Fratto, B.E., Roby, L.J., Guz, N., and Katz, E. (2014). *Chem. Commun.* 50: 12043–12046.
74 Verma, A., Fratto, B.E., Privman, V., and Katz, E. (2016). *Sensors* 16: 1042.
75 Pike, D.J., Kapur, N., Millner, P.A., and Stewart, D.I. (2013). *Sensors* 13: 58–70.
76 Zavalov, O., Bocharova, V., Privman, V., and Katz, E. (2012). *J. Phys. Chem. B* 116: 9683–9689.
77 Halámek, J., Zavalov, O., Halámková, L. et al. (2012). *J. Phys. Chem. B* 116: 4457–4464.
78 Halámek, J., Zhou, J., Halámková, L. et al. (2011). *Anal. Chem.* 83: 8383–8386.
79 Pita, M., Privman, V., Arugula, M.A. et al. (2011). *PhysChemChemPhys* 13: 4507–4513.
80 Pedrosa, V., Melnikov, D., Pita, M. et al. (2010). *Int. J. Unconventional Computing* 6: 451–460.
81 Privman, V., Halámek, J., Arugula, M.A. et al. (2010). *J. Phys. Chem. B* 114: 14103–14109.
82 Privman, V., Domanskyi, S., Mailloux, S. et al. (2014). *J. Phys. Chem. B* 118: 12435–12443.
83 Privman, V., Zavalov, O., Halámková, L. et al. (2013). *J. Phys. Chem. B* 117: 14928–14939.
84 Bakshi, S., Zavalov, O., Halámek, J. et al. (2013). *J. Phys. Chem. B* 117: 9857–9865.
85 Privman, V., Fratto, B.E., Zavalov, O. et al. (2013). *J. Phys. Chem. B* 117: 7559–7568.
86 Zavalov, O., Bocharova, V., Halámek, J. et al. (2012). *Int. J. Unconventional Computing* 8: 347–365.
87 Katz, E., Privman, V., and Zavalov, O. (2014). *Proceedings of The Eighth International Conference on Quantum, Nano/Bio, and Micro Technologies (ICQNM 2014)*, 22–27. Wilmington, DE: ThinkMind Online Publishing.
88 Privman, V. and Katz, E. (2015). *Physica Status Solidi A* 212: 219–228.

8

Enzyme-Based Memory Systems

8.1 Introduction

Molecular [1–4] and biomolecular [5] logic gates, particularly including enzyme-based logic systems [6, 7], perform sophisticated information processing operations, but do not store any actual information. Long-term use of these logic systems requires the addition of memory components, providing the ability to store an output in a stable form for later reference. Therefore, the development of memory systems is essential for the advancement of chemical and biochemical digital systems. The construction of these memory units, capable of multifunction information processing (read–write–erase), using biochemical components for analysis and storage, could potentially revolutionize the field of bioelectronics. By coupling these memory units with existing biocomputing ensembles, the functionality of switchable materials could be drastically improved, allowing for complex biocomputing devices, with built-in logic and long-term data storage, capable of dynamic response to their environment.

8.2 Enzyme-Based Flip-Flop Memory Elements

Of particular interest is the base-level integration of novel molecular memory units into chemical and biochemical logic systems for sophisticated information processing; however, the conceptualization and fabrication of these systems proves to be immensely challenging. Of the traditional silicon-based flip-flop memory circuits used in modern computers, such as the set/reset (SR) flip-flop, the delay (D) flip-flop, the toggle (T) flip-flop, and the Jack Kilby (JK) flip-flop, the SR flip-flop is, by far, the least challenging to reproduce using chemical/biochemical systems. This is primarily due to the fact that two antonymous reactions proceeding in one controlled direction, or the reverse, can serve as a viable method for the set or reset of a system. SR flip-flop memory units have been produced using a number of methods, such as chemical, electrochemical, and photochemical [8–13], as well as a number of biomolecular systems based on DNA reactions [14–16] and enzyme-catalyzed reactions [17]. Less common in chemical [18, 19] and biochemical [20–22] models are the remaining flip-flops (e.g., JK, D, and T). This is due primarily to the fact that, by their very design, they

Enzyme-Based Computing Systems, First Edition. Evgeny Katz.

exhibit higher complexity, making them difficult to mimic using biochemical systems. However, there have been successes in overcoming these complexities, resulting in the mimicking of both the D and T flip-flops [20–22]. Presented here are examples of these biochemical realizations of flip-flop memory systems based on enzyme-catalyzed reactions, as well as a review of the current state and perspectives for these memory units, progressing forward in this area of biocomputing.

8.2.1 Set/Reset (SR) Flip-Flop Memory Based on Enzyme-Catalyzed Reactions

A set-reset flip-flop memory unit is switchable between two distinct states defined digitally as **0** and **1** and is activated by two input signals, set (***S***) and reset (***R***), while each of them can be applied at two levels digitally defined as **0** and **1** [23]. Application of $\boldsymbol{S} = \mathbf{0}$ and $\boldsymbol{R} = \mathbf{0}$ should preserve the unit state regardless of its value (**0** or **1**). Application of $\boldsymbol{S} = \mathbf{1}$ should result in the unit state **1**, while application of $\boldsymbol{R} = \mathbf{1}$ should yield the unit state **0**, regardless of the unit state (**0** or **1**) prior to the signals. The simultaneous application of the signals $\boldsymbol{S} = \mathbf{1}$ and $\boldsymbol{R} = \mathbf{1}$ converts the system in the opposite directions; therefore, the resultant unstable situation is not permitted [23]. Usually, the memory unit operation is explained by a table showing the previous ($\boldsymbol{Q_t}$) and new ($\boldsymbol{Q_{t+1}}$) states of the unit upon application of different combinations of the ***S*** and ***R*** signals (**0**,**0**; **0**,**1**; and **1**,**0**, assuming that the **1**,**1** combination is not permitted) (Figure 8.1A).

(A)

Truth table for set/reset flip-flop memory

Set (S)	Reset (R)	Previous state, Q_t	New state, Q_{t+1}
0	0	Q_t (0 or 1)	Q_t (0 or 1) – unchanged
0	1	Q_t (0 or 1)	0
1	0	Q_t (0 or 1)	1
1	1	Q_t (0 or 1)	Unstable – not permitted

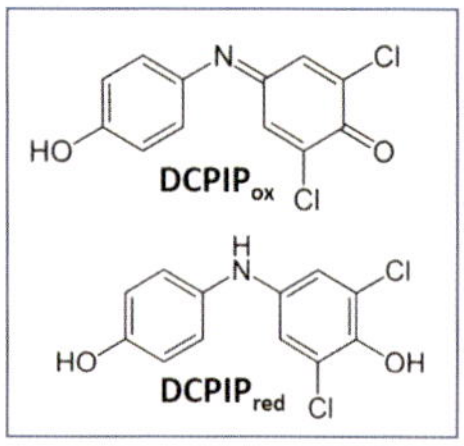

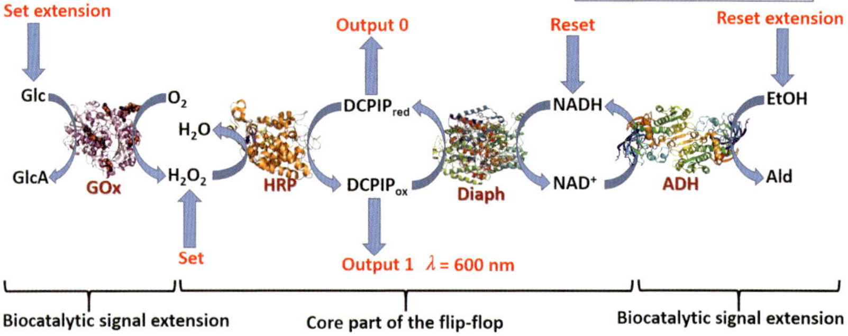

(B)

Figure 8.1 (A) The truth table for the set/reset (SR) flip-flop memory. (B) The enzyme biocatalytic system mimicking SR flip-flop memory operations using HRP and Diaph as the components of the core part and GOx and ADH as terminal biocatalysts converting primary SR signals (glucose and ethanol) to H_2O_2 and NADH inputs controlling the switchable core part. The structures of the used redox mediator (DCPIP$_{ox}$ and DCPIP$_{red}$) are shown in the frame.

In order to mimic such electronic devices with biochemical reactions, a simple biocatalytic system composed of two enzymes, horseradish peroxidase (HRP) and diaphorase (Diaph) catalyzing oxidation and reduction of redox species, respectively, was designed [17] (Figure 8.1B). The oxidative reaction biocatalyzed by HRP was activated by the addition of H_2O_2, while the reductive reaction biocatalyzed by Diaph was activated by NADH. The oxidative process resulted in the oxidized state of the redox species, operating as mediators between both enzymes. If the redox species already existed in the oxidized state prior to the addition of H_2O_2, the biocatalytic reaction did not proceed because of the lack of the oxidizable substrate. The reductive process resulted in the reduced state of the redox species. If the redox species already existed in the reduced state prior to the addition of NADH, the biocatalytic reaction did not proceed because of the lack of the reducible substrate; however, the oxidation reaction in the presence of H_2O_2 was obviously possible. The presence of the reduced or oxidized species in the system could be detected by optical or electrochemical means (also using other signal transduction methods; see Chapter 6), thus reporting on the state of the system. It is a matter of definition that state of the system (reduced or oxidized) is selected as the digital **0** or **1** state. This selection could be motivated by the used signal transduction method. For example, the oxidized state of the redox mediator was defined as digital **1**, while the reduced state as digital **0**, simply because of higher optical absorbance of the oxidized species used in the system discussed below. However, the binary definition of the states is obviously arbitrary. When the digital states are defined, the processes and signals resulting in these states become certain: the H_2O_2 input producing the oxidized state (**1**) is defined as set (***S***), and the NADH input yielding the reduced state (**0**) is defined as reset (***R***).

The core system composed of two enzymes (HRP and Diaph) mimics the operation of the electronic counterpart set-reset flip-flop memory [17] (Figure 8.1B). The purpose of the present study was to demonstrate that this core part of the enzyme-based flip-flop memory can be activated by various biochemical signals storing information. Indeed, the set and reset signals (H_2O_2 and NADH) can be produced *in situ* by numerous biochemical reactions extending the initial set/reset signals to many different biochemical inputs. To demonstrate the concept the flip-flop memory performance was illustrated with two biochemical examples.

In the first example [17], H_2O_2 was produced *in situ* in the presence of glucose oxidase (GOx) activated by the addition of glucose (Glc), being the set (***S***) input extended signal. NADH was also generated *in situ* in the presence of alcohol dehydrogenase (ADH) upon addition of ethanol (EtOH), being the reset (***R***) input extended signal (Figure 8.1B). A synthetic dye (dichlorophenolindophenol [DCPIP]) was used as the redox species defining the state of the system (**0** for its reduced and **1** for its oxidized form, $DCPIP_{red}$ and $DCPIP_{ox}$, respectively; see the molecule structures in the box [Figure 8.1]). When the initial state of the system, $\boldsymbol{Q_t}$, was defined as **0**, a "pretreatment" of NADH was added in order to reduce DCPIP in the presence of diaphorase (Diaph). The $\boldsymbol{Q_t} = \mathbf{0}$ demonstrated the absorbance spectrum with no absorbance at $\lambda = 600$ nm characteristic of the $DCPIP_{red}$ reduced form, while the NADH absorbance appeared at

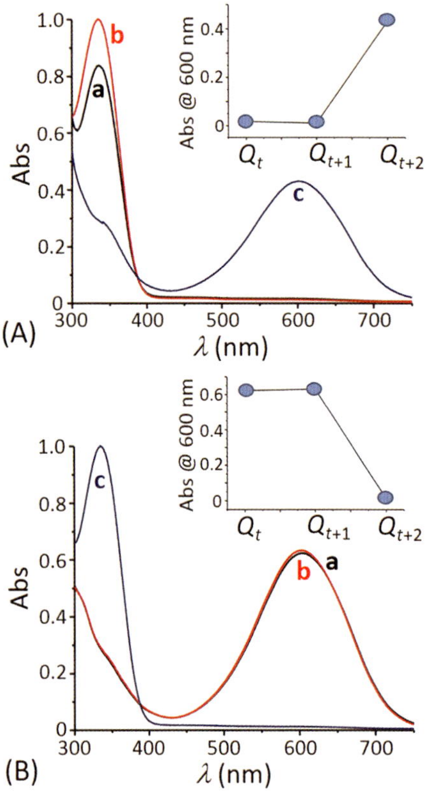

Figure 8.2 Absorbance spectra corresponding to the SR flip-flop system composed of HRP, Diaph, GOx, ADH, NAD^+, and DCPIP in 0.1 mM phosphate buffer, pH = 7.2. (A) The initial state of the system is reduced ($\boldsymbol{Q_t} = \mathbf{0}$): (a) the system pretreated with NADH, no signals applied ($\boldsymbol{Q_t} = \mathbf{0}$); (b) the system treated with the reset signal **1** (ethanol) keeping the system in **0** state ($\boldsymbol{Q_{t+1}} = \mathbf{0}$); and (c) the system treated with the set signal **1** (glucose) converting the system to **1** state ($\boldsymbol{Q_{t+2}} = \mathbf{1}$). Inset: Absorbance at $\lambda = 600$ nm at the initial **0** state, after **Reset** = **1** and after **Set** = **1** signals. (B) The initial state of the system is oxidized ($\boldsymbol{Q_t} = \mathbf{1}$): (a) no inputs applied ($\boldsymbol{Q_t} = \mathbf{1}$); (b) the system treated with the set signal **1** (glucose) preserving the system in **1** state ($\boldsymbol{Q_{t+1}} = \mathbf{1}$); and (c) the system treated with the reset signal **1** (ethanol) converting the system to **0** state ($\boldsymbol{Q_{t+2}} = \mathbf{0}$). Inset: Absorbance at $\lambda = 600$ nm at the initial **1** state, after **Set** = **1** and after **Reset** = **1** signals. Source: Pita et al. 2009 [17]. Adapted with permission of American Chemical Society.

$\lambda = 340$ nm (Figure 8.2A, curve a). The addition of ethanol ($\boldsymbol{R} = \mathbf{1}$) resulted in the further increase of NADH concentration reflected by the higher absorbance at $\lambda = 340$ nm; but it did not affect the state of the system readout at $\lambda = 600$ nm, thus keeping $\boldsymbol{Q_{t+1}} = \boldsymbol{Q_t}$ (Figure 8.2A, curve b). Conversely, the addition of glucose ($\boldsymbol{S} = \mathbf{1}$) resulted in the production of H_2O_2 and converted the system to its oxidized state ($\boldsymbol{Q_{t+2}} = \mathbf{1}$) reflected by the high absorbance of the oxidized $DCPIP_{ox}$ at $\lambda = 600$ nm (Figure 8.2A, curve c). Note that the absorbance of NADH at $\lambda = 340$ nm was also significantly decreased because of its reoxidation to NAD^+ biocatalyzed by Diaph. When the initial state of the system, $\boldsymbol{Q_t}$, was defined as **1**, the system presents in its oxidized state with the high absorbance at $\lambda = 600$ nm (Figure 8.2B, curve a). Note that the absorbance at $\lambda = 340$ nm was low, reflecting the presence of NAD^+ rather than NADH in the system. The addition of glucose ($\boldsymbol{S} = \mathbf{1}$) resulted in no changes, keeping $\boldsymbol{Q_{t+1}} = \boldsymbol{Q_t}$ (Figure 8.2B, curve b). The addition of ethanol ($\boldsymbol{R} = \mathbf{1}$) yielded NADH and induced the transformation of $DCPIP_{ox}$ to its reduced state $DCPIP_{red}$, thus increasing absorbance at $\lambda = 340$ nm and bleaching the absorbance band at 600 nm. This was considered as the system state $\boldsymbol{Q_{t+2}} = \mathbf{0}$. The stepwise response of the enzyme-based SR flip-flop system to the applied $\boldsymbol{S}/\boldsymbol{R}$ signals is shown in Figure 8.2A,B, insets. The system response to the set/reset signals mimics the flip-flop operation described by the truth table (Figure 8.1A).

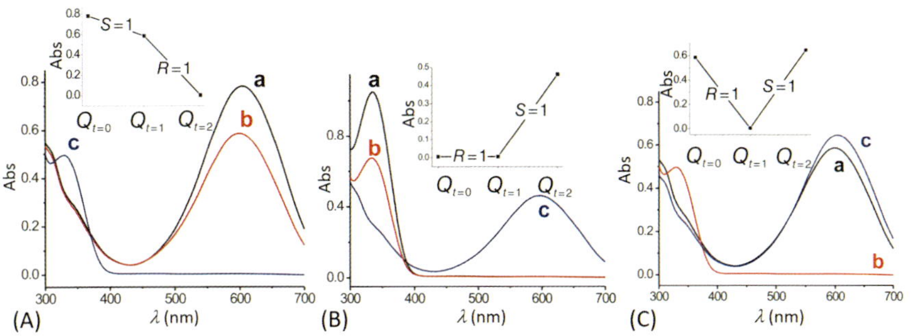

Figure 8.3 Absorbance spectra corresponding to the SR flip-flop system composed of HRP, Diaph, LOx, G6PDH, NAD$^+$, and DCPIP in 0.1 mM phosphate buffer, pH = 7.2. (A) The initial state of the system is oxidized ($\boldsymbol{Q_t}$ = **1**): (a) no inputs applied ($\boldsymbol{Q_t}$ = **1**); (b) the system treated with the set signal **1** (lactate) keeping the system in **1** state ($\boldsymbol{Q_{t+1}}$ = **1**); and (c) the system treated with the reset signal **1** (glucose 6-phosphate) converting the system to **0** state ($\boldsymbol{Q_{t+2}}$ = **0**). Inset: Absorbance at λ = 600 nm at the initial **0** state, after **Set** = **1** and after **Reset** = **1** signals. (B) The initial state of the system is reduced ($\boldsymbol{Q_t}$ = **0**): (a) the system was pretreated with NADH; no inputs applied ($\boldsymbol{Q_t}$ = **0**); (b) the system treated with the reset signal **1** (glucose 6-phosphate) preserving the system in **0** state ($\boldsymbol{Q_{t+1}}$ = **0**); and (c) the system treated with the set signal **1** (lactate) converting the system to **1** state ($\boldsymbol{Q_{t+2}}$ = **1**). Inset: Absorbance at λ = 600 nm at the initial **0** state, after **Reset** = **1** and after **Set** = **1** signals. (C) The initial state of the system is oxidized ($\boldsymbol{Q_t}$ = **1**): (a) no inputs applied ($\boldsymbol{Q_t}$ = **1**); (b) the system treated with the reset signal **1** (glucose 6-phosphate) converting the system to **0** state ($\boldsymbol{Q_{t+1}}$ = **0**); and (c) the system treated with the set signal **1** (lactate) returning the system to **1** state ($\boldsymbol{Q_{t+2}}$ = **1**). Source: Pita et al. 2009 [17]. Adapted with permission of American Chemical Society.

Another example [17] of the flip-flop performance was achieved when the NADH signal was produced *in situ* by glucose-6-phosphate dehydrogenase (G6PDH) in the presence of D-glucose-6-phosphate (Glc6P) as the extension of the reset input. In this example the H_2O_2 signal was generated *in situ* by lactate oxidase (LOx) in the presence of lactate as the extension of the set input. In this system, the core part of the flip-flop was the same as shown in Figure 8.1B, but the extensions used for producing the ***S***/***R*** signals were represented by different enzyme reactions. The optical responses of the system were similar to those in the previous example. Figure 8.3 shows the optical changes generated by this flip-flop system upon applying set-reset signals in various sequences. This experiment demonstrated the possibility to apply a large variety of biochemical signals for the activation of the enzyme-based memory unit. Eventually, the set/reset signals might be easily extended to long sequences of biocatalytic cascades starting from different substrates and finishing with NADH and H_2O_2.

The performance of the set-reset flip-flop system was also followed by electrochemical means by measuring the current corresponding to the redox state of the system. In this system $[Fe(CN)_6]^{3-}/[Fe(CN)_6]^{4-}$ redox pair was used as the redox mediator reporting on the system state. The reduced state of the redox species $[Fe(CN)_6]^{4-}$ was defined as **0** state, while the oxidized state $[Fe(CN)_6]^{3-}$ corresponded to **1** state of the system (this definition was also arbitrary). A gold electrode was modified with a self-assembled monolayer of 6-mercapto-1-hexanol to prevent the enzyme adsorption and to allow for a well-defined response from the redox species. The cyclic voltammogram of $[Fe(CN)_6]^{3-}/[Fe(CN)_6]^{4-}$ redox

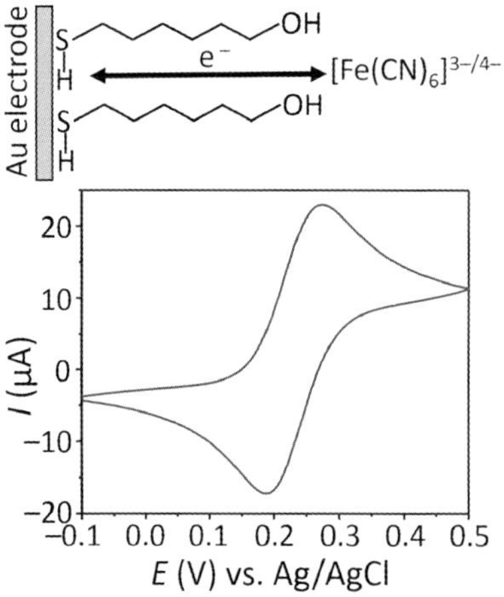

Figure 8.4 The cyclic voltammogram of $K_4[Fe(CN)_6]$ obtained in the presence of HRP, diaphorase at a Au electrode modified with a self-assembled monolayer of 6-mercapto-1-hexanol. The background solution was 0.1 mM phosphate buffer, pH = 7.2. The potential scan rate was 100 mV s^{-1}. The scheme above the plot shows the electrochemical process used to analyze the redox mediator oxidation state. Source: Pita et al. 2009 [17]. Adapted with permission of American Chemical Society.

pair measured with the modified electrode demonstrated the electrochemically reversible process with $E^0 = 230$ mV (vs. Ag/AgCl) (Figure 8.4). Selecting the constant potential at 0.26 V, corresponding to the anodic peak current, resulted in the large current response when the redox species existed in the reduced state (**0** state of the system). The system conversion to the oxidized state (**1**) resulted in the decrease of the current response. In this experiment, the core part of the flip-flop system consisting of HRP and Diaph was activated by H_2O_2 and NADH inputs (note that in this realization the SR flip-flop system was represented by the core part without biocatalytic extensions explained above). The experiment was started with the system state $\boldsymbol{Q_t} = \mathbf{0}$, when the anodic current reflected the presence of the reduced species $K_4[Fe(CN)_6]$ (Figure 8.5A). The addition of H_2O_2 ($\boldsymbol{S} = \mathbf{1}$) (observed on the curve as a current spike) resulted in the system transition to the oxidized state, $[Fe(CN)_6]^{3-}$, biocatalyzed by HRP. The resulting state of the system ($\boldsymbol{Q_{t+1}} = \mathbf{1}$) was observed as the cathodic current. Further addition of H_2O_2 ($\boldsymbol{S} = \mathbf{1}$) did not result in the state change, after a short current spike associated with the injection. The sequence of two set signals ($\boldsymbol{S} = \mathbf{1}$) applied to the system resulted in the transition from $\boldsymbol{Q_t} = \mathbf{0}$ to $\boldsymbol{Q_{t+1}} = \mathbf{1}$ upon the first set signal, then preserving $\boldsymbol{Q_{t+2}} = \mathbf{1}$ state after the second set signal (Figure 8.5A, inset). Starting again with the reduced species $[Fe(CN)_6]^{4-}$ ($\boldsymbol{Q_t} = \mathbf{0}$), the set input H_2O_2 ($\boldsymbol{S} = \mathbf{1}$) was applied, resulting in the system transition to $\boldsymbol{Q_{t+1}} = \mathbf{1}$ state characterized by the cathodic current (Figure 8.5B). Then the addition of NADH reset signal ($\boldsymbol{R} = \mathbf{1}$) returned the system to the initial state ($\boldsymbol{Q_{t+2}} = \mathbf{0}$) with the anodic current. The present sequence of set/reset signals is shown in Figure 8.5B, inset. Finally, the system at the state $\boldsymbol{Q_t} = \mathbf{0}$ was reacted with the NADH reset signal ($\boldsymbol{R} = \mathbf{1}$), resulting in no changes in the system ($\boldsymbol{Q_{t+1}} = \mathbf{0}$) (Figure 8.5C). The addition of H_2O_2 ($\boldsymbol{S} = \mathbf{1}$) resulted in the system transition to $\boldsymbol{Q_{t+2}} = \mathbf{1}$ state. The present sequence of set/reset signals is shown in Figure 8.5C, inset. It should be noted that the presence of H_2O_2 and NADH did not produce any effect on the measured current at the applied potential. The examples of the set/reset signals applied to the system demonstrated the possibility to read out the states of the SR flip-flop memory by electrochemical means. Further miniaturization of the

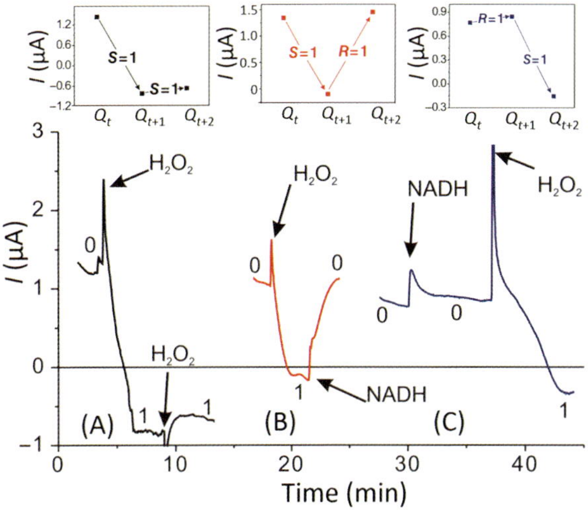

Figure 8.5 Chronoamperometric readout of the state changes in the enzyme-based SR flip-flop system starting from $Q_t = 0$. (A) Sequence of **S** = 1 and **S** = 1 signals. (B) Sequence of **S** = 1 and **R** = 1 signals. (C) Sequence of **R** = 1 and **S** = 1 signals. The system was composed of HRP, Diaph, and $K_4[Fe(CN)_6]$ in 0.1 mM phosphate buffer, pH = 7.2. The applied inputs were H_2O_2 for **S** = 1 and NADH for **R** = 1. The potential applied to the electrode was 0.26 V; note that this potential corresponds to the anodic peak in the cyclic voltammogram of the redox mediator; see Figure 8.4. Source: Pita et al. 2009 [17]. Adapted with permission of American Chemical Society.

electronic readout system will be possible, if microelectrodes [24] or field-effect transistors [25] are applied.

To demonstrate the applicability of the enzyme-based SR flip-flop memory for recording and storing biochemical signals, a multi-well microtiter plate was used as a reacting support, applying different combinations of the set/reset signals to each well [17]. For illustrative purposes, the words "**Clarkson**" and "**University**" with ASCII (American Standard Code for Information Interchange) were recorded and saved using the flip-flop memory system (Figure 8.6A,B), respectively. Each alphabetic character was represented by a column of eight wells, each containing the following biochemical "machinery": HRP, diaphorase, and $K_4[Fe(CN)_6]$ (obviously O_2 was included, while the solution was in equilibrium with air). The initial absorbance values at λ = 415 nm were below 0.2 units of optical density corresponding to the reduced state of the $[Fe(CN)_6]^{4-}$ redox species. Each digital output value, **0** or **1**, resulted from the addition of NADH or H_2O_2 to each well. When H_2O_2, considered as the set signal, was added to the well, ferrocyanide, $[Fe(CN)_6]^{4-}$, was biocatalytically oxidized to ferricyanide, $[Fe(CN)_6]^{3-}$, resulting in the increased absorbance at λ = 415 nm and yielding state **1**. The addition of NADH, considered as the reset signal, to the well resulted in a low value of optical density (state **0**) due to the biocatalytic reduction of $[Fe(CN)_6]^{3-}$. The biocatalytic reactions proceeding in 80 wells resulted in digital **0** or **1** values with eight values representing one of the eight letters in the word **Clarkson** using the ASCII encoding system (Figure 8.6A). The second set of

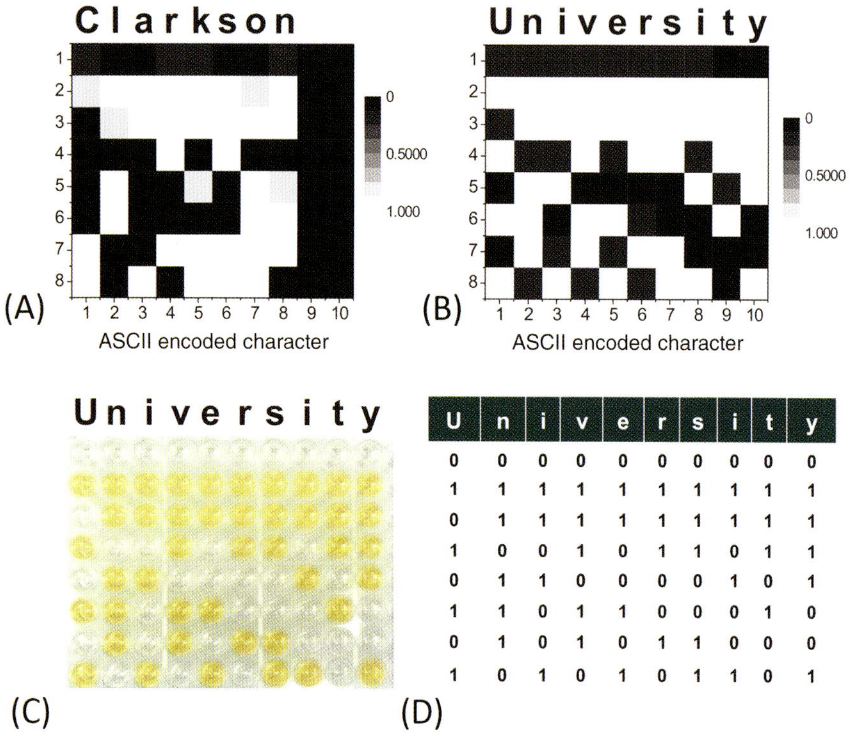

Figure 8.6 Words encoded using ASCII character codes upon application of different set/reset signals (H_2O_2–NADH, respectively) to the multi-well SR flip-flop system composed of HRP and Diaph as the biocatalysts and $K_4[Fe(CN)_6]$ as the redox species. Optical readout of the words **Clarkson** (A) and **University** (B) at $\lambda = 415$ nm. Color photo of the multi-well reactor with the encoded word **University** (C) and the respective ASCII codes for the used characters (D). Source: Pita et al. 2009 [17]. Adapted with permission of American Chemical Society.

inputs, H_2O_2 and NADH, applied to the same wells changed the output values to encode the word **University** (Figure 8.6B). Figure 8.6C shows the color photo of the plate with the encoded word **University**, and Figure 8.6D represents the corresponding table of the encoded states.

The enzyme-based flip-flop system was stable as long as the enzymes preserved their activities (hours–days depending on the applied temperature), allowing for multiple set-reset cycles. It should be noted that the system composed of soluble enzymes was diluted upon each signal application. This problem can be easily solved by application of immobilized enzymes. The signals recorded by the system were stable for several days (particularly in the case of $[Fe(CN)_6]^{4-}/[Fe(CN)_6]^{3-}$ redox species revealing better stability than DCPIP).

8.2.2 Delay (D) Flip-Flop Memory Based on Enzyme-Catalyzed Reactions

In electronics the D flip-flop is realized as a circuitry composed of four interconnected NAND gates with the Clock and Data inputs [26] (Figure 8.7). Taking

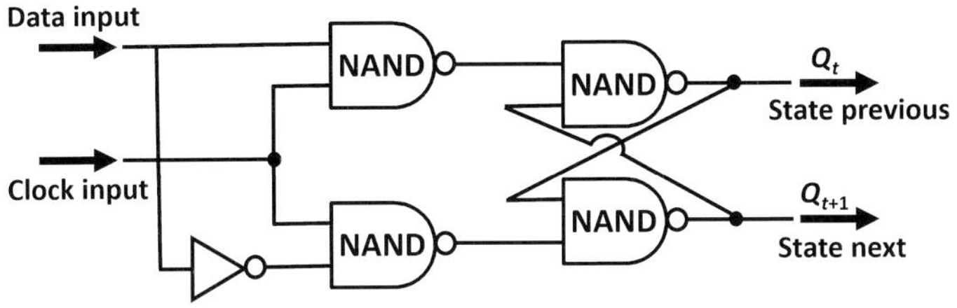

Figure 8.7 Schematics of the D flip-flop used in electronics. Note that this design was not used in the enzyme-based systems because of its complexity.

into account that realization of NAND gates by chemical means (e.g., using enzymatic reactions [27]) is not simple, direct copying of this circuitry with molecular systems implying NAND logic gates with the chemical pathways between them would be extremely difficult. This was the reason why very few chemical systems were designed for mimicking D flip-flop memory [18, 19]. However, the chemical systems do not have to follow exactly the same hardware design, which is used in electronics, and can be based on alternative concepts to result in the same logic operation [21]. To realize the different approach to the D flip-flop, one needs to consider the logic truth table characteristic of its operation (Figure 8.8A). When the Clock input has logic value **0**, the chemical system should not respond to any value applied for the Data input, thus preserving the $\boldsymbol{Q_t}$ state regardless of the Data input value (reactions A–D in Figure 8.8A). In other words, the chemical reactions should be inhibited by the Clock input **0**. On the other hand, when the Clock input has value **1**, the system should be active, and $\boldsymbol{Q_{t+1}}$ should be set according to the value of the Data input. In this case the logic value of $\boldsymbol{Q_{t+1}}$ should be equal to the value of the Data input regardless of the value of $\boldsymbol{Q_t}$ (reactions E–H in Figure 8.8A). As soon as we understand that the Clock input **0** means in chemical term *inhibition*, the setup of the enzyme-based D flip-flop becomes very simple, and its operation requires only the correct definitions of the Clock and Data input **0**/**1** values (Figure 8.9). The chemical "machinery" of the D flip-flop was a mixture of two enzymes, lactate dehydrogenase (LDH) and alcohol dehydrogenase (ADH), with the β-nicotinamide adenine dinucleotide cofactor being in one of two possible states: NAD^+, corresponding to $\boldsymbol{Q} = \mathbf{0}$, or NADH, corresponding to $\boldsymbol{Q} = \mathbf{1}$. The "machinery" system was dissolved in 0.1 mM phosphate buffer, pH 7.2. The state of the system was read out by optical means measuring the NADH absorbance at $\lambda = 340$ nm and defining the absorbance value above 0.3 as the logic **1** and below 0.3 as the logic **0** states. The Clock input was formulated as the absence (logic value **1**) or presence (logic value **0**) of the enzyme inhibitors: oxaloacetate, glutamate, and butyramide. Note that oxaloacetic acid and glutamate are known reversible inhibitors of LDH, while butyramide reversibly inhibits ADH [28]. The concentrations of the inhibitors were optimized experimentally to allow the complete inhibition of both enzymes in their presence while recovering the enzyme activity upon removing the inhibitors from the system. The Data input was defined as the presence of pyruvate (Pyr) and absence of ethanol (EtOH), logic value **0**, and as the presence of ethanol and absence of pyruvate, logic

(A)

Chemical reaction	Clock input	Data input	Q_t	Q_{t+1}
A	0	0	0 (0.07)	0 (0.09)
B	0	0	1 (0.59)	1 (0.57)
C	0	1	0 (0.05)	0 (0.05)
D	0	1	1 (0.57)	1 (0.51)
E	1	0	0 (0.01)	0 (0.05)
F	1	0	1 (0.56)	0 (0.05)
G	1	1	0 (0.01)	1 (0.54)
H	1	1	1 (0.57)	1 (0.56)

(B)

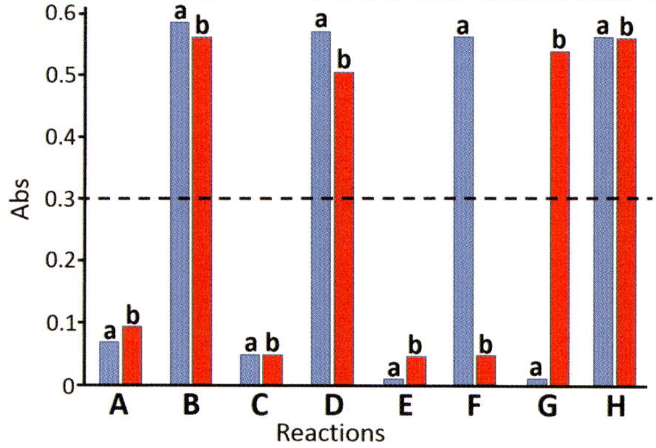

Figure 8.8 (A) Truth table of the D flip-flop based on the enzyme reactions. For the detailed definition of the Clock/Data inputs and optical output signals, see the text. Q_t and Q_{t+1} describe the Q output before and after the application of a given Clock/Data input combination, respectively. The values in parentheses in the last two columns are the average experimental optical absorbance values ($\lambda = 340$ nm) extracted from the spectral measurements. (B) Optical responses of the enzyme D flip-flop system measured at $\lambda = 340$ nm upon application of various combinations of the Clock/Data inputs corresponding to the truth table (A). Source: MacVittie et al. 2012 [21]. Adapted with permission of Royal Society of Chemistry.

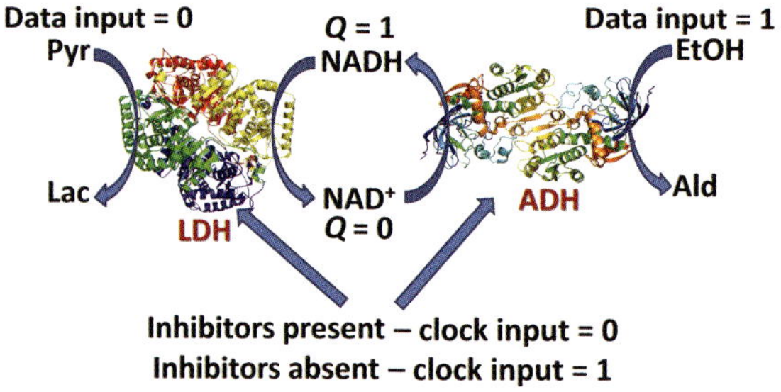

Figure 8.9 Schematics of the D flip-flop realized with the enzyme system. Source: MacVittie et al. 2012 [21]. Adapted with permission of Royal Society of Chemistry.

value **1**. The concentrations of Pyr and EtOH were optimized experimentally to allow the complete transitions between $\boldsymbol{Q} = \mathbf{1}$ (NADH present) and $\boldsymbol{Q} = \mathbf{0}$ (NAD^+ present) states in the same time interval (to have similar kinetics of the reactions biocatalyzed by LDH and ADH). The system operation can be demonstrated by reactions A–H (Figure 8.8A). When the inhibitors are added to the system (Clock input **0**), the enzymes, LDH and ADH, are not active, and the addition of the enzyme substrates Pyr (Data input **0**) or EtOH (Data input **1**) does not result in any chemical reaction and preserves the system in its current state regardless the presence of NADH ($\boldsymbol{Q_t} = \mathbf{1}$ and $\boldsymbol{Q_{t+1}} = \mathbf{1}$) or NAD^+ ($\boldsymbol{Q_t} = \mathbf{0}$ and $\boldsymbol{Q_{t+1}} = \mathbf{0}$) (reactions A–D, Figure 8.8A). Opposite to that, when the inhibitors are not added (Clock input **1**), both enzymes are active and able to respond to the corresponding substrates. In this case the addition of Pyr (Data input **0**) results in the oxidation of NADH to NAD^+ biocatalyzed by LDH, thus moving the system from $\boldsymbol{Q_t} = \mathbf{1}$ to $\boldsymbol{Q_{t+1}} = \mathbf{0}$ (reaction F, Figure 8.8A). If the system is already in $\boldsymbol{Q_t} = \mathbf{0}$, the biocatalytic reaction does not proceed because the enzyme does not have the cofactor in the reduced state and the system state is preserved, $\boldsymbol{Q_{t+1}} = \mathbf{0}$ (reaction E, Figure 8.8A). The addition of EtOH (Data input **1**) results in the reduction of NAD^+ to NADH biocatalyzed by ADH, thus moving the system from $\boldsymbol{Q_t} = \mathbf{0}$ to $\boldsymbol{Q_{t+1}} = \mathbf{1}$ (reaction G, Figure 8.8A). If the system is already in $\boldsymbol{Q_t} = \mathbf{1}$, the biocatalytic reaction does not proceed because the enzyme does not have the cofactor in the oxidized state and the system state is preserved, $\boldsymbol{Q_{t+1}} = \mathbf{1}$ (reaction H, Figure 8.8A). The experimental data demonstrating the system responses to each combination of the Clock/Data inputs (corresponding to reactions A–H, Figure 8.8A) are shown in Figure 8.8B. The data shown in Figure 8.8B was collected in the systems prepared separately, each time dissolving the new samples of the enzymes and NAD^+/NADH and applying different combinations of the Clock/Data inputs. Random sequence of the Clock/Data inputs applied to the same enzyme sample and resulting in the various transitions from the $\boldsymbol{Q_t}$ and $\boldsymbol{Q_{t+1}}$ is shown in Figure 8.10. In summary, the present system offers a new approach to the realization of the D flip-flop system, which is easy for integration with sophisticated biocatalytic cascades operating as the logic networks performing various computing steps.

8.2.3 Toggle (T) Flip-Flop Memory Based on Enzyme-Catalyzed Reactions

A T flip-flop is a single input device that does not change the system state ($\boldsymbol{Q_{t+1}} = \boldsymbol{Q_t}$) when **T-input** = **0**. On the other hand, **T-input** = **1** should change the system state to the opposite value (changing $\boldsymbol{Q_t} = \mathbf{0}$ to $\boldsymbol{Q_{t+1}} = \mathbf{1}$ and $\boldsymbol{Q_t} = \mathbf{1}$ to

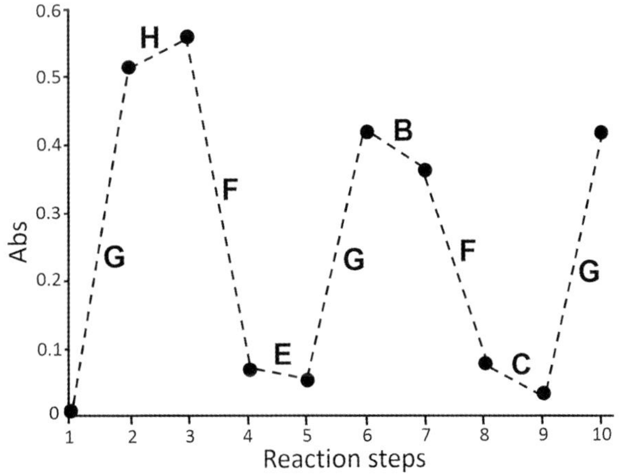

Figure 8.10 Optical responses of the enzyme D flip-flop system measured at $\lambda = 340$ nm upon application of random sequence of the Clock/Data inputs corresponding to the reactions in Figure 8.8. Source: MacVittie et al. 2012 [21]. Adapted with permission of Royal Society of Chemistry.

$\boldsymbol{Q_{t+1}} = \mathbf{0}$) (Figure 8.11A). Keeping the system state unchanged is obviously easy in chemical systems if the input signal **0** is defined as the absence of reacting species. However, it is very difficult to use the application of the same reactant (input signal **1**) to change the state depending on the initial state of the system. The first approach to the chemical model of the T flip-flop [22] is illustrated with Figure 8.11B. In this system horseradish peroxidase (HRP) was activated by H_2O_2 oxidizing 3,3′,5,5′-tetramethylbenzidine (TMB_{red}) to yield the first oxidized product (single oxidized; TMB_{sox}) with the absorbance peak at $\lambda_{max} = 652$ nm. Then the presence of the H_2O_2 excess produced the double-oxidized product (TMB_{dox}) with the absorbance peak at $\lambda_{max} = 459$ nm [29]. The system state $\boldsymbol{Q} = \mathbf{0}$ was defined as low optical absorbance (O.D. < 0.05) at 652 nm, while system state $\boldsymbol{Q} = \mathbf{1}$ was defined as high absorbance (O.D. > 0.25) at 652 nm. It should be noted that the low absorbance value ($\boldsymbol{Q} = \mathbf{0}$) can be achieved in this system in the presence of either TMB_{red} or TMB_{dox}, while the high absorbance value ($\boldsymbol{Q} = \mathbf{1}$) corresponds specifically to the presence of TMB_{sox}. The **T-input 0** was defined as the absence of H_2O_2, thus resulting in no reaction and no changes in the system state. If the system state was defined by the presence of TMB_{red}, it was not oxidized, thus preserving the $\boldsymbol{Q_t} = \mathbf{0}$ state. If the system state was defined by the presence of TMB_{sox}, it was not oxidized to the next oxidation state, thus preserving the $\boldsymbol{Q_t} = \mathbf{1}$ state. On the other hand, the application of **T-input 1** defined as the presence of H_2O_2 resulted in the system switch from $\boldsymbol{Q_t} = \mathbf{0}$ state to $\boldsymbol{Q_{t+1}} = \mathbf{1}$ state by oxidizing TMB_{red} to TMB_{sox} or from $\boldsymbol{Q_{t+1}} = \mathbf{1}$ state to $\boldsymbol{Q_{t+2}} = \mathbf{0}$ state by oxidizing TMB_{sox} to TMB_{dox}. The H_2O_2 concentration corresponding to the **T-input 1** was optimized experimentally to allow for the complete oxidation. Figure 8.11C shows the absorbance spectra changes measured experimentally upon application of the **T-input** with **0**/**1** values to the system with the $\boldsymbol{Q} = \mathbf{0}/\mathbf{1}$. In summary, the present system offers a new approach to the realization of the T flip-flop system, which is easy for integration with sophisticated biocatalytic cascades operating as the logic networks performing various computing steps.

8.2.4 Enzyme-Based Flip-Flop Memory Systems: Conclusions and Perspectives

With the multitude of fundamental advancements in the chemical and biochemical realizations of memory systems, the challenges approaching are becoming based more in the macro-development of these intelligent systems: focusing on their concatenation and integration into existing and emerging paradigms. By building upon fundamental biocomputing concepts, such as the manipulation of enzyme pathways, the SR, D, and T flip-flops presented here can be easily integrated with existing biocomputing ensembles. The coupling of memory systems with molecular "devices" capable of processing biochemical inputs would lead to new biomedical applications, particularly in the frame of the novel theranostic concept. Taking these "sense-and-treat" conceptualizations one step further results in an interesting approach: the integration of implanted bioelectronic devices [30], including implantable biofuel cells [31], with chemical and biochemical logic systems equipped with built-in memory capabilities. The development

(A)

Chemical reaction	T-input	Q_t	Q_{t+1}
a	0	0 (0.004)	0 (0.004)
b	0	1 (0.306)	1 (0.306)
c	1	1 (0.296)	0 (0.009)
d	1	0 (0.005)	1 (0.290)

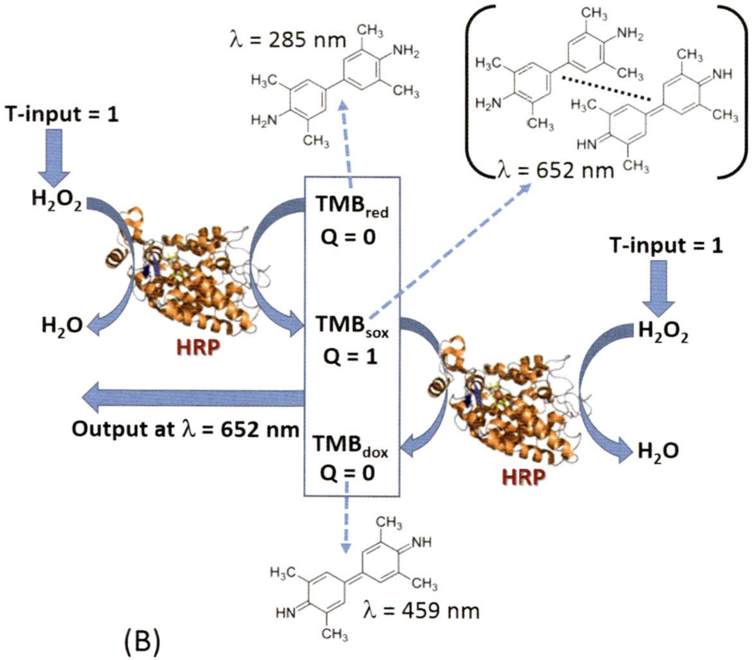

(B)

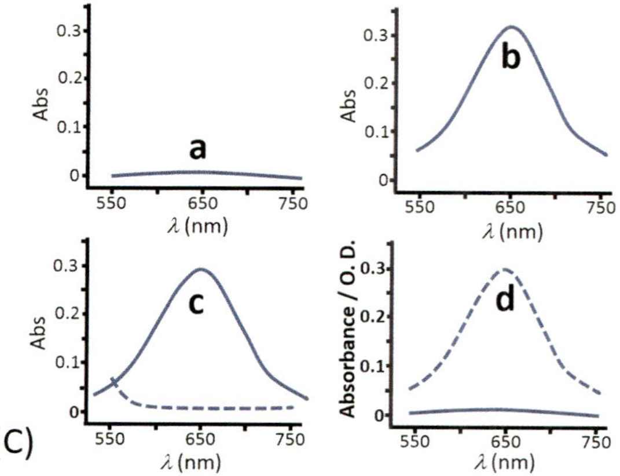

(C)

Figure 8.11 (A) The truth table corresponding to the T flip-flop operation. The numbers shown in the brackets correspond to the experimentally measured optical absorbance at $\lambda = 652$ nm (see also the spectra in C). (B) The reactions biocatalyzed by HRP used for mimicking the T flip-flop operation. Note that the TMB_{sox} is a charge transfer complex of the oxidized and reduced TMB species. (C) The optical spectra measured after performing the biocatalytic reactions ("a" through "d" steps are specified in the table (A)). The solid line signifies the initial state ($\boldsymbol{Q_t}$) of the system, with the dashed line signifying the state of the system after the corresponding reactions ($\boldsymbol{Q_{t+1}}$). Note that in "a" and "b" both lines (solid and dashed) coincided (meaning no changes in the system state).

of self-powered bioelectronic devices operating with complete autonomy would thoroughly revolutionize the field of medicine. These autonomous physiological diagnostic devices, or more aptly put "physiodiagnostics," could be capable of real-time monitoring of biomarkers, performing complex biocomputations, and responding accordingly to stresses, variations, or pathogens in the body while simultaneously recording all necessary information using built-in enzyme-based flip-flop memory systems. While each individual component would be essential, the ability to have constant and reliable long-term storage of information would be imperative. A sufficiently advanced device would have the ability to record a complete collection of system states, saving information on both the biological activity of the host and subsequent actions performed by the device (i.e., administration of medicine). This concept of constant monitoring at the molecular level would likely result in early-stage diagnosis of diseases; however, this has little benefit if the system has no way to store the information processed by the implantable bioelectronic device.

With the continued development of increasingly complex chemical and biochemical systems mimicking the logic and operation of modern silicon computers, additional components must also be developed to maximize their potential applications. The addition of sophisticated memory motifs to existing biocomputing systems has the potential to revolutionize a number of fields, the most prominent being medicine. The prospect for real-time physiological monitoring with reliable, long-term memory storage via sophisticated "physiodiagnostic" devices could result in a future where diseases are diagnosed and treated before they even present detectable symptoms.

8.3 Memristor Based on Enzyme Biocatalytic Reactions

8.3.1 Memristors: From Semiconductor Devices to Soft Matter and Biomolecular Materials

While resistors, capacitors, and inductors are long-time known electrical devices relating current–voltage, voltage–charge, and flux–current, a memristor is a novel electrical element relating charge and flux (Figure 8.12). A **memristor** (**mem***ory-capable* ***res*****istors**; a portmanteau of memory resistor) was envisioned, and its name coined in 1971 by circuit theorist Leon Chua [32]. Memristors retain memory states even when no current is being passed through them, and they can be used not just as a memory device but also to perform computations. That means computation could be performed on chips where data is stored, rather than on a specialized processing unit, thus accelerating the information processing. A drastic leap in the design of chemical information processing/memory systems could be achieved if the concept of memory-capable electronics – developed in physics [32–34] and mostly realized in semiconductor devices [35–37] – is brought to soft matter materials and potentially to biomaterials. Presently, only a few examples of polymer-based memristors have been realized; however, being developed as solid-state electronic devices, they are hardly compatible with biomolecular systems [38].

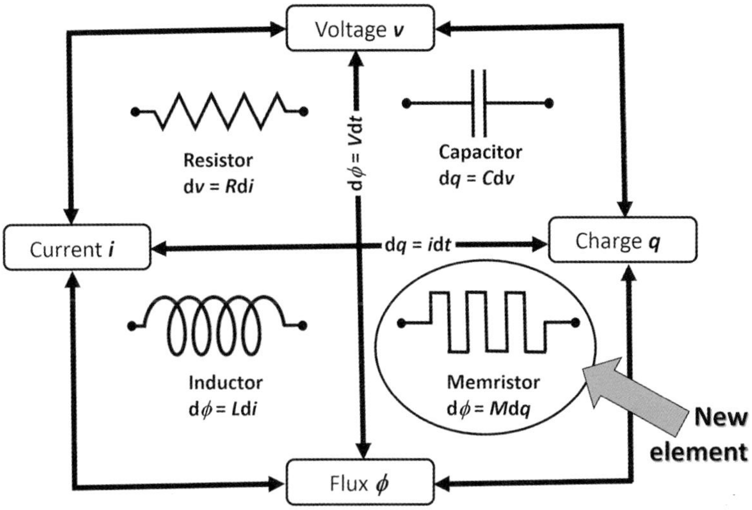

Figure 8.12 Illustration of the memristor in electrical network theory. The memristor and its relation with the four fundamental electric variables. Variables: v, voltage; i, current; q, charge; and ϕ, flux.

On the other hand, switchable electrode interfaces controlled by biomolecular systems have been recently realized [39, 40], keeping the promise for the development of memristive devices that include biomolecular components. It should be also noted that two kinds of memory-capable electronic systems, such as memristors and memcapacitors [41], can be easily and naturally realized in one device using switchable electrode interfaces, since such interfaces change the electron transfer resistance and interfacial capacitance upon transition between different states [42]. Recently a new approach to the electrochemical realization of memristor/memcapacitor (memimpedance) devices has been reported [43] based on pH-switchable electrode interfaces. Integration of the switchable electrodes in pairs to design a signal-switchable biofuel cell is easy [44–47]. Then, interfacing the biofuel cell demonstrating memristive features with enzyme catalytic systems is straightforward, thus allowing enzyme logic control of the electrochemical memristor.

8.3.2 The Memristor Device Based on a Biofuel Cell

Biofuel cells, traditionally considered for use as power sources [48–50], particularly as implantable devices operating *in vivo* [31, 51, 52], have been utilized extensively for nontraditional information processing applications [47], including self-powered biosensors [53–58], biocomputing devices [59, 60], keypad lock systems [61, 62], generation of electrochemically readable code [63], signal-controlled "sense–act–treat" drug-release systems [64], etc. The system discussed below reports on the first application of a biofuel cell for controlling an electrochemical memristor. The developed approach can be extended to systems where multi-input enzyme logic networks control the memristor device, thus allowing for intelligent interfacing between bioelectronic and biocomputing systems.

The simplest definition of the memristor's properties states that the resistance of the device depends on the direction of the flow of electrical current passing through the device, while in the absence of the current, the device remembers its last resistance [32, 33]. In order to realize these features in an electrochemical system, a pH-switchable polymer-brush-modified electrode, which demonstrated a low and high interfacial resistance in acidic and neutral solutions, respectively, was used [65]. An indium tin oxide (ITO) electrode was modified with poly(4-vinyl pyridine) (P4VP) brush, which is known for changing its ionic permeability upon reversible protonation–deprotonation of pyridine moieties [66] (Figure 8.13A). When the pH value is below 4.5, the pyridine moieties in the P4VP brush are protonated, and the positively charged polymer is swollen, hydrophilic, and permeable for the negatively charged redox probe, $[Fe(CN)_6]^{3-}$, resulting in a reversible electrochemical process in cyclic voltammetry measurements (Figure 8.13B, curve a). In the case of pH values above 5.5, the pyridine units are deprotonated, and the neutral polymer is collapsed to the hydrophobic state, which is not permeable for the anionic redox probe, thus inhibiting the electron transfer process (Figure 8.13B, curve b) (note that the extent of the polymer brush nonpermeability is increasing with the pH increase from 3 to 6, thus resulting in the current decrease) (Figure 8.13B, inset). This reversible restructuring of the thin film polymer brush on the electrode surface

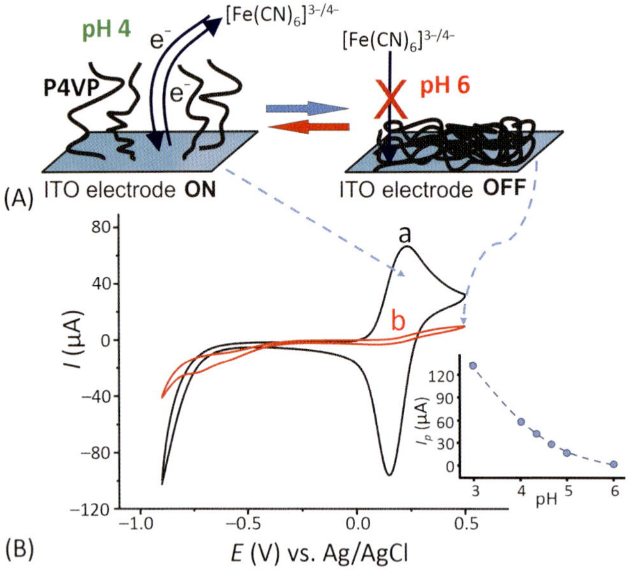

Figure 8.13 (A) pH-controlled restructuring of the P4VP brush at the ITO electrode surface resulting in the active electrode interface for the electrochemical process of $[Fe(CN)_6]^{3-/4-}$ redox probe at pH 4 and the inhibited electrochemical process at pH 6. (B) The cyclic voltammograms of the $[Fe(CN)_6]^{3-/4-}$ redox probe obtained at the P4VP-modified ITO electrode at pH = 4 (a) and pH = 6 (b). Inset: Peak current dependence on the pH value of the electrolyte solution.

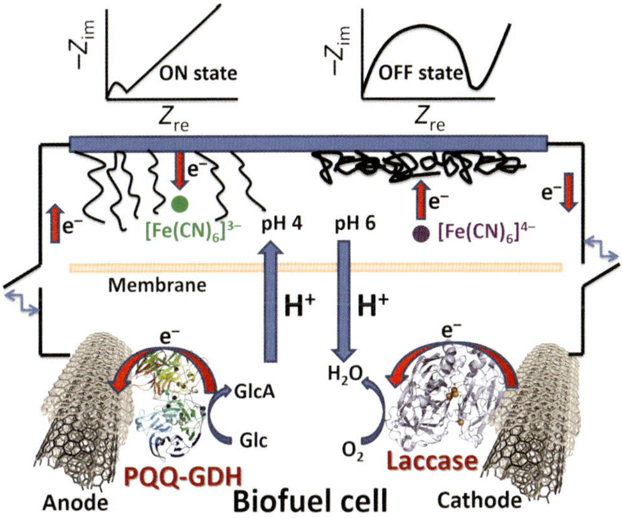

Figure 8.14 Schematics of the bioelectrochemical memristor composition and operation. The impedance spectra on the top show schematically low and high electron transfer resistances characteristic of the "open" and "closed" states of the pH-switchable electrode. Source: MacVittie and Katz 2014 [65]. Adapted with permission of Royal Society of Chemistry.

resulted in controllable changes of the interfacial resistance. Importantly, the current passing through the memristor device and the concomitant pH changes switching the device between the low and high resistance were produced by chemical signals (glucose and O_2) powering the biofuel cell.

The device was constructed [65] as a thin-layer electrochemical cell with three electrodes: one was a pH-switchable P4VP-brush-modified ITO electrode, while two other electrodes comprising a biofuel cell were located on the opposite side of the cell being separated from the switchable electrode with a Nafion® proton exchange membrane (Figure 8.14). The solutions between the electrodes and membrane were confined by two rubber O-rings. Note the small volume (ca. 250 μl) of the electrolyte solution in the cell, as this was necessary to obtain the bulk pH changes needed for controlling the resistance of the pH-switchable electrode upon operation of the biocatalytic electrodes (Figure 8.15). Both biocatalytic electrodes used in the biofuel cell were based on a buckypaper conducting support (composed of compressed multiwalled carbon nanotubes [MWCNTs]) (Figure 8.16A). The glucose-oxidizing anode was modified with pyrroloquinoline quinone (PQQ)-dependent glucose dehydrogenase (PQQ-GDH), and the oxygen-reducing cathode was modified with laccase (Figure 8.14). The enzymes were attached to the MWCNTs with the help of the heterobifunctional reagent 1-pyrenebutanoic acid succinimidyl ester (PBSE), which included an aromatic entity for the π–π stacking on the MWCNT walls and an active ester group for formation of an amide bond with amino groups of the enzymes [52] (Figure 8.16B,C). PQQ-GDH and laccase enzymes are known for direct electron transfer with an electrode support composed of MWCNTs [52, 67]. The biocatalytic electrodes were activated in the presence of glucose

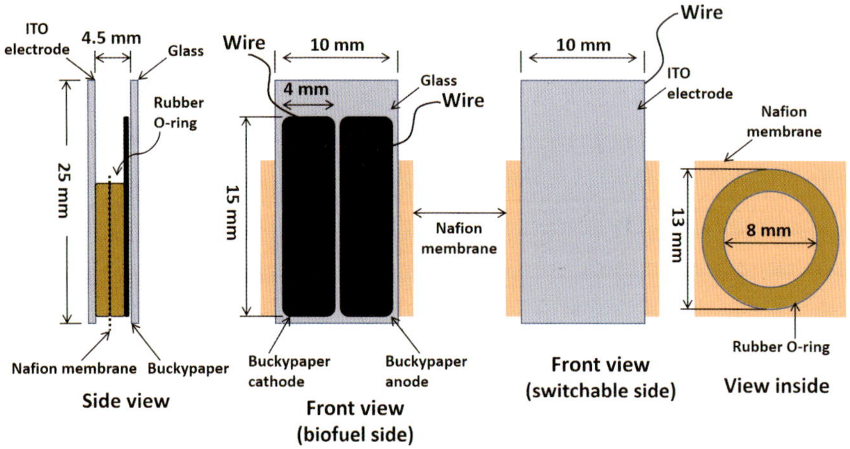

Figure 8.15 Schematics of the electrochemical cell. Note that the ITO electrode was modified with P4VP polymer brush and the buckypaper electrodes were modified with PQQ-GDH and laccase (on the biofuel cell anode and cathode, respectively). Source: MacVittie and Katz 2014 [65]. Adapted with permission of Royal Society of Chemistry.

and O_2 in the cell compartment facing the biofuel side. The opposite side of the cell facing the pH-switchable electrode (separated from the biofuel cell with the membrane) contained $[Fe(CN)_6]^{3-}/[Fe(CN)_6]^{4-}$ (1 : 1 mixture) redox probe. Figure 8.14 outlines the biofuel cell/memristor composition and operation. When the biocatalytic anode oxidizing glucose was electrically connected to the P4VP-modified electrode, the current was passing from the anode to the P4VP electrode, and glucose oxidation resulted in the formation of gluconic acid, thus decreasing the pH value and yielding the "open" state of the pH-switchable electrode with the low interfacial resistance. When the biocatalytic cathode reducing O_2 was electrically connected to the P4VP-modified electrode, the current was passing to the cathode from the P4VP electrode, and oxygen reduction resulted in the formation of water (note the proton consumption in the process), thus increasing the pH value and yielding the "closed" state of the pH-switchable electrode with the high interfacial resistance. In both processes, the electrochemical transformations at the P4VP electrode were the reduction/oxidation (depending on the current direction) of the included $[Fe(CN)_6]^{3-}/[Fe(CN)_6]^{4-}$ redox probe.

In order to illustrate the memristor's features, the following set of experiments was performed. The experiments were started from the electrolyte solution with pH 4.0 when the P4VP electrode is "open." The P4VP electrode was connected to the laccase electrode, and the electrochemical impedance spectrum was obtained to derive the electron transfer resistance, R_{et}, ca. 0.4 kΩ (Figure 8.17A, inset). Note that the measurement was performed between two electrodes in the mode typically used to characterize biofuel cell performance; a reference electrode was not used in this measurement [68]. It should be also noted that the electron transfer resistance of the cell represents the electrochemical kinetics of the redox processes rather than the ohmic resistances of the electrodes. The

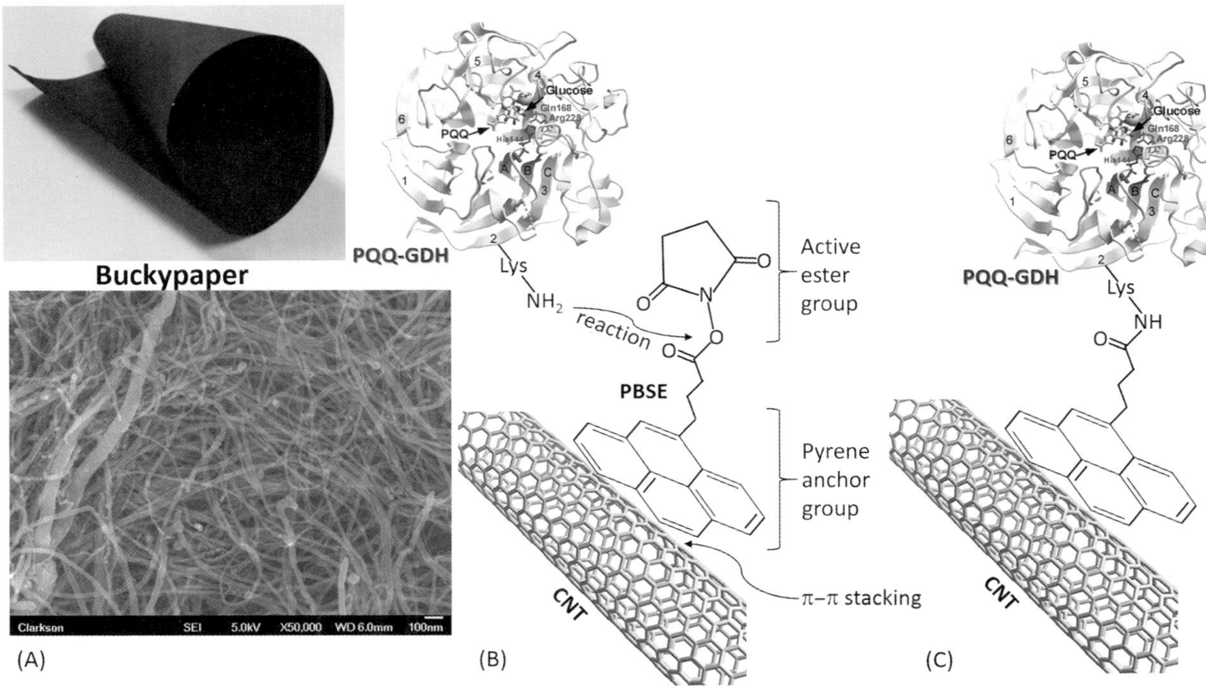

Figure 8.16 (A) Photo and SEM image of the buckypaper used as the conductive support for the enzyme-modified electrodes. (B) Immobilization of the PQQ-GDH on MWCNTs with the help of the heterobifunctional linker 1-pyrenebutanoic acid succinimidyl ester (PBSE), which provides covalent binding with amino groups of protein lysine residues through the formation of amide bonds and also interacts with MWCNTs via π–π stacking of the polyaromatic pyrenyl moiety. (C) Schematic structure of the enzyme-modified electrode. Note that laccase was attached to the buckypaper in the same way. Source: MacVittie and Katz 2014 [65]. Adapted with permission of Royal Society of Chemistry.

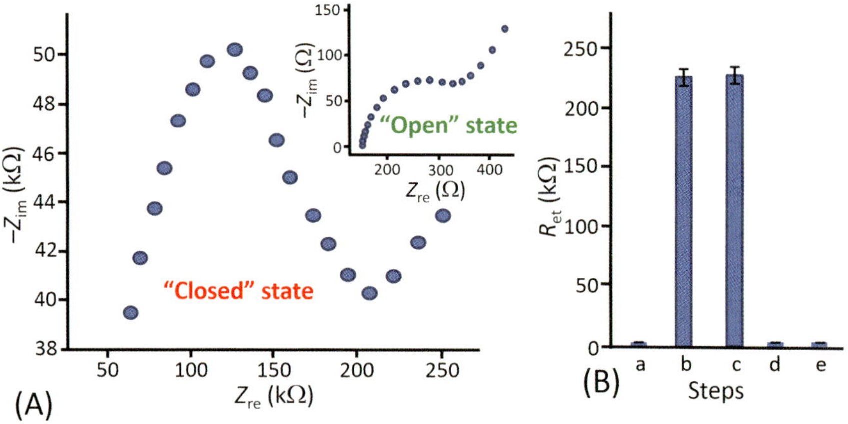

Figure 8.17 (A) The impedance spectrum obtained on the "closed" state of the P4VP-modified electrode connected to the biofuel cell cathode (laccase electrode), pH 6. Inset: The impedance spectrum obtained on the "open" state of the P4VP-modified electrode connected to the biofuel cell anode (PQQ-GDH electrode), pH 4 (note the scale difference in the main plot and inset). Bias potential of 350 mV was applied on the P4VP electrode (vs. the enzyme electrode), frequency range 100 mHz–10 kHz; note the absence of a reference electrode due to microscale cell setup. (B) Switchable R_{et} of the memristor device: (a) initial "open" state of the modified electrode, pH 4.0; (b) "closed" state of the modified electrode, pH ca. 6, produced by passing current from the P4VP electrode to the laccase electrode for 30 minutes; (c) repeat of the measurement after 30 minutes without passing any current; (d) "reopen" of the electrode, pH ca. 4, produced by passing current to the P4VP electrode from the PQQ-GDH electrode for 30 minutes; and (e) repeat of the measurement after 30 minutes without passing any current. The R_{et} values were extracted from the impedance spectra similar to those shown in part A. Source: MacVittie and Katz 2014 [65]. Adapted with permission of Royal Society of Chemistry.

low resistance value is shown as step "a" in Figure 8.17B. Continuous current resulted in $Fe(CN)_6]^{4-}$ oxidation at the P4VP electrode (note anodic current at this electrode) and O_2 reduction at the laccase electrode. The later process resulted in pH increase in the compartment connected to the laccase electrode (Figure 8.18A), which rapidly propagated to the P4VP electrode through the Nafion® proton exchange membrane, reaching pH ca. 6 in 30–35 minutes. This pH value resulted in a partially "closed" state of the P4VP electrode with much higher electron transfer resistance, R_{et}, value of ca. 225 kΩ measured by the impedance spectroscopy (Figure 8.17A). It should be noted that the pH value was optimized to still allow redox process to occur at the P4VP electrode, even though it proceeded with a high electron transfer resistance at the partially "closed" electrode interface. Thus, the anodic current at the P4VP electrode yielded the high resistance shown as step "b" in Figure 8.17B. Then the P4VP electrode was disconnected from the laccase electrode, and the cell was left without any current passing between the electrodes. In the absence of current, the system preserved both its chemical composition and the pH obtained in the previous step. Impedance spectroscopy was repeated after a 30 minutes time interval and resulted in the spectrum similar to the one shown in Figure 8.17A with R_{et} ca. 225 kΩ, thus demonstrating the system memory preserving the previous resistance value in the absence of current, shown as step

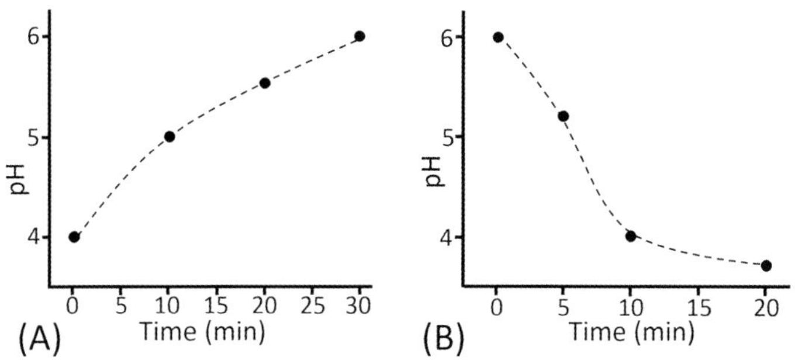

Figure 8.18 Changes of the solution pH generated *in situ* upon connecting the P4VP-ITO electrode with the biofuel cell cathode (laccase electrode) and with the biofuel cell anode (PQQ-GDH electrode), panels A and B, respectively. Source: MacVittie and Katz 2014 [65]. Adapted with permission of Royal Society of Chemistry.

"c" in Figure 8.17B. Next, the P4VP electrode was connected to the PQQ-GDH electrode, thus resulting in the opposite current direction, corresponding to $[Fe(CN)_6]^{3-}$ reduction and glucose oxidation. The latter process resulted in the pH decrease in the compartment of the PQQ-GDH electrode (note the formation of gluconic acid at the biocatalytic electrode) (Figure 8.18B), rapidly propagating to the P4VP electrode through the proton exchange membrane and reaching pH ca. 4 in approximately 20 minutes. This resulted in restoring the "open" state of the P4VP electrode with the R_{et} value of ca. 0.4 kΩ measured between the P4VP electrode and PQQ-GDH electrode. Thus, the opposite current direction (cathodic at the P4VP electrode) resulted in a low resistance value shown as step "d" in Figure 8.17B. Disconnecting electrodes and leaving the cell without current for 30 minutes resulted in the same resistance of ca. 0.4 kΩ remeasured by impedance spectroscopy and shown as step "e" in Figure 8.17B, thus demonstrating the cell memory at its low resistance value. The experimental steps (Figure 8.17B), steps "a–e," demonstrated the expected memristor's features, when the resistance value is dependent on the current direction and the cell remembers the previous resistance value in the absence of the current.

One of the features of the classical solid-state memristor devices is the hysteresis loop in the current–voltage function [69, 70] (Figure 8.19, inset). This feature was also nicely confirmed for the studied electrochemical device [43] (Figure 8.19). It should be noted that the I–V curve of the memristor device is recorded on the pH-switchable polymer-modified electrode and it is independent on the presence or absence of the biofuel cell in the system, thus representing features of the pH-switchable modified electrode. Also, the present electrochemical realization of the memristor device allows memory properties for the interfacial capacitance of the modified electrode, thus representing a memcapacitor operating in parallel with the memristor. In other words, the discussed electrochemical device demonstrates "memimpedance" features. Similar features have been reported previously for solid-state semiconductor devices [71], but the present electrochemical realization is the very first example of the memimpedance device based on (bio)organic materials.

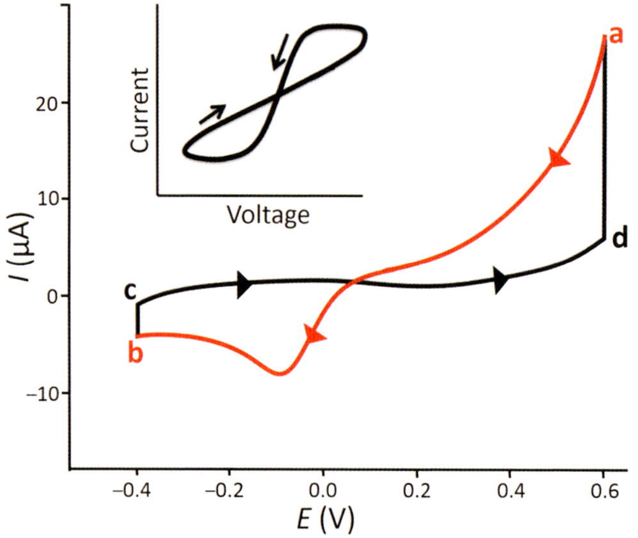

Figure 8.19 Cyclic voltammogram demonstrating the hysteresis loop in the current–voltage function: (a) "open" electrode at initial pH 4.0, (b) the potential of −1.0 V was applied for 20 minutes to allow switching from the "open" to "closed" state, (c) "closed" state of the electrode at pH ca. 6 produced electrochemically, and (d) the potential of 1.0 V was applied for 20 minutes to allow switching from the "closed" to "open" state. Potential scan rate between points "a–b" and "c–d" was 50 mV s^{-1}. Inset: Schematic hysteresis loop in the current–voltage function characteristic of a memristor device. Source: MacVittie and Katz 2013 [43]. Adapted with permission of American Chemical Society.

8.3.3 The Memristor Device Controlled by Logically Processed Biomolecular Signals

The important difference of the present bioelectrocatalytic realization of the memristor's function from previously published electrochemical system with memristic features [43] is the dependence of the system on the presence of the biofuel (glucose) and the oxidizer (O_2). In the absence of glucose and oxygen, the current cannot be established between the electrodes, and the system cannot switch between the low and high resistance values. Even more importantly, the system does not need the application of any external potential to activate the current and is self-activating in the presence of glucose and oxygen. The only operation needed to switch the system between two states is connecting/reconnecting the pH-switchable electrode with the anode or cathode of the biofuel cell. In another mode of operation, the pH-switchable electrode can be continuously electrically connected to both biofuel cell electrodes (the anode and cathode), and the switching between the low and high resistance value can be obtained by applying solutions containing either glucose or oxygen to the biofuel cell compartment. In this operation mode the current direction and the pH changes switching the resistance are controlled by the presence or absence of glucose and O_2. Note that this operation mode does not allow simultaneous presence of glucose and O_2 in the solution because they will drive

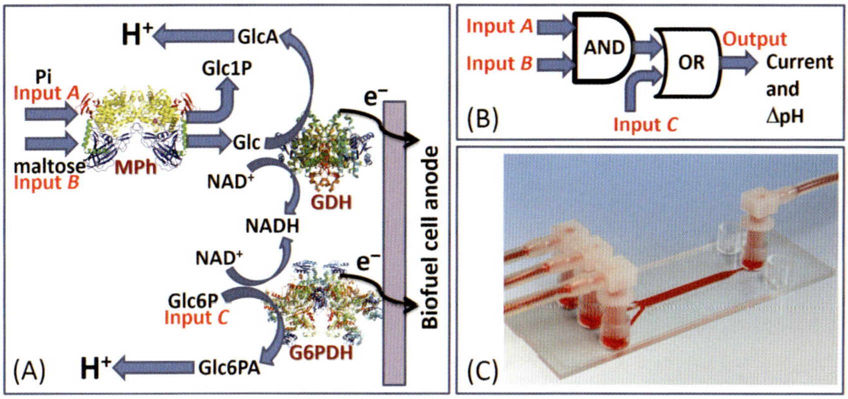

Figure 8.20 (A) An example of a biocatalytic cascade that can be used to control the switchable memristor device. (B) The equivalent logic network composed of concatenated AND/OR gates corresponding to the biocatalytic cascade. (C) Fluidic device for performing biocatalytic reactions mimicking logic gates (the red color in the cell does not correspond to the real experimental solutions). Source: MacVittie and Katz 2014 [65]. Adapted with permission of Royal Society of Chemistry.

the pH changes in the opposite directions, thus producing unstable operation of the bioelectronics device. The designed memristor can be easily interfaced with enzyme-based biocomputing systems producing glucose and/or oxygen as the result of biocatalytic cascades performing logic operations. Thus, the memristor can operate as a memory element activated by logically processed biomolecular signals and read out electronically (in the present example by impedance spectroscopy), allowing for the direct interfacing between biocomputing steps and electronic devices.

Figure 8.20A outlines an example of a biocatalytic cascade to activate the memristor device via bioelectrocatalytic reactions at an enzyme-modified electrode. In this example inorganic phosphate (Pi) and maltose (used as Input signals *A* and *B*, respectively) generate glucose (Glc) and glucose-1-phosphate (Glc1P) by-product in the reaction biocatalyzed by maltose phosphorylase (MPh). Glucose, produced only in the presence of both inputs (logic combination **1**,**1** of the AND gate; note that logic **0** and logic **1** values correspond to the absence and presence of the chemicals), can be oxidized through the reaction biocatalyzed by glucose dehydrogenase (GDH) in the presence of NAD^+ to yield gluconic acid (GlcA). Glucose-6-phosphate (Glc6P) (Input *C*) can be oxidized by glucose-6-phosphate dehydrogenase (G6PDH) in the presence of NAD^+ to yield gluconate-6-phosphate acid (Glc6PA). Both oxidizing enzymes (GDH and G6PDH) can be connected to a biofuel cell anode similarly to the recently reported approach [72, 73]. The anodic current generated by the biocatalytic reactions will be accompanied by the concomitant production of acidic pH values (note the production of GlcA and Glc6PA), thus switching the memristor to its low resistance state. Figure 8.20B shows the logic network composed of concatenated AND/OR logic gates equivalent to the biocatalytic cascade shown in Figure 8.20A. In this scheme the AND logic gate represents the biocatalytic reaction of MPh, which is activated in the presence of both Inputs *A* and *B*. The OR gate represents the bioelectrocatalytic process at the electrode (the biofuel cell anode) modified with GDH and G6PDH activated in the presence of either Glc or Glc6P. The anodic current at the enzyme-modified electrode and the concomitant pH change will occur upon application of the Input signals (*A*, *B*, *C*) in the following combinations: **1,1,0** (through the GDH biocatalytic pathway); **1,1,1** (through both the GDH and G6PDH biocatalytic pathways); and **0,1,1**; **1,0,1**; **0,0,1** (through the G6PDH biocatalytic pathway). On the other hand, the following signal combinations, **0,1,0**; **1,0,0**; and **0,0,0**, will not result in the current formation, and pH changes because both biocatalytic cascades will be mute. The biocatalytic reactions processing biomolecular signals prior to their reaction in the memristor device can be performed in a fluidic system (Figure 8.20C), similarly to the recently designed reversible enzyme-based logic gates [74–78] (see Chapter 5).

8.3.4 Enzyme-Based Memristors: Conclusions and Perspectives

The device discussed above is the first experimental realization of the electrochemical system with memimpedance properties, thus demonstrating the memristor and memcapacitance in one device. It should be noted that the slow processes resulting in the transition of the electrochemical device from the open to the closed states and back originate from the bulk pH changes produced by biocatalytic reactions performed at enzyme-modified electrode surfaces. Further miniaturization of the device, using nano-size biofuel cells, including those operating *in vivo* [79], will decrease the waiting time for the switching processes. However, this switching process will never be as fast as in semiconductor solid-state devices because of the time required for the restructuring of the polymer brush on the electrode surface. The present system does not pretend to be a competitor for solid-state devices, but opens future options for combining electrochemical memory devices with unconventional molecular/biomolecular information processing systems. The biocatalytic system shown in Figure 8.20A is only one of the numerous possible logic networks for processing various combinations of (bio)chemical input signals and activation of the electrochemical memristor device. A vast array of future applications in (bio)sensing, and other medicinal subfields, rather than in electronic systems, becomes increasingly feasible.

8.4 Enzyme-Based Associative Memory Systems

Development of processes with functionalities such as memory by mechanisms suggested by nature, but based on a simple set of biomolecular processes rather than a full complexity of living systems, is of interest for many reasons. New paradigms for information processing might be possible beyond the presently used approaches of analog and digital electronics. Presently there is no complete understanding of how natural systems manage complex, large-scale information processing [80]. Therefore, researching model biomolecular systems with memory can contribute to understanding how "complexity" is handled in nature. Indeed, memory properties are among the basic mechanisms of natural networks [80]. Near-term, new network elements with memory will result in novel computing paradigms [81] and add to the emerging "toolbox" of biocomputing and multi-signal-biosensing systems and components [82]. One of the interesting examples of biological memory features, particularly challenging for realization in nonbiological model systems, is associative memory [83], which is discussed below. While electronic realization of the associative memory, using, for example, memristor devices [84], became possible, model systems based on biomolecular reactions are only at the very beginning of investigation. In this section we consider new model systems activated by enzymes as signals and accomplishing associative memory [85, 86]. It is important to emphasize that associative memory [83] is a concept that has been studied for a long time [87] and in diverse fields [88] but at the same time it has been a subject of active recent research [89–92] driven by conceptual interest and also by the promise of adding adaptive capabilities to information processing systems. However, despite active research in the area, very few biomolecular systems based on proteins [85, 93] or DNA [94–96] mimicking associative memory *in vitro* have been realized.

8.4.1 Associative Memory: Biological Origin and Function

The most famous and widely known experiments of the Nobel Prize-winning Russian scientist Ivan Pavlov (1849–1936) Figure 8.21A [87] with dogs responding to food and to other non-related to food signals laid foundation for research on associative memory. In these experiments Pavlov observed dog's salivation in response to given food (Figure 8.21B), which was considered as the "correct" signal. Obviously, the dog was not responding in this way to the "wrong" signal (e.g., sound), which was not related to the food. However, by training the dog by applying the "wrong" signal together with the "correct" signal, the dog learned

(A) (B)

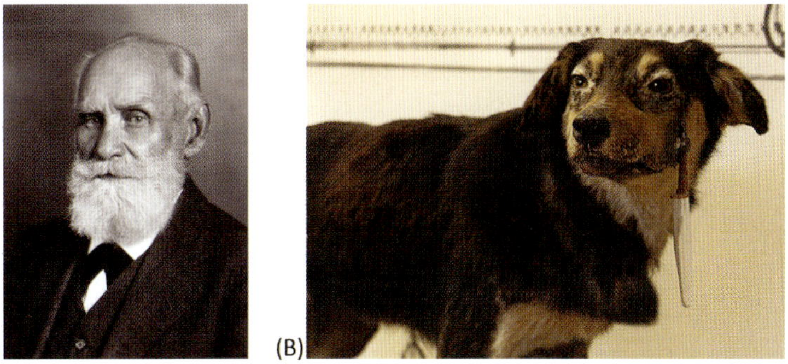

Figure 8.21 (A) Ivan Petrovich Pavlov (1849–1936). (B) One of Pavlov's dogs with a surgically implanted cannula to measure salivation used in the experiments related to associative memory (see also Figure 8.22 for the experiment explanations). Source: Adapted from https://en.wikipedia.org/wiki/Ivan_Pavlov. Licensed Under CC BY 3.0.

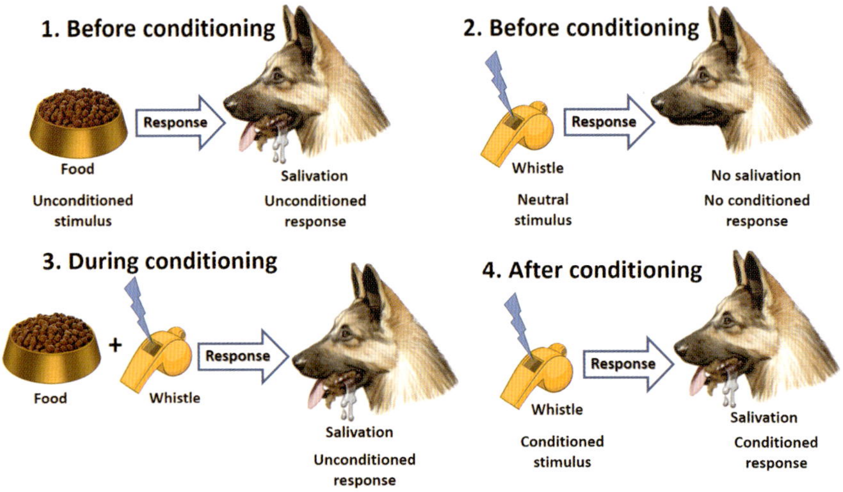

Figure 8.22 Pavlov's experiments with a dog, leaning to associate a "wrong" sound signal with a "correct" food signal and then responding to the "wrong" signal as it would be the "correct" signal. Note that these biological experiments are not directly related to the chemical model mimicking the associative memory – they are discussed here only for the problem formulation and comparison with the model shown in Figure 8.23.

to respond by salivation to the "wrong" signal alone (Figure 8.22). In this very brief explanation, the biological terminology is purposely avoided to simplify the concept and to keep the discussion in terms more related to chemistry and computer science; the biological description of Pavlov's experiments and their more recent advances can be found in specialized literature [97]. The learning process observed in Pavlov's experiments is based on very complex physiological mechanisms; however, in a simple way it could be reproduced with relatively simple chemical reactions.

Various definitions of the process are possible [83]. In the enzyme-based associative memory system discussed below, application of the "correct" input signal activates reactions to yield the output signal, while the "wrong" signal results in no chemical changes. A simultaneous application of the "correct" and "wrong" signals ("training" step) results in the memory effect after which the system reacts to the application of the "wrong" signal as it would react to the "correct" signal, by producing the output signal. Later the system can "forget" the "training" and stop reacting to the "wrong" signal applied alone. Here we focus on a system [86] in which the processes of conversion of the input signal to output are entirely those biocatalyzed by enzymes in two parallel cascades (Figure 8.23), allowing a simple biochemical system to mimic the Pavlov's dog behavior without complex biological mechanisms. It should be noted that inputting signals, detecting the output, and resetting the system are steps that will typically involve physical or chemical processes interfacing with the main biochemical steps. These enable spatial and temporal ("clocking") control of chemical concentrations by transport, separation, compartmentalization, and other physical or chemical means.

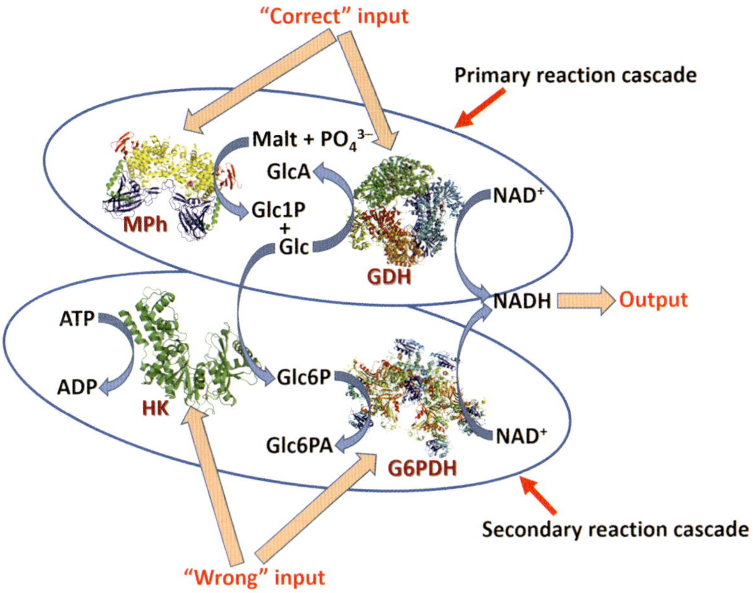

Figure 8.23 The biocatalytic process steps of the associative memory system based on a two-pathway reaction cascade. Source: MacVittie et al. 2013 [86]. Adapted with permission of Royal Society of Chemistry.

8.4.2 Realization of the Associative Memory with a Multienzyme Biocatalytic Cascade

The system "machinery" (the constant part of the reacting system) was composed of NAD^+, ATP, and maltose in phosphate buffer saline, pH = 7.0. The "correct" signal was defined as the presence of two enzymes, maltose phosphorylase (MPh) and glucose dehydrogenase (GDH), which activate the primary pathway converting maltose to glucose and then use glucose for the reduction of NAD^+ to NADH (Figure 8.23). The optical absorbance increase at $\lambda = 340$ nm corresponding to the formation of NADH was defined as the output signal. The "wrong" signal was defined as the presence of two other enzymes, hexokinase (HK) and glucose-6-phosphate dehydrogenase (G6PDH), which does not result in any reaction in the system if glucose is not present in the solution. Simultaneous application of the "correct" and "wrong" signals (i.e., all four enzymes) results in the activation of both pathways, where the supply of glucose produced in the presence of MPh is split in two parallel reactions, with the secondary pathway including glucose conversion to glucose-6-phosphate (Glc6P) catalyzed by HK, and then Glc6P is used for the reduction of NAD^+ to yield NADH. Note that the output signal is produced by both pathways (Figure 8.23). The system composition was optimized in such way that the "training" step (application of the "correct" and "wrong" signals together) resulted in accumulation of the intermediate product Glc6P representing the system memory. In order to achieve this, the activity of G6PDH was fivefold smaller than the activity of HK, thus providing the production of Glc6P faster than its consumption. Later application of the

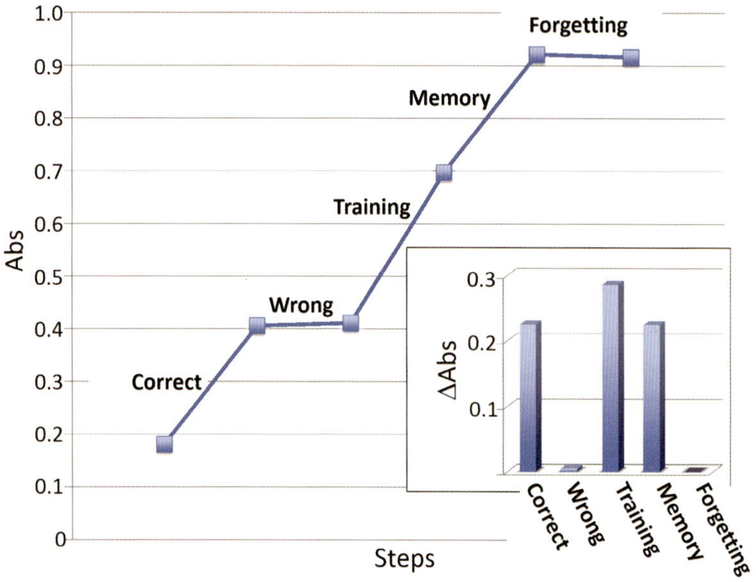

Figure 8.24 Optical measurements at 340 nm of the solution after the reaction with various input applications, as described in the text. Inset: Change in the absorbance measured after each 10 minutes input step, indicating the quantity of the output product. Source: MacVittie et al. 2013 [86]. Adapted with permission of Royal Society of Chemistry.

"wrong" signal alone results in the output signal (production of NADH) using Glc6P as a reductant accumulated in the "training" step. Note that the activities of all enzymes (particularly of HK and GDH) were optimized experimentally to split glucose into two pathways in such a way that the output signal is preserved almost at the same level at the reaction time. After each signal application step, the enzymes were removed from the solution by ultrafiltration, and NAD^+, ATP, and maltose were added to the reacting solution in order to have them at the same concentrations for the next signal application step.

Figure 8.24 shows a succession of steps involving applications of different inputs. The "correct" input produced finite output signal, whereas during the application of the "wrong" input only, no signal is produced. However, when both the "correct" and "wrong" inputs were applied together (in the "training" step), the signal was produced again through both biocatalytic cascades. As mentioned, the system was tailored to allow for a surplus of Glc6P to be produced *in situ* during the "training" step. Therefore, in a later application of the "wrong" input only, during the step termed "memory," NAD^+ was reduced by Glc6P, producing the output signal. Subsequent additions of the "wrong" input produce no signal, because practically all of the Glc6P-based memory is used up during the first "memory" step (if all the process steps are properly timed). This demonstrates that the memory realized here is not "self-reinforcing" – it "forgets" the training. Figure 8.25A demonstrates the possibility to "reset" the system after various step sequences. The biocatalytically produced NADH was recycled to NAD^+ by a photochemical reaction [98] in the presence of thionine and oxygen upon

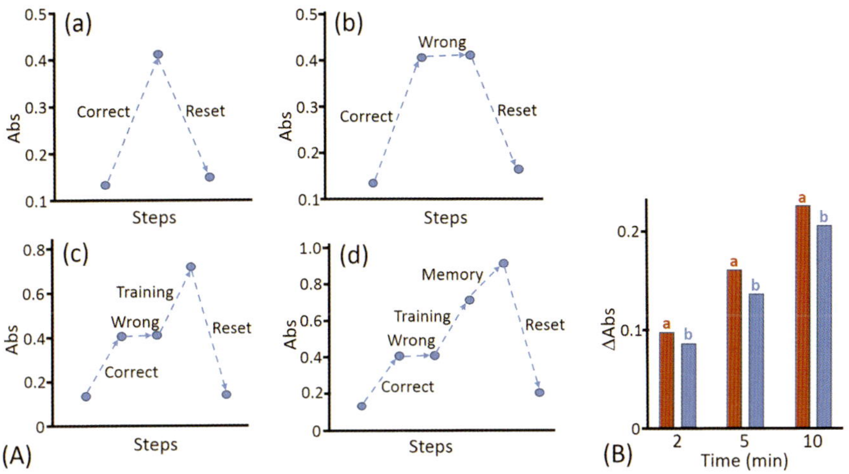

Figure 8.25 (A) Experiments demonstrating that the system can be "reset," i.e., returned to a minimum absorbance level, at any point in the succession of reactions steps. (B) Time dependence of "training" (a) and "memory" (b) for varying "training" times: 2, 5, and 10 minutes. The "memory" step was carried out for the same amount of time, 10 minutes, in each case. Source: MacVittie et al. 2013 [86]. Adapted with permission of Royal Society of Chemistry.

illumination. This was done after each step sequence to illustrate that the system can be returned to its initial state at any point.

The effect of the "training" step duration time on "memory" was also tested, as shown in Figure 8.25B. This was done by varying the initial incubation time of the "correct" and "wrong" inputs jointly, allowed to proceed for 2, 5, and 10 minutes. Measurements were then taken with the "wrong" input only, as the "memory" step for 10 minutes. Notably the longer "training" step results in more pronounced "memory" step due to a larger accumulation of Glc6P.

8.4.3 Enzyme-Based Associative Memory: Challenges and Perspectives

The results obtained for this system illustrate rather general properties specific to biocatalytic systems used as memory processes. It could be expected that such systems will be utilized in multi-input and/or multistep setups with digital, binary signal information processing steps temporally controlled ("clocked") and spatially controlled/compartmentalized for sensor and signal/information processing applications. Each of the process steps will then be timed, and signal levels will have to be maintained at well-defined values. The present system functions under rather stringent conditions, limiting its flexibility. Networking of memory elements for useful information processing is a rather new, recently emerging topic in electronics [99, 100]. An associative system should preferably be devised in such a way that in the "training" step, the added "wrong" input does not significantly modify this output intensity (as seen in Figure 8.24). Furthermore, after the "training," application of the "wrong" output alone should result in an output rate comparable with the original rate (cf. Figure 8.24). However, the accumulated chemical will eventually be exhausted, and the system will "forget" the training, and after one or more fixed-duration time steps, the product will no longer be generated. Based on the experiments shown in Figures 8.24 and 8.25, one can conclude that the discussed system shows all the aspects of this expected pattern of behavior, including fast "forgetting" practically after a single "memory" step.

However, the demonstrated realization of a purely enzymatic associative memory is limited to a rather specific set of conditions. Future work should focus on ideas of improving flexibility and versatility of enzymatic and other bio-inspired memory elements for enabling their networking. Networking concepts for useful information processing are rather recent even in electronics and will have to be developed for biomolecular systems. The "toolbox" of future gates should also include some modularity, i.e., flexibility in which compounds to use and what is their role. For example, in the system discussed above, the accumulating compound is Glc6P. Changing this, for instance, to have glucose accumulate instead, for the memory effect in an otherwise similar system, would require complete redesign and use of different operating regimes. The time evolution of memory systems realized as purely enzymatic biocatalytic cascades should also be studied for kinetic regimes involved. In typical kinetic studies of enzymatic reactions, the systems are kept under steady-state conditions for the time intervals of the experiment. In memory systems, however, accumulation/depletion of intermediate chemical(s) suggests non-steady-state kinetics for both the "learning" and "forgetting" processes. The development of novel biomolecular systems operating as associative memory system can be extended to the use of biorecognition processes in addition to the biocatalytic reactions [85], thus broadening their composition and operation for various applications.

In summary, it should be noted that the system discussed above is the first demonstration of the associative memory as a bio-inspired memory-involving signal processing system functioning with the internal dynamics entirely biocatalyzed by enzymes, activated by enzymes as inputs. This model system shows general characteristics expected of memory "devices" based on chemical compound accumulation. It is hoped that such systems can offer new "network elements" and novel bio-inspired design ideas for biomolecular logic being developed for biosensors [101] and biocomputing [5–7] applications.

8.5 Enzyme-Based Memory Systems: Challenges, Perspectives, and Limitations

As it is overviewed in the chapter, the enzyme-based memory systems have been designed with very broad functionality, including various flip-flop devices, memristors/memcapacitors, and even systems mimicking sophisticated biological functions, such as associative memory. At the present stage of their designing and characterization, they are mostly operating as stand-alone chemical devices being not integrated yet in computing networks. The first attempts to store information produced by logic networks composed of concatenated logic gates and activated by multiple inputs are still very preliminary and exemplified with limited systems (see, for example, Figure 8.20 [65] showing a reaction cascade mimicking two concatenated AND/OR logic gates connected to a biomolecular memristor). Some of the systems described above could be easily connected to logic networks. For example, the signals activating the Set-Reset flip-flop (Figure 8.1) could be produced *in situ* through an extended biocatalytic cascade [17]. On the other hand, the input signals represented by enzymes used for activating the associate memory (Figure 8.23) cannot be produced by any chemical means [86], thus not allowing integration of the designed system into biocomputing networks. This is certainly a big drawback of the developed approach. Some of the developed memory systems represent examples of general concepts, which can be easily reformulated to different chemical compositions performing the same function. For example, the SR flip-flop [17] (Figure 8.1) based on the competing catalytic reactions running in the opposite directions can be easily realized with different enzymes and different reversibly reduced–oxidized mediators. This property of the system is highly important for connecting the memory to different inputs generated by different biocatalytic cascades. However, some other memory systems have been represented by unique designs using very specific reactions. For example, the T flip-flop [22] (Figure 8.11) is based on the reactions resulting in two different oxidation states of the redox mediator, thus representing a very specific system, which is hard to adopt to different chemical species. The systems mimicking the associative memory based on biocatalytic reactions [86] or biorecognition processes [85] use very specific biomolecules for their operation and require complete redesign, if the composition needs to be changed. The most important limitation of the enzyme-based memory systems is very limited amount of the saved information, which is incomparably smaller than the information stored in DNA molecules. This limitation cannot be overcome by any redesign of the enzyme systems, and it is a conceptual limitation characteristic of the entire approach. This limitation should be taken into account when the enzyme systems are planned for their use as a memory storage. Also, the information saved by the enzyme reactions is preserved in the changed states of reactive molecules, for example, in the form of different redox states of reactive mediators. These states cannot be preserved for a long time to allow long-term storage of the information. This is completely different from the information saved in the sequence of DNA bases, which could be preserved for very long time. Obviously, the enzyme systems should not be considered for storing large amounts of information, particularly for long time, but should be used instead for temporary saving small amount of information produced by enzymatic cascades.

Overall, the future work on the enzyme-based memory systems should be directed to the design of generic, rather than specific, systems easily adaptable to different input signals and used for different applications, mostly meaning biosensor/bioanalytical applications. Another interesting direction in the study would be the kinetic analysis of the memory systems to allow optimization of their transitions from one to another state. Design of very sophisticated memory systems mimicking features of biological systems but based on relatively simple processes may help in understanding of biological processes and their biological evolution from simple prebiological systems. It is interesting to note that biological membranes [102] and plant tissues, e.g., in flowers, apples, pumpkin seeds, etc. [103–108], demonstrate some memristive features. The memristive behavior has been obtained in model biochemical systems based on various biomolecules, e.g., using DNA aptasensors for biomedical applications [109, 110]. A DNA-based biomolecular memcapacitor and memristor has been realized at a nanoscale [111], thus allowing for miniaturization of these devices.

The memory systems based on enzyme-catalyzed reactions should be considered in the general content of logic/computing systems, including logic gates and different non-Boolean operators, thus allowing multistep processes and multitask operations, including the temporary storage of the processed information in the catalytically changed composition of the reacting molecules. Finally, the research on memory systems based on biological molecules is presently in its infancy, and the huge variety of possibilities is waiting for future research, particularly involving contribution from experts in biochemistry, computer science, biology, and engineering.

References

1 Szacilowski, K. (2012). *Infochemistry – Information Processing at the Nanoscale*. Chichester: Wiley.
2 de Silva, A.P. (2013). *Molecular Logic-Based Computation*. Cambridge: Royal Society of Chemistry.
3 Sienko, T., Adamatzky, A., Rambidi, N.G., and Conrad, M. (eds.) (2003). *Molecular Computing*. Cambridge, MA: MIT Press.
4 Katz, E. (ed.) (2012). *Molecular and Supramolecular Information Processing – From Molecular Switches to Logic systems*. Weinheim: Wiley-VCH.
5 Katz, E. (ed.) (2012). *Biomolecular Computing – From Logic Systems to Smart Sensors and Actuators*. Weinheim: Wiley-VCH.
6 Katz, E. and Privman, V. (2010). *Chem. Soc. Rev.* 39: 1835–1857.
7 Katz, E. (2015). *Curr. Opin. Biotechnol.* 34: 202–208.
8 De Ruiter, G., Tartakovsky, E., Oded, N., and Van der Boom, M.E. (2010). *Angew. Chem. Int. Ed.* 49: 169–172.
9 De Ruiter, G., Motiei, L., Choudhury, J. et al. (2010). *Angew. Chem. Int. Ed.* 49: 4780–4783.
10 Periyasamy, G., Collin, J.-P., Sauvage, J.-P. et al. (2009). *Chem. Eur. J.* 15: 1310–1313.
11 Pischel, U. and Andréasson, J. (2010). *New J. Chem.* 34: 2701–2703.
12 Baron, R., Onopriyenko, A., Katz, E. et al. (2006). *Chem. Commun.* 2147–2149.
13 De Ruiter, G. and Van der Boom, M.E. (2011). *J. Mater. Chem.* 21: 17575–17581.
14 Elbaz, J., Wang, Z.-G., Orbach, R., and Willner, I. (2009). *Nano Lett.* 9: 4510–4514.
15 Elbaz, J., Moshe, M., and Willner, I. (2009). *Angew. Chem. Int. Ed.* 48: 3834–3837.

16 Wang, Z.-G., Elbaz, J., Remacle, F. et al. (2010). *Proc. Nat. Acad. Sci. USA* 107: 21996–22001.
17 Pita, M., Strack, G., MacVittie, K. et al. (2009). *J. Phys. Chem. B* 113: 16071–16076.
18 Remón, P., Bälter, M., Li, S. et al. (2011). *J. Am. Chem. Soc.* 133: 20742–20745.
19 Puntoriero, F., Nastasi, F., Bura, T. et al. (2011). *New J. Chem.* 35: 948–952.
20 Hoteit, I., Kharma, N., and Varin, L. (2012). *BioSystems* 109: 57–71.
21 MacVittie, K., Halámek, J., and Katz, E. (2012). *Chem. Commun.* 48: 11742–11744.
22 MacVittie, K., Halámek, J., and Katz, E. (2012). *Int. J. Unconventional Computing* 2012 (8): 383–389.
23 Shiva, S.G. (1998). *Introduction to Logic Design*, 2e. New York: Marcel Dekker.
24 Polsky, R., Harper, J.C., Wheeler, D.R., and Brozik, S.M. (2008). *Electroanalysis* 20: 671–679.
25 Krämer, M., Pita, M., Zhou, J. et al. (2009). *J. Phys. Chem. C* 113: 2573–2579.
26 Mano, M.M. and Kime, C.R. (2000). *Logic and Computer Design Fundamentals*. Upper Saddle River, NJ: Prentice Hall.
27 Zhou, J., Arugula, M.A., Halámek, J. et al. (2009). *J. Phys. Chem. B* 113: 16065–16070.
28 BRENDA - The comprehensive enzyme information system. http://www.brenda-enzymes.org/ (accessed 4 March 2019).
29 Josephy, P.D., Eling, T., and Mason, R.P. (1982). *J. Biol. Chem.* 257: 3669–3675.
30 Katz, E. (ed.) (2014). *Implantable Bioelectronics – Devices, Materials and Applications*. Weinheim: Wiley-VCH.
31 Katz, E. (2015). *Bioelectronic Medicine* 2: 1–12.
32 Chua, L.O. (1971). *IEEE Trans. Circuit Theory* 18: 507–519.
33 Chua, L.O. (1976). *Proc. IEEE* 64: 209–223.
34 Pershin, Y.V. and Di Ventra, M. (2019). *J. Phys. D: Appl. Phys.* 52: 01LT01.
35 Struckov, D.B., Snider, G.S., Stewart, D.R., and Williams, R.S. (2008). *Nature* 453: 80–83.
36 Yang, J.J., Pickett, M.D., Li, X. et al. (2008). *Nat. Nanotechnol.* 3: 429–433.
37 Szot, K., Rogala, M., Speier, W. et al. (2011). *Nanotechnology* 22: 254001.
38 Erokhin, V. and Fontana, M.P. (2011). *J. Comp. Theor. Nanosci.* 8: 313–330.
39 Katz, E., Bocharova, V., and Privman, M. (2012). *J. Mater. Chem.* 22: 8171–8178.
40 Privman, M., Tam, T.K., Pita, M., and Katz, E. (2009). *J. Am. Chem. Soc.* 131: 1314–1321.
41 Di Ventra, M., Pershin, Y.V., and Chua, L.O. (2009). *Proc. IEEE* 97: 1717–1724.
42 Tam, T.K., Pita, M., Motornov, M. et al. (2010). *Electroanalysis* 22: 35–40.
43 MacVittie, K. and Katz, E. (2013). *J. Phys. Chem. C* 117: 24943–24947.

44 Katz, E., Bocharova, V., and Halámek, J. (2014). *Enzymatic Fuel Cells: From Fundamentals to Applications*, Chapter 18 (ed. P. Atanassov, G. Johnson and H. Luckarift), 374–421. Weinheim: Wiley-VCH.
45 Katz, E., Minko, S., Halámek, J. et al. (2013). *Anal. Bioanal. Chem.* 405: 3659–3672.
46 Katz, E. (2010). *Electroanalysis* 22: 744–756.
47 Katz, E. and Pita, M. (2009). *Chem. Eur. J.* 15: 12554–12564.
48 Atanassov, P., Johnson, G., and Luckarift, H. (eds.) (2014). *Enzymatic Fuel Cells: From Fundamentals to Applications*. Weinheim: Wiley-VCH.
49 Davis, F. and Higson, S.P.J. (2007). *Biosens. Bioelectron.* 22: 1224–1235.
50 Cracknell, J.A., Vincent, K.A., and Armstrong, F.A. (2008). *Chem. Rev.* 108: 2439–2461.
51 Katz, E. and MacVittie, K. (2013). *Energy Environ. Sci.* 6: 2791–2803.
52 Halámková, L., Halámek, J., Bocharova, V. et al. (2012). *J. Am. Chem. Soc.* 134: 5040–5043.
53 Katz, E., Bückmann, A.F., and Willner, I. (2001). *J. Am. Chem. Soc.* 123: 10752–10753.
54 Zhang, L.L., Zhou, M., and Dong, S.J. (2012). *Anal. Chem.* 84: 10345–10349.
55 Zhou, M. and Wang, J. (2012). *Electroanalysis* 24: 197–209.
56 Zhou, M., Kuralay, F., Windmiller, J.R., and Wang, J. (2012). *Chem. Commun.* 48: 3815–3817.
57 Zhou, M. and Dong, S.J. (2011). *Acc. Chem. Res.* 44: 1232–1243.
58 Zhou, M., Du, Y., Chen, C.G. et al. (2010). *J. Am. Chem. Soc.* 132: 2172–2174.
59 Zhou, M., Wang, F.A., and Dong, S.J. (2011). *Electrochim. Acta* 56: 4112–4118.
60 Zhou, M., Zheng, X.L., Wang, J., and Dong, S.J. (2011). *Bioinformatics* 27: 399–404.
61 Halámek, J., Tam, T.K., Chinnapareddy, S. et al. (2010). *J. Phys. Chem. Lett.* 1: 973–977.
62 Halámek, J., Tam, T.K., Strack, G. et al. (2010). *Chem. Commun.* 46: 2405–2407.
63 Strack, G., Luckarift, H.R., Nichols, R. et al. (2011). *Chem. Commun.* 47: 7662–7664.
64 Zhou, M., Zhou, N., Kuralay, F. et al. (2012). *Angew. Chem. Int. Ed.* 51: 2686–2689.
65 MacVittie, K. and Katz, E. (2014). *Chem. Commun.* 50: 4816–4819.
66 Tam, T.K., Pita, M., Trotsenko, O. et al. (2010). *Langmuir* 26: 4506–4513.
67 Parimi, N.S., Umasankar, Y., Atanassov, P., and Ramasamy, R.P. (2012). *ACS Catal.* 2: 38–44.
68 Katz, E. and Willner, I. (2003). *J. Am. Chem. Soc.* 125: 6803–6813.
69 Pershin, Y.V., Martinez-Rincon, J., and Di Ventra, M. (2011). *Comput. Theoret. Nanosci.* 8: 441–448.
70 Pershin, Y.V. and Shevchenko, S.N. (2017). *Nanotechnology* 28: 075204.
71 Wakrim, T., Vallée, C., Gonon, P. et al. (2016). *Appl. Phys. Lett.* 108: 053502.
72 Kar, P., Wen, H., Li, H.Z. et al. (2011). *J. Electrochem. Soc.* 158: B580–B586.

73 Moehlenbrock, M.J., Toby, T.K., Waheed, A., and Minteer, S.D. (2010). *J. Am. Chem. Soc.* 132: 6288–6289.
74 Fratto, B.E. and Katz, E. (2017). *Int. J. Parallel, Emergent and Distributed Systems* 32: 139–156.
75 Fratto, B.E. and Katz, E. (2016). *ChemPhysChem* 17: 1046–1053.
76 Katz, E. and Fratto, B.E. (2017). *Advances in Unconventional Computing*, vol. 2, Chapter 2 (ed. A. Adamatzky), 29–59. Springer.
77 Fratto, B.E. and Katz, E. (2015). *ChemPhysChem* 16: 1405–1415.
78 Fratto, B.E., Guz, N., and Katz, E. (2015). *Parallel Process. Lett.* 25 (1): 1540001.
79 Hansen, B.J., Liu, Y., Yang, R., and Wang, Z.L. (2010). *ACS Nano* 4: 3647–3652.
80 Alon, U. (2007). *An Introduction to Systems Biology. Design Principles of Biological Circuits.* Boca Raton, FL: Chapman & Hall/CRC Press.
81 Di Ventra, M. and Pershin, Y.V. (2013). *Nat. Phys.* 9: 200–202.
82 Privman, V. (2011). *Israel J. Chem.* 51: 118–131.
83 Kohonen, T. (2012). *Associative Memory*. Berlin: Springer-Verlag.
84 Yang, L., Zeng, Z., and Wen, S. (2018). *Neurocomputing* 272: 513–519.
85 Bocharova, V., MacVittie, K., Chinnapareddy, S. et al. (2012). *J. Phys. Chem. Lett.* 3: 1234–1237.
86 MacVittie, K., Halámek, J., Privman, V., and Katz, E. (2013). *Chem. Commun.* 49: 6962–6964.
87 Pavlov, I.P. (1927). *Conditioned Reflexes: An Investigation of the Physiological Activity of the Cerebral Cortex.* Translated and Edited by (ed. G.V. Anrep). London: Oxford University Press.
88 Ziegler, M., Soni, R., Patelczyk, T. et al. (2012). *Adv. Funct. Mater.* 22: 2744–2749.
89 Gandhi, N., Ashkenasy, G., and Tannenbaum, E. (2007). *J. Theor. Biol.* 249: 58–66.
90 Fernando, C.T., Liekens, A.M.L., Bingle, L.E.H. et al. (2009). *J. Royal Soc. Interface* 6: 463–469.
91 Qadir, O., Liu, J., Tempesti, G. et al. (2011). *Artif. Intell.* 175: 673–693.
92 Pershin, Y.V. and Di Ventra, M. (2010). *Neural Networks* 23: 881–886.
93 Birge, R.R., Gillespie, N.B., Izaguirre, E.W. et al. (1999). *J. Phys. Chem. B* 103: 10746–10766.
94 von Schmiedeberg, S., Soebel, C., Cjleichmann, E., and Uetrecht, J. (1995). *Science* 268: 583–565.
95 Chen, J.H., Deaton, R., and Wang, Y.Z. (2004). *DNA Computing; Lecture Notes in Computer Science*, vol. 2943 (ed. J. Chen and J. Reif), 145–156. Berlin: Springer-Verlag.
96 Nowak, R.M., Mulawka, J.J., and Pucienniczak, A. (2006). *Proceedings of SPIE*, vol. 6159 (ed. R.S. Romaniuk), S1593.
97 Mackintosh, N.J. (1983). *Conditioning and Associative Learning*. Oxford: Clarendon Press.
98 Rickus, J.L., Chang, P.L., Tobin, A.J. et al. (2004). *J. Phys. Chem. B* 108: 9325–9332.
99 Pershin, Y.V. and Di Ventra, M. (2011). *Phys. Rev. E* 84: 046703.

100 Pershin, Y.V. and Di Ventra, M. (2012). *Proc. IEEE* 100: 2071–2080.
101 Katz, E., Wang, J., Privman, M., and Halámek, J. (2012). *Anal. Chem.* 84: 5463–5469.
102 Volkov, A.G. and Markin, V.S. (2016). Memristors in biomembranes. In: *Advances in Biomembranes and Lipid Self-Assembly*, vol. 24 (ed. A. Iglic, C.V. Kulkarni and M. Rappolt), 91–117.
103 Volkov, A.G. (2018). *Memory and Learning in Plants*, Book Series: Signaling and Communication in Plants (ed. F. Baluska, M. Gagliano and G. Witzany), 139–161.
104 Volkov, A.G. and Nyasani, E.K. (2018). *Funct. Plant Biol.* 45: 222–227.
105 Volkov, A.G. and Markin, V.S. (2017). *Russian J. Electrochem.* 53: 1011–1018.
106 Volkov, A.G., Nyasani, E.K., Tuckett, C. et al. (2016). *Bioelectrochemistry* 112: 9–15.
107 Volkov, A.G., Nyasani, E.K., Tuckett, C. et al. (2016). *Plant Signaling Behavior* 11: e1151600.
108 Volkov, A.G., Nyasani, E.K., Blockmon, A.L., and Volkova, M.I. (2015). *Plant Signaling Behavior* 10: e1071750.
109 Tzouvadaki, I., Jolly, P., Lu, X.L. et al. (2016). *Nano Lett.* 16: 4472–4476.
110 Tzouvadaki, I., Aliakbarinodehi, N., De Micheli, G., and Carrara, S. (2017). *Nanoscale* 9: 9676–9684.
111 Chu, H.L., Lai, J.J., Wu, L.Y. et al. (2017). *NPG Asia Mater* 9: e430.

9

Arithmetic Functions Realized with Enzyme-Catalyzed Reactions

9.1 Molecular and Biomolecular Arithmetic Systems: Introduction and Motivation

Fast progress in the design of molecular [1–3] and biomolecular [4] information processing systems has resulted in the formulation of various basic Boolean logic gates and their small networks. Integration of logic operations in sophisticated circuitries allowed for their use as simple computing elements [2], including half-adders/half-subtractors and full-adders/full-subtractors [5]. These molecular computing "devices" were designed using synthetic molecules [6–15] or nanoparticles [16] with switchable properties controlled by chemical or optical signals. The majority of the designed molecular arithmetic "devices" were based on optical readout of the generated signals, while very few systems utilized electrochemical methods for the output signal analysis [17].

Another much more powerful approach to the design of computing elements performing basic arithmetic functions, particularly half-adders/half-subtractors and full-adders/full-subtractors, was implemented using biomolecular species, mostly based on the use of DNA [18–33], DNAzymes [34–36], or aptamers [37]. Additional functionalities have been added by using magnetic nanoparticles [38] and quantum dots [39] in combination with the DNA-based computing circuitries. The use of biomolecular systems, particularly based on DNA/RNA molecules, allowed for computational operations in a biological environment, with the ultimate goal being to encompass the creation of a biocomputer [40]. Particularly, DNA origami robots operating in living cockroaches performed the half-adder computing function [41], while RNA-based computation in mammalian cells realized half-adder and half-subtractor operations [42]. The use of protein-based systems for realization of arithmetic functions was limited to very few examples, including light-activated bacteriorhodopsin [43, 44], verotoxin [45], actin filaments [46, 47], and proteins for computation in mammalian cells based on bacterial protein-triggered cytotoxicity [48].

Realization of computing functions (limited to simple arithmetic operations) with enzyme-catalyzed reactions is difficult in one-pot systems where all biocatalytic reactions proceed in the same solution. The only enzyme-based system performing the half-adder operation in a homogeneous solution included concerted operation of four enzymes; the half-subtractor was claimed to be

Enzyme-Based Computing Systems, First Edition. Evgeny Katz.

possible in a similar way, but not experimentally realized [49]. The motivation behind enzyme-based computing systems is their potential integration into multi-input binary-operating (YES/NOT format) biosensors [50], which are particularly useful in biomedical sensing [51] as well as in diagnostic applications [52, 53]. Indeed, while DNA-based information processing systems could be rather powerful for computational operations, they do not communicate with physiological biomarkers and cannot logically respond to variable biomedical conditions. On the other hand, the enzyme-based logic systems are the most adaptive to physiological changes and can report on them in the YES/NO sensing format, operating simultaneously with several physiologically relevant input signals. Unfortunately, despite the fact that the enzyme logic systems can be assembled in biocatalytic cascades to perform concatenated logic operations [54, 55], the complexity of these systems is limited by cross-reactivity of the enzyme-biocatalyzed reactions, thus making the arithmetic operations in the enzyme reactions very difficult for experimental realization. This is why the number of studies on the enzyme-based arithmetic operations is very limited, particularly compared with the substantially larger body of research performed on DNA systems. Recent work on modular design of enzyme logic systems, where each reaction step proceeds in a separate reacting cell and all cells are connected in a flow device, allowed for design of very sophisticated logic systems [56–58] (see Chapter 5), thus providing background for realization of arithmetic operations in similar modular flow cell systems described below.

Half-adder and half-subtractor devices represent two double input–double output systems [59] composed of XOR and AND logic gates operating in parallel. The only difference between them is the presence of an Inverter (NOT) gate located in one of the input channels of the AND gate in the half-subtractor device. Thus, their biochemical realization could be similar and particularly easy if the problems of cross-reactivity that plague enzymatic reactions in homogeneous systems can be resolved. A resolution was achieved by separating all of the reaction steps by localizing them in different cells that were then assembled in a flow circuitry [60].

9.2 Half-Adder

The half-adder adds two single binary digits A and B and generates two outputs, sum (S) and carry (C) [59]. The carry signal represents an overflow into the next digit of a multi-digit addition. The value of the sum is $2C + S$ (Figure 9.1A). The simplest realization of this device can be achieved by the parallel operation of XOR and AND gates (Figure 9.1B). Note that two inputs A and B are directed to two different gates – the inputs to the XOR gate are also the inputs to the AND gate. The truth table of the system operation is shown in Figure 9.1C.

The biochemical realization of the XOR gate was achieved by the balanced reduction–oxidation of redox species biocatalyzed by two enzymes operating in the opposite directions and activated by two input signals A and B [61] (Figure 9.2). The reductive signal, NADH, (Input A) passing through the cell

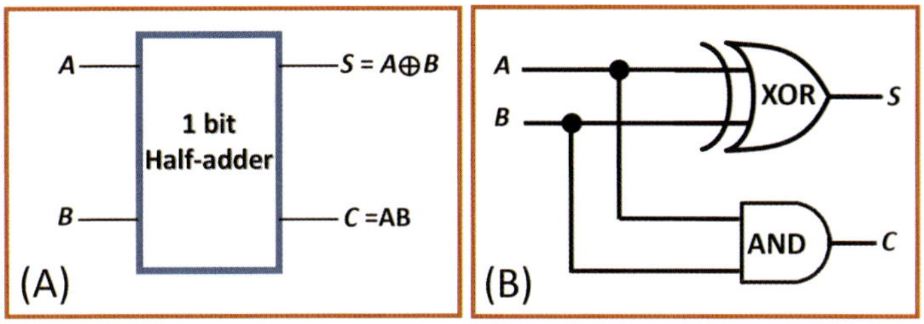

(C)

Input *A*	Input *B*	Carry digit (*C*)	Sum digit (*S*)
0	0	0	0
0	1	0	1
1	0	0	1
1	1	1	0

Figure 9.1 The block diagram (A) and equivalent electronic circuitry (B) for the half-adder system. (C) Half-adder truth table. Standard Boolean notations are used in the schemes [59]. Source: Fratto et al. 2016 [60]. Adapted with permission of John Wiley and Sons.

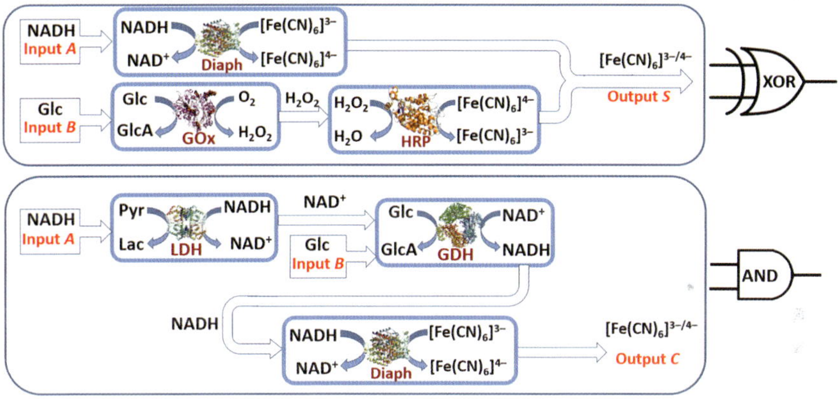

Figure 9.2 Experimental realization of the biocatalytic half-adder system in the flow device. Source: Fratto et al. 2016 [60]. Adapted with permission of John Wiley and Sons.

modified with diaphorase (Diaph) results in the reduction of $[Fe(CN)_6]^{3-}$ present in the background ("machinery") solution to yield the reduced species $[Fe(CN)_6]^{4-}$. The opposite process was activated by the combined operation of glucose oxidase (GOx)- and horseradish peroxidase (HRP)-modified flow cells. The second signal, glucose (Glc), (Input *B*) passing through the GOx-modified cell resulted in the production of H_2O_2 (note the presence of oxygen in the background solution). The generated H_2O_2 was directed to the next flow cell modified with HRP, where the reaction resulted in the oxidation of $[Fe(CN)_6]^{4-}$ to yield $[Fe(CN)_6]^{3-}$. Overall, when Input *A* appeared at the logic value **1** (meaning the presence of NADH), the biocatalytic process resulted in the consumption of $[Fe(CN)_6]^{3-}$ and formation of $[Fe(CN)_6]^{4-}$. On the other hand, when Input *B* appeared at the logic value **1** (meaning the presence of Glc), the opposite process was activated resulting in consumption of $[Fe(CN)_6]^{4-}$ and formation of $[Fe(CN)_6]^{3-}$. The solutions produced in the reacting flow cells then flowed into a mixing cell before being analyzed optically. The changes in the oxidation state of $[Fe(CN)_6]^{4-/3-}$ were followed by the optical absorbance measurements at $\lambda = 420$ nm where the oxidized species $[Fe(CN)_6]^{3-}$ have specific absorbance. The biocatalytic reductive process activated by Input *A* resulted in the decreasing absorbance at $\lambda = 420$ nm corresponding to the consumption of the light-absorbing $[Fe(CN)_6]^{3-}$ species. The biocatalytic oxidation process activated by Input *B* resulted in the opposite absorbance change due to the increasing concentration of $[Fe(CN)_6]^{3-}$. The balanced application of both inputs resulted in no changes in the absorbance. Note that the output signal measured optically was defined as the absolute value of the absorbance change, $|\Delta Abs|$, regardless the direction (decreasing or increasing) of the absorbance. The observed absorbance changes upon application of Inputs *A* and *B* in various logic combinations (**0**,**0**; **0**,**1**; **1**,**0**; and **1**,**1**) corresponded to the XOR function of the half-adder.

The AND function of the half-adder was realized by the concerted operation of three biocatalytic flow cells modified with lactate dehydrogenase (LDH), glucose dehydrogenase (GDH), and Diaph – all connected in a serial (Figure 9.2). Input *A* represented by NADH (logic value **1**), when passing through the LDH-modified cell, resulted in the formation of NAD^+ (note that the oxidative species, pyruvate (Pyr), required for this reaction were always present in high excess in the background solution as a part of the system "machinery"). Importantly, this reaction resulted in the complete consumption of NADH. In the next reaction step biocatalyzed by GDH, NAD^+ was returned back to the reduced form, NADH, if Input *B* (Glc) appeared at the logic value **1**. It should be noted that this operation already performed the AND logic function. Indeed, in the absence of NADH (Input *A* equal to **0**), NAD^+ cannot be produced. On the other hand, in the absence of Glc (Input *B* equal to **0**), the NADH will not be formed. Overall, the NADH formation at this reaction step is only possible when both input signals (*A* and *B*) appear at their logic values **1**, thus performing the AND logic gate operation. Notably, the AND gate operation needed NAD^+ as one of the inputs; however, the input should be consistent with one used in the XOR gate. This is why the first reaction in this cascade converted NADH to NAD^+. The last reaction step performed in the flow cell modified with Diaph converted NADH intermediate signal to the final output signal of $[Fe(CN)_6]^{4-}$ to be consistent with the output generated by the XOR gate. Finally, the $[Fe(CN)_6]^{3-/4-}$ absorbance variation was considered as the output signals from the XOR and AND gates, being *S* and *C* outputs, respectively (Figure 9.1B).

Figure 9.3 shows the experimental results for the XOR/AND gates operation explained above. Figure 9.3A shows the kinetics of the absorbance changes in the Output *S* processed through the XOR gate for different combinations of the input signals. Inputs *A* and *B* applied in **0**,**1** and **1**,**0** combinations resulted in the increase and decrease of the absorbance corresponding to $[Fe(CN)_6]^{3-}$ formation/consumption, respectively. It should be noted that Output *S* was defined as the *absolute value* of the absorbance change, thus considering the absorbance

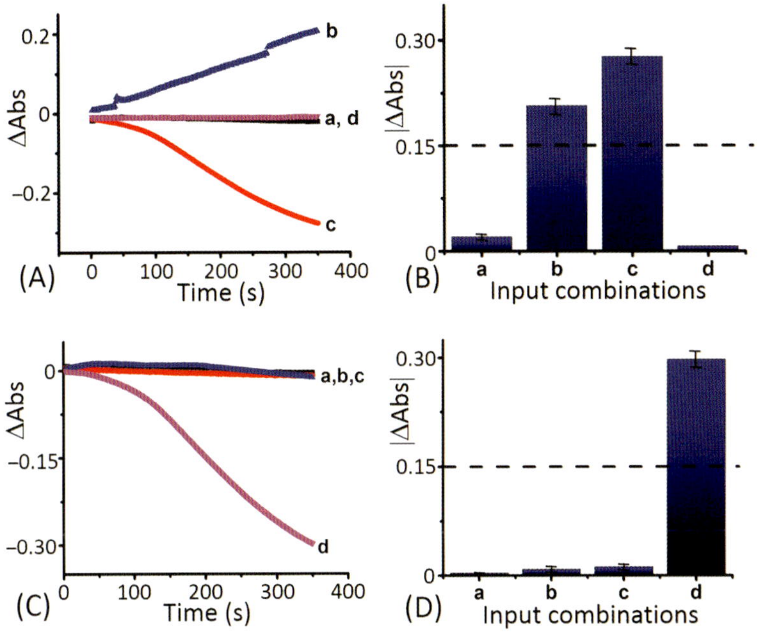

Figure 9.3 Optical responses of the half-adder system to various combinations of the input signals: (a) **0**,**0**; (b) **0**,**1**; (c) **1**,**0**; (d) **1**,**1**. (Note that the logic values are shown for the input signals in the following order: *A*, *B*.) (A) and (B) panels show Output *S* corresponding to the XOR logic operation. (C) and (D) panels show Output *C* corresponding to the AND logic operation. Output signals were measured optically as absolute values of the absorbance change at $\lambda = 420$ nm. Bar charts show the output signals produced after 350 seconds of the reaction time. Absolute value of the absorbance change, |ΔAbs|, of 0.15 corresponded to the threshold, shown as a dashed line, separating logic **0** and **1** values of the output signals. The data shown in the bar charts are average of three independent experiments; the overall distribution of the independent experiments was ca. ±5%. Source: Fratto et al. 2016 [60]. Adapted with permission of John Wiley and Sons.

increase and decrease as the same output. When Inputs *A* and *B* were applied in **0**,**0** and **1**,**1** combinations, the absorbance was not changed due to the absence of the reacting species (**0**,**0** combination) or balanced reactions proceeding in the opposite directions (**1**,**1** combination). Figure 9.3B summarizes the absorbance changes in Output *S* for all combinations of the input signals achieved at the reaction time of 350 seconds. Figure 9.3C shows the kinetics of the absorbance changes in Output *C* processed through the AND gate. As expected for the Boolean AND function, input combinations **0**,**0**; **0**,**1**; **1**,**0** do not result in any absorbance changes, while the input combination **1**,**1** results in the absorbance decrease consistent with the consumption of $[Fe(CN)_6]^{3-}$. Considering Output *C* as the absolute value of the absorbance changes, Figure 9.3D summarizes the output signals measured after 350 seconds of the reactions proceeding in the connected flow cells. Finally, Outputs *S* and *C* represented by the bar charts (Figure 9.3B,D) correspond to the half-adder operation.

9.3 Half-Subtractor

Half-subtractor is used for subtracting one single bit binary digit from another single bit binary digit [59]. Unlike the binary adder that produces a sum output, *S*, and a carry output, *C*, when two binary numbers are added together, the binary subtractor produces difference, *D*, and borrow, *Bo*, outputs (Figure 9.4A). The logic circuitry operating as a half-subtractor can be represented by XOR and AND logic gates operating in parallel, similarly to the half-adder, but with inversion (NOT function) of Input *A* applied to the AND gate (Figure 9.4B). The truth table of the system operation is shown in Figure 9.4C. It should be noted that the NOT–AND pathway (the NOT function was applied to Input *A* channel) can be also represented by the INHIBIT-*A* logic gate, which has the same logic function and identical truth table; however, in the electronic realization of half-subtractor, this function is usually represented by NOT–AND operation (sometimes called "single inversion AND" ["SAND"]). While the electronic realizations of NOT–AND and INHIBIT-*A* logic functions are technically substantially different, the biomolecular system is essentially identical for both logic operations; thus the selection of the equivalent logic circuitry used when modeling the chemical reactions is flexible, allowing both logic representations to be possible. When considering the consistency of discussed system to the electronic counterpart, the NOT–AND logic function has been selected.

The XOR logic operation was performed by two biocatalytic reactions operating in the opposite directions (Figure 9.5). Input *A* represented by Glc (logic value **1**) was processed by the flow cell modified with GDH and resulted in the production of the reduced NADH species. On the other hand, Input *B* represented by Pyr/ATP mixture (logic value **1**) was processed by the cell functionalized with LDH and resulted in the formation of the oxidized NAD^+

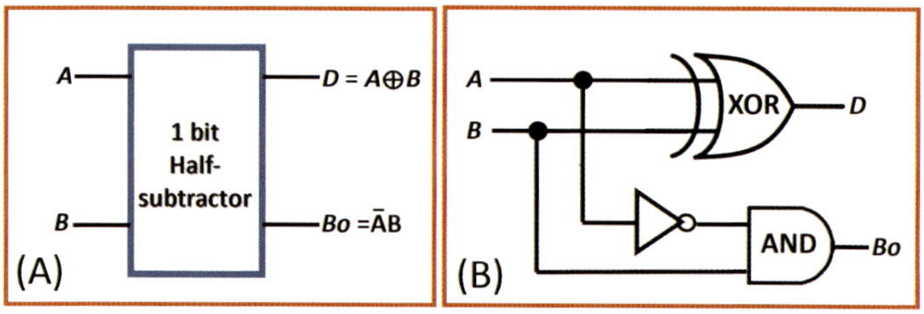

(C)

Input *A*	Input *B*	Borrow digit (Bo)	Difference digit (D)
0	0	0	0
0	1	1	1
1	0	0	1
1	1	0	0

Figure 9.4 The block diagram (A) and equivalent electronic circuitry (B) for the half-subtractor system. (C) Half-subtractor truth table. Standard Boolean notations are used in the schemes [59]. Source: Fratto et al. 2016 [60]. Adapted with permission of John Wiley and Sons.

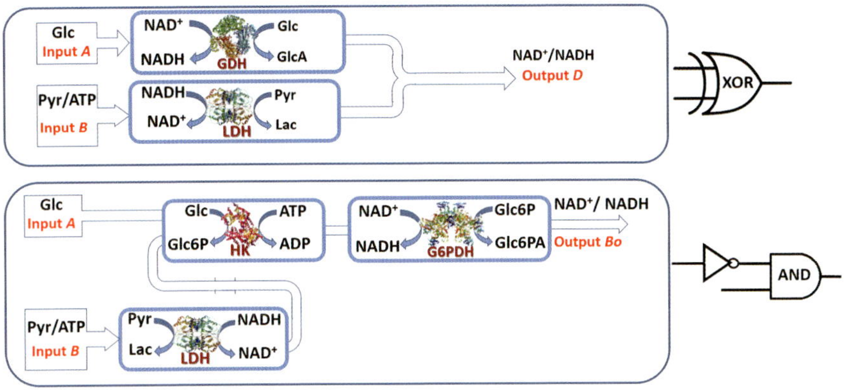

Figure 9.5 Experimental realization of the biocatalytic half-subtractor system in the flow device. Source: Fratto et al. 2016 [60]. Adapted with permission of John Wiley and Sons.

species. These reactions resulted in the decreased or increased absorbance (λ = 340 nm) corresponding to NADH, when input signals were applied in **0**,**1** and **1**,**0** combinations, respectively. Obviously, the absence of the reacting species (Glc and Pyr/ATP) in Inputs *A* and *B* (meaning the input combination **0**,**0**) resulted in no reactions and no absorbance changes in the system. When Inputs *A* and *B* were applied at **1**,**1** combination, the output signal *D* was also at the logic value **0** due to careful balancing of the reductive and oxidative reactions in the cells functionalized with GDH and LDH. It should be noted that the presence of ATP in Input *B* is needed only for its operation in the NOT–AND channel operating in parallel with the XOR gate.

The NOT–AND concatenated logic gates processing Inputs *A* and *B* in parallel to the XOR gate were composed of flow cells modified with LDH, hexokinase (HK), and glucose 6-phosphate dehydrogenase (G6PDH) (Figure 9.5). The inversion of Input *A* (NOT operation) prior to the AND gate resulted in the Output *Bo* being at the logic value **1** only for the input combination **0**,**1**, while all other input combinations (**0**,**0**; **1**,**0**; **1**,**1**) should yield the output value **0**. This is easy to understand, since Input *A* is converted by the NOT gate from **0** to **1** and from **1** to **0**; see Figure 9.4B,C. The input combination **0**,**1** corresponds to the absence of Glc in Input *A* (logic value **0**) and presence of Pyr/ATP mixture in Input *B* (logic value **1**). In the absence of Glc, the reaction biocatalyzed by HK is not activated, regardless of the presence or absence of ATP, which is the second co-substrate of HK. Then gluconate-6-phosphate (Glc6P) is not produced in the HK-modified cell, thus keeping the next cell modified with G6PDH also in an inactive state. At the same time, pyruvate (Pyr) present in Input *B* (logic value **1**) results in consumption of NADH in the cell modified with LDH. The produced NAD^+ flows through all connected cells and finally results in the absorbance decrease in Output *Bo*, which is considered as logic output **1** (note that the output signal is considered as the *absolute* value of the absorbance change). All other input combinations should not change the NADH absorbance, thus producing Output *Bo* at the logic value **0**. Let us consider the reactions proceeding for all other input combinations (Figure 9.5). In case of the input combination **0,0** (absence of Glc, Pyr, and ATP), all reactions in all enzyme-modified flow cells are inhibited, and the NADH concentration is not changed (logic output **0**). When input combination **1,0** (presence of Glc, but absence of Pyr and ATP) is realized, again all reactions are inhibited, because ATP required for the HK-catalyzed reaction is missing and Pyr required for the LDH-catalyzed reaction is also missing, thus producing the output signal **0**. Finally, when input combination **1,1** is applied, NADH is consumed in the cell modified with LDH (note the presence of Pyr in Input *B*). Also, Glc and ATP reacting together with HK produce glucose-6-phosphate (Glc6P), which in the next cell modified with G6PDH reacts producing NADH. The balanced processes (consumption of NADH in the LDH cell and production of NADH in G6PDH cell) result in no changes in the NADH concentration, thus yielding **0** output. Overall, the XOR gate and operating in parallel NOT–AND gates mimic the half-subtractor performance (Figure 9.6).

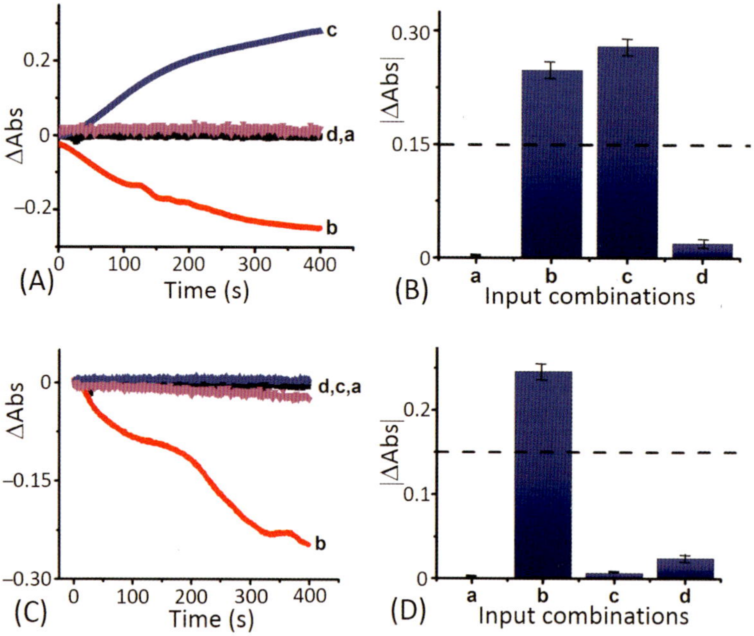

Figure 9.6 Optical responses of the half-subtractor system to various combinations of the input signals: (a) **0,0**; (b) **0,1**; (c) **1,0**; (d) **1,1**. (Note that the logic values are shown for the input signals in the following order: *A*, *B*.) (A) and (B) panels show Output *D* corresponding to the XOR logic operation. (C) and (D) panels show Output *Bo* corresponding to the NOT–AND logic operation. Output signals were measured optically as absolute values of the absorbance change at $\lambda = 340$ nm. Bar charts show the output signals produced after 350 seconds of the reaction time. Absolute value of the absorbance change, |ΔAbs|, of 0.15 corresponded to the threshold, shown as a dashed line, separating logic **0** and **1** values of the output signals. The data shown in the bar charts are average of three independent experiments; the overall distribution of the independent experiments was ca. ±5%. Source: Fratto et al. 2016 [60]. Adapted with permission of John Wiley and Sons.

9.4 Conclusions and Perspectives

The modular design of the enzyme-based system allowed for very simple realization of the half-adder and half-subtractor. This approach allows for easy change of the system configuration, replacing flow cells and alternating the logic/arithmetic function. It should be noted that the same chemical species ($[Fe(CN)_6]^{3-/4-}$ in the half-adder and NADH/NAD^+ in the half-subtractor) were applied in both output channels, producing chemically consistent outputs. While they are easily detected separately in physically different flow channels, the same species cannot be applied for different output signals in a single solution system. In addition to the optical detection of the output signals delivering results of the arithmetic operations, one can expect the use of the output species for triggering various actuation reactions, e.g., substance release processes [62]. Keeping in mind this extension of the present project, different kinds of the redox species ($[Fe(CN)_6]^{3-/4-}$ and NADH/NAD^+) in the half-adder and half-subtractor systems have been applied assuming that they can be used in different actuation processes. The option to extend the arithmetic operations made on metabolite signals to actuation processes (e.g., drug release) controlled by the processed signals is another important feature of the present system that differentiates this system from other molecular/biomolecular half-adder/half-subtractor system. Overall, the extensively applied modular approach to biomolecular computing systems, which was recently used for assembling various reversible logic gates with high complexity [56–58], was presently extended to the arithmetic functions performed by the half-adder and half-subtractor.

Despite the fact that the half-adder and half-subtractor have been successfully realized using enzyme-catalyzed reactions in flow systems, the complexity of full-adder and full-subtractor is much above the limits placed by the present technology; thus they require a different approach, possibly using a combination of different kinds of biomolecules including DNA, DNAzymes, enzymes, aptamers, etc.

References

1 Szacilowski, K. (2012). *Infochemistry – Information Processing at the Nanoscale.* Chichester: Wiley.

2 de Silva, A.P. (2013). *Molecular Logic-Based Computation.* Cambridge: Royal Society of Chemistry.

3 Katz, E. (ed.) (2012). *Molecular and Supramolecular Information Processing – From Molecular Switches to Logic Systems.* Weinheim: Wiley-VCH.

4 Katz, E. (ed.) (2012). *Biomolecular Computing – From Logic Systems to Smart Sensors and Actuators.* Weinheim: Wiley-VCH.

5 Pischel, U. (2007). *Angew. Chem. Int. Ed.* 46: 4026–4040.

6 Pérez-Inestrosa, E., Montenegro, J.-M., Collado, D. et al. (2007). *J. Phys. Chem. C* 111: 6904–6909.

7 Margulies, D., Melman, G., and Shanzer, A. (2006). *J. Am. Chem. Soc.* 128: 4865–4871.
8 Qu, D.-H., Wang, Q.-C., and Tian, H. (2005). *Angew. Chem. Int. Ed.* 44: 5296–5299.
9 Andréasson, J., Straight, S.D., Kodis, G. et al. (2006). *J. Am. Chem. Soc.* 128: 16259–16265.
10 Andréasson, J., Kodis, G., Terazono, Y. et al. (2004). *J. Am. Chem. Soc.* 126: 15926–15927.
11 López, M.V., Vázquez, M.E., Gómez-Reino, C. et al. (2008). *New J. Chem.* 32: 1473–1477.
12 Kuznetz, O., Salman, H., Shakkour, N. et al. (2008). *Chem. Phys. Lett.* 451: 63–67.
13 Brown, G.J., de Silva, A.P., and Pagliari, S. (2002). *Chem. Commun.* 2461–2463.
14 Chen, K. and Schmittel, M. (2016). *Anal. Bioanal. Chem.* 408: 7077–7083.
15 Mondal, D., Bar, M., Maity, D., and Baitalik, S. (2015). *J. Phys. Chem. C* 119: 25429–25441.
16 Li, Y., Han, X., and Deng, Z. (2011). *Langmuir* 27: 9666–9670.
17 Meng, T.T., Xue, L.X., Wang, H. et al. (2017). *J. Mater. Chem. C* 5: 3390–3396.
18 Wu, Z., Dong, B., Zhou, X. et al. (2015). *Chem. Eur. J.* 21: 14301–14304.
19 Wang, J. and Huang, Y. (2014). *IEEE Trans. Nanobiosci.* 13: 146–151.
20 Xu, S., Li, H., Miao, Y. et al. (2013). *NPG Asia Materials* (5): e76.
21 Stojanović, M.N., Stefanović, D., and Rudchenko, S. (2014). *Acc. Chem. Res.* 47: 1845–1852.
22 Yang, C.-N., Hsu, C.-Y., and Chuang, Y.-C. (2012). *Chem. Commun.* 48: 112–114.
23 Yang, C.-N., Chen, Y.-L., Lin, H.-Y., and Hsu, C.-Y. (2013). *Chem. Commun.* 49: 8860–8862.
24 Liu, Y., Dong, B., Wu, Z. et al. (2014). *Chem. Commun.* 50: 12026–12029.
25 Wang, K., Ren, J., Fan, D. et al. (2014). *Chem. Commun.* 50: 14390–14393.
26 Stojanović, M.N. and Stefanović, D. (2003). *J. Am. Chem. Soc.* 125: 6673–6676.
27 Lin, H.-Y., Chen, J.-Z., Li, H.-Y., and Yang, C.-N. (2015). *Sci. Rep.* 5: 10686.
28 Voelcker, N.H., Guckian, K.M., Saghatelian, A., and Ghadiri, M.R. (2008). *Small* 4: 427–431.
29 Wang, Y.F., Sun, J.W., Zhang, X.C., and Cui, G.Z. (2011). *J. Comput. Theor. Nanosci.* 8: 1288–1295.
30 Sun, J.W., Li, X., Cui, G.Z., and Wang, Y.F. (2017). *J. Nanoelectronics Optoelectronics* 12: 375–380.
31 Wang, K., He, M.Q., Wang, J. et al. (2016). *Sci. Rep.* 6: 34810.
32 Li, W., Zhang, F., Yan, H., and Liu, Y. (2016). *Nanoscale* 8: 3775–3784.
33 Li, H.L., Guo, S.J., Liu, Q.H. et al. (2015). *Adv. Sci.* 2: 1500054.
34 Lederman, H., Macdonald, J., Stefanovic, D., and Stojanovic, M.N. (2006). *Biochemistry* 45: 1194–1199.
35 Orbach, R., Wang, F., Lioubashevski, O. et al. (2014). *Chem. Sci.* 5: 3381–3387.
36 Elbaz, J., Lioubashevski, O., Wang, F. et al. (2010). *Nature Nanotechnol.* 5: 417–422.

37 Hu, X.Y., Liu, Y.Q., Qu, X.J., and Sun, Q.J. (2017). *Chem. Commun.* 53: 11181–11184.
38 Zhang, S., Wang, K., Huang, C., and Sun, T. (2015). *Nanoscale* 7: 20749–20756.
39 He, X., Li, Z., Chen, M., and Ma, N. (2014). *Angew. Chem. Int. Ed.* 53: 14447–14450.
40 Benenson, Y. (2009). *Mol. BioSyst.* 5: 675–685.
41 Amir, Y., Ben-Ishay, E., Levner, D. et al. (2014). *Nature Nanotechnol.* 9: 353–357.
42 Auslaender, S., Auslaender, D., Mueller, M. et al. (2012). *Nature* 487: 123–127.
43 Roy, S. and Prasad, M. (2011). *IEEE Trans. Nanobiosci.* 10: 160–171.
44 Roy, S., Prasad, M., Topolancik, J., and Vollmer, F. (2010). *J. Appl. Phys.* 107: 053115.
45 Adamatzky, A. (2017). *ChemPhysChem* 18: 1822–1830.
46 Siccardi, S., Tuszynski, J.A., and Adamatzky, A. (2016). *Phys. Lett. A* 380: 88–97.
47 Adamatzky, A. (2017). *Sci. Rep.* 7: 11755.
48 Zhu, K., Shen, J.Z., Dietrich, R. et al. (2014). *Chem. Commun.* 50: 676–678.
49 Baron, R., Lioubashevski, O., Katz, E. et al. (2006). *Angew. Chem. Int. Ed.* 45: 1572–1576.
50 Katz, E., Wang, J., Privman, M., and Halámek, J. (2012). *Anal. Chem.* 84: 5463–5469.
51 Wang, J. and Katz, E. (2011). *Israel J. Chem.* 51: 141–150.
52 Halámková, L., Halámek, J., Bocharova, V. et al. (2012). *Analyst* 137: 1768–1770.
53 Halámek, J., Windmiller, J.R., Zhou, J. et al. (2010). *Analyst* 135: 2249–2259.
54 Katz, E. and Privman, V. (2010). *Chem. Soc. Rev.* 39: 1835–1857.
55 Katz, E. (2015). *Curr. Opin. Biotechnol.* 34: 202–208.
56 Fratto, B.E. and Katz, E. (2017). *Int. J. Parallel, Emergent Distributed Systems* 32: 139–156.
57 Katz, E. and Fratto, B.E. (2017). *Advances in Unconventional Computing*, vol. 2, Chapter 2 (ed. A. Adamatzky), 29–59. Springer.
58 Fratto, B.E. and Katz, E. (2015). *ChemPhysChem* 16: 1405–1415.
59 Roth, C.H. Jr. and Kinney, L.L. (2010). *Fundamentals of Logic Design*. Cengage Learning.
60 Fratto, B.E., Lewer, J.M., and Katz, E. (2016). *ChemPhysChem* 17: 2210–2217.
61 Privman, V., Zhou, J., Halámek, J., and Katz, E. (2010). *J. Phys. Chem. B* 114: 13601–13608.
62 Katz, E., Pingarrón, J.M., Mailloux, S. et al. (2015). *J. Phys. Chem. Lett.* 6: 1340–1347.

10

Information Security Applications Based on Enzyme Logic Systems

10.1 Keypad Lock Devices as Examples of Electronic Information Security Systems

The previously discussed enzyme-based logic gates [1, 2] and their logic networks [3–5] have been already used in various information processing systems including binary operating (YES/NO) biosensors [6–8], signal-switchable materials [9–11], and bioelectronic devices [12–14]. While operating with different logic and complexity, all these systems have been processing input signals according to their applied combinations, but regardless of the input order. In other words, the generated output signal was not different if Input *A* was applied prior to Input *B*, if the order of the signal application was opposite, or if both signals were applied simultaneously. This property of the biomolecular logic systems excluded the parameter of time in the signal processing. While for many applications overviewed in other chapters the order of the signal application was not important, some special applications, particularly related to the information security systems, require the time parameter as an additional factor controlling the final output generated by the system. The present chapter outlines a few examples of the information processing systems, particularly including parameter of the time, for assembling biomolecular analogs mimicking operation of electronic keypad lock devices [15–17]. Notably, when using a keypad lock device (e.g., in ATM), one should introduce a pin code (e.g., composed of four digits) in specific order (Figure 10.1A). Notably, the input of 1,2,3,4 is not the same as 2,1,3,4 and all other permutations because of the different order of the inputs introduced. In electronic realization of the keypad lock (Figure 10.1B), the input signals are processed through elements performing AND logic operations, which produce high electronic output signal (**1**, **True**) only when each gate receives the high input (**1**) from the preceding gate accompanied with the high input (**1**) applied to this gate. If any input applied to the gate or received from the preceding gate is low (**0**), the final output is **0** (**False**).

Enzyme-Based Computing Systems, First Edition. Evgeny Katz.

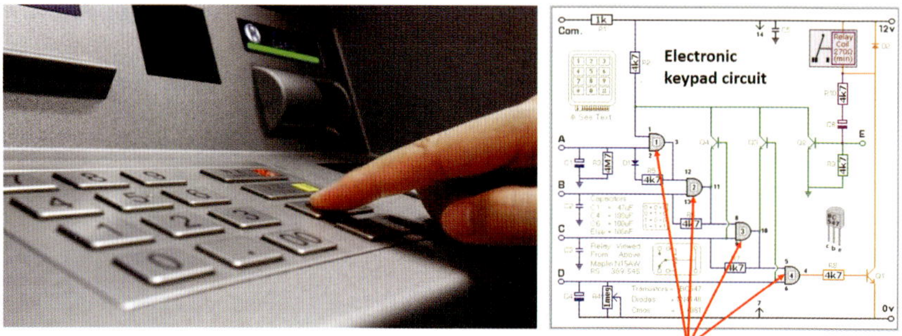

Input in the order 1234 is not the same as 2134 and all other permutations. Electronic elements operating as AND logic gates.

(A) (B)

Figure 10.1 (A) Conventional keypad lock system shown here as a part of an ATM. (B) Generic electronic circuit operating as a keypad lock electronic system – note interconnected AND logic gates.

10.2 Keypad Lock Systems Based on Biocatalytic Cascades

Figure 10.2A shows schematically a biocatalytic cascade activated with three input signals – all applied in the correct order (*A*, *B*, *C*) to generate the output signal **1** (**True**), thus mimicking the electronic version of the keypad lock device (note that in this example only three inputs are used) [15]. The input signals were represented with three enzymes bound chemically to small (1 mm diameter) glass beads: Input *A* was defined as the presence of invertase (Inv), Input *B* was glucose oxidase (GOx), and Input *C* was microperoxidase-11 (MP-11). The enzyme immobilization on the glass beads allowed their separation from the reacting solution, which included a "soup" of all reactants (sucrose, O_2, and ABTS) except the enzymes. When the beads functionalized with Inv were added to the solution, glucose (Glc) was produced from sucrose in the reaction catalyzed by Inv. After the reaction was finished, the reacting solution was separated from the Inv-functionalized beads, and another set of beads modified with GOx was added. This resulted in production of H_2O_2 in the reaction catalyzed by GOx. Then, the GOx-functionalized beads were removed from the solution, and the beads modified with MP-11 were added. This resulted in oxidation of ABTS catalyzed by MP-11 yielding optical absorbance corresponding to the oxidized ABTS ($ABTS_{ox}$) measured at 415 nm and defined as the output signal **1**. The biocatalytic cascade operated as three networked AND logic gates (Figure 10.2B), generating the output **1** only when the input signals (immobilized enzymes) were added in the correct order *ABC* (Figure 10.2C). If the enzyme inputs are added in a different (wrong) order, the reaction cascade cannot proceed till the end, and, thus, the output **0** (**False**) is generated. Indeed, if the beads holding GOx are added prior to the beads with Inv (the input order *BAC*), the reaction cannot proceed in the absence of glucose. All other input permutations (input signals applied in other wrong orders) do not allow production of $ABTS_{ox}$ and thus generate the output **0** (Figure 10.2C). It should

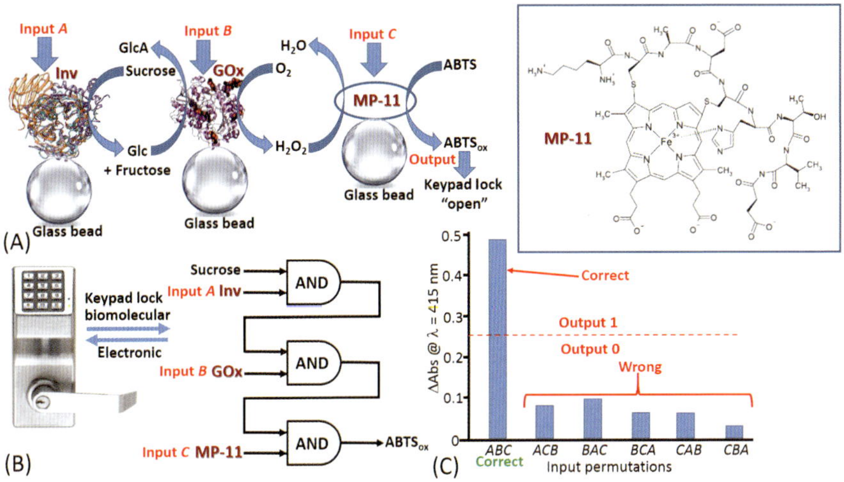

Figure 10.2 (A) Biocatalytic cascade activated with three input signals represented by enzymes, invertase (Inv), glucose oxidase (GOx), and microperoxidase-11 (MP-11; the structure is shown in the frame), all immobilized on glass beads. Note that the signals were removed from the reaction solution prior to application of the next signal. (B) The electronic keypad lock device and the equivalent logic network composed of three concatenated AND logic gates activated with the enzyme signals. (C) Optical responses measured for the biocatalytically produced $ABTS_{ox}$ when the enzyme signals were applied in different order. Note that only *ABC* input order resulted in the high absorbance (Output **1**) corresponding to the produced $ABTS_{ox}$, while all other input permutations did not result in the high output value (**0**).

be noted that the enzymes must be removed from the reacting solution after each signal application step – this is exactly why they were immobilized on glass beads allowing their easy separation from the solution. In case the enzymes are not removed and accumulated in the reacting solution, the system is not sensitive to the order of the signal application. The biomolecular model of a keypad lock device produces the optical output signal, which is considered as **1** (**True**) when the absorbance increases above a threshold level; otherwise the output **0** (**False**) is generated.

Another more advanced version of the enzyme-based keypad lock system was designed to produce an electronic (electrochemical) output signal [16] (Figure 10.3). In order to produce electrical output signal, the biocatalytic cascade processing input signals was connected to a biofuel cell producing the electrical output. A biocatalytic system composed of four enzyme-catalyzed reaction steps was designed to produce pH changes only at the very last biocatalytic step. Three enzymes involved in the biocatalytic cascade were considered as logic input signals and applied in different combinations. When the enzyme inputs were applied in the correct order (Inputs *A*, *B*, and then *C*), the biocatalytic cascade – starting from starch and finishing with gluconic acid – was in succession (Figure 10.3A). In the first reaction step, β-amylase (βAmy, Input *A*) resulted in the biocatalytic hydrolysis of starch and yielded β-maltose. In the second step, maltose phosphorylase (MPh; Input *B*), coupled with acid phosphatase (AcidP; being a part of the "machinery" composition),

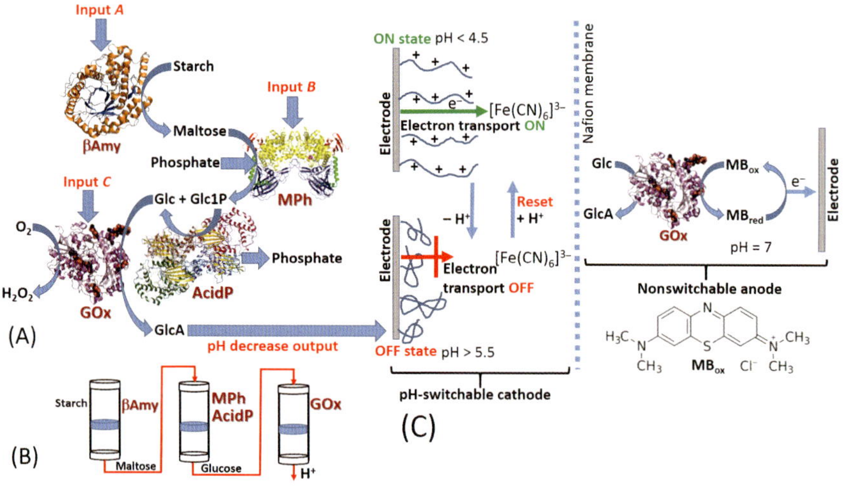

Figure 10.3 (A) Biocatalytic cascade activated with three input signals represented by enzymes, β-amylase (βAmy), maltose phosphorylase (MPh) coupled with acid phosphatase (AcidP), and glucose oxidase (GOx). The final output was represented by gluconic acid (GlcA) decreasing pH value in the reaction solution. (B) The reaction/separation setup allowing application of the enzyme inputs in different permutations (shown in the correct order *A*–*B*–*C*). (C) The pH-switchable biofuel cell for "reading" the output signals generated by the keypad lock system. The cathode was modified with P4VP brush to allow the pH-switchable electrode operation. Note that the electrode was active only at pH < 4.5 when the polymer brush is protonated and swollen, while being mute at pH > 5.5 when the brush is deprotonated and collapsed on the electrode surface. $[Fe(CN)_6]^{3-}$ was used as the oxidizer in the cathodic reaction at the pH-switchable electrode. MB_{ox} and MB_{red} are the oxidized and reduced states of methylene blue mediator transporting electrons from GOx to the anode.

produced glucose from maltose generated in the previous step. In this reaction step, MPh produced glucose (Glc) and glucose-1-phosphate (Glc1P) from maltose consuming phosphate, while AcidP converted Glc1P to the second glucose molecule releasing phosphate. This resulted in the increased yield of glucose and recirculation of phosphate ions in the solution, thus allowing to keep phosphate concentration low, which is important for having low buffer capacity in the solution. In the final step, glucose oxidase (GOx; Input *C*) produced gluconic acid (GlcA) from glucose, thus resulting in the pH decrease. The initial solution, pH 6.7, before applying the enzyme inputs included soluble starch as the primary substrate, inorganic phosphate as a co-substrate for the second enzymatic reaction step, O_2 in equilibrium with air as a co-substrate for the last reaction step, and $K_3[Fe(CN)_6]$ as a redox probe for contacting a modified electrode after all enzyme inputs are applied. The solution also included Na_2SO_4 as a background electrolyte to perform electrochemical measurements. To allow the stepwise application of the enzymes without their mixing, the reactions were performed in ultrafiltration tubes (cutoff 10 kDa), allowing separation of the low molecular products generated at each biocatalytic step from the enzyme inputs (Figure 10.3B). The product generated in the tube was applied to the next reaction step, while the enzyme input was preserved in the reaction volume. This setup allowed easy exchange of the reaction volumes applying the enzyme inputs in six different permutations, *ABC*, *CBA*, *ACB*, *BCA*, *BAC*, and *CAB*, where only the first one corresponded to the correct sequence of the biocatalytic reactions. All other permutations of the enzyme inputs did not facilitate the whole biocatalytic cascade since the correct sequence of the products–reactants was not achieved. Thus, the correct input sequence *ABC* represented the "password" for the biocatalytic cascade to yield gluconic acid as the final product decreasing the solution pH value, while all other input combinations did not result in significant pH changes (Figure 10.4A). Note that the biocatalytic cascade was designed to have the acid as the final product, while all intermediate products do not have acidic properties. The product solution generated at the end of the biocatalytic cascade was applied to switch the interfacial activity of a polymer-brush-modified electrode. As it was demonstrated earlier [18], an indium tin oxide (ITO) electrode functionalized with poly(4-vinyl pyridine) (P4VP) reveals electrochemical activity for soluble anionic redox species switchable by the solution pH values. The polymer brush being in the nonprotonated hydrophobic shrunk state at pH > 5.5 is impermeable for soluble redox species, thus inhibiting their electrochemical reactions. When pH < 4.5 is applied, the pyridine groups in the polymer brush are protonated yielding a positively charged swollen hydrophilic thin film permeable for anionic redox species, thus allowing their electrochemical reactions.

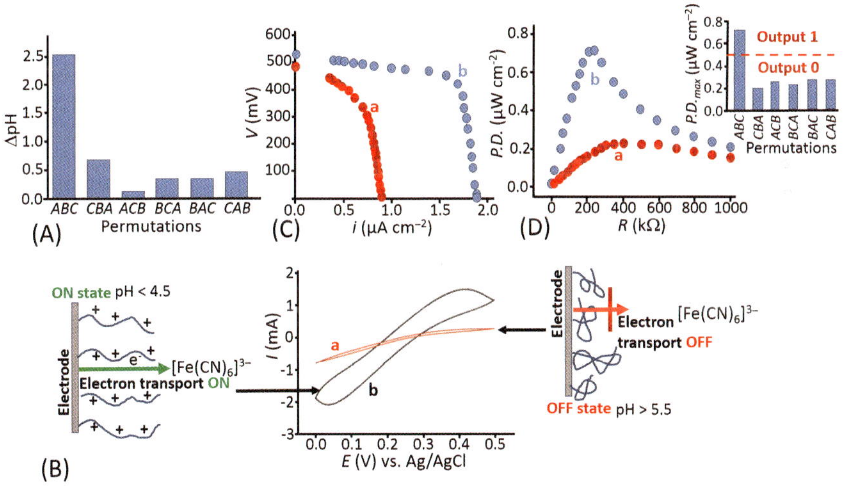

Figure 10.4 (A) pH changes generated at the final step of the reaction cascade (shown in Figure 10.3A) for different permutations of the enzyme inputs. (B) Cyclic voltammograms obtained on the P4VP-ITO electrode in the presence of 10 mM $[Fe(CN)_6]^{3-}$ at different pH values: (a) initial pH 6.7 and (b) pH 4.2 generated upon *A–B–C* sequence of the enzyme inputs. Potential scan rate, 100 mV s^{-1}. The polarization functions (C) and power density output (D) of the biofuel cell obtained at different pH values: (a) initial pH 6.7 and (b) pH 4.2 generated upon *A–B–C* sequence of the enzyme inputs. Inset in (D) shows the maximum power density output generated by the biofuel cell after applying the input signals in different permutations. Source: Halámek et al. 2010 [16]. Adapted with permission of Royal Society of Chemistry.

The electrochemical measurements on the P4VP-brush-modified electrode started at pH 6.7 when the electrode was in the OFF state for the soluble anionic redox probe, $[Fe(CN)_6]^{3-}$, thus showing no waves in the cyclic voltammogram (Figure 10.4B, curve a). When the enzyme inputs were applied in the correct order *ABC* and gluconic acid was produced, the pH value in the last reacting solution was reaching ca. 4.2. This pH value corresponded to the ON state of the P4VP-brush-modified electrode. When this solution was applied to the electrode, a cyclic voltammogram showing the redox process of $[Fe(CN)_6]^{3-}$ was obtained (Figure 10.4, curve b). The waves in the cyclic voltammogram were obtained only when the correct "password" *ABC* was used in the biocatalytic cascade, while all other permutations of the enzyme inputs did not change the initial cyclic voltammogram of the electrode being in the OFF state. After "reading" the "answer" of the enzyme-based keypad lock system by cyclic voltammetry, a biofuel cell was designed as a self-powered "reading" device. A simple model biofuel cell was composed of two ITO electrodes (Figure 10.3C). The cathode was modified with a pH-switchable P4VP polymer brush operating in the presence of $K_3[Fe(CN)_6]$ used as a model oxidizer. The anode was an unmodified ITO electrode operating in the presence of soluble GOx, which oxidized the glucose fuel with the help of a diffusional redox mediator methylene blue (MB) under Ar. The electrodes were separated with a Nafion membrane that allowed the pH changes only in the cathodic compartment. The oversimplified design of the biofuel was specially selected to clearly demonstrate the power control by the keypad lock system without any complications from secondary effects related to the bioelectrocatalytic reactions. The experiment was started at pH 6.7 in the cathodic solution when the switchable cathode was in the OFF state and the entire biofuel cell was "mute" demonstrating low current–voltage (Figure 10.4C, curve a) and low power (Figure 10.4D, curve a). The final solution from the last reaction step in the keypad lock was applied to the cathodic compartment of the biofuel cell. When the correct "password" *ABC* was used, the produced acidic solution, pH ca. 4.2, resulted in the activation of the cathode, thus switching the biofuel cell ON. This resulted in the increasing current and power output generated by the biofuel cell (Figure 10.4C,D, curves b). All other "wrong" permutations of the enzyme inputs did not result in switching "ON" the biofuel cell, thus, keeping the power output low (Figure 10.4D, inset). It should be noted that the difference between the current and power outputs generated by the biofuel cell in the OFF and ON states can be increased upon appropriate optimization of the polymer brush surface density. Since the reacting cocktail is consumed after application of the enzyme inputs, the additional inputs are not possible.

The designed enzyme-based keypad lock system demonstrated the IMPLICATION logic function generating the final output YES only when the correct order of the input signals is applied (the *ABC* "password"). The "answer" YES was obtained in the form of the electrical power produced by the biofuel cell. All other permutations of the enzyme inputs did not result in the activation of the biofuel cell, preserving it in the initial "mute" state. The designed security system operated without the need of any external power source producing electrical power itself when the correct "password" is applied. It should be noted that the present example device was aiming at concept demonstration only, while the real operating system should be based on a microfluidic system (lab-on-a-chip), allowing for its miniaturized design. The number of biocatalytic steps and the respective number of the enzyme inputs might be increased up to 10 without significant noise in the biocomputing network [19]. A different approach to the biomolecular keypad lock system, also connected to a self-powered biofuel cell, has been realized using stepwise immune complex formation upon applying antibody signals in the specific order [17] (see Section 13.3). This approach based on the bioaffinity complex formation, rather than on a biocatalytic cascade, allows the system operation without cross-reactivity of biocatalytic reactions.

10.3 Other Biomolecular Information Security Systems

10.3.1 Steganography and Encryption Methods Based on Bioaffinity Complex Formation Followed by a Biocatalytic Reaction

Given the tailored specificity of antibodies, particularly those labeled with biocatalytic enzymes, the biorecognition process could be exploited not only for mimicking keypad lock devices but also for information encryption and steganography. This application was demonstrated by using a mixture of proteins adsorbed to a polystyrene ELISA plate [20]. The randomly selected masking proteins were mixed in a ratio of 1 : 1 : 1, with the target protein applied in a specific pattern that encoded a hidden message. Application of enzyme-labeled antibodies to the spots containing only masking proteins, which are not encoded with the corresponding target protein, does not result in antibody binding and does not yield any color-producing reaction (Figure 10.5A). The only way to expose the pre-patterned proteins was to add the labeled antibody specific to the target protein, along with a redox dye reacting biocatalytically with the enzyme tag. When the enzyme-labeled antibody is applied, the enzyme will oxidize the dye, resulting in color development (Figure 10.5B), thus exposing the pattern (Figure 10.6). The described encryption technique relies on a digital pattern on a surface. When the enzyme–antibody conjugate binds to the target protein on the surface, the oxidized dye will reveal the hidden digital information (digital **1**), whereas areas without the oxidized dye are considered as digital **0**. The digital information revealed by the immune assay coupled to the biocatalytic reaction (actually performed as a standard ELISA technique) can be read using various encoding methods, in the easy way using ASCII code [20]. The level of encryption can be enhanced by controlling the protein ratio used such that the quantity of the oxidized dye will correspond to the amount of target protein applied at the surface [20]. Thus, the assigned value of digital **1** or **0** can also be controlled and introduces an additional level of information security (or a hierarchy); the correct antibody along with knowledge of the digital threshold will decode the hidden text.

Another variation on the ELISA-inspired encryption is the conjugation of various enzyme labels to the antibodies [20]. In this way, multiple codes can be encrypted on the same ELISA plate; the target is revealed only when the

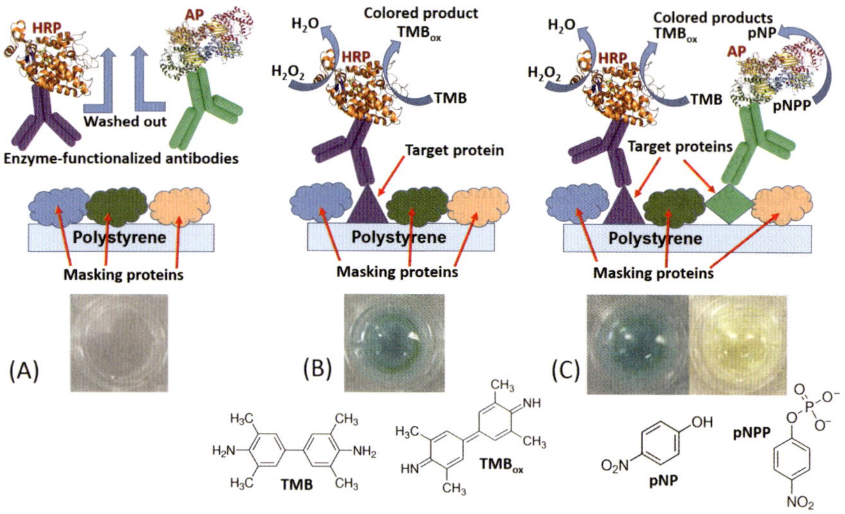

Figure 10.5 Steganography based on the immuno-specific interactions resulting in the color development by the enzyme-labeled complementary antibody similarly to ELISA technique. (A) There is no binding of the enzyme-labeled antibodies to a mixture of masking proteins. (B) The enzyme-labeled antibody (HRP-labeled anti-mouse IgG) binds to the corresponding target protein antigen (mouse anti-human IgG) in the presence of a mixture of masking proteins, thus resulting in the biocatalytic oxidation of TMB. (C) Two enzyme-labeled antibodies (HRP-labeled anti-mouse IgG and AP-labeled anti-rabbit IgG) bind to the corresponding target protein antigens (mouse anti-human IgG and rabbit anti-rat IgG) in the presence of a mixture of masking proteins, thus resulting in the biocatalytic oxidation of TMB and formation of pNP. Inset: Images show protein-modified wells corresponding to the biorecognition systems shown schematically above. The following abbreviations are used: AP, alkaline phosphatase; HRP, horseradish peroxidase; TMB, 3,3′,5,5′-tetramethylbenzidine; TMB$_{ox}$, the oxidized form of TMB; *p*NP, *p*-nitrophenol; and *p*NPP, *p*-nitrophenyl phosphate (see structures at the bottom).

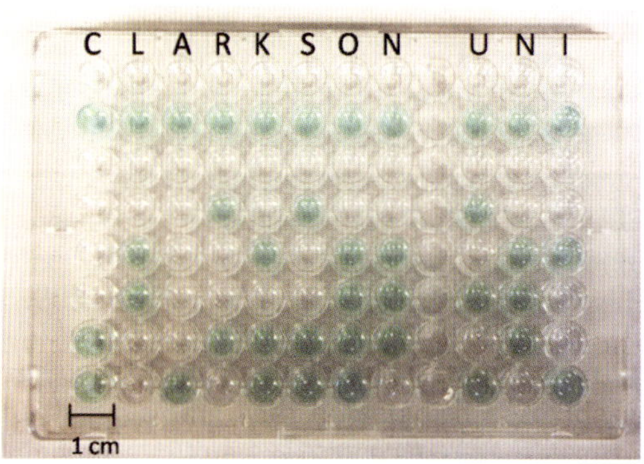

Figure 10.6 Photo image of the ELISA plate with text "CLARKSON UNI" encoded with mouse anti-human IgG in the presence of two masking proteins (goat anti-cat IgG and donkey anti-chicken IgG) after the treatment with HRP-labeled anti-mouse IgG and the color-developing solution for 30 minutes. Source: Kim et al. 2011 [20]. Adapted with permission of John Wiley and Sons.

enzyme-linked antibody adheres and only if the appropriate assay is performed to develop the antibody signal. An ELISA plate modified with several proteins, for example, may include two different target proteins placed in specific patterns that encode two overlapping patterns (Figure 10.5C). After the corresponding enzyme-linked antibodies are added, both patterns can be visualized only if subjected to two separate colorimetric assays, each tailored for the catalytic specificity of the enzyme tag. Therefore, multiple lines of text can be encoded in one device as information or to provide disinformation that distracts from the hidden text.

To advance the antibody encryption technique toward the practical realm, protein spots were deposited on nitrocellulose paper as an "invisible ink" to pattern code onto seemingly blank paper [20]. In this way, steganographic applications can be combined with cryptography. The protein spots can be applied with an inkjet printer for miniaturization and precision [20]. Scaling down the technique to micro-size can also be extended to the detection mode [21]. Recently, atomic force microscopy (AFM) studies demonstrated that the spot size needed to detect a specifically bound antibody layer can be reduced to approximately 3 μm^2, thus scaling down the required spot size by a factor of ca. 100. The molecular and biomolecular systems described are essentially conceptual demonstrations that have inherent limitations, but nevertheless exhibit specificity that may prove valuable in specific niche operational conditions.

The proteins used in the preliminary study (IgG antibodies) already provide a broad spectrum of combinations of the target and masking protein components. However, the future use of antibodies of an individual person or animal toward specific fractions, namely, MHC I class structures, will make the immune steganography absolutely unbreakable due to the absence of detection techniques able to distinguish between the target and mask proteins, which can even come from the same biological species. It should be noted that a similar approach based on DNA molecules [22] instead of proteins provides much less information security because DNA, even in a very small quantity, can be analyzed and read by a well-developed technique.

10.3.2 Barcodes Produced by Bioelectrocatalytic Reactions

Most conventional encryption uses logic and binary functions for essential processing; an example is barcoding. Using biocatalysts to generate barcodes provides an interesting mechanism of encryption processing that is based entirely on the catalytic mechanism of an enzyme. Not all examples of unconventional barcodes employ biogenerated digital sequences [23–30]; encryption tags, for example, are fabricated from specific alignment of metallic nanowires and quantum dots [23, 24]. These types of unconventional barcodes include medical diagnostic models and multiplexed bioanalytical assays that can detect specific proteins or molecules of DNA [24–27, 29]. Generating a barcode in time domain, however, creates significant amounts of binary code, which implies a complementary, reversible on–off toggle. Tailoring this kind of response requires the integration of a biocatalyst with a reliable and sustainable bioelectronic interface. Given that the addition of the inputs into chemical information processing systems is essentially a fluctuating environment, the response integrity, sustainability, and longevity of the bioelectronic interface must be addressed. In other words, an electrode capable of processing relatively large amounts of information must withstand constant input flow and flux over a considerable period of time. As a demonstration, a laccase-functionalized carbon electrode was fabricated, which exhibited sustainable bioelectrocatalytic functionality for 20 days under continuous input fluctuation [31]. The catalytic response can be programmed in such a way that output is dependent on the catalysis of the enzyme. By varying the substrate input under continuous conditions, the first unconventional barcode generated was demonstrated by continuous relay of digital information through a bioelectronic interface (Figure 10.7). The biocatalytic electrode was composed of carbon nanotubes functionalized with laccase to yield cathodic current in the presence of O_2 while being mute under N_2 atmosphere (Figure 10.7D). The electrode longevity was tested by programing an ON/OFF change in the substrate range [31]. In this manner, the enzyme interface responds to an abrupt transition between N_2 (OFF) and O_2 (ON) substrate feeds and forms a series of "bars" observed as an amperometric output that are read immediately without further treatment or manipulation (Figure 10.7A). When plotted by area, the output data of current vs. time was visualized as a series of black or white lines (Figure 10.7B). The barcode was generated by implementing a series of binary codes (Figure 10.7C). In this case, ASCII 12-bit code 39 was chosen to generate a readable code currently available on barcode reader cell phone applications (Figure 10.7E).

The bioelectronic barcode method was demonstrated using simple input sequences and electrode fabrication techniques. Several levels of encryption can be generated, including traditional electronic encryption of the binary sequence before it is translated by the electrode. Furthermore, the bioelectronic interface can be prepared with other biomolecules capable of biocatalytically responding

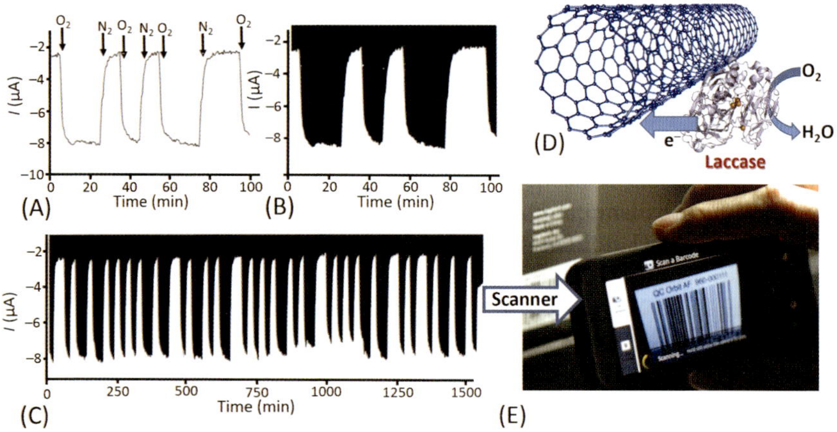

Figure 10.7 Generation of a barcode from direct amperometric output of immobilized laccase. (A) Input of nitrogen- or oxygen-saturated electrolyte generates an amperometric output, which when plotted by area reveals a series of white (nitrogen) or black (oxygen) bars (B). The process can be programmed to generate a specific barcode by varying the residence time, 10 minutes (thin line) and 20 minutes (thick line) as demonstrated for the word TEST (C). The barcode (C) can be directly scanned using a conventional barcode scanner (E) available as a cell phone application. (D) The cartoon showing the enzyme (laccase) bound to an electrode made of carbon nanotubes (buckypaper) and operating as a biocatalyst for O_2 reduction.

to other chemical inputs and biological cascades; in this manner, the electrode can serve as a tailored biological key used to unlock a series of chemical inputs.

10.4 Summary

In conclusion, the inherent selectivity of enzyme biocatalysts and enzyme-functionalized antibodies allows for tailored reactions, in which unique functionalities, unobtainable using existing *in silico* computing techniques, can be realized. Security systems, cryptography, steganography, and code relay have now been demonstrated using unconventional biomolecular methods. Biomolecule-based information processing will not only aid development of medical sensors capable of injury or disease diagnosis but also advance encryption, security systems, and code relay. The design and optimization of such systems could lead to a wealth of tools and technologies, such as logically controlled biofuel cells, programmable signal-responsive transducers, biocompatible security apparatuses, and high-fidelity biosensors.

References

1 Katz, E. (2019). *ChemPhysChem* 20: 9–22.
2 Katz, E. and Privman, V. (2010). *Chem. Soc. Rev.* 39: 1835–1857.
3 Niazov, T., Baron, R., Katz, E. et al. (2006). *Proc. Natl. Acad. USA* 103: 17160–17163.
4 Privman, V., Arugula, M.A., Halámek, J. et al. (2009). *J. Phys. Chem. B* 113: 5301–5310.
5 Privman, V., Zavalov, O., Halámková, L. et al. (2013). *J. Phys. Chem. B* 117: 14928–14939.
6 Katz, E., Wang, J., Privman, M., and Halámek, J. (2012). *Anal. Chem.* 84: 5463–5469.
7 Wang, J. and Katz, E. (2011). *Israel J. Chem.* 51: 141–150.
8 Wang, J. and Katz, E. (2010). *Anal. Bioanal. Chem.* 398: 1591–1603.
9 Katz, E. and Minko, S. (2015). *Chem. Commun.* 51: 3493–3500.
10 Minko, S., Katz, E., Motornov, M. et al. (2011). *J. Comput. Theor. Nanoscience* 8: 356–364.
11 Pita, M., Minko, S., and Katz, E. (2009). *J. Mater. Sci.: Materials in Medicine* 20: 457–462.
12 Katz, E., Bocharova, V., and Privman, M. (2012). *J. Mater. Chem.* 22: 8171–8178.
13 Amir, L., Tam, T.K., Pita, M. et al. (2009). *J. Am. Chem. Soc.* 131: 826–832.
14 Tam, T.K., Pita, M., Ornatska, M., and Katz, E. (2009). *Bioelectrochemistry* 76: 4–9.
15 Strack, G., Ornatska, M., Pita, M., and Katz, E. (2008). *J. Am. Chem. Soc.* 130: 4234–4235.
16 Halámek, J., Tam, T.K., Strack, G. et al. (2010). *Chem. Commun.* 46: 2405–2407.

17 Halámek, J., Tam, T.K., Chinnapareddy, S. et al. (2010). *J. Phys. Chem. Lett.* 1: 973–977.
18 Privman, M., Tam, T.K., Pita, M., and Katz, E. (2009). *J. Am. Chem. Soc.* 131: 1314–1321.
19 Privman, V., Strack, G., Solenov, D. et al. (2008). *J. Phys. Chem. B* 112: 11777–11784.
20 Kim, K.-W., Bocharova, V., Halámek, J. et al. (2011). *Biotechnol. Bioengin.* 105: 1100–1107.
21 Volkov, D., Strack, G., Halámek, J. et al. (2010). *Nanotechnology* 21: 145503.
22 Clelland, C.T., Risca, V., and Bancroft, C. (1999). *Nature* 399: 533–534.
23 Xiang, Y., Zhang, Y., Chang, Y. et al. (2010). *Anal. Chem.* 82: 1138–1141.
24 Kim, J., Seo, K., and Wang, J. (2004). *Conf. Proc. IEEE Eng. Med. Biol. Soc.* 1: 137–140.
25 Demirok, U.K., Burdick, J., and Wang, J. (2009). *J. Am. Chem. Soc.* 131: 22–23.
26 Thaxton, C.S., Elghanian, R., Thomas, A.D. et al. (2009). *Proc. Natl. Acad. Sci. USA* 106: 18437–18442.
27 Appleyard, D.C., Chapin, S.C., and Doyle, P.S. (2011). *Anal. Chem.* 83: 193–199.
28 White, K.A., Chengelis, D.A., Gogick, K.A. et al. (2009). *J. Am. Chem. Soc.* 131: 18069–18071.
29 Gunnarsson, A., Sjovall, P., and Hook, F. (2010). *Nano Lett.* 10: 732–737.
30 Dejneka, M.J., Streltsov, A., Pal, S. et al. (2003). *Proc. Natl. Acad. Sci. USA* 100: 389–393.
31 Strack, G., Luckarift, H.R., Nichols, R. et al. (2011). *Chem. Commun.* 47: 7662–7664.

11

Enzyme Logic Digital Biosensors for Biomedical, Forensic, and Security Applications

11.1 Introduction: Short Overview

The chapter overviews recent advances in digital biosensors operating in the binary (YES/NO) mode, mostly emphasizing biomedical applications of enzyme-based logic systems [1], particularly for the analysis of pathophysiological conditions associated with various injuries. Novel biosensors digitally processing multiple biomarker signals produce a final output in the form of YES/NO response through Boolean logic networks composed of biomolecular systems. The biocomputing approach applied to biosensors leads to a high-fidelity biosensing compared to traditional single-analyte sensing devices. By processing complex patterns of multiple physiological biomarkers, such multi-signal digital biosensors should have a profound impact on the rapid diagnosis and treatment of diseases and particularly can provide timely detection and alert of medical emergencies along with immediate therapeutic intervention. The novel biosensing concept has been exemplified with the systems for logic analysis of various injuries, including soft tissue injury, traumatic brain injury, liver injury, abdominal trauma, hemorrhagic shock and oxidative stress. The biocomputing systems analyzing injury biomarkers were realized in serum solutions mimicking *in vivo* operation. Biochemical filter systems were designed to convert a convex-shaped concentration response typical of biochemical reactions to a sigmoidal function characteristic of electronic digital devices, thus suppressing noise at the binary **0** and **1** logic points and improving digitalization of the output signals. The attractive performance of such new enzyme logic biosensors holds considerable promise for diverse practical and important future applications, including novel forensic and security applications, which are also briefly discussed here.

11.2 From Traditional Analog Biosensors to Novel Binary Biosensors Based on the Biocomputing Concept

Biosensors, from the very first prototype pioneered by Clark in 1962 [2] to sophisticated modern devices [3], are all based on the same general concept: a biomolecular reaction with analyte species followed by transduction of a

Enzyme-Based Computing Systems, First Edition. Evgeny Katz.

chemical signal to an electronic one. Suitable biomolecular reactions can be based on enzyme-catalyzed processes [4] or biorecognition/bioaffinity events [5] using immune-specific [6], DNA-specific [7], or bioreceptor-specific [8] interactions. The major advantage of all these processes is the high specificity in the biorecognition of a single selected analyte. The results of the biochemical reactions are usually transduced to electronic signals by electrochemical [9], optical [10], or other [11] physical means. The entire biosensor assembly includes a biosensing interface integrated with an electronic transduction instrument, a power supply, and an electronic (frequently computerized) signal-processing unit (Figure 11.1A). Usually, the output signal is linearly proportional to the analyte concentration and can be characterized by its dynamic concentration range, where the linear dependence on the analyte concentration is preserved, and selectivity (being independent of various interfering species). Typically, a biosensor provides quantitative information on the concentration of a single analyte, e.g., glucose [12]. Many different bioanalytical assays, e.g., enzyme-linked immunosorbent assay (ELISA) [13], while relying on different recognition events, function in a similar manner. Sometimes the differences between biosensors and bioanalytical assays are merely technical, e.g., when the biomolecular reaction is not fully integrated with the transduction interface of the instrument converting the chemical signal to electronic one. Simultaneous analysis of several different species is carried out by biosensor or bioassay arrays. Such arrays generate signals from multiple analytical channels working in parallel, each channel providing quantitative information on one specific analyte [14]. The resulting signals can then be processed by a computer (Figure 11.1B). In "field" situations, when extensive use of computers or human involvement is not practical, drawing quantitative conclusions from the results obtained by multichannel analysis can be challenging, because the concentrations of different analytes typically span wide ranges of values, and their direct inspection is not definitive. Presently, very few applications generate a bioanalytical result with a qualitative (binary) YES/NO conclusion. This format can be useful when there is no need for precise,

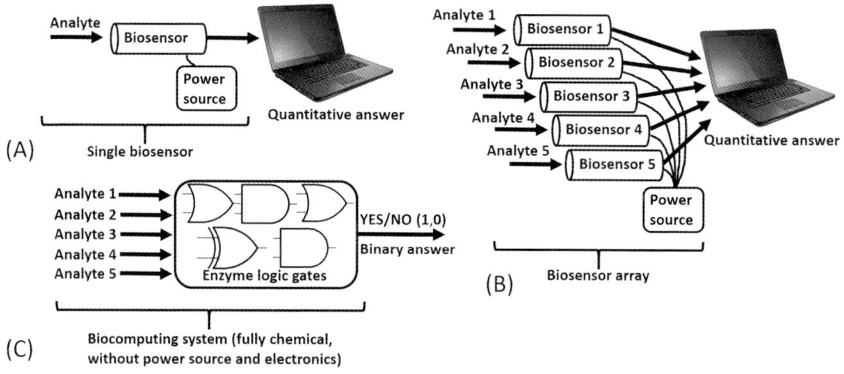

Figure 11.1 Different approaches to biosensing: (A) a single-analyte biosensor, (B) an array of parallel biosensors for multi-analyte analysis with the computer-processed multichannel signals, and (C) multi-analyte analysis carried out by a biomolecular logic system with the input signals chemically processed and the final output generated in the binary YES/NO form.

quantitative measurements, e.g., in a pregnancy test [15] or for detection of chemical weapons on a battlefield [16] when a rapid answer is important. Novel approaches to qualitative analysis providing the final result in the YES/NO form would be beneficial for various end users and point-of-care applications [17] in medicine, homeland security, or military applications, requiring rapid simultaneous analysis of multiple analyte species.

A successful approach has recently been advanced based on unconventional chemical computing [18] specifically on recently pioneered biomolecular computing systems [19]. Such biocomputing systems based on proteins/enzymes [20], DNA [21], RNA [22], DNAzymes [23], and whole cells [24] can perform logic operations processing multiple biochemical input signals. On the basis of biomolecular systems, various Boolean logic operations, such as AND, OR, XOR, NOR, NAND, INHIB, and XNOR, were realized [19]. Sophisticated networks composed of several concatenated biomolecular logic gates performing complex logic operations were designed for unconventional computing applications [19]. Particularly, rapid progress was recently achieved in the area of enzyme-based biocomputing systems, allowing their assembling in the form of concatenated logic networks composed of many gates performing various Boolean logic operations and including non-Boolean elements, such as enzyme-based multiplexer and demultiplexer, amplifier, and filter, mimicking their electronic counterparts [20]. Although chemical information processing systems were originally considered exclusively for computational applications [25], it has recently been realized that they have features suitable for analytical/bioanalytical use [26]. These systems can analyze several biochemical signals according to a predesigned "program" and generate a binary "YES/NO" answer without using a computer. Such chemical testing, analogous to a pregnancy test, might be convenient for certain end user and point-of-care applications (Figure 11.1C). Networks with computational steps that involve only biochemical processes [19, 20] are being investigated for new technological capabilities that include multi-input biosensors with new functionalities [26]. New approaches are being explored, allowing one to reduce the use of batteries, inorganic leads, and electrical power supply for those stages of information processing that occur during biomedical testing, in implantable devices, and toward fast decision-making steps (e.g., therapeutic intervention). The next section of this chapter describes how the biomedical analysis can benefit from the use of biomolecular information processing (biocomputing) systems. The following sections overview the application of binary biosensing methods based on the enzyme logic gates to various types of injury biomarkers, particularly concentrating on battlefield injuries. This selection is mostly based on the practical need of rapid analysis of injury conditions in the absence of hospital facilities, thus providing fast preliminary evaluation to allow immediate therapeutic action. Obviously, the developed approach can be used for analysis of regular patients suffering from different health problems; however, one should take into account that the discussed methods provide only preliminary limited information, which might be not enough for comprehensive medical analysis. In other words, the approach discussed here is beneficial mostly in the conditions when hospital facilities are not available and rapid analysis is critically important. The methods primary developed for biomedical analysis have been successfully extended to novel forensic and security applications of such binary biosensors/bioassays. These advances will be briefly presented at the end of the chapter.

11.3 How Binary Operating Biosensors Can Benefit Biomedical Analysis: Requirements, Challenges, and First Applications

When a biomedical analysis aimed at recognizing a pathophysiological dysfunction is performed with a biosensor, each specific biomarker should be analyzed separately according to a traditional approach. The search for biomarkers signaling various medical problems has become a very important area of medical research [27]. However, diagnostic conclusions can rarely be based on the analysis of just a single highly specific biomarker. Moreover, biomarkers may appear at low concentrations in a complex mixture with many other biomolecular species of similar structures and properties. For example, neuron-specific enolase is considered as a specific biomarker for traumatic brain injury (TBI) [28]. However, its concentration in physiological fluids is much lower than that of generic enolases with similar biocatalytic properties. Furthermore, elevated levels of neuron-specific enolase might result from some other brain dysfunctions, which are not directly related to TBI [29]. Therefore, even the species, which is considered as a biomarker, might not be specific enough to draw a reliable biomedical conclusion based on a single-species analysis. A standard solution of the specificity problem has been to analyze a set of less specific biomarkers present simultaneously with overlapping specificity (Figure 11.2), and in the biocomputing approach, such analysis is realized via a cascade of biocatalytic reactions.

A biocomputing approach can reduce technical problems in the analysis of highly specific biomarkers by analyzing species appearing at much higher concentrations. For example, glutamate and lactate dehydrogenase, analyzed together for their pathophysiologically elevated concentrations, could provide an attractive alternative for the challenging analysis of neuron-specific enolase to diagnose TBI [30]. Analysis of several biomarkers, each with limited selectivity, requires a proper design of a logic network for processing of the analyte signals. For example, analysis of two biomarkers for their simultaneous presence requires a biochemical reaction mimicking the Boolean AND logic operation [31]. In case of multiple biomarkers, more sophisticated logic networks composed of several logic gates performing various logic operations might be needed [32]. For example, a multienzyme/multi-input logic network composed of many concatenated AND/OR gates equipped with different biomolecular switches controlling the pathways involved in the input processing was designed for the analysis of different injury biomarkers [32]. These logic operations can be performed as a sequence of biochemical reactions without the use of electronic computers [19, 20]. Both optical and electrochemical transduction modes can be used for detecting the products of such logic operations. Biomedical use of logic gates

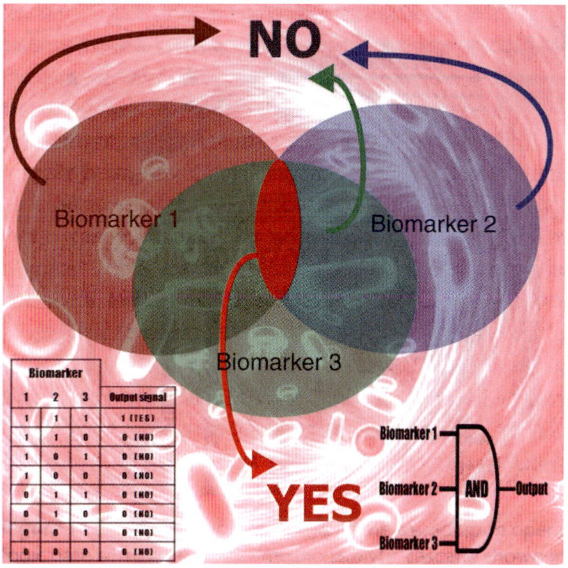

Figure 11.2 Simultaneous presence of biomarkers with limited specificity indicates a definitive YES conclusion, schematically represented by the overlapping region in the diagram. In the biochemical computing approach, this conclusion is realized as the output YES (or **1**) of a multi-input AND logic gate, obtained only when all the three inputs are in the ranges corresponding to **1**. The output NO (or **0**) is obtained for all the other combinations of inputs. The "truth table" for such a three-input AND gate and its logic diagram are also shown. Source: Katz et al. 2012 [1]. Adapted with permission of American Chemical Society.

and sophisticated logic networks is a rather new and undeveloped research area, being advanced in parallel with similar developments, such as logic operations realized with synthetic organic molecules aimed at computational applications [18, 33].

Biomolecular information processing systems have already been successfully applied to analyze protein libraries associated with multiple sclerosis [34]. Biosensor systems for detection of genetic modifications in avian influenza were developed based on the DNA computing principles, involving various oligonucleotide signals being processed by a DNA logic network [35]. Coupling enzyme logic systems with controlled self-assembly of nanoparticles allowed AND/OR logic responses to matrix metalloproteinases MMP2 and MMP7 [36]. In this study, the enzymes used as the input signals were important cancer biomarkers. MMP2 is overexpressed in many cancers, including breast cancers, and is an indicator of cancer invasiveness, metastasis, and angiogenesis, while MMP7, a protease with broader substrate specificity, is thought to facilitate early stages of mammary carcinoma progression. The use of input signals with obvious biomedical relevance has represented a significant step in the development of enzyme logic systems aiming at their practical applications. However, the following important issues have required additional attention: (i) In model studies, the logic **0** values of the input signals were frequently taken as the complete absence of the biomaterial, whereas for practical applications they should be defined as normal physiological concentrations. (ii) The logic **1** values of the input signals did not always correspond to the concentrations expected *in vivo*. Instead, they were selected as convenient concentrations, which sometimes significantly exceeded pathophysiological levels. (iii) Processing of the input signals according to different logic schemes did not always correspond to their diagnostic uses. For example, the proteases MMP2 and MMP7 were applied to activate the AND and OR logic gates without justification of the logic operation needed for the appropriate biomedical conclusion [36]. Some of these issues, which are particularly important for practical biomedical applications, were addressed in recent research and explained in the following sections. Logic systems for the analysis of biomarkers characteristic of various battlefield injuries, which are important for immediate field-based decision-making and therapeutic action (in the absence of hospital facilities), have been developed [32, 37, 38] and then theoretically modeled and optimized [39].

11.4 Binary (YES/NO) Analysis of Liver Injury Biomarkers: From Test Tube Probes to Animal Research

Let us illustrate how logic gates can be optimized for practical biomedical analytical applications and interfaced to electronic transducing devices amplifying the chemical signal. For this discussion, in order to keep it simple, we will consider only one example system performing a single AND logic operation (many other examples are briefly explained in the next section). The selected example will allow one to follow the development of the system from its biosensoric use to bio-actuation and from analysis of model solutions artificially spiked with biomarkers to the assay of real biological samples containing naturally produced biomarkers. It should be noted, however, that much more sophisticated logic networks composed of many concatenated logic gates and processing more than two input signals were designed for computational and biosensoric applications [19, 20].

We consider the system [38] activated by two biomarkers mimicking an AND logic gate and signaling liver injury (LI) when the concentrations of both biomarkers are elevated from the normal to pathophysiological levels. The two inputs are enzymes, alanine transaminase (ALT) and lactate dehydrogenase (LDH), which are biomarkers characteristic of LI [40]. It should be noted that each of them alone is not specific enough to indicate LI; however, their simultaneous increase in concentration, from normal to pathophysiological levels, provides an unambiguous evidence of LI [41]. Logic **0** and **1** levels of ALT (0.02 and $2\,U\,ml^{-1}$) and LDH (0.15 and $1\,U\,ml^{-1}$) input signals were selected in order to mimic meaningful circulating levels of these biomarkers under normal and pathophysiological conditions, respectively [40–43]. The input signals (ALT and LDH enzymes) were applied in four combinations (**0**,**0**; **0**,**1**; **1**,**0**; and **1**,**1**) to a solution containing all needed substrates: glutamate (Glu), alanine (Ala), and NADH with experimentally optimized concentrations. The biocatalytic reactions allowed for combining both enzyme signals in one reaction cascade

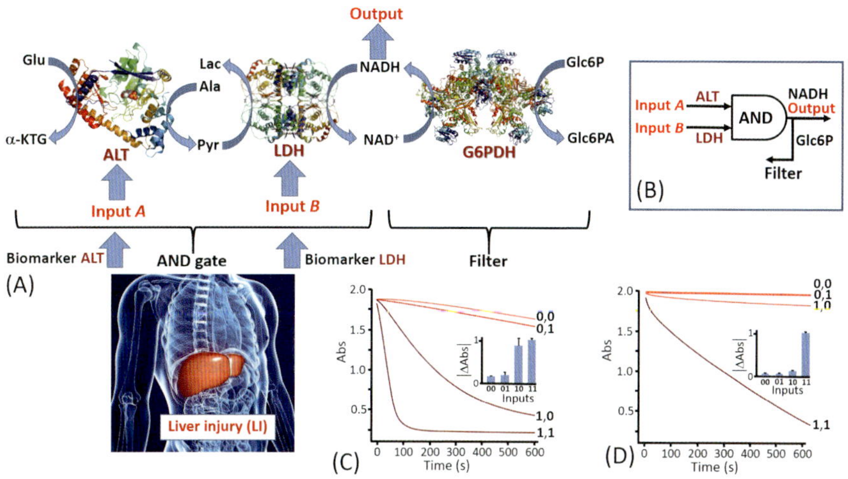

Figure 11.3 (A) The biocatalytic cascade for analysis of LI, activated by enzymes ALT and LDH as inputs for the AND logic gate. (B) Schematically shown AND logic gate for the analysis of the input signals. Note that the output signal was defined in this system as the absolute values of the absorbance change, thus the logic operation corresponds to the AND gate. (C) Optical absorbance changes corresponding to the decreasing concentration of NADH upon application of different combinations of input signals in the absence of the "filter" system. The inset shows the normalized output signals measured at 600 seconds from the reaction initiation. (D) The absorbance changes for the biocatalytic system shown in (A) operating with the "filter" resetting the output signal, back to nearly its zero level, as long as the filter-activating substrate is not consumed. The inset in (D) shows the normalized output signals measured at 600 seconds with the improved separation between the logic outputs **0** and **1**. The insets in (C) and (D) show the absolute values of the absorbance changes, |ΔAbs|.

with the cooperative catalysis provided by both enzymes. On the basis of the sequence of the biochemical reactions (Figure 11.3A), the final result, oxidation of NADH that causes the decrease in the optical absorbance (Figure 11.3C), should be obtained only upon concerted work of the two enzyme biomarker inputs, thus representing the AND logic gate (Figure 11.3B). However, it should be remembered that logic **0** values in their present definition are not the absence of the enzymes but rather their presence at normal physiological levels. Therefore, the decrease in absorbance is observed not only for the **1**,**1** input combination but also to some extent for inputs **0**,**0**; **0**,**1**; and **1**,**0** (Figure 11.3C). When the readout time interval is limited to 50–200 seconds, the absorbance decrease is measurably larger for the **1**,**1** input combination, defining output **1**. The other three input combinations yield smaller absorbance changes, defining output **0** (Figure 11.3C). However, for reaction times exceeding 200 seconds, which are relevant for recently investigated actuation applications [44], the absorbance decrease for the **1**,**0** input combination becomes comparable with one for **1**,**1** inputs (Figure 11.3C). At sufficiently long reaction times, the results for the **1**,**0** and **1**,**1** input combinations will become indistinguishable, and the AND gate will no longer be realized (Figure 11.3C, inset).

In order to increase the gap separating output signals **0** and **1**, a "filter" process [45] was added, consuming the chemical product NAD^+, converting it back to

NADH for small input concentrations (Figure 11.3A). This has allowed to achieve high quality signal separation for the reaction times as long as 600 seconds and beyond (Figure 11.3D). It is likely that such "filter" processes can potentially be implemented with any so-called NAD^+-dependent dehydrogenase [46], e.g., glucose dehydrogenase activated by physiological amounts of glucose. However, aiming at the eventual application of the system in a physiological environment, glucose-6-phosphate dehydrogenase (G6PDH) was selected as the filter enzyme. It is activated by D-glucose-6-phosphate (Glc6P), which does not interfere with glucose naturally existing in blood, thus allowing one to tune the Glc6P concentration independently on the physiological glucose concentration. The filter system works in the following way: In the presence of G6PDH and Glc6P, the biocatalytically produced NAD^+ is converted back to NADH. Thus, absorbance changes are prevented until Glc6P is totally consumed. Only then the depletion of NADH can fully set in, resulting in the absorbance decrease. The delay in the biocatalytic oxidation of NADH is controlled by the amount of the added Glc6P and can be optimized. A comprehensive approach to the filter performance optimization could include detailed analysis of the reactions kinetics [45]. However, a simple experimental optimization might suffice. Addition of the G6PDH–Glc6P (enzyme substrate) filter to the biocatalytic cascade activated by ALT–LDH biomarker inputs (Figure 11.3A) has allowed a much better separation of the output signals generated by the system for the **1**,**1** vs. all the other combinations of the inputs (Figure 11.3D). However, while improving the binary signal separation, such filtering can decrease the overall signal strength that could be an added source of relative noise [47]. Thus, filtering is useful at sufficiently large times, such that the decrease in the absorbance reaches its saturation, of relevance in actuation applications [44]. When the output signals were measured at 600 seconds, the desired system operation corresponding to the high-tolerance AND logic realization was obtained in the presence of the filter (Figure 11.3D, inset). Good quality separation of the **0** and **1** output signals was found to persist at much larger times as well, up to three hours. The robustness of this analytical system has allowed its use in human serum solutions [38].

A system similar to that described in Figure 11.3A, but with glucose dehydrogenase (GDH) as the filter enzyme, was used for electrochemical transduction/amplification of the signals generated by the AND logic gate activated by ALT and LDH biomarkers [44] (Figure 11.4). The reaction biocatalyzed by the filter enzyme provided improved resolution between the output **0** and **1** signals as described above. In addition, glucose was oxidized in the "filter" biocatalytic step resulting in the formation of gluconic acid and lowering the pH value. A pH-switchable ITO electrode modified with poly(4-vinyl pyridine) (P4VP) was activated for electrochemical reactions only when the solution pH value reached the pK_a of the P4VP polymer brush. In the pH range below pK_a, protonation of the polymer brush produced a swollen state permeable for anionic redox species. The pH changes resulting in the electrode activation

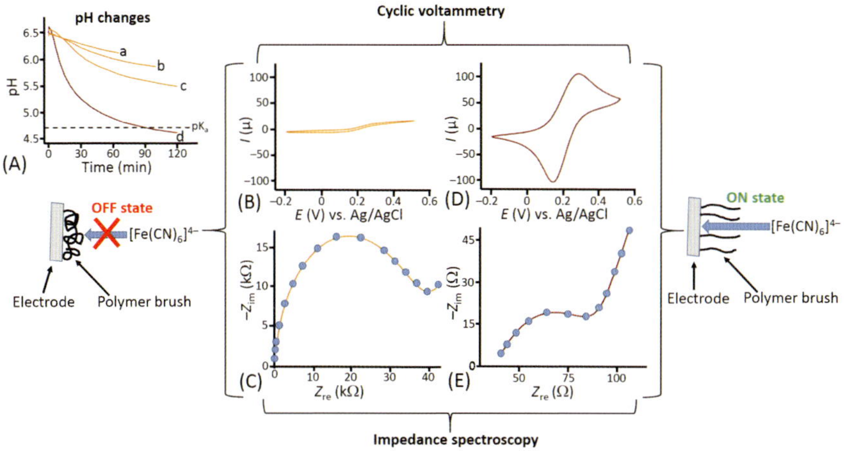

Figure 11.4 Electrochemical transduction of the signals generated by the AND logic gate for analysis of LI activated by ALT and LDH input signals: (A) pH changes generated *in situ* by the biocatalytic cascade activated with various combinations of the ALT/LDH signals: (a) **0,0**; (b) **0,1**; (c) **1,0**; and (d) **1,1**. The dashed line corresponds to the pK_a value of the P4VP-brush immobilized on the electrode surface. (B, D) Cyclic voltammograms, 10 mM $K_4[Fe(CN)_6]$, 100 mV s^{-1}, obtained at the electrode in the OFF and ON states, respectively. (C, E) Impedance spectra, 10 mM $K_4[Fe(CN)_6]$, bias potential of 0.17 V, obtained at the electrode in the OFF and ON states, respectively. Note the different scales in the impedance spectra shown in (C) and (E) panels. A P4VP-brush-modified ITO electrode was used in all electrochemical measurements. The biocatalytic cascade is essentially the same as in Figure 11.3A, except the filter enzyme which was here glucose dehydrogenase (GDH), the substrate and product operating with GDH were glucose and gluconic acid, respectively. Source: Katz et al. 2012 [1]. Adapted with permission of American Chemical Society.

were achieved only when both biomarker inputs, ALT and LDH, were applied at their **1,1** logic values corresponding to the conditions of LI (Figure 11.4A). Any other combinations of the input signals (**0,0**; **0,1**; and **1,0**) did not produce pH changes reaching the polymer pK_a value and thus did not result in the electrode activation (Figure 11.4A). The inactive state of the pH-switchable electrode was characterized by cyclic voltammetry and Faradaic impedance spectroscopy, which demonstrated no peaks and a large electron transfer resistance, respectively (Figure 11.4B,C). This is consistent with the properties of the P4VP polymer brush in the neutral (nonprotonated) state, which is hydrophobic and nonpermeable to ionic redox species. The active state of the electrode was obtained only upon application of the **1,1** combination of the input signals and showed peaks in the cyclic voltammogram typical for the reversible electrochemical reaction of $[Fe(CN)_6]^{3-}$ redox probe (Figure 11.4D), while the impedance spectrum showed the corresponding decrease in the electron transfer resistance (Figure 11.4E) (note the difference in the impedance scales of the (C) and (E) panels). Utilization of a pH-sensitive electrode allowed not only electrochemical transduction of the biochemical output signal generated by the biocatalytic cascade but also its significant amplification. Small concentration change of NAD^+, which is required for glucose oxidation, sufficed for the electrode activation for the redox probe used at a large concentration.

It should be noted that the filter reaction increasing the separation between the output signal **0** and **1** logic values is critically important for achieving the results described previously. In the described systems, ALT and LDH were systematically added to the analyte solutions in four different combinations: **0**,**0**; **0**,**1**; **1**,**0**; and **1**,**1**, where **0** and **1** input values corresponded to the normal and pathophysiological concentrations of the biomarkers. The input combinations **0**,**0** and **1**,**1** correspond to the normal physiological and liver-injury conditions, respectively, while the combinations **0**,**1** and **1**,**0** have medical meanings unrelated to the liver injury. In order to analyze performance of the system under real physiological conditions, the samples containing biomarkers should be obtained from biological sources rather than prepared in a laboratory. A well-established [48] model for porcine injury was utilized to obtain samples mimicking physiological conditions of human liver injury. Assuming that porcine samples contain the biomarker concentrations similar to the human physiology, time-dependent decrease of NADH absorbance during the reaction of the analytical system with the porcine serum samples has been studied [49]. The bottom section of the absorbance decay (red curves) corresponds to the porcine samples originating from the liver injured animals (elevated concentrations of the biomarker enzymes), while the top part (blue curves) was obtained for control animals without the liver injury (with the biomarker enzymes present at the normal physiological concentrations) (Figure 11.5). These responses are statistically different from each other (Figure 11.6). The established difference between normal physiological and pathophysiological levels of the biomarkers allows one to distinguish liver-injured animals from the control group.

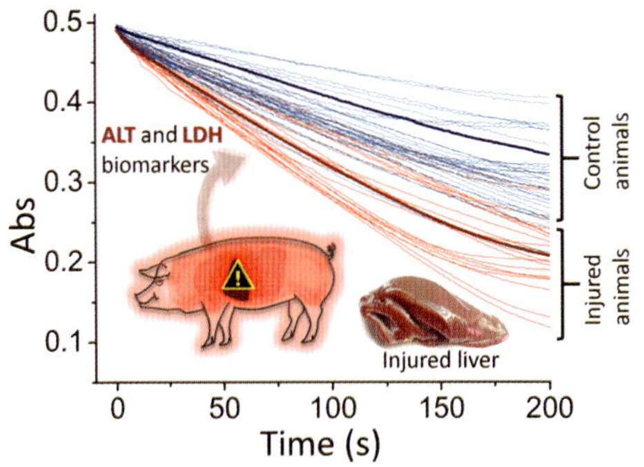

Figure 11.5 Absorbance changes corresponding to the consumption of NADH upon operation of the analytical system activated by porcine samples naturally containing ALT and LDH biomarkers. The bottom (red) traces correspond to the application of porcine samples from the liver-injured animals, while the top (blue) traces correspond to the control group of animals without liver injury. The experimentally measured optical responses correspond to the natural distribution of the injury biomarkers in different animals. Bold solid curves show the average responses for both groups. Source: Katz et al. 2012 [1]. Adapted with permission of American Chemical Society.

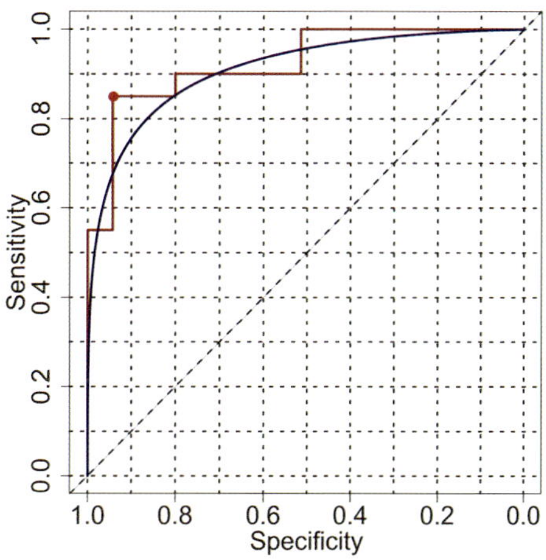

Figure 11.6 Receiver operating characteristic (ROC) empirical (red) and smoothed (blue) curves for the analytical system. Random choice is denoted by the gray diagonal line. The red point on the plot corresponds to the best sensitivity/specificity pair (the best tradeoff between them) giving the most accurate cut-off point for discrimination between pathological and non-pathological porcine serum samples. The statistical analysis of the data obtained confirmed that the injured animals can be distinguished from the control (healthy) animals with probability of 95%. Source: Halámková et al. 2012 [49]. Adapted with permission of Royal Society of Chemistry.

11.5 Further Examples of Injury Biomarker Analysis Using AND/NAND Logic Gates

While the liver injury biomarker analysis explained earlier required the additional filter step, particularly at long gate operation time, some other similar systems aiming at the analysis of other injury biomarkers may operate well enough without filter steps (this does not mean that the filter steps should not be considered for improving the gate operation). The examples briefly discussed below are based on the logic analysis of two biomarkers for each kind of injury, providing positive answer (logic **1** output for the AND logic gate or **0** output for the NAND logic gate) when both biomarkers appear at their elevated pathophysiological concentrations (input combination **1**,**1**). The difference between AND/NAND logic gates realized in these systems is purely based on the definition of the output signals. When the output signal measured (optically or electrochemically) decreased upon increasing concentrations of the input signals, the logic operation was considered as the NAND gate; otherwise it was not different from the AND gate. If the definition of the output signal is changed to the absolute value of the signal change, all discussed systems can be simply considered as the AND logic gates. As it was explained earlier, the advantage of the use of two biomarkers instead of one is the increase of the confidence in the diagnostic conclusion. It should be noted that the analyzed biomedical problems exemplified below are related to combat injuries; however, the same approach can be adopted to many other biomarkers signaling on other biomedical dysfunctions. The reason to concentrate on the combat injuries originated from the practical needs to have rapid conclusion on the pathological conditions without access to standard hospital facilities. Among pervasive battlefield injuries, soft tissue injury (STI), traumatic brain injury (TBI), acute liver injury (LI – already discussed in the previous section), abdominal trauma (ABT), hemorrhagic shock (HS), oxidative stress (OS), and radiation injury (RI) are the most common sustained by soldiers in combat [50–53]. Thus, they have been selected [31] as convenient examples for the rapid analysis of the corresponding biomarkers using the discussed biocomputing approach. While considering the examples given in the following sections, one should remember that the logic values **0** and **1** for the input signals are set by their biomedical (physiological) concentrations. In other words, the logic input **0** corresponds to the normal physiological concentration of the used biomarker (which is not physically zero concentration) and the logic input **1** corresponds to the most typical abnormal (pathophysiological) concentration of the biomarker corresponding to the specific injury being analyzed. Overall, this biomedical definition of the input signals results in the biocatalytic process being always activated even at the **0,0** input combination. The difference between the logic output **0** and **1** values originated from different rates of the biocatalytic cascades for different concentrations of the input signals applied.

11.5.1 Soft Tissue Injury (STI) Logic Analysis

Soft tissue injuries (STI) are among the most pervasive injuries sustained in combat and can be difficult to identify in numerous circumstances [54]. Among clinically established indicators of STI, serum creatine kinase (CK) and lactate dehydrogenase (LDH) have been routinely employed in the assessment of muscular exertion, fatigue, injury, and trauma [42]. Logic **0** and **1** levels of CK (0.1 and 0.71 U ml^{-1}) and LDH (0.15 and 1 U ml^{-1}) input signals were applied to the biocomputing/bioanalytical system in order to realize meaningful circulating levels of these biomarkers [31]. The input signals, being represented by the enzymes, were applied to the biocatalytic cascade (Figure 11.7A) which also included some other non-variable ("machinery") components, such as pyruvate kinase (PK), creatine, adenosine 5′-triphosphate (ATP), phospho(enol)pyruvate (PEP), β-nicotinamide adenine dinucleotide reduced (NADH), and some additional components (salts, buffer, etc.). The PK enzyme is essential for combining both enzyme signals in one biocatalytic cascade. The biocatalytic reaction activated by the enzyme input signals, applied in four different combinations (**0,0**; **0,1**; **1,0**; and **1,1**), resulted in the consumption of NADH, thus reducing the output signal corresponding to the NADH concentration measured optically (Figure 11.8A) or electrochemically (Figure 11.8B). The overall process (decreasing the output signal upon increasing the concentrations of the input signals) corresponded to the NAND logic gate operation (Figure 11.7B) with the corresponding truth table shown in Figure 11.7C. The response values (the output signals) were measured after the specific gate time and then digitized by applying a threshold separating logic **0** and **1** output values (Figure 11.8C,D). The measured low signal (**0** output) was considered as the diagnostic conclusion on the STI based on the binary analysis of two biomarkers following the NAND logic operation (Figure 11.7C). It should be noted that the logic output **1** (meaning the high output signal corresponding to the low reaction rate) could originate from many reasons including normal physiological conditions as well as various physiological dysfunctions not related to the STI (Figure 11.7C). This means

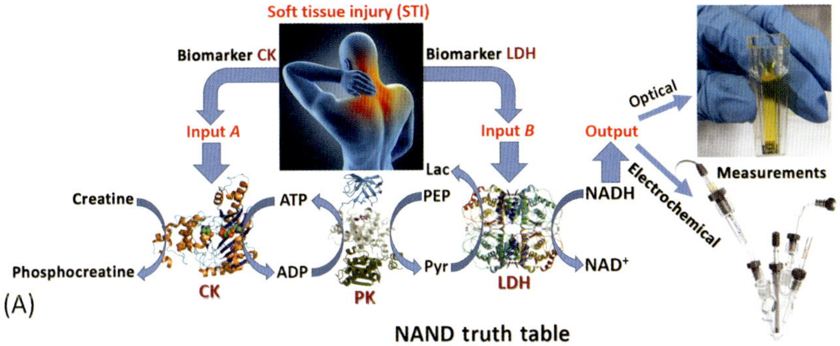

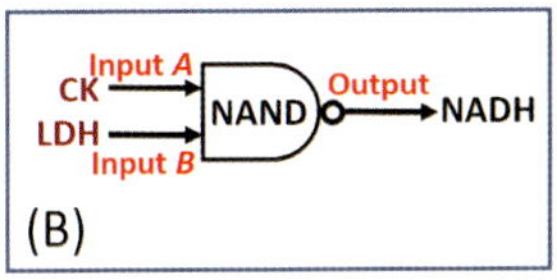

NAND truth table

Input *A* (CK)	Input *B* (LDH)	Output	Biomedical conclusion
0	0	1	Normal
0	1	1	Not STI
1	0	1	Not STI
1	1	0	Soft tissue injury

(C)

Figure 11.7 (A) The biocatalytic cascade realized for the logic analysis of the STI biomarkers. The biocatalytic reactions are activated by CK and LDH inputs, which are biomarkers of the STI. The biocatalytically consumed NADH, being the final output signal, is analyzed optically and electrochemically. (B) The NAND logic gate corresponding to the biocatalytic cascade. (C) The truth table of the NAND gate. The output signal **1** corresponds to the physiologically normal conditions or abnormal conditions, which are not related to the STI. The output signal **0** corresponds to the STI condition.

that the binary (YES/NO) diagnostic approach does not allow to distinguish the normal physiological conditions from pathological conditions that are not related to the STI. However, the obtained results can clearly indicate the STI based on the logic analysis of two biomarkers.

11.5.2 Traumatic Brain Injury (TBI) Logic Analysis

Traumatic brain injuries (TBI) are among the most debilitating injuries suffered in the battlefield [55] and have garnered much recent attention due to the wide range of symptoms and characteristics presented by individuals suffering from the condition [56, 57]. The TBI is also possible during various sport activities (particularly in skiers and snowboarders) [58, 59] and as a result of car accidents [60]. Thus, the rapid diagnostics of the TBI conditions is very important in many different circumstances. Norepinephrine (NE), a catecholamine hormone neurotransmitter, and L-glutamic acid (Glu; glutamate), an amino acid excitatory neurotransmitter, are among the most widely employed indicators of neurological damage and trauma [61, 62]. Therefore, NE (with **0** and **1** logic values of 2.2 nM and 3.5 μM) and Glu (with **0** and **1** logic values of 40 and 140 μM) were applied as input signals activating the biocatalytic cascade [31] (Figure 11.9A). The "machinery" in this cascade included glutamate oxidase (GluOx) and microperoxidase-11 (MP-11) with some salts, buffer, and oxygen

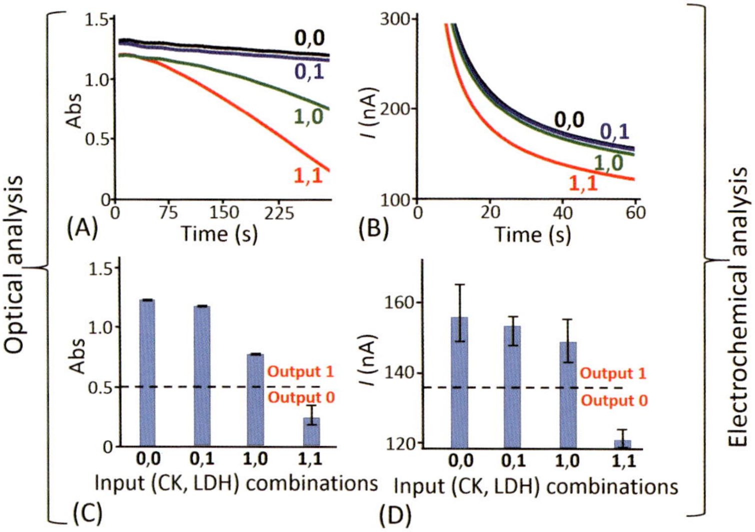

Figure 11.8 Optical (A) and electrochemical (B) responses generated by the NAND logic gate corresponding to the biocatalytic cascade shown in Figure 11.7A. The biocatalytic reactions were activated by the CK and LDH biomarkers of STI applied in different combinations. Optical absorbance measurements were performed at $\lambda = 340$ nm and electrochemical chronoamperograms corresponding to the NADH oxidation (in the presence of methylene green mediator) were performed at $E = 0.0$ V (vs. Ag/AgCl). Both optical and electrochemical measurements demonstrated the decreasing concentration of NADH in the course of the biocatalytic reactions. Optical (C) and electrochemical (D) bar charts obtained by sampling the output of the STI NAND gate upon various combinations of the input biomarkers. Optical absorbance measurements were extracted at $t = 300$ seconds. Electrochemical chronoamperograms were sampled at $t = 60$ seconds. Dashed lines indicate the decision threshold for the realization of Boolean logic gate operation. Technical details on the optical and electrochemical measurements can be found in Ref. [31]. Source: Halámek et al. 2010 [31]. Adapted with permission of Royal Society of Chemistry.

dissolved. In the presence of Glu (Input *B*), oxygen is reduced to H_2O_2 in the process catalyzed by GluOx. Then, the biocatalytically produced H_2O_2 results in oxidation of NE (Input *A*) resulting in the formation of norepiquinone that was detected optically (Figure 11.10A) and electrochemically (Figure 11.10B) as the final output. The output signal appeared at the high level (logic value **1**) only when both inputs appeared at their **1,1** input combination (Figure 11.10C,D), thus representing the AND logic gate (Figure 11.9B). The corresponding truth table (Figure 11.9C) was used to derive the biomedical conclusions from the logic output values, which corresponded to the TBI conditions only in the case of both inputs appearing at their elevated pathophysiological concentrations (**1,1** combination). The low value of the output signal (logic **0**) corresponded to the normal physiological conditions or physiological dysfunctions not related to TBI. The bioanalytical systems operating in the binary YES/NO (**1,0**) mode allowed rapid and high-fidelity diagnostics for the TBI conditions, which is important for the out-of-hospital usage.

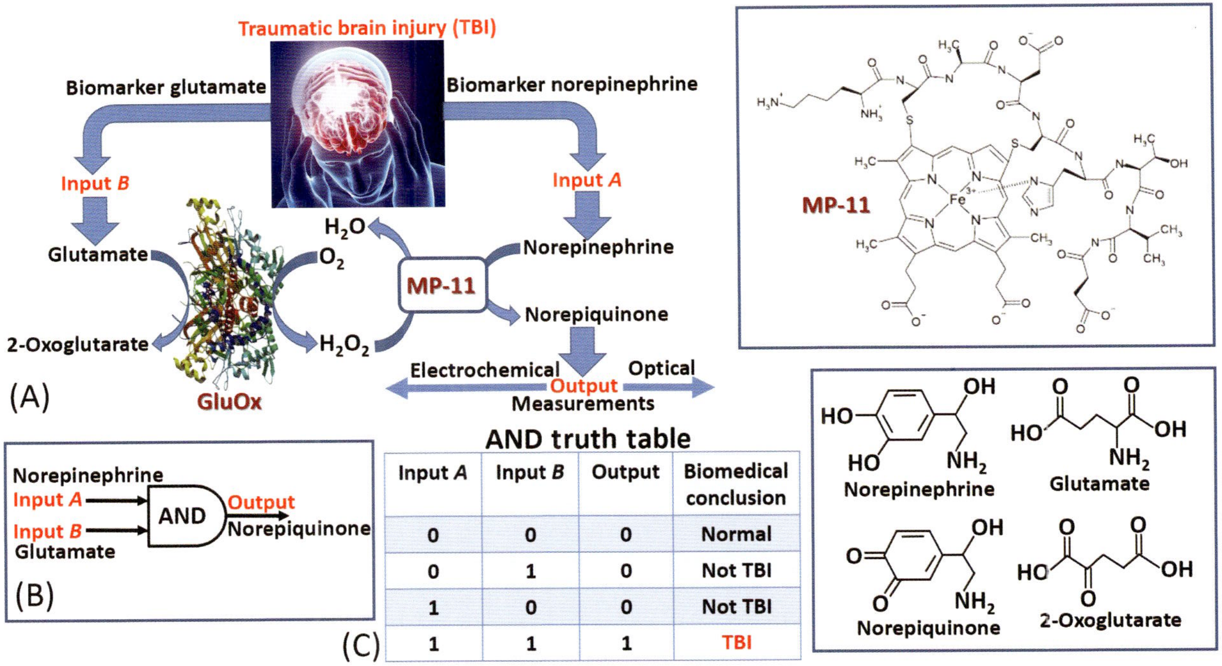

AND truth table

Input *A*	Input *B*	Output	Biomedical conclusion
0	0	0	Normal
0	1	0	Not TBI
1	0	0	Not TBI
1	1	1	TBI

Figure 11.9 (A) The biocatalytic cascade realized for the logic analysis of the TBI biomarkers. The biocatalytic reactions are activated by norepinephrine and glutamate inputs, which are biomarkers of the TBI. The biocatalytically produced norepiquinone, being the final output signal is analyzed optically and electrochemically. (B) The AND logic gate corresponding to the biocatalytic cascade. (C) The truth table of the AND gate. The output signal **0** corresponds to the physiologically normal conditions or abnormal conditions, which are not related to the TBI. The output signal **1** corresponds to the TBI. Chemical structures of microperoxidase-11 (MP-11), norepinephrine, norepiquinone, glutamate (in the acidic form) and 2-oxoglutarate (in the acidic form) are shown in the frames.

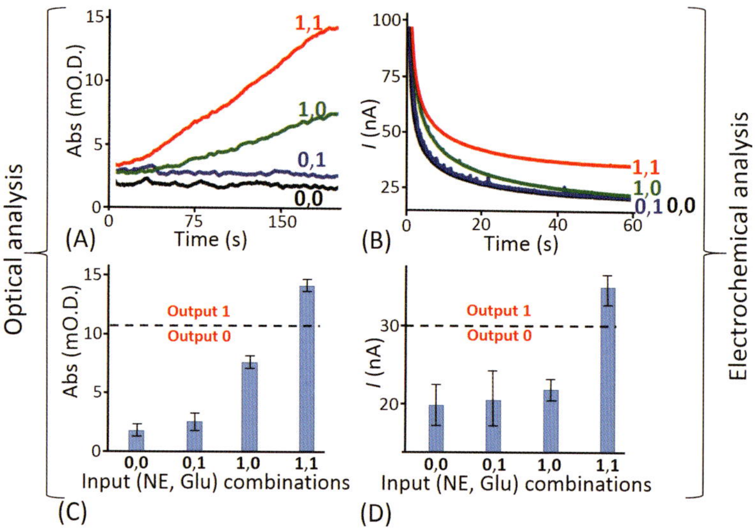

Figure 11.10 Optical (A) and electrochemical (B) responses generated by the AND logic gate corresponding to the biocatalytic cascade shown in Figure 11.9A. The biocatalytic reactions were activated by the norepinephrine and glutamate biomarkers of TBI applied in different combinations. Optical absorbance measurements were performed at $\lambda = 487$ nm and electrochemical chronoamperograms corresponding to the norepiquinone reduction were performed at $E = -0.4$ V (vs. Ag/AgCl). Both optical and electrochemical measurements demonstrated the increasing concentration of norepiquinone in the course of the biocatalytic reactions. Optical (C) and electrochemical (D) bar charts obtained by sampling the output of the TBI AND gate upon various combinations of the input biomarkers. Optical absorbance measurements were extracted at $t = 200$ seconds. Electrochemical chronoamperograms were sampled at $t = 60$ seconds. Dashed lines indicate the decision threshold for the realization of Boolean logic gate operation. Technical details on the optical and electrochemical measurements can be found in Ref. [31]. Source: Halámek et al. 2010 [31]. Adapted with permission of Royal Society of Chemistry.

11.5.3 Abdominal Trauma (ABT) Logic Analysis

As with liver injury discussed earlier, severe abdominal trauma (ABT) frequently results in sepsis and must be addressed with little or no delay in order to improve survival [63]. In addition, ABT is another example of a common battlefield injury, which has been linked to high mortality rates [64]. Serum lactate (Lac) and lactate dehydrogenase (LDH) are well-established biomarkers of such injury [42, 43, 65, 66] and they are frequently used in corresponding assays [67]. Thus, Lac and LDH were used as the biomarker input signals (Input A and Input B, respectively) activating the biocatalytic cascade [31] (Figure 11.11A). NAD^+ and some other system components (salts, buffer, etc.) were used as the non-variable "machinery" parts of the biocatalytic system. Logic **0** and **1** levels of LDH (0.15 and 1.0 U ml^{-1}) and lactate (1.6 and 6.0 mM) input signals were applied to the biocatalytic system in order to realize meaningful circulating levels of these biomarkers corresponding to the normal physiological and

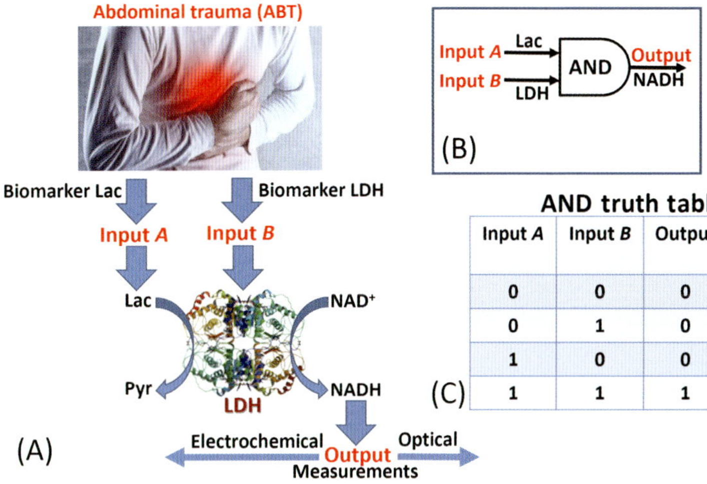

AND truth table

Input *A*	Input *B*	Output	Biomedical conclusion
0	0	0	Normal
0	1	0	Not ABT
1	0	0	Not ABT
1	1	1	ABT

Figure 11.11 (A) The biocatalytic reaction realized for the logic analysis of the ABT biomarkers. The biocatalytic reaction is activated by Lac and LDH inputs, which are biomarkers of the ABT. The biocatalytically produced NADH, being the final output signal, is analyzed optically and electrochemically. (B) The AND logic gate corresponding to the biocatalytic reaction. (C) The truth table of the AND gate. The output signal **0** corresponds to the physiologically normal conditions or abnormal conditions, which are not related to the ABT. The output signal **1** corresponds to the ABT.

elevated pathophysiological concentrations, respectively [31]. Notably, one of the biomarkers (LDH) was represented by the enzyme, while the second (Lac) was the corresponding substrate operating with the enzyme (Figure 11.11A). In the presence of both biomarkers the NAD^+ cofactor was reduced biocatalytically to NADH, which was detected optically (Figure 11.12A) and electrochemically (Figure 11.12B). The high output signal level (logic **1**) was generated when both inputs appeared with the logic combination **1**,**1** corresponding to both biomarkers at the elevated pathophysiological concentrations (Figure 11.12C,D), thus resembling the AND logic gate (Figure 11.11B). The corresponding truth table (Figure 11.11C) was used to derive the biomedical conclusions from the logic output values, which corresponded to the ABT conditions only in the case of both inputs appearing at their elevated pathophysiological concentrations (**1**,**1** combination). The low value of the output signal (logic **0**) corresponded to the normal physiological conditions or physiological dysfunctions not related to ABT.

11.5.4 Hemorrhagic Shock (HS) Logic Analysis

Hemorrhagic shock, a condition that arises due to uncontrolled bleeding, is another pervasive example of a high-mortality combat injury and frequently occurs as a result of the infliction of gunshot wounds and blast injuries [68]. This critical condition must be assessed before the individual who has sustained

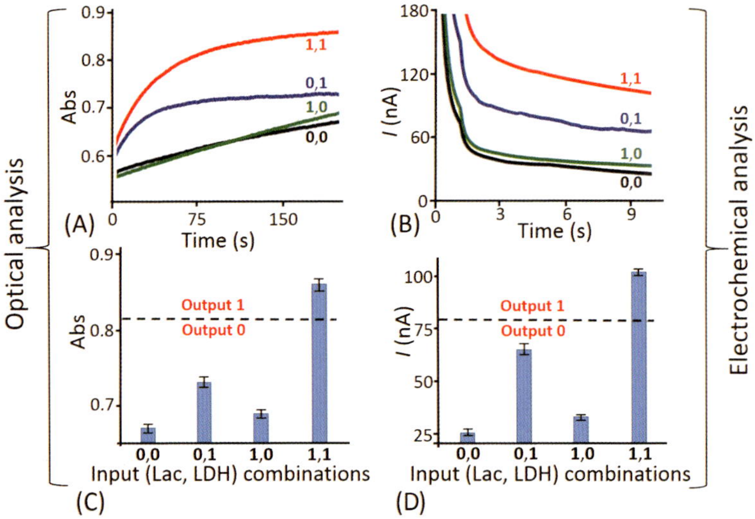

Figure 11.12 Optical (A) and electrochemical (B) responses generated by the AND logic gate corresponding to the biocatalytic reaction shown in Figure 11.11A. The biocatalytic reaction was activated by Lac and LDH biomarkers of ABT applied in different combinations. Optical absorbance measurements were performed at $\lambda = 340$ nm and electrochemical chronoamperograms corresponding to the NADH oxidation (in the presence of methylene green mediator) were performed at $E = 0.1$ V (vs. Ag/AgCl). Both optical and electrochemical measurements demonstrated the increasing concentration of NADH in the course of the biocatalytic reaction. Optical (C) and electrochemical (D) bar charts obtained by sampling the output of the ABT AND gate upon various combinations of the input biomarkers. Optical absorbance measurements were extracted at $t = 200$ seconds. Electrochemical chronoamperograms were sampled at $t = 10$ seconds. Dashed lines indicate the decision threshold for the realization of Boolean logic gate operation. Technical details on the optical and electrochemical measurements can be found in Ref. [31]. Source: Halámek et al. 2010 [31]. Adapted with permission of Royal Society of Chemistry.

the injury bleeds to death [69]. Norepinephrine (NE) and Glc were selected as typical biomarkers of the HS conditions (Inputs *A* and *B*, respectively), which significantly increase their concentrations in the case of HS [31]. The NE and Glc biomarker inputs were applied to the biocatalytic system outlined in Figure 11.13A [31]. Logic **0** and **1** levels of NE (2.2 nM and 3.5 μM) and Glc (4 and 26 mM) input signals were applied to the biocatalytic system mimicking the normal physiological and elevated pathophysiological biomarker concentrations, respectively. Two biocatalysts (GOx and MP-11) as well as some additional chemical components (salts, buffer, etc.) were added as non-variable components of the system. While in the presence of glucose the reaction biocatalyzed by glucose oxidase (GOx) resulted in the production of H_2O_2, the second reaction catalyzed by MP-11 consumed the *in situ* produced H_2O_2 and oxidized NE to yield norepiquinone, which was the final output product analyzed by optical (Figure 11.14A) and electrochemical (Figure 11.14B) methods. The high output concentration (logic **1**) was obtained only when both inputs were applied in their **1,1** logic input values (Figure 11.14C,D), thus mimicking the AND logic gate (Figure 11.13B). The logic output **1** was considered as the diagnostic conclusion

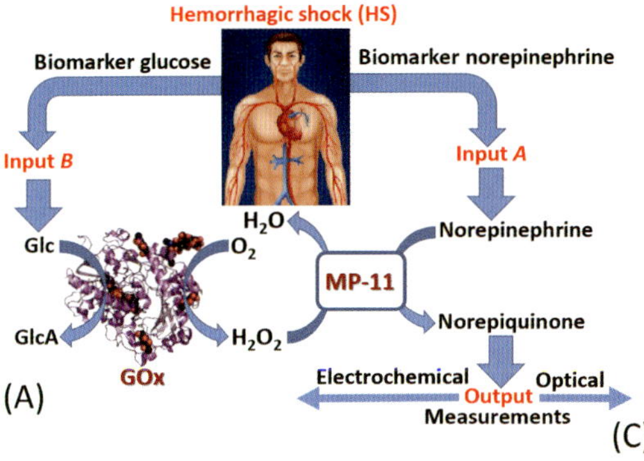

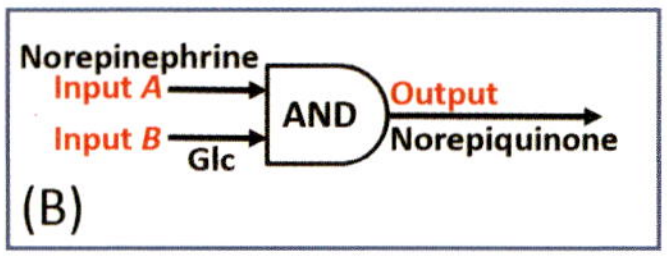

AND truth table

Input *A*	Input *B*	Output	Biomedical conclusion
0	0	0	Normal
0	1	0	Not HS
1	0	0	Not HS
1	1	1	HS

(C)

Figure 11.13 (A) The biocatalytic cascade realized for the logic analysis of the HS biomarkers. The biocatalytic reactions are activated by glucose (Glc) and norepinephrine inputs, which are biomarkers of HS. The biocatalytically produced norepiquinone, being the final output signal, is analyzed optically and electrochemically. (B) The AND logic gate corresponding to the biocatalytic cascade. (C) The truth table of the AND gate. The output signal **0** corresponds to the physiologically normal conditions or abnormal conditions that are not related to HS. The output signal **1** corresponds to HS.

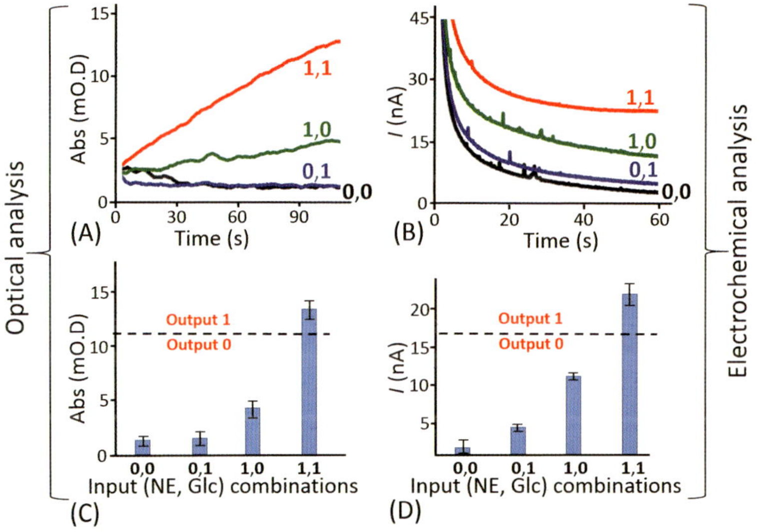

Figure 11.14 Optical (A) and electrochemical (B) responses generated by the AND logic gate corresponding to the biocatalytic cascade shown in Figure 11.13A. The biocatalytic reactions were activated by glucose (Glc) and norepinephrine biomarkers of HS applied in different combinations. Optical absorbance measurements were performed at $\lambda = 487$ nm and electrochemical chronoamperograms corresponding to the norepiquinone reduction were performed at $E = -0.4$ V (vs. Ag/AgCl). Both optical and electrochemical measurements demonstrated the increasing concentration of norepiquinone in the course of the biocatalytic reactions. Optical (C) and electrochemical (D) bar charts obtained by sampling the output of the HS AND gate upon various combinations of the input biomarkers. Optical absorbance measurements were extracted at $t = 110$ seconds. Electrochemical chronoamperograms were sampled at $t = 60$ seconds. Dashed lines indicate the decision threshold for the realization of Boolean logic gate operation. Technical details on the optical and electrochemical measurements can be found in Ref. [31]. Source: Adapted from Halámek et al. 2010 [31].

about the HS conditions, while logic value **0** was related to normal physiological or some other conditions not related to the HS (Figure 11.13C).

11.5.5 Oxidative Stress (OS) Logic Analysis

Oxidative stress (OS) refers to a broad scope of pathological states and accompanies nearly all forms of physical stress or strain experienced by the body [70, 71], including those acquired in a battle. The OS is characterized by elevated levels of reactive oxygen species (ROS) and reactive nitrogen species (RNS) resulting in cell dysfunction, cell apoptosis, and tissue necrosis. Measurement and real-time monitoring of oxidative species produced under OS are difficult and in general impossible or even meaningless. This is because the extremely active radical species (ROS and RNS) appear at very low and unsteady concentrations and the damage produced by them to biological tissues and physiological processes can be hardly estimated based on direct measurements of these labile species. Instead of direct measurements of the radical species, it should be possible to measure much more stable biological redox (reduced or corresponding oxidized) species appearing at concentrations related to the physiological response to the primary oxidative species.

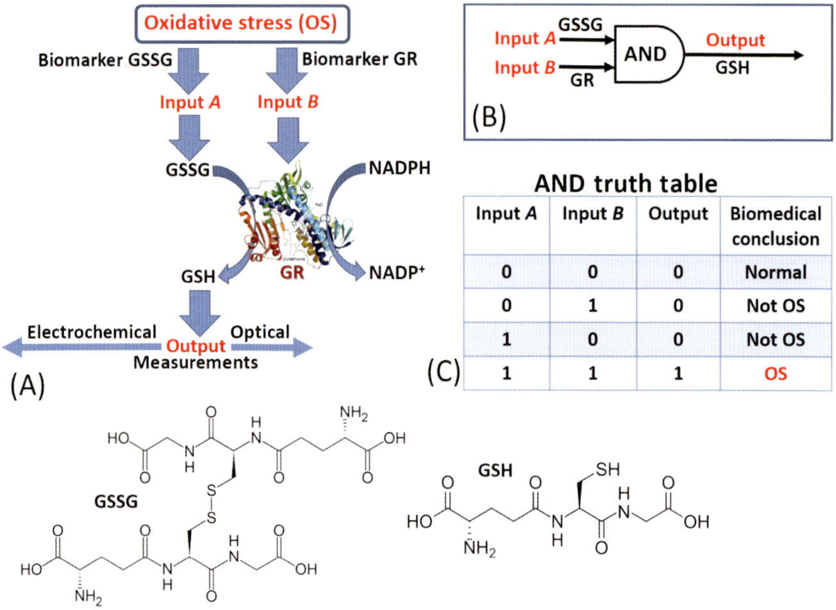

AND truth table

Input *A*	Input *B*	Output	Biomedical conclusion
0	0	0	Normal
0	1	0	Not OS
1	0	0	Not OS
1	1	1	OS

Figure 11.15 (A) The biocatalytic reaction realized for the logic analysis of the OS biomarkers. The biocatalytic reaction is activated by glutathione oxidized (GSSG) and GR inputs, which are biomarkers of the OS. The biocatalytically produced glutathione reduced (GSH), being the final output signal, is analyzed optically and electrochemically. (B) The AND logic gate corresponding to the biocatalytic reaction. (C) The truth table of the AND gate. The output signal **0** corresponds to the physiologically normal conditions or abnormal conditions, which are not related to the OS. The output signal **1** corresponds to the OS.

While many different reduced/oxidized biochemical species and enzymes regulating their redox state could serve as oxidative stress biomarkers, glutathione (GSSG, oxidized form) and glutathione reductase (GR) have been selected as the simplest example biomarkers of OS [31]. Logic **0** and **1** levels of GSSG (150 and 400 μM) and GR (0.55 and 0.65 U ml^{-1}) input signals were applied to the biocatalytic system in order to realize meaningful circulating levels of these biomarkers [31]. Some additional non-variable components (particularly NADPH) were included in the biocatalytic system to support its operation (Figure 11.15A). The glutathione reduced (GSH) species produced by the biocatalytic system in the presence of GSSG and GR were analyzed optically (Figure 11.16A) and electrochemically (Figure 11.16B). The high concentration level (logic value **1**) for the GSH output signal was obtained only upon application of the **1,1** input combination (Figure 11.16C,D), thus resembling the Boolean AND logic gate (Figure 11.15B). The logic output **1** allowed high-confidence conclusion on the OS conditions, while the logic **0** value originated from normal

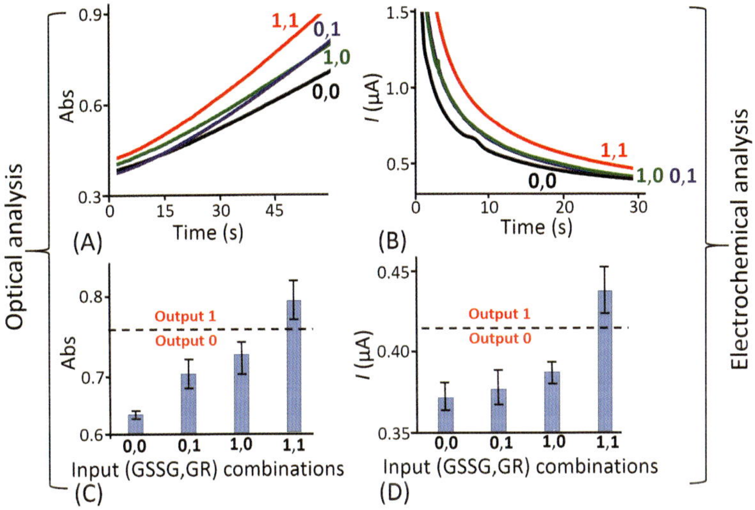

Figure 11.16 Optical (A) and electrochemical (B) responses generated by the AND logic gate corresponding to the biocatalytic cascade shown in Figure 11.15A. The biocatalytic reactions were activated by glutathione oxidized (GSSG) and GR biomarkers of OS applied in different combinations. Optical absorbance measurements (using standard procedure with Ellman's reagent) were performed at $\lambda = 412$ nm and electrochemical chronoamperograms corresponding to the glutathione reduced (GSH) oxidation were performed at $E = 0.5$ V (vs. Ag/AgCl). Both optical and electrochemical measurements demonstrated the increasing concentration of norepiquinone in the course of the biocatalytic reactions. Optical (C) and electrochemical (D) bar charts obtained by sampling the output of the HS AND gate upon various combinations of the input biomarkers. Optical absorbance measurements were extracted at $t = 60$ seconds. Electrochemical chronoamperograms (measured at the electrode modified with cobalt(II)-phthalocyanine) were sampled at $t = 30$ seconds. Dashed lines indicate the decision threshold for the realization of Boolean logic gate operation. Technical details on the optical and electrochemical measurements can be found in Ref. [31]. Source: Adapted from Halámek et al. 2010 [31].

physiological conditions or some other physiological dysfunctions not related to the OS (Figure 11.15C).

The example system for the OS analysis based on two biomarkers (GSSG and GR) outlined above can be extended to the analysis of many more biomarkers including participation of various reduced biomolecular species (Figure 11.17). An extended collection of the analyzed biomarkers can include glutathione reduced (GSH), methionine, NAD(P)H, xanthine, etc. These reduced species can be biocatalytically oxidized with the help of the corresponding enzymes immobilized at an electrode surface. All oxidation reactions biocatalyzed by the oxidase enzymes result in the production of H_2O_2 upon O_2 reduction (Figure 11.17). The enzymes can be attached to SiO_2 nanoparticles assembled at the electrode surface to provide a high-surface platform for the enzyme immobilization. The *in situ* produced H_2O_2 will be catalytically reduced by hemin immobilized at a graphene-functionalized carbon fiber electrode surface [72, 73] (Figure 11.18), thus yielding current dependent on the total concentration of the analyzed reduced biomolecules. Their concentration decrease measured electrochemically will signal on the OS. The present system will simplify the analytical process providing the total response for many reduced biomarkers instead of the separate analysis of each component individually. The major advantage of this briefly explained approach is in the standard bioanalytical platform that includes various biomarkers and different enzymes operating in the same concert way – all producing H_2O_2 analyzed electrochemically. The joint operation of all enzymatic systems would allow analysis of the total effect of OS on the physiological processes reflected by the concentration change of different reduced species used as the OS biomarkers. The biocatalytic/bioanalytic reactions can be performed in parallel mimicking Boolean OR gates (Figure 11.19A) and some other biocatalytic cascades can operate in sequence, thus mimicking AND logic gates (Figure 11.19B). The reactions schematically shown in Figure 11.19 can be particularly useful for analyzing biomarkers that cannot operate directly with oxidases (the corresponding enzymes may not be available), instead the reduced biomarkers can be first reacted to yield NAD(P)H, which can be then used to produce H_2O_2 through the reactions catalyzed by NAD(P)H oxidase. This would

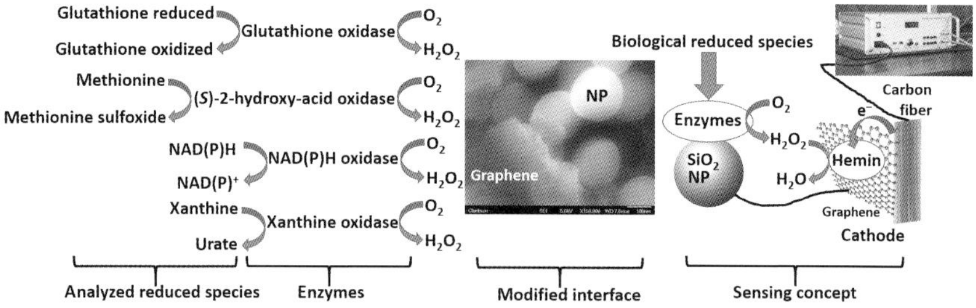

Figure 11.17 Electrochemical analysis of various biological reduced species with the electrode modified with different enzymes. Note that different reduced species are all converted to the same electrochemically analyzed product (H_2O_2), thus providing a general platform for analyzing multiple OS biomarkers.

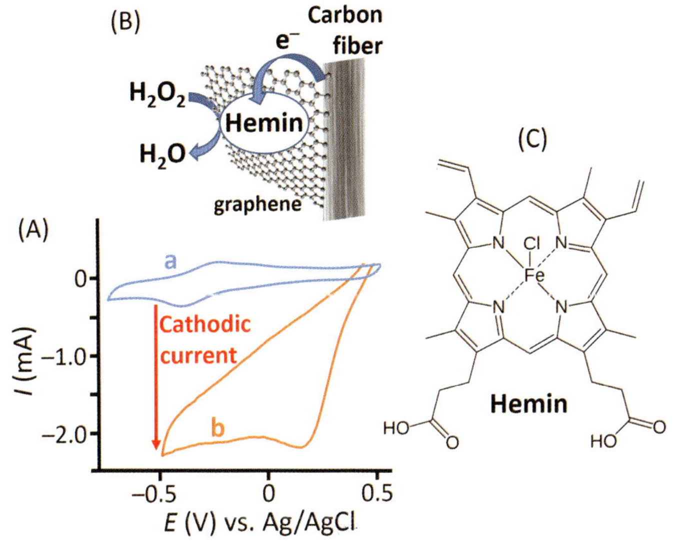

Figure 11.18 (A) Cyclic voltammograms obtained with a graphene-functionalized carbon fiber electrode modified with adsorbed hemin in the absence (a) and presence of H_2O_2, 5 mM (b). Similar electrochemical processes can be realized for the *in situ* produced H_2O_2 upon operation of the system shown in Figure 11.17. (B) A scheme shows H_2O_2 reduction catalyzed by hemin. (C) The hemin structure is shown. The background solution was composed of 0.1 M phosphate buffer, pH 6.5. The solution was deaerated with Ar. Potential scan rate, 2 mV s^{-1}. Source: Koushanpour et al. 2017 [72]. Adapted with permission of John Wiley and Sons.

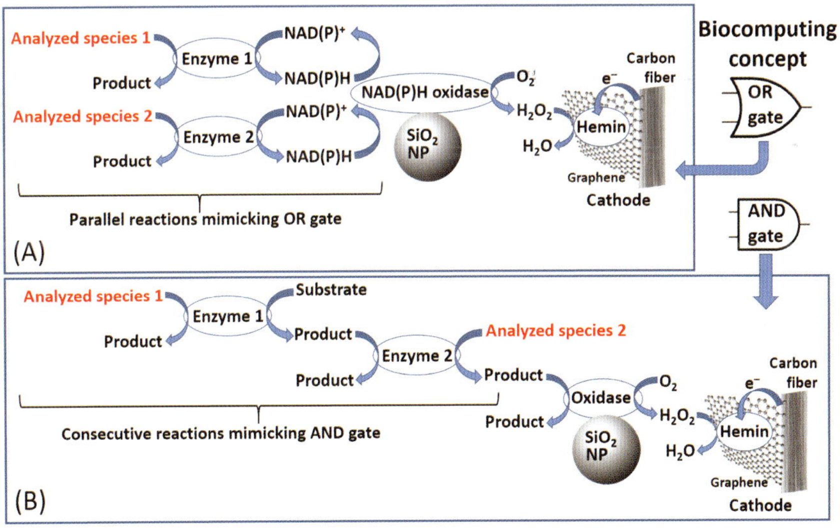

Figure 11.19 Biocomputing approach to simultaneous sensing of various biological species with logically processed input signals: OR/AND Boolean logic gates: (A) Realizing parallel biocatalytic reactions mimicking the OR gate. (B) Realizing consecutive biocatalytic reactions mimicking the AND gate. Both processes can be particularly useful for the analysis of the OS biomarkers, which cannot be used directly for production of H_2O_2.

allow to include additional biomarkers in the analytical assay even in the case they cannot produce H_2O_2 directly.

11.5.6 Radiation Injury (RI) Logic Analysis

Radiation injury (RI), a life threatening form of organ tissue damage, is caused by exposure to ionizing radiation. A short intensive exposure results in acute radiation syndrome [74], while chronic radiation syndrome [75] is caused by a prolonged exposure to lower intensity radiation. Immediate and effective medical treatment of persons suspected of radiation exposure requires a reliable detection approach to provide adequate diagnostic information. Traditional instrumentation such as physical dosimeters does not provide sufficient information about the physiological condition in a human organism following radiation exposure. Variations in age, gender, health, and genetic background can also affect biological responses, which cannot be quantified by a stand-alone physical dosimeter. Biological dosimetry [76, 77] based on the analysis of biomarkers in a body exposed to radiation, rather than physical measure of a radiation dose, is considered as an important tool for the effective medical management of radiation therapy for cancer [78–82], occupational radiation exposure [83] (including astronauts in manned space missions) [84–86], accidents similar to *Chernobyl* disaster [87], and possible radiological terror attacks [88]. Identification of biomarkers characteristic of a radiation injury is a challenging goal since radiation can result in different physiological processes, many of which are not specifically related to the direct instantaneous damage caused by radiation [89]. Potentially important radiation injury biomarkers include various metabolites, proteins, RNA, and DNA [89–92]. Thus, different biomedical analytical approaches have been developed to analyze their variations; among them, the most important methods are chromosome [90, 91] and proteomic analysis [92]. Unfortunately, none of the presently available methods are based on a single specific biomarker, thus requiring massive analysis of multiple biomarkers with low specificity. Technically, such methods are based on sophisticated and expensive equipment, e.g., mass spectrometry for the proteomic analysis [92], which cannot be used for rapid on-site field detection of pathophysiological changes caused by radiation. Biocomputing approach to the logic analysis of RI biomarkers allows rapid and unambiguous conclusion on radiation-induced physiological processes, thus operating as an alert-type biosensing device for preliminary on-site detecting of the RI.

Among many known biomarkers [89] characteristic of radiation-caused damage, enzyme inputs able to operate in a biocatalytic cascade performing a logic operation should be selected. An increased concentration of α-amylase (αAmy) in blood, urine, and saliva was reported as a result of irradiation exposure [89, 93–97]. αAmy release originates mainly from gland tissue damaged by radiation. Another imminent effect of irradiation is tissue breakdown and hemolysis causing massive release of lactate dehydrogenase (LDH) into blood stream [98–100]. Since αAmy and LDH can be combined in a biocatalytic cascade resulting in the consumption of NADH, these two enzymes are convenient biomarkers of the radiation damage [101]. The biocatalytic cascade (Figure 11.20A) includes

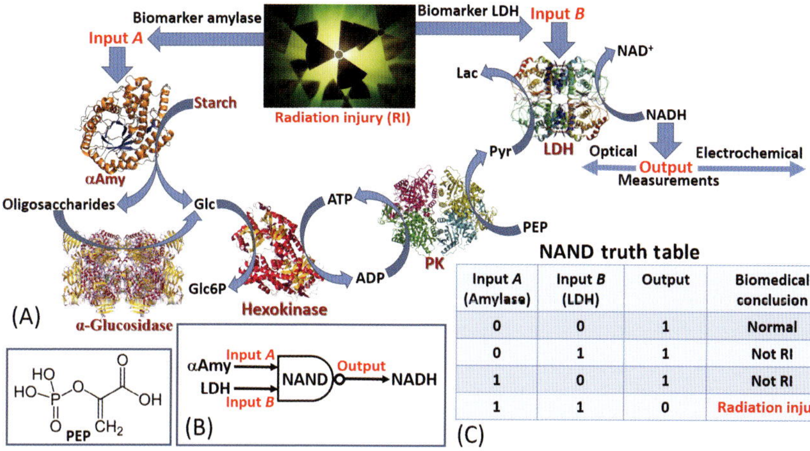

Input A (Amylase)	Input B (LDH)	Output	Biomedical conclusion
0	0	1	Normal
0	1	1	Not RI
1	0	1	Not RI
1	1	0	Radiation injury

Figure 11.20 (A) The biocatalytic cascade realized for the logic analysis of the RI biomarkers. The biocatalytic reactions are activated by αAmy and LDH inputs, which are biomarkers of the RI. The biocatalytically consumed NADH, being the final output signal, is analyzed optically and electrochemically. (B) The NAND logic gate corresponding to the biocatalytic cascade. (C) The truth table of the NAND gate. The output signal **1** corresponds to the physiologically normal conditions or abnormal conditions, which are not related to the RI. The output signal **0** corresponds to the RI. The chemical structure of PEP is shown in the frame.

a multistep biocatalytic reaction, which can be completed only in the presence of all biocatalytic species, including both the biomarkers – αAmy and LDH. The reaction cascade starts with the decomposition of starch upon a cooperative biocatalytic effect of αAmy and α-glucosidase yielding glucose (Glc) as a product. Then, Glc is converted to glucose-6-phosphate (Glc6P) in the presence of hexokinase (HK) and ATP, resulting in the formation of adenosine 5′-diphosphate (ADP), which allows the next reaction step producing pyruvate (Pyr) from phospho (enol)pyruvate (PEP) in the presence of pyruvate kinase (PK). Finally, the *in situ* produced Pyr allows oxidation of NADH to NAD^+ biocatalyzed by lactate dehydrogenase (LDH) to yield decrease of the optical absorbance at $\lambda = 340$ nm as the final output signal (obviously the electrochemical analysis of NADH is also possible). The reactions biocatalyzed by HK and PK are needed to couple the αAmy and LDH biomarker input signals in a single biocatalytic cascade. Therefore, the NADH concentration decreases as a result of the completed reaction and should be observed only in the presence of both biomarkers, αAmy and LDH (meaning their **1**,**1** logic combination). The absence of both or either input signals (**0**,**0**; **0**,**1**; and **1**,**0** combinations) should result in the incomplete biocatalytic chain, thus inhibiting the NADH consumption. For the digital operation of the enzyme logic system, the presence of αAmy and LDH at normal physiological concentrations in blood was defined as logic input **0**, while their elevated pathological concentrations were considered as logic input **1**. Logic **0** and **1** levels of Amy (0.25 and 2.5 U ml^{-1}) [93] and LDH (0.15 and 1.0 U ml^{-1}) [43] input signals were applied in serum to realize meaningful circulating levels of these biomarkers under normal physiological and pathological conditions caused by radiation, respectively. Therefore, the **1**,**1** (αAmy, LDH) input signal combination resulted in the highest activity of the biocatalytic cascade with the rapid decrease of NADH concentration. It should be noted however, that **0** logic concentrations of the input signals correspond to the normal physiological concentrations of αAmy and LDH rather than their complete absence. Thus, **0,0**; **0,1**; and **1,0** combinations of the input signals still allow the biocatalytic reaction to proceed, however, with a lower rate compared to the **1,1** combination of the biomarker inputs. Figure 11.21 shows the NADH absorbance decrease upon application of four different combinations of the input signals: **0,0**; **0,1**; **1,0**; and **1,1**. Figure 11.21 (inset) shows a bar chart for the optical output signals generated by the enzyme logic system obtained after 350 seconds of the biocatalytic reaction in the presence of various combinations of the biomarker input signals. A threshold absorbance value of 1.9 allows clear separation of a low output signal generated upon the cooperative effect of elevated concentrations of αAmy and LDH (**1,1** inputs) from a high signal for the inhibited reaction when either or both input signals were applied at the logic **0** level (**0,0**; **0,1**; **1,0** inputs). The output signal pattern corresponds to the Boolean logic operation NAND, which results in the output signal **0** (low signal) only for the **1,1** combination of the inputs (Figure 11.20B,C).

Numerous compounds present in the serum samples can potentially interfere with the enzymatic "machinery" of the developed analytical system. Particularly, glucose, pyruvate and ADP are intermediate products in the biocatalytic cascade and their natural presence in serum could result in a false positive signal even in the absence of the αAmy input signal. It should be noted that the depletion

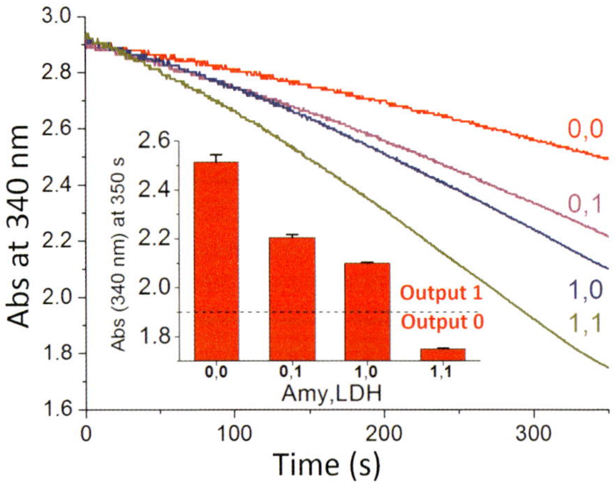

Figure 11.21 Optical detection of the output signal generated by the logic system shown in Figure 11.20A obtained upon application of different combinations of the biomarker input signals (αAmy, LDH). Inset: Bar chart for the optical output signals generated by the enzyme logic system obtained after 350 seconds of the biocatalytic reaction. The radiation damage diagnosis corresponds to the output signal below the decision threshold (dashed line). Technical details on the optical and electrochemical measurements can be found in Ref. [101]. Source: Bocharova et al. 2011 [101]. Adapted with permission of Elsevier.

of naturally existing serum glucose prior to the measurements was essential for the discrimination between the outputs generated by **0,1** and **1,1** (αAmy, LDH) signals [101]. On the other hand, the system operation was possible in the presence of pyruvate and ADP naturally existing in serum. The system operation was examined using different samples of human serum. Some minor sample-to-sample absorbance variations were observed, mostly due to the difference in the transparency of the serum samples. However, the robust operation of the bioanalytical system always allowed convenient discrimination of **0** and **1** output signals, thus providing reliable diagnostics of the pathological conditions [101]. Practical future applications would benefit from further miniaturization and/or integration into a "reagentless" solid-state optical or electrochemical device.

11.6 Multienzyme Logic Network Architectures for Assessing Injuries: Aiming at the Increased Complexity of the Biocomputing–Bioanalytic Systems

The present section describes a high complexity multienzyme biocatalytic cascade processing simultaneously five biomarkers characteristic of traumatic brain injury (TBI) and soft tissue injury (STI) [32]. The system operates as a digital biosensor based on concerted function of eight Boolean AND logic gates, resulting in the decision on physiological conditions based on the logic analysis of complex patterns of the biomarkers. The system represents the first example of a multistep/multienzyme biosensor with the built-in logic for the analysis of complex combinations of biochemical inputs. The approach is based on recent advances in enzyme-based biocomputing systems and the present example [32] demonstrates the potential applicability of biocomputing for developing novel digital biosensor networks. This conceptually novel approach to biosensors aims at increasing further the complexity of biomarker-analyzing systems by concatenating many logic operations represented by a multienzyme system capable to process many variable biomarker signals. Bimodal way of action is enabled by a built-in SWITCH feature activated by the presence or absence of switching inputs in specific gates. Switching between two different subsystems provides a solution for controlling the biochemical pathways and performance correlation for all multiple AND gates in the system. The built-in SWITCH feature enables analysis of different physiological conditions within a biochemical logical system operating in a "single-pot" solution.

11.6.1 The System Structure Based on the Complex Biocatalytic Cascade

The assessment of two common battlefield injuries, soft tissue injury (STI) and traumatic brain injury (TBI), was performed by a multienzyme biocatalytic cascade (Figure 11.22), which can be described as a comprehensive concatenated

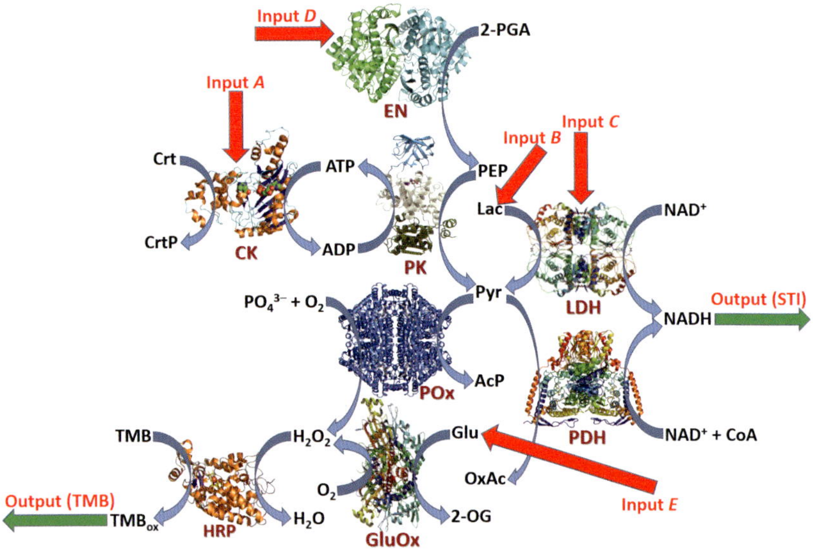

Figure 11.22 Multienzyme biocatalytic cascade for the analysis of STI and TBI. Biomarker-inputs for STI (CK, Lac, LDH) and for TBI (EN and Glu) are shown with the red arrows. Output signals for STI and for TBI are NADH and TMB_{ox}, respectively. Note that for simplicity the scheme does not include some reacting cofactors, promoters and by-products – for the full composition of the system refer to Ref. [32]. Abbreviations used in the scheme are specified in the list of abbreviations, while some of the reacting species and products are not mentioned in the text for brevity.

logic system (Figure 11.23). The logic network represented in Figure 11.23 is actually identical to the biocatalytic cascade depicted in Figure 11.22, but it offers another way of its description, which is more convenient for the discussion of the system operation. The system architecture includes eight networked logic gates of the AND type. The biocatalytic system operated in two different modes: one for the analysis of TBI and another for the analysis of STI with the possibility to switch between them. The system was designed to analyze five different biomarkers appearing in different combinations and to lead to a logic conclusion about the presence or absence of TBI or STI conditions. Three biomarkers: creatine kinase (CK), lactate dehydrogenase (LDH), and lactate (Lac) corresponded to the physiological conditions characteristic of STI, while two other biomarkers – enolase (EN) and glutamate (Glu) – were reporting on the TBI diagnosis. All five biomarkers were applied as biochemical input signals (labeled red in Figure 11.23) activating the logic network at two different concentration levels: logic **0** corresponded to the normal physiological concentrations of the biomarkers, while logic **1** was selected at the elevated pathophysiological concentrations corresponding to the respective injuries (Table 11.1). Simultaneous processing of many biomarkers through the complex biocatalytic cascade required optimization of the biocatalytic reactions by tuning the reaction rates in order to have comparable output signals for various combinations of the

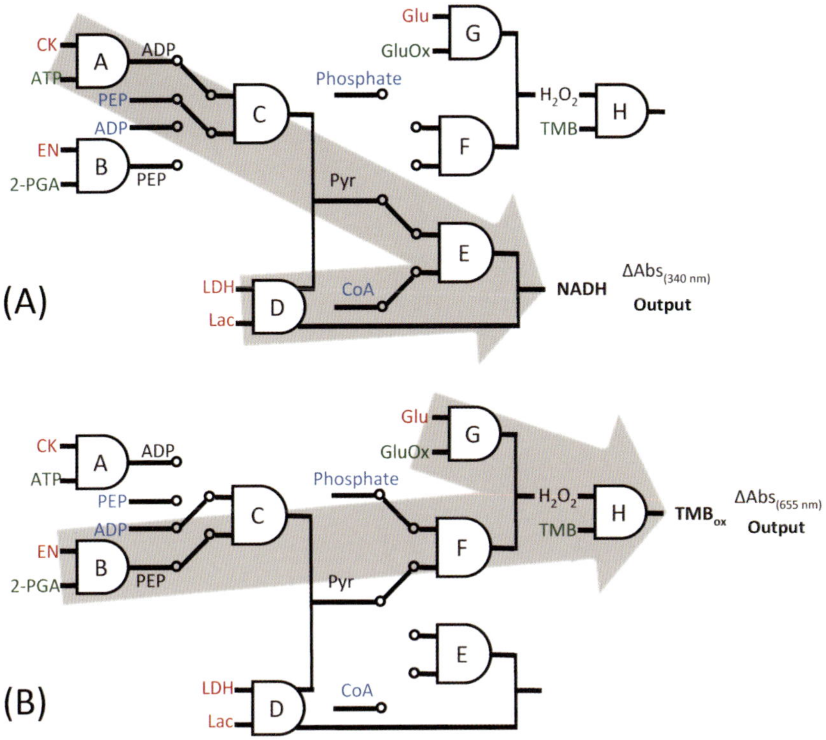

Figure 11.23 Equivalent logic schemes for the concatenated logic gates analyzing STI and TBI corresponding to the biocatalytic cascade shown in Figure 11.22. (A) The system switched to the STI analysis mode. (B) The system switched to the TBI analysis mode. Biomarker inputs are red labeled, auxiliary inputs are green labeled, and switching inputs are blue labeled. The switch-regulated pathways for the STI and TBI operational modes are indicated by arrows. Source: Halámek et al. 2010 [32]. Adapted with permission of Royal Society of Chemistry.

biomarker inputs. The optimization was achieved by careful selection of the auxiliary inputs (labeled green in Figure 11.23): adenosine 5′-triphosphate (ATP), D(+)2-phosphoglyceric acid sodium salt (2-PGA), glutamate oxidase (GluOx), and 3,3′,5,5′-tetramethylbenzidine dihydrochloride (TMB). In order to digitize the logic network operation, the auxiliary inputs were applied at two levels: logic **0** corresponded to the physical zero concentration, while logic **1** was selected experimentally upon optimization of the system (Table 11.1). The auxiliary inputs did not provide any information about physiological conditions related to the injuries, but they were needed for optimal processing of the biomarker inputs. When they were applied at **0** levels, the system was mute and insensitive to the biomarker inputs, while their application at **1** levels provided optimized processing of the biomarker inputs applied in different combinations of **0** and **1** logic levels. Four additional inputs (labeled blue in Figure 11.23) – phospho (enol)pyruvate (PEP), adenosine 5′-diphosphate (ADP), phosphate, and coenzyme A sodium salt (CoA) – were used to switch the system

Table 11.1 Concentrations of the inputs activating the biocatalytic cascade for the logic analysis of STI and TBI.

Inputs	Logic 0	Logic 1	References
CK[a)]	$100\,U\,l^{-1}$[b)]	$710\,U\,l^{-1}$[c)]	[42, 43]
LDH[a)]	$150\,U\,l^{-1}$[b)]	$1000\,U\,l^{-1}$[c)]	[42, 43]
Lac[a)]	1.6 mM[b)]	6 mM[c)]	[102]
EN[a)]	$0.42\,U\,l^{-1}$[b)]	$1.2\,U\,l^{-1}$[c)]	[103]
Glu[a)]	40 μM[b)]	140 μM[c)]	[61, 62]
ATP[d)]	0	2 mM	e)
2-PGA[d)]	0	8 mM	e)
GluOx[d)]	0	$0.3\,U\,l^{-1}$	e)
TMB[d)]	0	0.45 mM	e)
PEP[f)]	0	2 mM	e)
ADP[f)]	0	2 mM	e)
Phosphate[f)]	0	31 mM	e)
CoA[f)]	0	0.5 mM	e)

a) Biomarker inputs.
b) Corresponds to the normal physiological concentration.
c) Corresponds to pathophysiological elevated concentration.
d) Auxiliary inputs.
e) Optimized experimentally.
f) Switching inputs.

operation between the TBI and STI analytical modes. These inputs were applied at logic **0** levels corresponding to the physical zero concentrations and logic **1** levels being experimentally optimized (Table 11.1).

11.6.2 STI Operation Mode of the Logic Network

Since none of the used biomarkers is specific enough for the STI diagnosis, only simultaneous appearance of all three STI-related biomarkers (CK, LDH, and Lac) at logic **1** values would provide the reliable conclusion about the STI conditions. The system was operating in the following way (see Figure 11.22 for the biochemical representation and Figure 11.23A for the logic equivalent circuitry – the STI pathway is highlighted): The biocatalytic reaction of CK (STI biomarker) and ATP (gate A) resulted in the production of ADP. Further reaction of ADP with PEP biocatalyzed by PK (gate C) resulted in the formation of Pyr, which then reacted in the presence of CoA and pyruvate dehydrogenase (PDH) (gate E) to yield the reduced NADH considered as the output signal. Note that the pathway composed of A–C–E gates was activated only when the biomarker input CK, auxiliary input ATP and switching inputs PEP and CoA appeared at logic **1** values. Simultaneous application of the LDH and Lac inputs (STI biomarkers) at logic **1** (gate D) resulted in the reduction of NAD^+ and further increase of the NADH output. It should

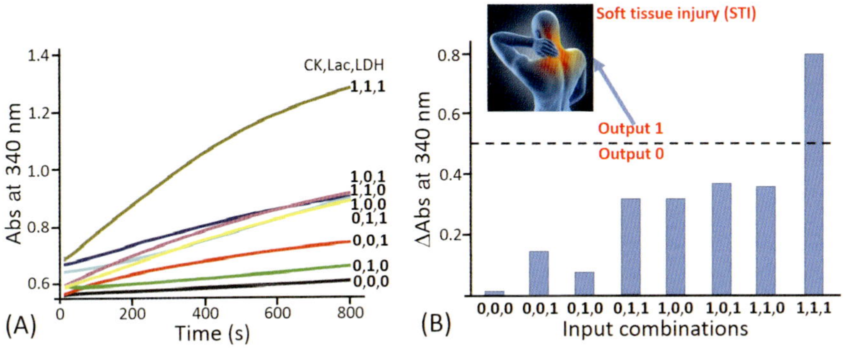

Figure 11.24 (A) Optical detection of the output signal (NADH) generated by the logic system operating for the STI analysis obtained upon application of different combinations of the injury biomarker input signals (CK, Lac, LDH). (B) Bar chart for the output signals generated by the enzyme logic system for the analysis of STI at 800 seconds. The STI diagnosis corresponds to the output signal above the decision threshold (dashed line). The logic system composition is detailed in Ref. [32] and the biomarker input concentrations corresponding to the logic **0** and **1** values are specified in Table 11.1. The auxiliary (ATP) and switching (PEP and CoA) inputs were at logic **1** for all combinations of the biomarker inputs. Source: Adapted from Halámek et al. 2010 [32].

be noted that this pathway resulted in the concomitant production of Pyr, which was passing through gate E producing one more equivalent of NADH, thus further amplifying the output signal. Figure 11.24A shows the optical changes in the system measured at $\lambda = 340$ nm (NADH absorbance) for different combinations of the biomarker inputs. Only simultaneous application of all three biomarker inputs at logic **1** values (input combination **1**,**1**,**1**) resulted in the high optical absorbance changes allowing an unambiguous conclusion about the STI condition. The experimentally derived threshold level of 0.5 O.D. allowed perfect separation of the logic **0** and **1** levels for the output signal being different at least by twofold (Figure 11.24B). It should be noted that in all measurements shown in Figure 11.24, the auxiliary (ATP) and switching (PEP and CoA) inputs were applied at logic **1** values to allow optimal performance of the analytical pathway. At the same time, the switching inputs ADP and phosphate were applied at logic **0** levels to inhibit the alternative TBI pathway.

11.6.3 TBI Operation Mode of the Logic Network

Opposite to STI, the TBI biomarkers (EN and Glu) are rather specific and can report on the injury presence even appearing alone. (It should be noted that a generic EN enzyme was used in the model experiments instead of neuron-specific enolase released from damaged brain.) The system was operating in the following way (see Figure 11.22 for the traditional biochemical representation and Figure 11.23B for the logic equivalent circuitry – the TBI pathway is highlighted): The biocatalytic reaction of EN (TBI biomarker) and 2-PGA (gate B) resulted in the formation of PEP. The next step included the reaction of PEP with ADP (switching input) biocatalyzed by PK (gate C) and resulted in the production of Pyr. Further reaction of Pyr with inorganic phosphate (switching input) biocatalyzed by pyruvate oxidase (POx) (gate F) yielded H_2O_2, which reacted with TMB in the presence of HRP (gate H). This reaction resulted in the oxidation of TMB and produced the absorbance increase at $\lambda = 655$ nm characteristic of the TMB oxidized form (TMB_{ox}) being considered as the final output signal from the pathway composed of B–C–F–H gates. The same signal was generated through another pathway composed of G and H gates: The biocatalytic reaction of Glu (TBI biomarker) and GluOx resulted in O_2 reduction and yielded H_2O_2 (G gate). Then H_2O_2 reacted with TMB in the presence of HRP to yield TMB_{ox} with the respective absorbance changes (H gate). Careful optimization of the system (by tuning the concentrations of the auxiliary inputs) allowed comparable output signals produced in the both pathways. Figure 11.25A shows the optical changes in the system measured at $\lambda = 655$ nm (TMB_{ox} absorbance) for different combinations of the biomarker inputs. Any or both biomarkers appearing at logic **1** level resulted in high absorbance changes signaling about TBI conditions. The experimentally derived threshold level of 0.02 O.D. allowed perfect separation of the logic **0** and **1** levels for the output signal being significantly different for the input combination **0**,**0** and all other combinations (**0**,**1**; **1**,**0**; **1**,**1**) (Figure 11.25B). It should be noted that in all measurements shown in Figure 11.25 the auxiliary (2-PGA, TMB, and GluOx) and switching (ADP and phosphate) inputs were applied at logic **1** values to allow optimized performance of the analytical pathway. At the same

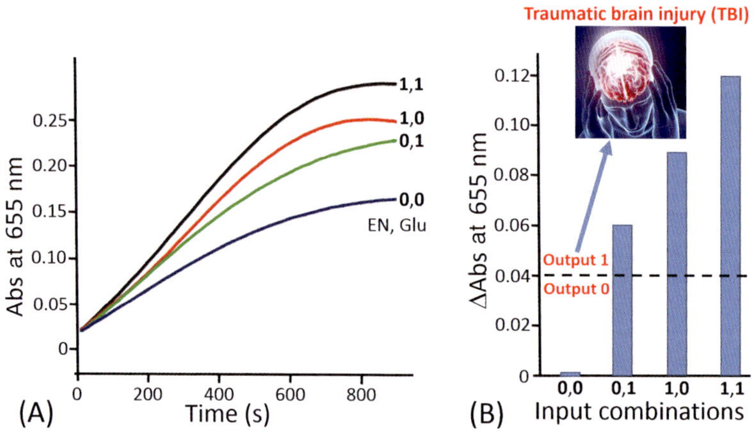

Figure 11.25 (A) Optical detection of the output signal (TMB_{ox}) generated by the logic system operating for the TBI analysis obtained upon application of different combinations of the biomarker input signals (EN, Glu). (B) Bar chart for the output signals generated by the enzyme logic system for the analysis of TBI at 800 seconds. The TBI diagnosis corresponds to the output signals above the decision threshold (dashed line). The logic system composition is detailed in Ref. [32] and the input concentrations corresponding to the logic **0** and **1** values are specified in Table 11.1. The auxiliary (2-PGA, GluOx, TMB) and switching (ADP and inorganic phosphate) inputs were at logic **1** for all combinations of the biomarker inputs. Source: Halámek et al. 2010 [32]. Adapted with permission of Royal Society of Chemistry.

time the switching inputs PEP and CoA were applied at logic **0** levels to inhibit the alternative STI pathway.

11.6.4 Switching Between the STI and TBI Modes and General Comments on the System

Operation of the logic system in the STI mode results in production of ADP as a product of the biocatalytic reaction at gate A. Further reaction of ADP in gate C requires the presence of PEP, which is not produced by the system when it operates in the STI mode (note that the gate B is mute because of the absence of EN input). Therefore PEP should be added artificially in order to activate gate C. Opposite to this, when the system operates in the TBI mode, PEP is produced *in situ* in gate B, while ADP, which is also needed for the operation of gate C, is missing (note that gate A is mute because of the absence of CK). Therefore, in this case ADP should be added artificially in order to activate gate C. Finally, for activation of the network in the STI mode the switching inputs PEP and ADP should be applied at the logic **1** and **0** values, respectively, while for the TBI mode they should be at the opposite **0** and **1** values (Figure 11.23). This switch allowed the use of gates C and F in two different modes of operation. Similarly, in order to switch between two operational modes, gates F and E should be selectively activated by the correct concentrations of phosphate and CoA. Specifically, phosphate and CoA were applied at logic **0** and **1** levels in the STI mode and at logic **1** and **0** levels in the TBI mode.

The present study demonstrated that even a very sophisticated multienzyme/multistep biocatalytic cascade can provide reliable diagnostic of physiological conditions upon logic analysis of complex patterns of various biomarkers. The obtained results experimentally confirmed the recent theoretical prediction that enzyme logic networks with up to 10 concatenated logic gates should be able to process biochemical information within a reasonable noise level [104]. The designed system exemplify the novel approach to multi-signal processing biosensors mimicking natural biochemical pathways and operating according to the biocomputing concept [20]. Further studies will be needed to transfer this approach from a conceptual demonstration to real-life biosensor applications. The future biosensor devices will be based on electrochemical methods rather than optical analysis used in the present study. This will require a lot of scientific and engineering work to integrate multienzyme systems in a rational design with mini-invasive electrodes before a real practically applicable biosensor becomes possible. It should be noted that in addition to the biocomputing and biosensor challenges, additional biomedical studies will be needed to formulate analyzed biomarkers and their normal and pathophysiological concentrations reflecting specific medical problems. The broadening of the possible applications of this concept will result in the design of various bioelectronic devices and bioactuators controlled by complex patterns of multiple inputs. Microrobotics and bioimplantable computing systems are among the most likely applications to benefit from advances in biomolecular computing. Future progress in these areas will depend on the development of novel computing concepts and design of new signal-responsive and information processing materials contributing to molecular information technology [105, 106].

11.7 New Approach in Forensic Analysis: Biomolecular Computing-Based Analysis of Forensic Biomarkers

Forensic analysis of biofluids can benefit from novel bioanalytical/biosensing methods, including those based on the biocomputing approach [107–109]. One of the new applications based on biocatalytic cascades mimicking Boolean logic gates and analyzing simultaneous presence of creatine kinase (CK) and lactate

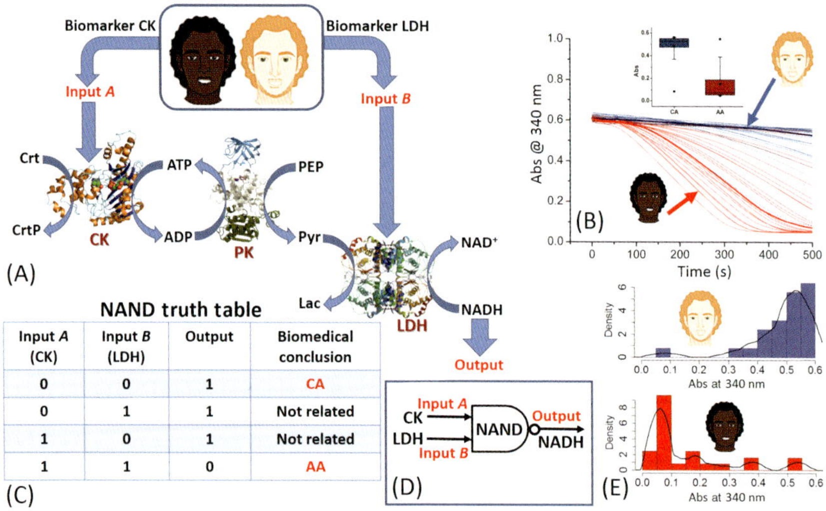

Input *A* (CK)	Input *B* (LDH)	Output	Biomedical conclusion
0	0	1	CA
0	1	1	Not related
1	0	1	Not related
1	1	0	AA

Figure 11.26 (A) The biocatalytic cascade for the two-enzyme CK–LDH assay analyzing ethnic origin of biofluids. The exact information on the system is given in Ref. [110]. (B) Absorbance ($\lambda = 340$ nm) corresponding to the consumption of NADH upon operation of the CK–LDH assay. The bottom (red) and top (blue) traces correspond to the application of samples with CK and LDH concentrations mimicking AA and CA groups, respectively. The bold solid curves show the median responses for both groups. Inset: box and whisker plot of Abs in AA and CA groups. The median value for each group is noted with the horizontal line in a box, the boxes represent the range of values from the 25th percentile to the 75th percentile, the ends of the whiskers represent the 5th and the 95th percentile of values, and the dots are the mean, maximum, and minimum values. (C) Truth table corresponding to the logic operation of the biocatalytic cascade shown in (A). Logic combination **0,0** corresponds to the biofluid with the CA origin, while logic combination **1,1** corresponds to the biofluid with the AA origin. Note that the logic combinations **0,1** and **1,0** cannot be realized in natural biofluids (they can be only prepared in artificial model solutions). (D) The NAND logic gate corresponding to the system operation. (E) Density histograms of the output signal (absorbance) obtained for CA group (blue color) and AA group (red color) using the CK–LDH assay. The histograms were derived from the experimental data shown in (B). Superimposed is the kernel density curve (black line). Note well separated data distribution for the CA and AA samples.

dehydrogenase (LDH) was developed aiming at the recognition of biofluids of different ethnic origin for forensic analysis [110] (Figure 11.26A). Knowing the difference in concentrations of CK and LDH in the blood of healthy adults of two ethnical groups, Caucasian (CA) and African American (AA), and taking into account the human population distribution patterns, samples of different ethnic origin with various CK/LDH concentrations were mimicked and analyzed [110]. The analysis was performed using a multienzyme/multistep biocatalytic cascade where the differences in both included enzymes resulted in an amplified difference in the final analytical response (Figure 11.26A). Higher concentration of the biomarkers (CK and LDH) in the biofluids of the AA origin, comparing with CA origin, resulted in faster reactions in the biocatalytic cascade, thus resulting in lower concentration of NADH, which was considered as the final output (Figure 11.26B). The biocatalytic process mimicked the NAND Boolean gate, where the low input concentrations (**0**,**0** combination) resulted in the higher output value (logic **1** output), while the higher inputs (**1**,**1** combination) produced the smaller output (logic **0** output) (Figure 11.26C,D). The output **0** corresponded to the CA origin of the analyzed biofluid and the output **1** corresponded to the AA origin. It should be noted that the "mixed" input combinations (**0**,**1** and **1**,**0**) could be artificially prepared in model solutions but cannot be realized in natural biofluids (unless their concentrations are affected by other than ethnic origin physiological reasons). The statistically established analytical results confirmed excellent probability to distinguish samples of different ethnic origin (CA vs. AA) (Figure 11.26E). The results obtained on the model solutions were confirmed by the analysis of real serum samples collected from human subjects of different ethnic origin.

Another biocatalytic assay analyzing simultaneous presence of creatine kinase (CK) and alanine transaminase (ALT) was developed aiming at the recognition of biofluids of different gender for forensic applications [111]. Knowing the difference in concentrations of CK and ALT enzymes in the blood of healthy adults of male and female groups, the samples of different gender with various CK/ALT concentrations were mimicked. The analysis was performed using a multienzyme/multistep biocatalytic cascade where the differences in both included enzymes resulted in an amplified difference in the final analytical response. The analysis performed in human serum solutions allowed discrimination of samples corresponding to male/female groups. The robustness of the developed analysis allowed determination of the gender for serum solutions after their drying and ageing at least for one hour. Importantly for forensic application, reaction with a chromogenic reactant nitroblue tetrazolium allowed qualitative discrimination of the "male" and "female" samples with a naked eye.

Overall, implementation of the forensic analytical methods based on the logic processing of various biomarkers keeps great potential for the fast on-site analysis of biofluids found directly on the crime scene without sophisticated equipment and with no delay. While more detailed analysis can be performed later in a specialized lab, the fast preliminary results can be highly important for immediate action of a police team.

11.8 Logic Analysis of Security Threats (Explosives and Nerve Agents) Based on Biocatalytic Cascades

The ability to assess rapidly diverse security threats using biochemical logic network systems is an important task. A new biocatalytic cascade, emulating a NOR logic gate, capable to identify the presence of explosive compounds or/and nerve agents by providing a simple and rapid "YES/NO" alert has been developed recently [112], following the biocomputing approach designed earlier for biomedical/bioanalytical applications. Two model compounds, trinitrotoluene (TNT) and paraoxon (PAX) have been used as input signals for activating a biocatalytic cascade (Figure 11.27A) operating as an NOR logic gate (Figure 11.27B). While TNT is a real explosive material, PAX is an acetylcholinesterase inhibitor with a toxicity lower than highly toxic nerve gases (such as VX-gas, sarin, soman, etc.), thus allowing its use in a lab as a model compound (*note that PAX is still toxic and requires very careful handling*). The biocatalytic cascade included two reaction branches activated/inhibited by TNT and PAX, respectively. The reaction catalyzed by nitroreductase (NRd) (Figure 11.27A, left part of the scheme) was activated by TNT (Input *A*, logic value **1**). This reaction resulted in the oxidizable hydroxylamine (ArNHOH), which further reacted with horseradish peroxidase (HRP) consuming H_2O_2 and reducing its concentration. Another reaction pathway (Figure 11.27A, right part of the scheme) catalyzed by acetylcholinesterase (AChE) and choline oxidase (ChOx) produced H_2O_2. However, this pathway was inhibited in the presence of PAX (Input *B*, logic value **1**). Thus the H_2O_2 production was decreased in the presence of PAX. Overall, both input signals (TNT and PAX), reacting in a different way, resulted in the decreasing concentration of H_2O_2, which was considered as the final output signal analyzed electrochemically with a

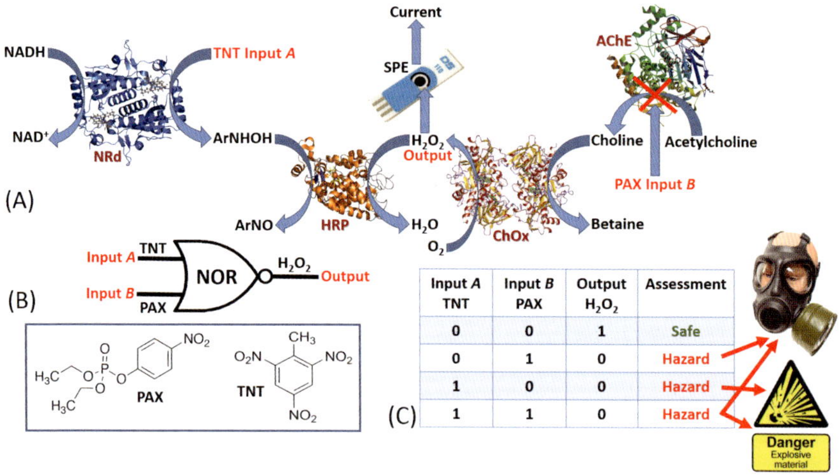

Input *A* TNT	Input *B* PAX	Output H_2O_2	Assessment
0	0	1	Safe
0	1	0	Hazard
1	0	0	Hazard
1	1	0	Hazard

Figure 11.27 (A) Biocatalytic cascade used to perform NOR logic operation in connection to trinitrotoluene (TNT) and paraoxon (PAX) inputs. (B) The equivalent logic scheme and (C) the corresponding truth table with assessment drawn from the combinations of the input signals.

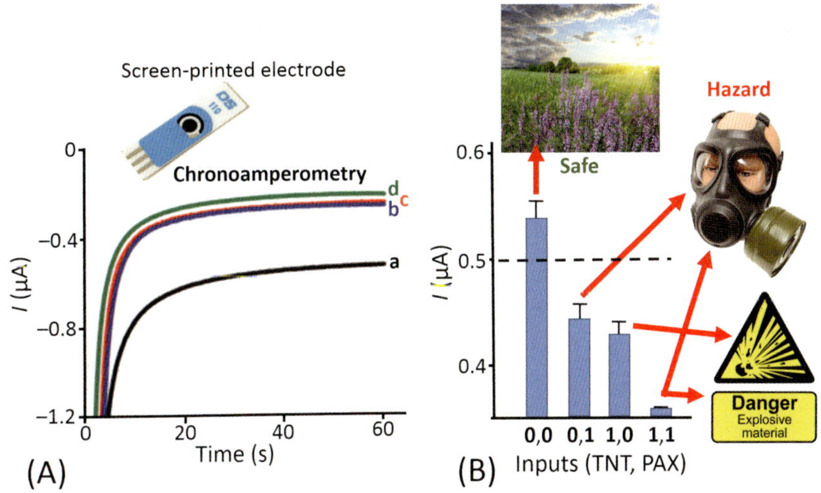

Figure 11.28 (A) Chronoamperograms measured upon operation of the enzyme-based NOR logic gate corresponding to the biocatalytic cascade shown in Figure 11.27A. The input signals (TNT and PAX) were applied in four logic combinations: **0,0**; **0,1**; **1,0**; **1,1**. (B) Bar chart comparing the magnitude of the response measured at 60 seconds after applying the potential step for the four combinations of the inputs. The bars correspond to the threat-free condition (**0,0** combination), toxic environment (**0,1** combination), explosives detected (**1,0** combination), and toxic material and explosives detected (**1,1** combination). The dashed line indicates the decision threshold separating output **0** and **1** values. Note that the alert-type binary (YES/NO) response does not allow to distinguish difference between **0,1**; **1,0**; **1,1** combinations, however the threat-free condition (**0,0** combination) is well separated from the dangerous conditions. Source: Chuang et al. 2011 [112]. Adapted with permission of Royal Society of Chemistry.

screen-printed electrode (SPE) modified with Prussian blue for effective and selective amperometric detection of H_2O_2 (Figure 11.28A). The measured current corresponding to the H_2O_2 concentration was dependent on the input combination applied in four different variants (**0,0**; **0,1**; **1,0**; **1,1**), while only the absence of the inputs (**0,0** combination) corresponded to the safe environmental conditions free of explosives and toxic materials modeled in the present system with TNT and PAX, respectively (Figure 11.28B). Figure 11.27C offers the truth table characteristic of the NOR logic gate providing the hazard assessment. The only safe (hazard-free) condition was detected with the high H_2O_2 concentration (output **1**) in the absence of both inputs (**0,0** input combination). The low H_2O_2 concentration (output **0**) originated from the presence of either or both inputs (TNT or/and PAX) (**0,1**; **1,0**; **1,1** input combinations), thus signaling on the presence of the toxic nerve agent or explosive or both of them (Figure 11.27C). The present design of the system does not allow separate detection of TNT and PAX, which is an acceptable feature for the fast alert system.

Figure 11.29 shows the response surface obtained experimentally upon variation of the input concentrations. The output signal (amperometric detection of H_2O_2) obtained in the absence of the TNT and PAX (**0,0** input combination) is sharply decreasing upon appearing of TNT and PAX in the analyte solution,

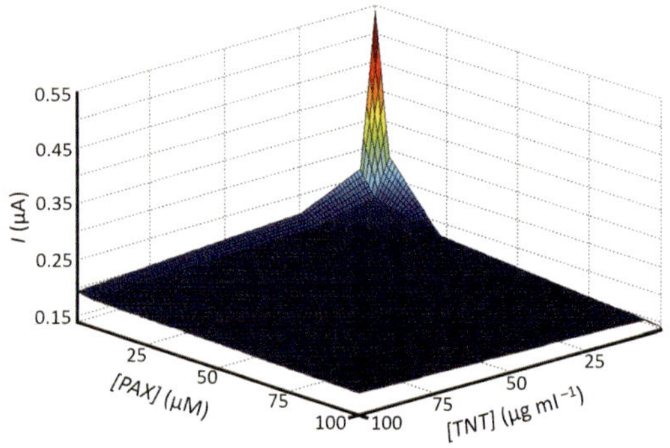

Figure 11.29 Gate mapping illustrating the functional dependence between the levels of the TNT and PAX inputs and the output current obtained. The output currents are sampled using chronoamperometry (30 seconds) with inputs levels of 0, 5, 25, 100 mg ml^{-1} and mM for TNT and PAX, respectively. The 2D linear interpolation was performed using a 20× oversampling algorithm in MATLAB. Source: Chuang et al. 2011 [112]. Adapted with permission of Royal Society of Chemistry.

getting to the flat response surface with increasing their concentrations. The flat response surface provides very small noise for the randomly variable TNT and PAX concentrations, thus allowing output signal **0** measured with the high confidence in the broad range of the input concentrations. This should provide a strong evidence of the hazard (TNT or/and PAX) presence even at their very small concentrations. On the other hand, the sharp decrease of the output signal at the **0,0** logic input combination could result in a false positive response because of the very high variability of the output signal at **0,0** logic point.

The presented concept could be readily expanded toward rapid warning of other chemical and biological threats in connection to different logic gates and should be coupled with a follow-up identification of the exact threat. The ability to assess the presence of different types of hazards holds considerable promise for enhancing and simplifying a variety of security screening protocols.

11.9 Integration of Biocatalytic Cascades with Microelectronics and Wearable Sensors

The enzyme logic gates and circuits can be analyzed using different detection tools [113] ranging from optical [37] and electrochemical [30] to nanotechnological methods, e.g., electrochemical scanning microscopy [114] and atomic force microscopy [115]. However, the majority of the studies have been directed to assembling novel logic functions using various biocatalytic and biorecognition processes, giving little attention to the transduction tools. While the biocomputing research is becoming more matured and aiming at practical applications, mostly in the area of binary biosensors for biomedical applications, the interests started to shift to integration of the signal processing biomolecular systems with microelectronics for reading the processed output signals. This novel research area has been facilitated by adopting research advances achieved recently with flexible electronics, allowing for small size wearable devices being in direct contact with human skin. The detailed discussion on this research subtopic is not in the scope of the book; thus, only a few examples of these devices will be outlined in this section.

Application of new polymeric materials and printing technologies resulted in flexible and stretchable electronic circuits (Figure 11.30A–C) and allowed for electronic circuits printed on polymeric support and then directly attached to human skin (Figure 11.30D). The developed printed circuits have been used for various wearable (bio)sensors, not necessary operating with biomolecular computing systems, but still providing a convenient platform for integration with biomolecular logic gates. One of the most exciting examples based on this new technology is represented by an "electronic ring" (Figure 11.31A), which operates as 3-input signal electrochemical sensor for monitoring of nitroaromatic and peroxide explosives as well as organophosphate nerve agents in aerosol, vapor, and solutions (Figure 11.31B) [116]. The miniaturized electronic device packed in a wearable ring included three independently operating electrochemical sensors. The device was tested for the analysis of model explosives, 2,4-dinitrotoluene (DNT) (Figure 11.32A), and H_2O_2 (Figure 11.32C), detected with square

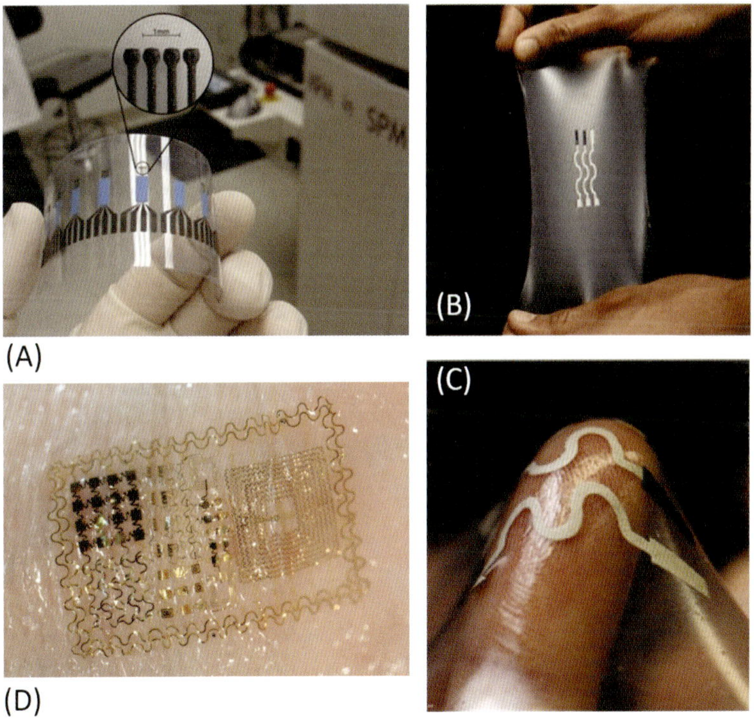

Figure 11.30 (A–C) Flexible and stretchable bioelectronic devices allow interfacing with biological tissue. (D) A new type of a biosensor uses flat, flexible electronics ("tattoo" bioelectronics) printed on a thin rubbery sheet, which can stick to human skin for at least 24 hours. Source: Courtesy of Dr. Joseph Wang, University California San Diego, USA, and "D" by Dr. John A. Rogers, University of Illinois at Urbana-Champaign, USA.

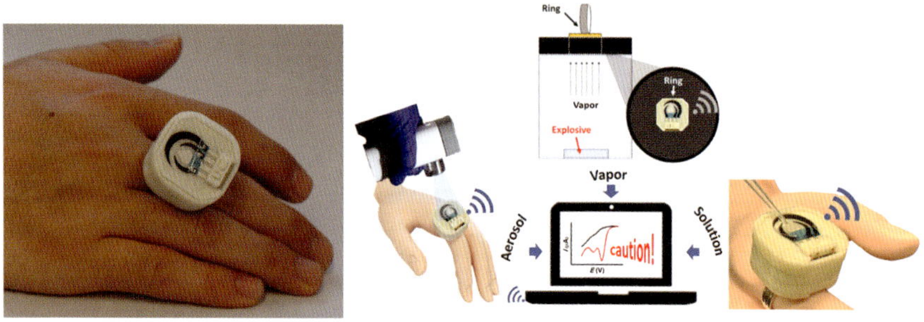

(A) (B)

Figure 11.31 (A) The three-input electrochemical sensor packed in a wearable ring. (B) The electrochemical analysis of model explosives and organophosphate nerve agent in different states: aerosol, vapor, and solution. Source: Adapted from Sempionatto et al. 2017 [116] with permission.

wave voltammetry (SWV) (Figure 11.32B), and chronoamperometry (CA) (Figure 11.32D) techniques, respectively. The DNT reduction was performed at a carbon electrode coated with solid-state electrolyte agarose hydrogel, and the H_2O_2 reduction was electrocatalyzed with Prussian blue layer deposited on the electrode surface. Organophosphate nerve agent (modeled with methyl paraoxon, MPAX) was first biocatalytically hydrolyzed with organophosphorous hydrolase (OPH), and then produced *p*-nitrophenol was electrochemically oxidized (Figure 11.32E), yielding current proportional to the MPAX concentration (Figure 11.32F). The engineered sensor was aimed at the goal similar to one based on the enzyme logic and shown in Figure 11.27. The main difference of the new design from the previous approach was in the use of three independent channels processing the input signals independently and providing quantitative information on concentrations of all three threats (two explosives and one toxic material). While the new design is not based on the enzyme logic gates and represents the traditional concept of biosensors, the novelty is in its high quality engineering. This device can be easily adapted to the analysis of signals produced by various enzyme logic gates, particularly benefitting from three independent read-out channels. The number of input signals obviously can exceed three upon appropriate organization of biocomputing system associated with the electronic device used as a convenient platform for the logic gates. Two other example devices (selected from many available in recent publications [116–120]) are wearable biosensors for alcohol [121] (Figure 11.33A) and glucose [122] (Figure 11.33C). Their electronic parts are based on temporary "tattoo" electrodes attached to human skin and the biocatalytic processes are similar with only difference in the used enzymes: alcohol oxidase (AOx) and glucose oxidase (GOx), respectively (Figure 11.33B,C). These biocatalytic reactions can be extended to multistep/multienzyme cascades, then converting the electrochemical biosensors to biocomputing systems assembled on the same platform. Overall, advances in microelectronics and wearable sensors can provide engineering solutions to integration of biocomputing elements with electronic parts located on human body or other biological substances.

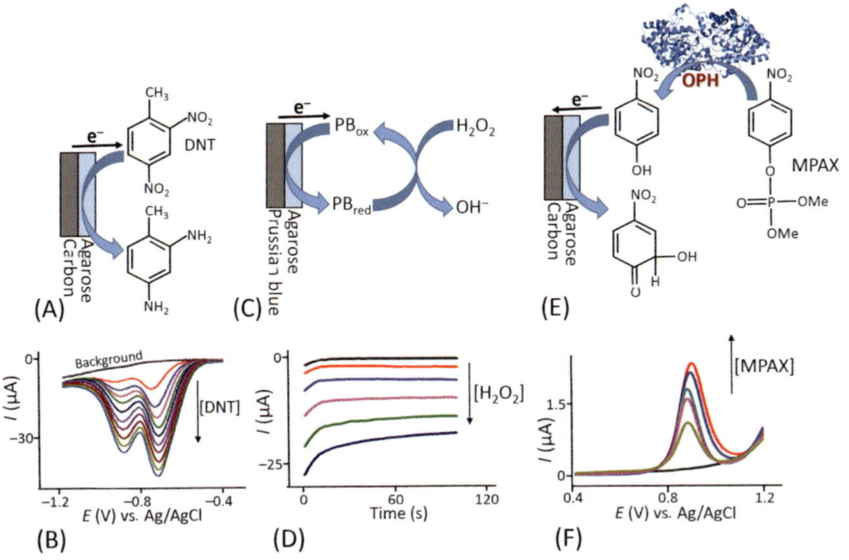

Figure 11.32 Electrochemical analysis of DNT shown schematically (A) and corresponding SWV traces recorded with increasing concentration of DNT (B). Electrochemical analysis of H_2O_2 shown schematically (C) and corresponding CA traces recorded with increasing concentration of H_2O_2 (D). Electrochemical analysis of MPAX shown schematically (E) and corresponding SWV traces recorded with increasing concentration of MPAX (F). Source: Parts B, D, and F are adapted from Sempionatto et al. 2017 [116] with permission.

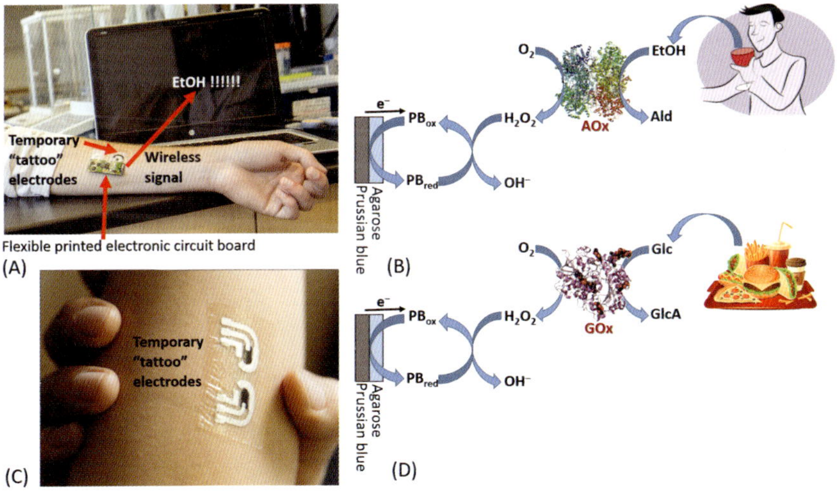

Figure 11.33 A wearable tattoo-based biosensing system for noninvasive alcohol monitoring in induced sweat (A) and the scheme of the biosensing process (B). A wearable tattoo-based system for noninvasive monitoring of interstitial glucose (C) and the scheme of the biosensing process (D). Source: Courtesy of Dr. Joseph Wang, University California San Diego, USA.

11.10 Conclusions and Perspectives

Applying novel concepts from the area of unconventional computing (specifically from biomolecular computing) [19, 20] to biosensing and bioanalytical assays [1, 26, 123] has resulted in the design of biomolecular systems logically processing several chemical signals and converting them to a single binary output in the format of YES/NO. Information processing in biomolecular systems does not require electronic computers and proceeds at the level of chemical reactions. The "program" for processing chemical inputs can be implemented in the composition of the biomolecular system and can include various logic operations applied in different combinations. The systems exemplified in this chapter demonstrated the simplest logic processing based mostly on logic AND/OR gates, applied to biochemical input signals. However, many other logic operations integrated in various logic circuitries are possible with the use of different enzymes and other biomolecules, to allow high-fidelity detection of various biomarkers. The resulting digital biosensors would thus benefit different important fields, being particularly useful in rapid on-field injury analysis and forensic and security analytical methods. An important challenge in developing this kind of the digital multi-signal biosensor system is obtaining a significant difference between the logic **1** and **0** output values (in other words, a well-defined YES/NO answer). One should remember that in case of forensic analysis the input signals (biomarkers) appear with small difference. In order to obtain significant difference in the output signals, thus being able to distinguish biological samples of different origin, the response function in the information processing steps should be sigmoid rather than linear. In other words, the system should demonstrate a nonlinearity with a sharp transition between the **0** (NO) and **1** (YES) states. The first steps in this direction have been already done experimentally and analyzed theoretically [38, 45, 124–129]. However, extensive research effort aimed at designing chemical "filter" systems similar to the electronic counterparts is needed. The thresholds separating the logic **0** and **1** values could be tailored for a specific application by following expected concentrations of biomarkers in the biological fluids. This will be an important step toward future on-site biochemical analysis.

References

1 Katz, E., Wang, J., Privman, M., and Halámek, J. (2012). *Anal. Chem.* 84: 5463–5469.
2 Clark, L.C. Jr. and Lyons, C. (1962). *Ann. N.Y. Acad. Sci.* 102: 29–45.
3 Renneberg, R. and Lisdat, F. (eds.) (2008). Biosensing for the 21st Century. In: *Series Advances in Biochemical Engineering/Biotechnology*, Scheper, T. (Series ed.). Berlin, Heidelberg: Springer.
4 Willner, I. and Katz, E. (2000). *Angew. Chem. Int. Ed.* 39: 1180–1218.
5 Leech, D. (1994). *Chem. Soc. Rev.* 23: 205–213.

6 van Emon, J.M. (ed.) (2006). *Immunoassay and Other Bioanalytical Techniques*. Boca Raton: CRC Press.
7 Sassolas, A., Leca-Bouvier, B.D., and Blum, L.J. (2008). *Chem. Rev.* 108: 109–139.
8 Zourob, M. (ed.) (2010). *Recognition Receptors in Biosensors*. New York: Springer.
9 Ronkainen, N.J., Halsall, H.B., and Heineman, W.R. (2010). *Chem. Soc. Rev.* 39: 1747–1763.
10 Borisov, S.M. and Wolfbeis, O.S. (2008). *Chem. Rev.* 108: 423–461.
11 Steinem, C. and Janshoff, A. (eds.) (2010). *Piezoelectric Sensors*. In: Series on chemical sensors and biosensors, Wolfbeis, O.S. (Series ed.). Berlin, Heidelberg: Springer.
12 Wang, J. (2008). *Chem. Rev.* 108: 814–825.
13 Crowther, J.R. (2010). *The ELISA Guidebook*. New York: Humana Press.
14 Schena, M. (2002). *Microarray Analysis*. Hoboken: Wiley-Liss.
15 Ehrenkranz, J.R.L. (2002). *Epidemiology* 13: S15–S18.
16 Kumar, V. and Kaushik, M.P. (2011). *Analyst* 136: 5151–5156.
17 Warsinke, A. (2009). *Anal. Bioanal. Chem.* 393: 1393–1405.
18 Katz, E. (ed.) (2012). *Molecular and Supramolecular Information Processing: From Molecular Switches to Logic Systems*. Weinheim: Wiley-VCH.
19 Katz, E. (ed.) (2012). *Biomolecular Information Processing – From Logic Systems to Smart Sensors and Actuators*. Weinheim: Wiley-VCH.
20 Katz, E. and Privman, V. (2010). *Chem. Soc. Rev.* 39: 1835–1857.
21 Stojanovic, M.N. and Stefanovic, D. (2003). *Nat. Biotechnol.* 21: 1069–1074.
22 Win, M.N. and Smolke, C.D. (2008). *Science* 322: 456–460.
23 Stojanovic, M.N., Mitchell, T.E., and Stefanovic, D. (2002). *J. Am. Chem. Soc.* 124: 3555–3561.
24 Li, Z., Rosenbaum, M.A., Venkataraman, A. et al. (2011). *Chem. Commun.* 47: 3060–3062.
25 Calude, C.S., Costa, J.F., Dershowitz, N. et al. (eds.) (2009). *Unconventional Computation*, Lecture notes in computer science, vol. 5715. Berlin: Springer.
26 Wang, J. and Katz, E. (2010). *Anal. Bioanal. Chem.* 398: 1591–1603.
27 Vaidya, V.S. and Bonventre, J.V. (eds.) (2010). *Biomarkers: In Medicine, Drug Discovery, and Environmental Health*. Hoboken: Wiley.
28 Honda, M., Tsuruta, R., Kaneko, T. et al. (2010). *J. Trauma Inj. Infect. Crit. Care* 69: 104–109.
29 Lee, S.-Y., Choi, Y.-C., Kim, J.-H., and Kim, W.-J. (2010). *J. Neurol.* 257: 1708–1712.
30 Zhou, N., Windmiller, J.R., Valdés-Ramírez, G. et al. (2011). *Anal. Chim. Acta* 703: 94–100.
31 Halámek, J., Windmiller, J.R., Zhou, J. et al. (2010). *Analyst* 135: 2249–2259.
32 Halámek, J., Bocharova, V., Chinnapareddy, S. et al. (2010). *Mol. Biosyst.* 6: 2554–2560.
33 de Silva, A.P. and Uchiyama, S. (2007). *Nat. Nanotechnol.* 2: 399–410.
34 Margulies, D. and Hamilton, A.D. (2009). *J. Am. Chem. Soc.* 131: 9142–9143.
35 May, E.E., Dolan, P.L., Crozier, P.S. et al. (2008). *IEEE Sens. J.* 8: 1011–1019.

36 von Maltzahn, G., Harris, T.J., Park, J.-H. et al. (2007). *J. Am. Chem. Soc.* 129: 6064–6065.
37 Zhou, J., Halámek, J., Bocharova, V. et al. (2011). *Talanta* 83: 955–959.
38 Halámek, J., Zhou, J., Halámková, L. et al. (2011). *Anal. Chem.* 83: 8383–8386.
39 Melnikov, D., Strack, G., Zhou, J. et al. (2010). *J. Phys. Chem. B* 114: 12166–12174.
40 Tan, K.-K., Bang, S.-L., Vijayan, A., and Chiu, M.-T. (2009). *Injury* 40: 978–983.
41 Khalili, H., Dayyeh, B.A., and Friedman, L.S. (2010). Assessment of liver function in clinical practice. In: *Clinical Gastroenterology: Chronic Liver Failure* (ed. P. Ginès, P.S. Kamath and V. Arroyo), 47–76. New York: Humana Press.
42 Olerud, J.E., Homer, L.D., and Carroll, H.W. (1976). *Arch. Intern. Med.* 136: 692–697.
43 Kratz, A., Ferraro, M., Sluss, P.M., and Lewandrowski, K.B. (2004). *N. Engl. J. Med.* 351: 1548–1563.
44 Privman, M., Tam, T.K., Bocharova, V. et al. (2011). *ACS Appl. Mater. Interfaces* 3: 1620–1623.
45 Privman, V., Halámek, J., Arugula, M.A. et al. (2010). *J. Phys. Chem. B* 114: 14103–14109.
46 Jeffery, J. (ed.) (1980). *Dehydrogenases: Requiring Nicotinamide Coenzymes.* Basel: Springer.
47 Privman, V. (2011). *Isr. J. Chem.* 51: 118–131.
48 Lusczek, E.R., Nelson, T., Lexcen, D. et al. (2011). *J. Bioanal. Biomed.* 3: 38–48.
49 Halámková, L., Halámek, J., Bocharova, V. et al. (2012). *Analyst* 137: 1768–1770.
50 Michaeli, D. (1979). *J. R. Soc. Med.* 72: 370–373.
51 Wilson, M., Davis, D.P., and Coimbra, R. (2003). *J. Emerg. Med.* 24: 413–422.
52 Champion, H.R., Bellamy, R.F., Roberts, C.P., and Leppaniemi, A. (2003). *J. Trauma* 54: 13–19.
53 Clark, M.E., Bair, M.J., Buckenmaier, C.C. et al. (2007). *J. Rehabil. Res. Dev.* 44: 179–194.
54 Bhasale, A. (1998). *Fam. Pract.* 15: 308–318.
55 Taber, K.H., Warden, D.L., and Hurley, R.A. (2006). *J. Neuropsychiatry Clin. Neurosci.* 18: 141–145.
56 Pineda, J.A., Wang, K.K.W., and Hayes, R.L. (2004). *Brain Pathol.* 14: 202–209.
57 Papa, L., Robinson, G., Oli, M. et al. (2008). *Expert Opin. Med. Diagn.* 2: 937–945.
58 Bailly, N., Afquir, S., Laporte, J.D. et al. (2017). *Med. Sci. Sports Exercise* 49: 1–10.
59 Haider, A.H., Saleem, T., Bilaniuk, J.W., and Barraco, R.D. (2012). *J. Trauma Acute Care Surg.* 73: 1340–1347.
60 Viano, D., von Holst, H., and Gordon, E. (1997). *Accid. Anal. Prev.* 29: 811–816.

61 Rosenberg, J.C., Lillehei, R.C., Longerbean, J., and Zini-Nierinann, B. (1961). *Ann. Surg.* 154: 611–627.
62 Prasad, M.R., Ramaiah, C., McIntosh, T.K. et al. (1994). *J. Neurochem.* 63: 1086–1094.
63 Kudsk, K.A., Croce, M.A., Fabian, T.C. et al. (1992). *Ann. Surg.* 215: 503–513.
64 Rignault, D.P. (1992). *World J. Surg.* 16: 940–946.
65 Freitas, A.D. and Franzon, O. (2015). *Arq. Bras. Cir. Dig.* 28: 163–166.
66 Laverde Sabogal, C.E., Correa Rivera, A.F., and Joya Higuera, A.Y. (2014). *Colomb. J. Anesthesiol.* 42: 60–64.
67 Sprules, S.D., Hart, J.P., Wring, S.A., and Pittson, R. (1995). *Anal. Chim. Acta* 304: 17–24.
68 Wedmore, I., McManus, J.G., Pusateri, A.E., and Holcomb, J.B. (2006). *J. Trauma* 60: 655–658.
69 Lakstein, D., Blumenfeld, A., Sokolov, T. et al. (2003). *J. Trauma* 54: 221–225.
70 Sies, H. (1997). *Exp. Physiol.* 82: 291–295.
71 Griendling, K.K. and FitzGerald, G.A. (2003). *Circulation* 108: 1912–1916.
72 Koushanpour, A., Gamella, M., and Katz, E. (2017). *Electroanalysis* 29: 1602–1611.
73 Koushanpour, A., Gamella, M., Guz, N., and Katz, E. (2017). *Electroanalysis* 29: 950–954.
74 Donnelly, E.H., Nemhauser, J.B., Smith, J.M. et al. (2010). *South. Med. J.* 103: 541–546.
75 Akleyev, A.V., Akushevich, I.V., Dimov, G.P. et al. (2010). *Radiat. Environ. Biophys.* 49: 281–291.
76 Swartz, H.M., Flood, A.B., Gougelet, R.M. et al. (2010). *Health Phys.* 98: 95–108.
77 Alexander, G.A., Swartz, H.M., Amundson, S.A. et al. (2007). *Radiat. Meas.* 42: 972–996.
78 Jack, C.I.A., Bottier, B., Jackson, M.J. et al. (1996). *Int. J. Radiat. Oncol. Biol. Phys.* 34: 149–154.
79 Wickremesekera, J.K., Chen, W., Cannan, R.J., and Stubbs, R.S. (2001). *Int. J. Radiat. Oncol. Biol. Phys.* 49: 1015–1021.
80 Hara, R., Itami, J., Komiyama, T. et al. (2004). *Chest* 125: 340–344.
81 Haffty, B.G. and Glazer, P.M. (2003). *Oncogene* 22: 5915–5925.
82 Bower, J.E., Ganz, P.A., Tao, M.L. et al. (2009). *Clin. Cancer Res.* 15: 5534–5540.
83 Schneider, J., Presek, P., Braun, A., and Woitowitz, H.J. (1999). *Am. J. Ind. Med.* 36: 602–609.
84 Horneck, G., Facius, R., Reitz, G. et al. (2003). *Adv. Space. Res.* 31: 87–95.
85 Horneck, G. (1992). *Int. J. Radiat. Appl. Instrum. D* 20: 185–2005.
86 Horneck, G. and Comet, B. (2006). *Adv. Space Res.* 37: 100–108.
87 Gusev, I., Guskova, A., and Mettler, F.A. (eds.) (2001). *Medical Management of Radiation Accidents*. Boca Raton: CRC Press.
88 Blakely, W.F., Salter, C.A., and Prasanna, P.G.S. (2005). *Health Phys.* 89: 494–504.

89 Okunieff, P., Chen, Y., Maguire, D.J., and Huser, A.K. (2008). *Cancer Metastasis Rev.* 27: 363–374.
90 Stephan, G., Oestreicher, U., and Romm, H. (2007). *Chromosomal Alterations. Methods, Results and Importance in Human Health*, Chapter 22 (ed. G. Obe and Vijayalaxmi), 341–350. Berlin-Heidelberg: Springer.
91 Beninson, D., Lloyd, D.C., Natarajan, A.T. et al. (1986). *Biological Dosimetry: Chromosomal Aberration Analysis for Dose Assessment*. Vienna: International Atomic Energy Agency.
92 Ménard, C., Johann, D., Lowenthal, M. et al. (2006). *Cancer Res.* 66: 1844–1850.
93 Becciolini, A., Porciani, A., Lanini, A. et al. (1987). *Acta Oncologica.* 26: 139–142.
94 Blakely, W.F., Ossetrova, N.I., Manglapus, G.L. et al. (2007). *Radiat. Meas.* 42: 1164–1170.
95 Chen, I.W., Kereiakes, J.G., Silberstein, E.B. et al. (1973). *Radiat. Res.* 54: 141–151.
96 Barrett, A., Jacobs, A., Kohn, J. et al. (1982). *Br. Med. J.* 285: 170–171.
97 Becciolini, A., Porciani, S., Lanini, A. et al. (1994). *Eur. J. Nucl. Med.* 21: 1121–1125.
98 Bloor, C.M., Leroy, E.P., Friedman, P.J., and Sobel, B.E. (1974). *Radiat. Res.* 57: 311–318.
99 Park, S.H., Pradeep, K., Choi, M.H. et al. (2010). *Res. J. Pharm. Biol. Chem. Sci.* 1: 165–177.
100 Horvath, M., Rode, I.L., Kralovanszky, J., and Volant, M. (1980). *Strahlentherapie* 156: 244–247.
101 Bocharova, V., Halámek, J., Zhou, J. et al. (2011). *Talanta* 85: 800–803.
102 Zink, B.J., Schultz, C.H., Wang, X. et al. (1999). *Brain Res.* 837: 1–7.
103 Berger, R.P., Hymel, K., and Gao, W.-M. (2006). *Clin. Pediatr. Emerg. Med.* 7: 186–193.
104 Privman, V., Strack, G., Solenov, D. et al. (2008). *J. Phys. Chem. B* 112: 11777–11784.
105 Zauner, K.-P. (2005). *Critical Rev. Solid State Mater. Sci.* 30: 33–69.
106 Fu, L., Cao, L.C., Liu, Y.Q., and Zhu, D.B. (2004). *Adv. Colloid Interf. Sci.* 111: 133–157.
107 Katz, E., Halámek, J., Halámková, L. et al. (2016). Biochemical analysis of biomarkers for forensic applications. In: *Forensic Science: A Multidisciplinary Approach*, Chapter 8 (ed. E. Katz), 151–176. Weinheim: Wiley-VCH.
108 Katz, E., Halámek, J., and Bakshi, S. (2015). *J. Forensic Legal Invest. Sci.* 1: 004.
109 Katz, E. and Halámek, J. (2014). *Ann. Forensic Res. Anal.* 1: 1002.
110 Kramer, F., Halámková, L., Poghossian, A. et al. (2013). *Analyst* 138: 6251–6257.
111 Bakshi, S., Halámková, L., Halámek, J., and Katz, E. (2014). *Analyst* 139: 559–563.
112 Chuang, M.-C., Windmiller, J.R., Santhosh, P. et al. (2011). *Chem. Commun.* 47: 3087–3089.
113 Katz, E. (2017). *ChemPhysChem* 18: 1688–1713.

114 Gdor, E., Katz, E., and Mandler, D. (2013). *J. Phys. Chem. B* 117: 16058–16065.
115 Strack, G., Chinnapareddy, S., Volkov, D. et al. (2009). *J. Phys. Chem. B* 113: 12154–12159.
116 Sempionatto, J.R., Mishra, R.K., Martín, A. et al. (2017). *ACS Sensors* 2: 1531–1538.
117 Kim, J., Campbell, A.S., and Wang, J. (2018). *Talanta* 177: 163–170.
118 Kim, J., Kumar, R., Bandodkar, A.J., and Wang, J. (2017). *Adv. Electron. Mater.* 3: 1600260.
119 Bandodkar, A.J., Jeerapan, I., and Wang, J. (2016). *ACS Sensors* 1: 464–482.
120 Bandodkar, A.J. and Wang, J. (2014). *Trends Biotechnol.* 32: 363–371.
121 Kim, J., Jeerapan, I., Imani, S. et al. (2016). *ACS Sensors* 1: 1011–1019.
122 Bandodkar, A.J., Jia, W., Yardímcí, C. et al. (2015). *Anal. Chem.* 87: 394–398.
123 Wang, J. and Katz, E. (2011). *Isr. J. Chem.* 51: 141–150.
124 Privman, V., Domanskyi, S., Mailloux, S. et al. (2014). *J. Phys. Chem. B* 118: 12435–12443.
125 Mailloux, S., Zavalov, O., Guz, N. et al. (2014). *Biomater. Sci.* 2: 184–191.
126 Privman, V., Zavalov, O., Halámková, L. et al. (2013). *J. Phys. Chem. B* 117: 14928–14939.
127 Bakshi, S., Zavalov, O., Halámek, J. et al. (2013). *J. Phys. Chem. B* 117: 9857–9865.
128 Zavalov, O., Bocharova, V., Privman, V., and Katz, E. (2012). *J. Phys. Chem. B* 116: 9683–9689.
129 Halámek, J., Zavalov, O., Halámková, L. et al. (2012). *J. Phys. Chem. B* 116: 4457–4464.

12

Release of Molecular Species Stimulated by Logically Processed Biomolecule Signals

Biomolecule signals logically processed through different gates based on enzyme-catalyzed reactions [1, 2] have been used for activating various downstream processes, including electrochemical reactions [3–6] (particularly in biofuel cells [7–11]), activating electronic chips [12–17], restructuring "smart" signal-responsive materials [18, 19] (such as nanoparticle ensembles [20, 21], polymer membranes [22], and molecule brushes [23]), stimulating light emission processes [24, 25], etc. One of the most intriguing applications is signal-stimulated release of biomolecule and nanostructured species from signal-sensitive matrices [26, 27]. Several examples of such processes, mostly limited to the release from alginate hydrogels [26–30], are overviewed in this chapter. The reasons for concentrating on alginate hydrogels are explained in the following sections.

12.1 Motivation and Experimental Background

Chemical systems based on stimuli-responsive materials designed to release loaded substances in response to external signals received great attention due to their potential applications in various areas [31–33], particularly for the delivery of bioactive species (e.g., drugs) [34–37]. The releasing processes stimulated by various physical and chemical signals, such as light [38], magnetic field [39], temperature change [40], and pH variation [41], have been extensively studied. Drug release triggered by biomolecular signals (particularly represented by biomarkers signaling on physiological dysfunctions) [42] are of special interest, being promising for the functional integration of releasing systems with biological processes [43, 44]. Such systems could be important for the development of closed-loop *sense-and-act* biomedical devices [45–48] in the general frame of a theranostic approach (a combination of diagnostics and therapy) [49].

While the experimental background for the signal-triggered release processes is very broad, it is mostly based on the use of a single input activation. Combining the release systems with biomolecular logic gates processing multiple input signals can significantly increase the operational complexity of the release systems resulting in their functional integration with biological/physiological processes and allowing their adaptive behavior, when the release process is

Enzyme-Based Computing Systems, First Edition. Evgeny Katz.

activated upon receiving input signals in the logically correct combination. The "decision-making" logic system can be "programmed" to activate the release process according to different logic processes corresponding to the needs of biological systems.

12.2 Fe^{3+}-Cross-Linked Alginate Hydrogel is a Good Example of Matrix for Signal-Stimulated Release

While various polymeric matrices have been often used for biomolecule load-release processes [50–52], alginate-based assemblies are among the most attractive components of the controlled release systems due to their biocompatibility and versatility [53]. Alginate matrices have been extensively studied for encapsulation and, in some cases, release of many different biomolecular species, including proteins/enzymes [54–58], DNA [59–61], whole biological cells [62, 63], and other (bio)molecular substances, e.g., drugs [64, 65]. Alginate matrices loaded with biomolecular species are usually cross-linked with multi-charged metal cations, frequently using Ca^{2+} cations as the cross-linkers, to yield the alginate hydrogel for the slow release (leakage) of entrapped biomolecules [66]. Signal-triggered release of biomolecules loaded in alginate hydrogels is much more interesting for many applications but less explored experimentally [67]. In order to achieve the signal-triggered biomolecular release from alginate hydrogels, cross-linking of the alginate polymers is usually performed with Fe^{3+} cations [26]. These cations have an advantage over Ca^{2+} cross-linkers because the cross-linking can be controlled by the oxidation state of the iron cations. Fe^{2+} is a "soft" metal cation that tends to bind neutral ligands containing nitrogen and sulfur atoms, whereas Fe^{3+} cation is a typical example of a "hard" metal cation that preferentially binds oxygen atoms in negatively charged ligands, such as carboxylate groups [68], particularly in the alginate molecules. As the result of this difference, Fe^{3+} cations produce a hydrogel in the presence of alginate, while Fe^{2+} cations are not capable of the hydrogel formation (Figure 12.1). Using this feature of the iron cations, thin films of Fe^{3+}-cross-linked alginate hydrogels with various biomolecular species entrapped inside have been prepared electrochemically by oxidizing Fe^{2+} cations in the presence of soluble alginate (Figure 12.2) and then dissolved upon changing the potential applied on the modified electrode, resulting in the reduction process and formation of Fe^{2+} cations (note that these cations are not capable of alginate cross-linking). The electrochemical dissolution of these films resulted in the release of the species loaded in the hydrogel upon its formation. The reductive dissolution of the Fe^{3+}-cross-linked alginate hydrogel was performed by the direct application of the electrical potential on the modified electrode from an electrochemical instrument, thus resulting in the electrochemically stimulated release process [69, 70]. Similar reductive dissolution of Fe^{3+}-cross-linked alginate hydrogel was achieved by photochemical reduction of Fe^{3+}-cross-linking cations in the presence of lactate operating as a sacrificial electron donor [71]. This process allows light-stimulated release of species entrapped in the alginate hydrogel.

Figure 12.1 Photograph represents reversible formation of alginate hydrogel in the presence of Fe^{3+} cations capable of alginate cross-linking and hydrogel dissolution when Fe^{3+} is converted to Fe^{2+} cations, which have much weaker interactions with alginate and do not cross-link the alginate molecules. This figure aims only at the schematic illustration of different interactions of alginate with iron ions in different oxidation states. Note that in the real systems described in the chapter, the alginate matrix was produced on an electrode surface in the presence of electrochemically generated Fe^{3+} cations to yield a thin film, which was later dissolved upon electrochemical reduction of Fe^{3+} to Fe^{2+}. Source: Jin et al. [69]. Adapted with permission of American Chemical Society.

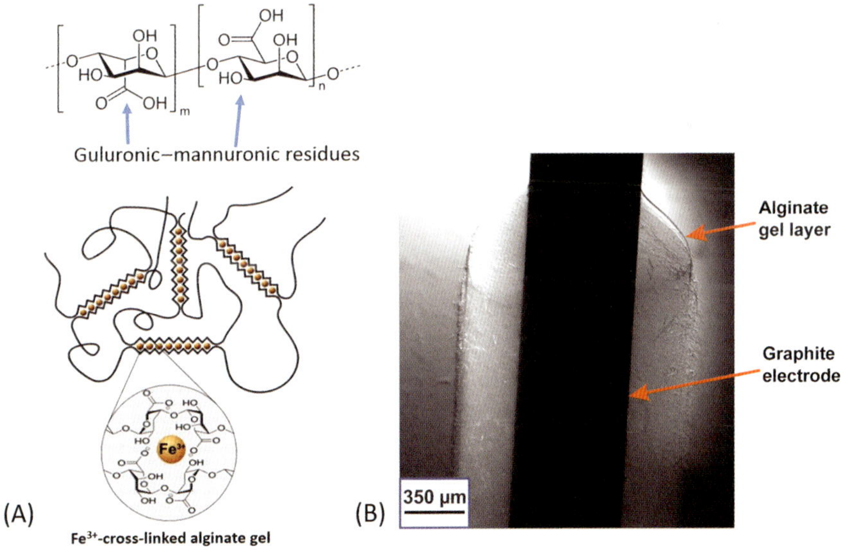

Figure 12.2 (A) Schematic structure of the Fe^{3+}-cross-linked alginate hydrogel deposited electrochemically on a graphite electrode surface. (B) Microscopic image of the alginate hydrogel film deposited on a graphite electrode surface. Note that the film thickness is controlled by the deposition time period and concentration of soluble alginate and Fe^{2+} cations during the deposition step. Source: Gamella et al. 2017 [28]. Adapted with permission of John Wiley and Sons.

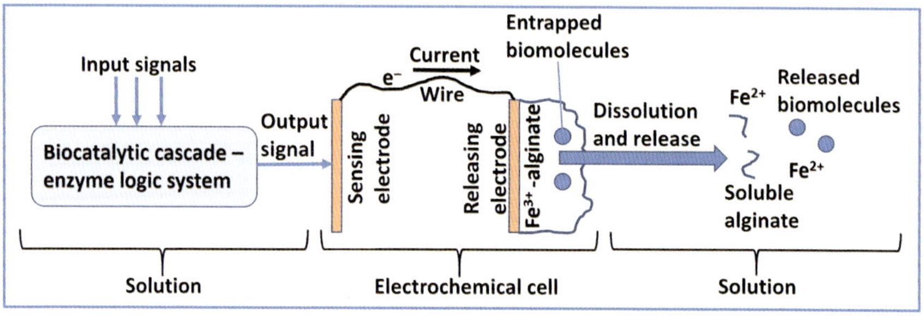

Figure 12.3 Electrochemical system composed of two electrically connected electrodes, one activated with output signals produced by an enzyme logic system and another coated with alginate hydrogel with entrapped biomolecules. The alginate film was dissolved, and the entrapped molecules were released when the sensing electrode was activated. Source: Gamella et al. 2017 [28]. Adapted with permission of John Wiley and Sons.

Much more interesting systems (detailed in the following sections) were composed of two modified electrodes connected to each other: one coated with the Fe^{3+}-cross-linked alginate film with the entrapped species, while the second electrode capable of producing the reductive potential in the presence of different biomolecular/biological signals, thus representing a biosensing electrode (Figure 12.3). The reductive potential on the biosensing electrode was produced in the presence of small biomolecules (e.g., glucose) [72], various combinations of biomolecules [73, 74], proteins [75], DNA [76], or microbial cells [77]. The most interesting examples were represented by biocatalytic cascades logically processing multiple input signals [73, 74] (Figure 12.3). The applied signals resulted in the current passing from the biosensing electrode to the alginate electrode, thus resulting in the reduction of iron cations, dissolution of alginate hydrogel, and release of the entrapped species. Biomolecular species released from the alginate hydrogel in the presence of the external signals included various drugs [78] (e.g., insulin [79]) for biomedical applications, enzymes for signal-activated biofuel cells [7], and oligonucleotides for activating DNA computing systems [80, 81]. While the developed approach based on the use of two interconnected electrodes [26], one for biosensing and another for the signal-triggered release of the entrapped biomolecules, has proven to be very powerful and has allowed for the logic processing of the incoming activating signals [73, 74], the disadvantage of this method is in the use of electronic interfaces in addition to the chemical signal-responsive materials. This problem might be partially overcome by using a bifunctional "Janus"-type electrode where a fraction of the electrode surface is used for biosensing of the incoming signals, while the second section of the same electrode is coated with the alginate film for the signal-stimulated release of the entrapped biomolecules [82]. This modified approach allows for miniaturization of the signal processing and molecule-releasing system. The detailed discussion on the alginate hydrogels used for the signal-stimulated molecule release is offered in the following sections.

12.3 DNA Release as an Example of Signal-Stimulated Biomolecule Release

While many different (bio)molecular and nano-size species can be entrapped into alginate hydrogel upon its electrochemical formation and then released when the alginate film is dissolved, many example systems discussed in the following sections are aimed for the DNA release from the alginate matrix. There are two major reasons for using DNA as the release substance in the discussed systems: one is scientific and another is technical. The scientific reason is the importance of the DNA release. Controlled DNA release from various biomolecular and bioelectronic systems [83, 84] stimulated by different signals [85–88] is highly important for many biomedical and biotechnological applications, including gene delivery therapy [83, 89], biosensors [90], biochips [91] and biomolecular computing [81]. Various methods of the signal-controlled DNA release have been reported recently, including irradiation of DNA-loaded nanoparticles with an alternating electromagnetic field [85], light-triggered capture and release of DNA [86], thermal release of DNA from nanostructured systems [87, 88], etc. Among different methods of the DNA controlled release, electrochemically stimulated release of DNA preloaded on an electrode surface is particularly convenient and interesting due to its simplicity and versatility of the method. The technical reason for studying DNA release from the alginate hydrogel is simply based on the experimental convenience. The DNA molecules demonstrated small uncontrolled leakage from the alginate matrix, thus allowing its signal-activated release with the kinetics significantly different from the leakage.

Electrochemical systems releasing DNA from electrode interfaces in response to electrical signals [92–107] represent particular interest due to possibility of their connection to biosensing systems triggering the DNA release. The electrical signals triggering DNA release can be generated *in situ* by bioelectrocatalytic electrodes in the presence of electron-donating species, e.g., NADH [81]. Connecting signal-producing bioelectrocatalytic electrodes and substance-releasing electrodes allows for close-loop self-activated systems responding to the external signals by the substance release [26, 47].

12.4 Bioelectrochemical Systems with Sensing and Releasing Electrodes

The bioelectronic system for signal-activated release of molecular species from alginate hydrogel was assembled with two electrodes, one for generating electrochemical signal and another coated with an alginate film with entrapped the species prepared for the release (Figure 12.3). The alginate film was produced at the electrode surface by electrochemical oxidation of Fe^{2+} cations in the presence of soluble alginate and the species to be entrapped. The oxidized Fe^{3+} cations cross-linked alginate polymer yielding a hydrogel film on the electrode surface with the species physically entrapped (Figure 12.2). The opposite electrochemical process resulted in the reduction of Fe^{3+} cations, alginate film dissolution, and the entrapped species release. This reductive dissolution process was stimulated by potential produced on the connected sensing electrode in the presence of oxidizable molecules generated *in situ* through biocatalytic reactions performing logic operations. Let us consider operation of both electrodes in details for different logic operations and the following release process.

12.4.1 Sensing Electrodes Activated with Single Input Identity Gate

In the simplest experimental realization, the biosensing electrode was functionalized with pyrroloquinoline quinone-dependent glucose dehydrogenase (PQQ-GDH) biocatalytically oxidizing glucose (Glc) to gluconic acid (GlcA). The material called "buckypaper" [108–110], made from compressed multiwalled carbon nanotubes (MWCNTs), was used as a conducting support for preparing the enzyme-modified electrode. PQQ-GDH was linked to the MWCNTs using a heterobifunctional cross-linker, 1-pyrenebutanoic acid succinimidyl ester (PBSE) (Figure 12.4A). Succinimidyl ester forms covalent amide bond with amino groups on protein lysine residues, while the other end of the linker, polyaromatic pyrenyl moieties, interacts with MWCNTs via π–π stacking. MWCNTs provided efficient direct non-mediated electron transfer from the PQQ-active center of the immobilized enzyme to the conducting support (Figure 12.4B), thus resulting in the formation of a bioelectrocatalytic anodic current (Figure 12.4C) and negative potential of ca. −150 mV when it is measured in open circuitry vs. a reference electrode in the presence of glucose (Figure 12.4D). It should be noted that similar glucose-oxidizing electrodes were successfully used in implantable biofuel cells operating *in vivo* [111–113], thus demonstrating the electrode operation in a biological environment. The reductive potential and current were produced on the modified electrode only in the presence of glucose (note that the reductive potential was produced on the electrode in a broad range of glucose concentrations), while in the absence of glucose, the electrode potential was ca. −5 mV. The operation of the sensing electrode mimics the Identity (YES) logic gate with a single input/single output signals. The output **1** was always produced in the presence of glucose (input **1**), while output **0** was generated in the absence of glucose (input **0**). Different configuration electrodes performing the same simple YES logic operation were activated with input signals represented with proteins [75], DNA [76], microbial cells [77], pH change [78], biomarkers [79], etc.

12.4.2 Sensing Electrodes Activated with Multi-input Logic Networks

Recent advances in biomolecular logic systems [114], particularly based on enzyme-biocatalyzed reactions [1, 2], allowed electrode activation by complex combinations of various biomolecular input signals [4–6]. Figure 12.5 shows two examples of biocatalytic cascades mimicking operation of a three-input OR

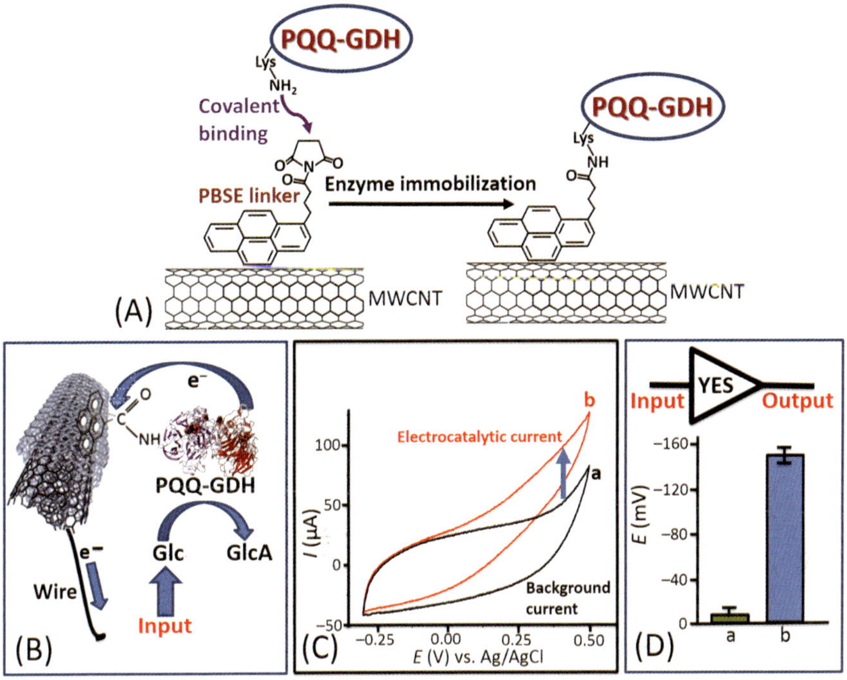

Figure 12.4 (A) Immobilization of PQQ-GDH enzyme on surface of the buckypaper electrode. (B) Direct (non-mediated) electron transfer from the PQQ-GDH enzyme to the electrode surface while glucose (Glc) is biocatalytically oxidized to gluconic acid (GlcA). (C) Cyclic voltammograms obtained with the PQQ-GDH electrode in the absence (a) and presence (b) of glucose (25 mM), scan rate 1 mV s^{-1}. (D) Potentials measured on the PQQ-GDH-modified electrode vs. Ag|AgCl|KCl reference electrode (open circuitry conditions) in the absence (a) and presence (b) of 25 mM glucose. Source: Katz et al. 2015 [26]. Adapted with permission of American Chemical Society.

logic gate and three concatenated AND logic gates for processing biomolecular signals. The three-input OR logic gate [73] was realized with reactions biocatalyzed by three NAD$^+$-dependent enzymes, glucose dehydrogenase (GDH), alcohol dehydrogenase (ADH), and glucose-6-phosphate dehydrogenase (G6PGH) operating in parallel (Figure 12.5A). The biocatalytic reactions were activated by input signals *A*, *B*, and *C*, represented by the corresponding substrates: glucose (Glc), ethanol (EtOH), and glucose-6-phosphate (Glc6P), respectively. Aiming at the simplest concept demonstration, logic value **0** for all input signals was defined as the absence of the substrates, while logic value **1** was defined as the experimentally convenient and optimized concentrations. All three reactions resulted in the generation of NADH upon reduction of the NAD$^+$ cofactor and concomitant oxidation of the substrate inputs. The NADH formation was activated in the presence of any of the input signals applied in any combination (**0,0,1**; **0,1,0**; **1,0,0**; **1,1,0**; **1,0,1**; **0,1,1**; **1,1,1**). The only situation when NADH was not produced was the complete absence of all input signals (**0,0,0**). When NADH was produced in the biocatalytic reactions it

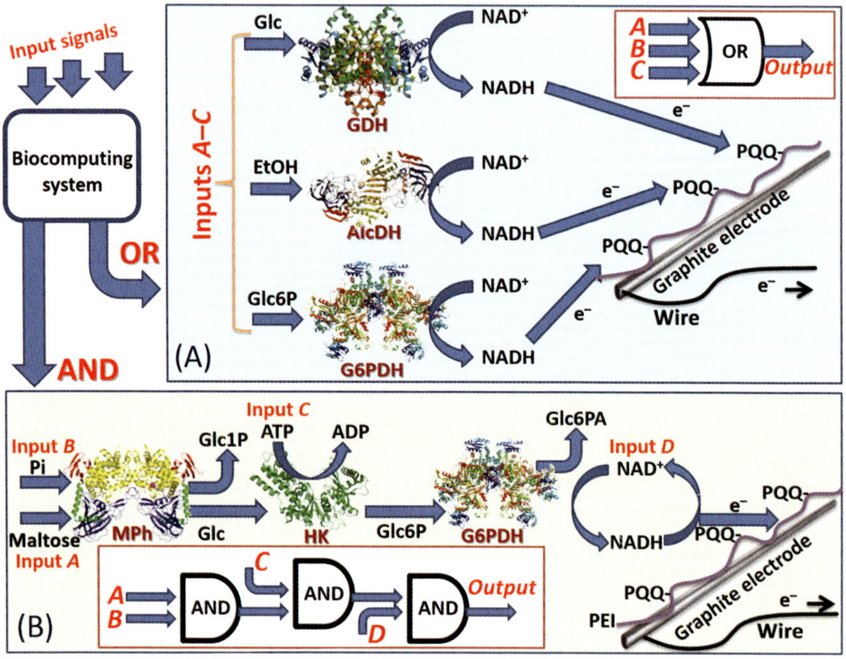

Figure 12.5 Biocatalytic cascades logically processing multiple biomolecular input signals. (A) The biocatalytic reactions operating in parallel and mimicking three-input OR logic gate. (B) The biocatalytic cascade including three consecutive biocatalytic reactions and mimicking three concatenated AND logic gates. The biocatalytically produced NADH was electrocatalytically reoxidized back to NAD^+ at the PQQ-modified graphite electrode. The NADH oxidation generated a negative potential and reductive current on the modified electrode. The electrode was electrically connected to an alginate-modified electrode to stimulate the molecule release from the alginate film. Source: Katz et al. 2015 [26]. Adapted with permission of American Chemical Society.

was reoxidized back to NAD^+ at the electrode modified with pyrroloquinoline quinone (PQQ), which is a well-known catalyst for NADH electrochemical oxidation [115] (PQQ immobilization was achieved by its covalent coupling to the adsorbed polyethyleneimine, PEI). The electrocatalytic process of NADH oxidation resulted in the formation of the reductive potential/current on the modified electrode.

Another example system shown in Figure 12.5B includes three concatenated AND gates [73] mimicked by the enzymatic cascade with three consecutive reactions activated by four input signals. The biocatalytic reaction of maltose phosphorylase (MPh) was activated in the presence of maltose (Input *A*) and inorganic phosphate (Pi, Input *B*) resulting in glucose (Glc) and glucose-1-phosphate (Glc1P) byproduct formation. In the next reaction step, biocatalyzed by hexokinase (HK), glucose was converted to glucose-6-phosphate (Glc6P) in the presence of adenosine 5′-triphosphate (ATP) (Input *C*). Finally, Glc6P reduced NAD^+ (Input *D*) to NADH in the process biocatalyzed by glucose-6-phosphate dehydrogenase (G6PDH). The NADH production was obtained only when all four input signals were applied at their logic **1** value (input combination **1,1,1,1**). The generated NADH was reoxidized and recycled to NAD^+ at the PQQ-modified electrode, resulting in the formation of an anodic current and a negative potential of ca. −80 mV on the electrode. If any of the input signals were applied at the logic **0** value (meaning the physical absence of the corresponding species; 15 different input combinations), the biocatalytic cascade did not continue to the very end, and NADH was not produced, thus inhibiting formation of the reductive potential on the electrode. It should be noted that the realized biocatalytic cascades and the input signals activating the processes do not have any specific biomedical meaning and only represent a convenient model where the final production of NADH and the potential formation on the electrode are controlled by the correct combination of multiple input signals. In the advanced system, the biocatalytic cascades can be activated by biomarkers signaling physiological conditions and their changes [5], thus resulting in the potential formation on the sensing electrode, reflecting biomedical conditions in real time.

Both example systems shown in Figure 12.5 illustrate operation of the biosensing electrode activated with several input signals applied in the correct combination according to the logic function "programmed" in the enzyme composition. Similarly, other different logic functions (e.g., majority and minority logic gates [74]) have been applied to activate the sensing electrode. Notably, the systems overviewed here represent logic networks of various composition performing different logic operations of different complexity. When the sensing electrode is activated and the reductive potential is produced, an alginate film on the connected electrode is dissolved due to Fe^{3+} reduction, and the entrapped species are released. This process is detailed in the next section.

12.4.3 Releasing Electrodes: Various Released Species for Different Applications

The negative potential and reductive current produced on the biocatalytic sensing electrode in the presence of biomolecular signals exemplified above were applied on another modified electrode coated with the Fe^{3+}-cross-linked alginate film (Figure 12.2) containing loaded substances of various composition and complexity. It was already explained above (Figure 12.1) that reduction of Fe^{3+} cations to Fe^{2+} state results in decomposition and dissolution of the alginate matrix, due to the reduced ability of Fe^{2+} cations to cross-link the alginate molecules [69]. Therefore, the electrochemically stimulated reduction process resulted in alginate matrix dissolution and release of the preloaded species from the dissolved thin film (Figure 12.6). The loaded and then released species of different composition were applied for various targets. In the simplest example [72], Au nanoparticles (Au-NPs) functionalized with enzyme molecules (e.g., horseradish peroxidase [HRP]) were used as model species mimicking drug release. The purpose of this experiment was the optimization of the loading/release processes and minimization of the non-controlled leakage from the alginate matrix. In a more advanced system [77], a real antibacterial drug, polymyxin B, was released in the process activated by

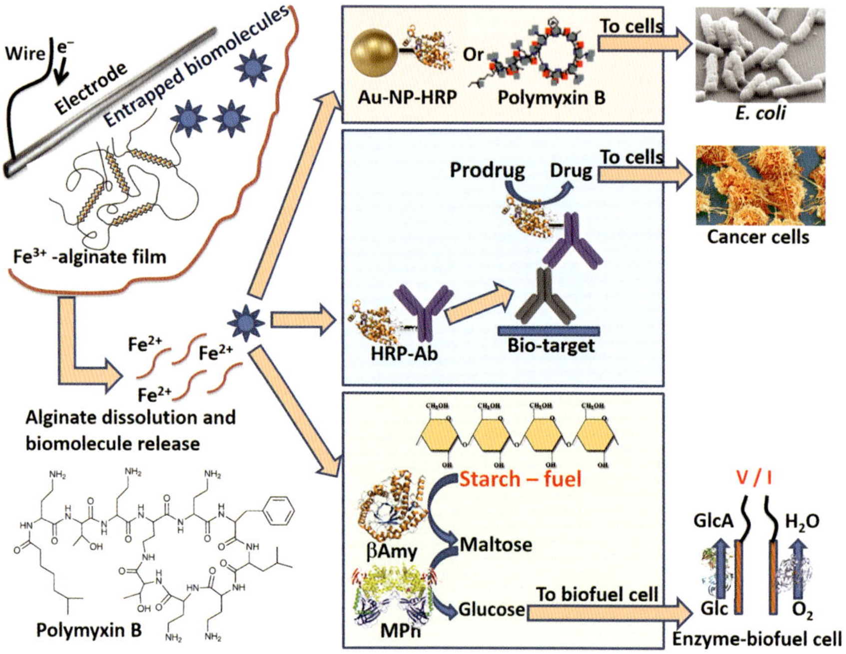

Figure 12.6 Electrochemically stimulated reductive dissolution of Fe^{3+}-cross-linked alginate and concomitant release of the loaded substances, drugs (e.g., polymyxin B), drug-mimicking species, and enzymes, aiming at different biomedical and biotechnological applications. Source: Katz et al. 2015 [26]. Adapted with permission of American Chemical Society.

the signal represented by bacterial cells and applied to the bacterial culture inhibiting its growth (Figure 12.7A). The released species were also represented by an enzyme–antibody conjugate (horseradish peroxidase-labeled anti-goat IgG-antibody from donkey; HRP-Ab) [75] (Figure 12.6). After HRP-Ab release and its complex formation with the target species, the complex was analyzed by an immune assay of the HRP-labeled assay (ELISA), where HRP-catalyzed biochemical transformations (Figure 12.7B). The non-controlled leakage of HRP-Ab from the alginate matrix (Figure 12.7B, curve a) was negligible compared to the signal-stimulated release (Figure 12.7B, curve b). In this case, the released species were able to target specific biomolecules (e.g., complementary polyclonal anti-rabbit IgG-antibody from goat), thus mimicking, for example, targeting cancer cells in future applications. The realized model system corresponds to the biomedical scenario where the "activator" (modeled with HRP-Ab) is released in response to a specific combination of biomarker signals and the "drug" is locally produced from the inactive "prodrug" with the help of the released "activator" (modeled with the reaction catalyzed by HRP).

In addition to future biomedical applications of the signal-triggered release processes (mostly related to the signal-stimulated drug release), another biotechnological application has been illustrated [7]. This system was releasing enzymes, β-amylase (βAmy) and maltose phosphorylase (MPh), decomposing starch and yielding glucose, which was used as a biomolecular "fuel" for a biofuel cell

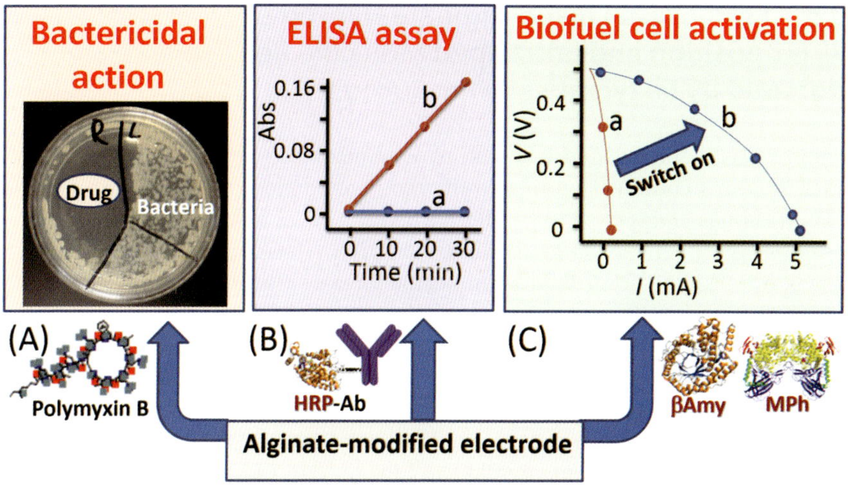

Figure 12.7 Different activity of the released substances. (A) Antibacterial action upon releasing of polymyxin B. The bacterial cell growth was inhibited in the area where the released drug was applied. (B) The released HRP-Ab conjugate was interacted with a surface containing complementary immune species, and then an ELSA assay was performed demonstrating low and high concentrations of the HRP-Ab conjugate upon its non-control leakage from the alginate matrix (a) and signal-stimulated release (b), respectively. (C) The biofuel cell activation upon increasing concentration of glucose in the solution. The glucose concentration was increased due to starch digestion in the presence of the enzymes (βAmy and MPh) released from the alginate matrix. The plot shows polarization curves before (a) and after (b) the biofuel cell activation. Source: Katz et al. 2015 [26]. Adapted with permission of American Chemical Society.

(Figure 12.6). When the enzymes were released from the alginate film and after they decomposed starch yielding glucose, the current–voltage output produced by the biofuel cell was significantly increased corresponding to the increased glucose concentration (Figure 12.7C). The present example demonstrated the application of the releasing system for the activation of a biofuel cell triggered by biomolecular signals.

It should be noted that only large biomolecules including DNA and enzymes (particularly enzymes bound to nanoparticles) demonstrated small non-controlled leakage from the alginate films. Smaller molecules (e.g., drugs) have significant leakage from the alginate matrices prior to their reductive dissolution triggered by logically processed signals. This problem requires attention and additional work to minimize the leakage process. The problem might be solved by optimizing alginate hydrogel density and cross-linking. Also, the drug molecules (or any other low molecular weight species) might be chemically (possibly covalently) bound to the alginate polymer, thus preventing their leakage prior to the alginate dissolution. Meanwhile, presently studied systems demonstrating signal-triggered release from alginate films are mostly applied for DNA release, as illustrated in the next sections. This selection of the entrapped and then released species is particularly motivated by the experimental convenience due to low leakage of DNA from alginate hydrogel.

12.5 Fe^{3+}-Cross-Linked Alginate Hydrogel Decomposition and Entrapped Molecule Release Triggered by Enzymatically Produced H_2O_2

The previous sections exemplified electrochemically stimulated release process, which requires combination of two interconnected electrodes, one for sensing of input signals and second for reductive dissolution of alginate film, resulting in release of entrapped molecules. While this approach has demonstrated successful realization of signal-controlled biomolecule release, it is rather complicated and difficult for miniaturization, particularly because of different composition of the anodic and cathodic solutions reacting with both modified electrodes.

Another approach has been developed to decompose alginate films and release entrapped molecules (DNA was selected as a convenient example) without electrochemical reactions. This approach aims at the signal-triggered release of DNA from alginate hydrogel, while the biocatalytic reactions resulting in the alginate film decomposition/dissolution proceed inside the polymer film. In order to achieve this goal, already known process of alginate degradation in the presence of H_2O_2 has been used [29, 116–121]. Notably, many enzyme-catalyzed reactions, particularly in the presence of various oxidases (e.g., glucose oxidase [122]), produce H_2O_2 that can be utilized for the controlled degradation of alginate and release of the entrapped biomolecules. These enzyme-catalyzed reactions can be easily used to mimic Boolean logic operations activated by various biomolecular signals and result in the production of H_2O_2 [1, 2]. In order to integrate signal processing enzyme systems with the DNA-releasing alginate hydrogel and to minimize the uncontrolled enzyme leakage, the enzymes were immobilized on silica nanoparticles (SiO_2-NPs, 200 nm diameter) (Figure 12.8) and included in the hydrogel matrix together with the entrapped DNA. Prior to their operation inside the hydrogel, the biocatalytic function and logic operation of the enzyme-functionalized SiO_2-NPs were studied in a solution by optical analysis of the produced species.

12.5.1 DNA Release from Fe^{3+}-Cross-Linked Alginate Hydrogel Stimulated by Signals Processed through OR, AND, and INHIB Logic Gates

Figure 12.9A shows two parallel reactions, both resulting in the production of H_2O_2 and catalyzed by glucose oxidase (GOx) and lactate oxidase (LOx) in the presence of glucose (Glc) and lactate (Lac), respectively. The produced H_2O_2 was further reacting with horseradish peroxidase (HRP) in the presence of ABTS yielding a colored product ($ABTS_{ox}$) absorbing light at λ_{max} = 420 nm [123] (Figure 12.9B). The reacting species, Glc and Lac, were applied at two concentration levels: high concentration (optimized experimentally) defined as logic value **1** and zero concentration (meaning their absence) defined as logic value **0**. Input *A* (Glc) and Input *B* (Lac) were added to the solution containing suspension of the GOx-SiO_2-NPs and LOx-SiO_2-NPs in four different combinations: **0,0**; **0,1**; **1,0**; and **1,1**. Figure 12.9C shows the truth table characteristic of the Boolean

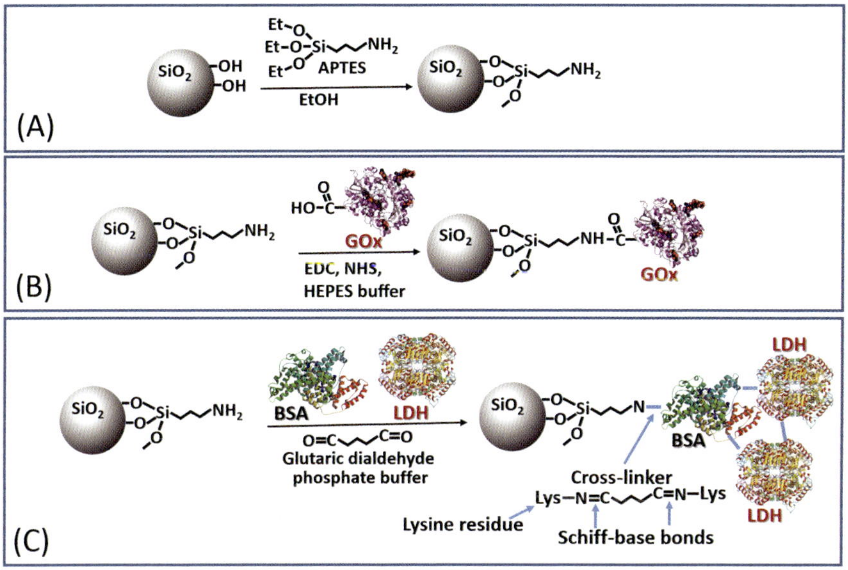

Figure 12.8 Functionalization of the SiO_2-NPs used in the enzyme logic gates operating in the alginate hydrogel. (A) Silanization of the SiO_2-NPs with (3-aminopropyl)triethoxysilane. (B) Carbodiimide coupling of GOx to the silanized SiO_2-NPs (note that attachment of LOx and Est was performed in the same way). (C) Cross-linking of BSA and LDH on the surface of the silanized SiO_2-NPs. BSA was added to increase the LDH load on the NPs.

OR logic gate, and Figure 12.9D shows the corresponding experimental result measured as the absorbance of $ABTS_{ox}$. The optically measured output signal demonstrated, as expected [1, 2], the H_2O_2 production mimicking the OR logic gate.

Figure 12.10A demonstrates another logic gate operation (AND gate) realized as a two-step biocatalytic reaction. Input *A* (NADH) was oxidized to NAD^+ in the presence of pyruvate (Pyr) (Input *B*) in the reaction catalyzed by lactate dehydrogenase (LDH). The concomitantly produced Lac reacted with O_2 in the reaction catalyzed by LOx to yield H_2O_2. The production of H_2O_2 was analyzed optically through the HRP-ABTS assay, resulting in the absorbance measured at 420 nm (Figure 12.10B). The H_2O_2 production and the absorbance change were only possible when both input reactants (NADH and Pyr) were present (applied at **1,1** combination), while all other input combinations did not allow the reaction to proceed to the end. Figure 12.10C shows the truth table of the AND logic gate, and Figure 12.10D demonstrates the experimentally measured output signal obtained for different combinations of the inputs and mimicking the AND gate logic operation.

The OR and AND logic gates realized in the enzyme-catalyzed reactions are rather trivial. Much more sophisticated logic networks composed of many concatenated logic operations can be realized with enzyme reactions [1, 2]. However, the logic gates demonstrated here can be effectively used to illustrate the concept of the signal-controlled DNA release from the alginate hydrogel. Figure 12.11

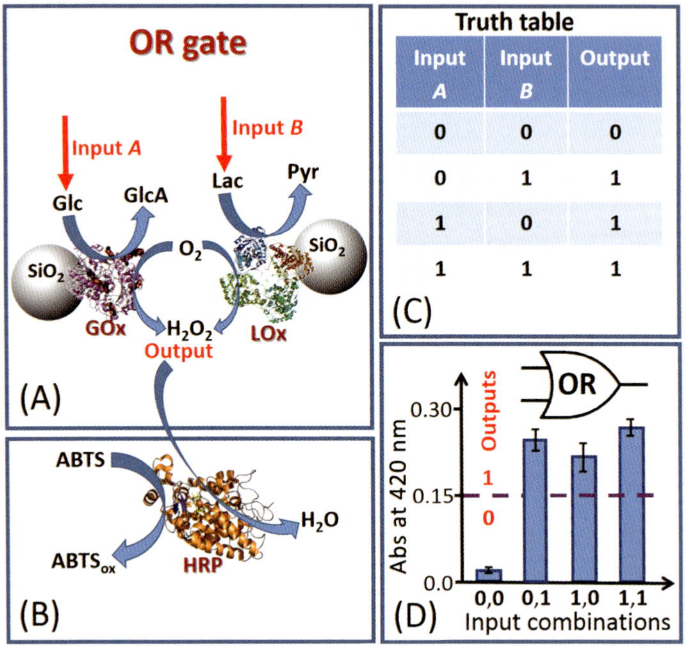

Figure 12.9 Realization of the enzyme-based OR logic gate in a solution with the optical analysis of the output signals. (A) The biocatalytic cascade mimicking the OR gate producing H_2O_2 as the final output. (B) The biocatalytic reaction converting H_2O_2 to the optically readable product $ABTS_{ox}$. (C) The truth table of the Boolean OR logic gate. (D) Optical absorbance corresponding to the formation of $ABTS_{ox}$ and measured as the final output signal after applying the input signals in different combinations. The absorbance was measured after 10 minutes of the biocatalytic process. The dashed line shows the threshold separating **0** and **1** logic values of the output signal. Source: Gamella et al. 2017 [28]. Adapted with permission of John Wiley and Sons.

shows schematically the whole system composition and operation. The alginate hydrogel cross-linked with Fe^{3+} cations was deposited electrochemically on a graphite electrode, and two kinds of enzyme-functionalized SiO_2-NPs were physically entrapped into the hydrogel upon its formation. DNA-FAM (oligonucleotide labeled with a fluorescent dye) was also entrapped into the hydrogel. In the presence of the input signals, the enzyme system produced H_2O_2 following the Boolean logic operations discussed above. The *in situ* produced H_2O_2 was catalytically decomposed to free radicals in the Fenton-type reaction catalyzed by iron cations [124]. This reaction is known to proceed in alginate hydrogels in the presence of iron cations, finally resulting in degradation/depolymerization of the alginate hydrogel [121]. In the system discussed this process stimulated the DNA-FAM release from the alginate hydrogel, thus increasing fluorescence in the solution.

Figure 12.12A shows the time-dependent fluorescence increase in the solution upon the DNA-FAM release from the alginate hydrogel containing the enzyme system mimicking the OR gate (GOx-SiO_2-NPs and LOx-SiO_2-NPs activated with Glc and Lac applied in four different combinations: **0**,**0**; **0**,**1**; **1**,**0**; **1**,**1**).

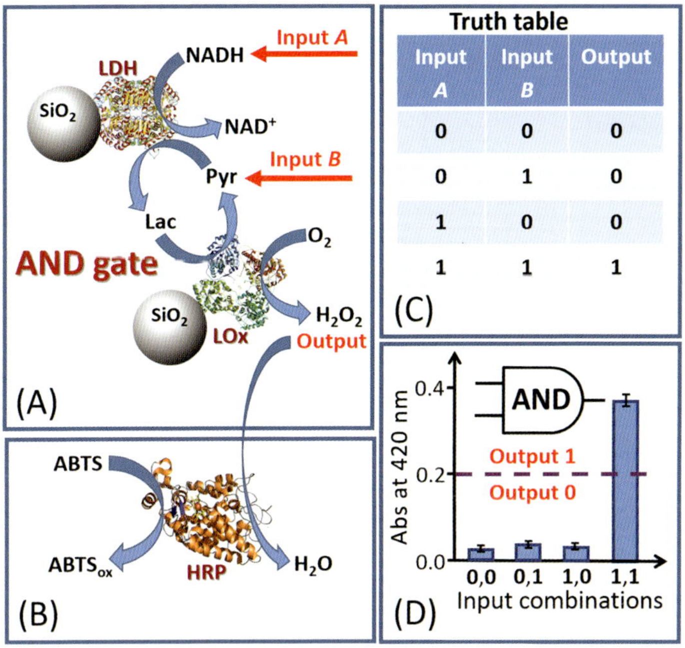

Figure 12.10 Realization of the enzyme-based AND logic gate in a solution with the optical analysis of the output signals. (A) The biocatalytic cascade mimicking the AND gate producing H_2O_2 as the final output. (B) The biocatalytic reaction converting H_2O_2 to the optically readable product $ABTS_{ox}$. (C) The truth table of the Boolean AND logic gate. (D) Optical absorbance corresponding to the formation of $ABTS_{ox}$ and measured as the final output signal after applying the input signals in different combinations. The absorbance was measured after 30 minutes of the biocatalytic process. The dashed line shows the threshold separating **0** and **1** logic values of the output signal. Source: Gamella et al. 2017 [28]. Adapted with permission of John Wiley and Sons.

Notably, the only input combination not resulting in the fluorescence increase was the absence of both input signals (**0,0** combination) typical for the OR gate (a minor fluorescence increase for the **0,0** combination corresponds to the uncontrolled leakage of the entrapped DNA-FAM). The faster release process in case of **1,1** input combination results from the H_2O_2 production by both biocatalytic reactions; however, the final result was almost the same for the operation of either or both biocatalytic reactions, being limited by the amount of the entrapped DNA-FAM species. The low and high fluorescence signals obtained in the solution at the end of the release process were defined as the logic output **0** and **1**, respectively, mimicking the OR logic gate (Figure 12.12B).

Figure 12.13A shows the time-dependent fluorescence increase in the solution upon the DNA-FAM release from the alginate hydrogel containing the enzyme system mimicking the AND gate (LDH-SiO_2-NPs and LOx-SiO_2-NPs activated with NADH and Pyr applied in four different combinations: **0,0**; **0,1**; **1,0**; **1,1**). Only one combination of the input signals (**1,1**) resulted in the significant fluorescence output signal, while all other input combinations demonstrated only minor fluorescence increase due to the uncontrolled leakage of DNA-FAM,

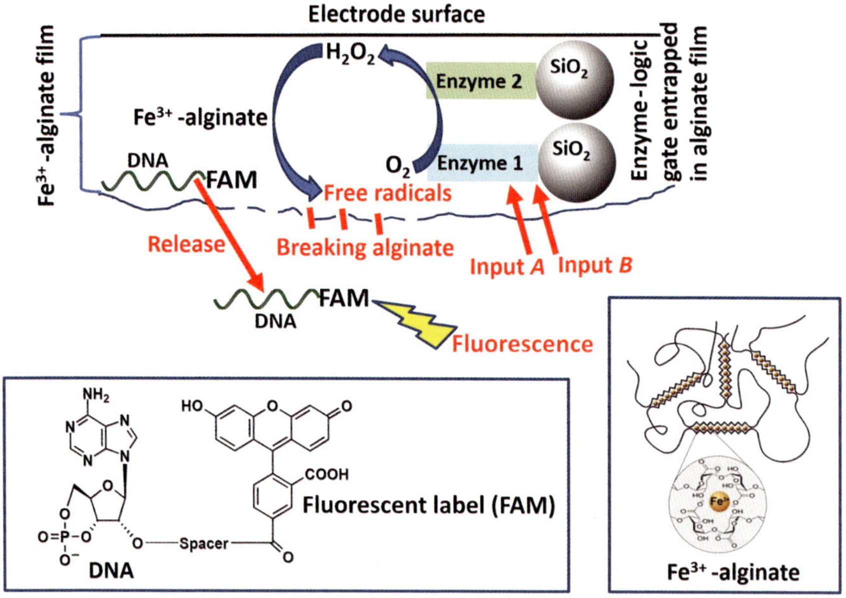

Figure 12.11 The scheme showing the signal-triggered release of the DNA-FAM from the Fe^{3+}-cross-linked alginate film on the electrode surface. The primary input signals are logically processed through the enzyme reactions resulting in the H_2O_2 production (OR and AND logic gates are realized using different enzymes and input chemicals). The produced H_2O_2 is decomposed to yield free radicals in the Fenton-type reaction catalyzed by iron cations. The generated free radicals damage the alginate film stimulating the DNA-FAM release. The released DNA-FAM is analyzed in the solution by measuring fluorescence of the dye label. Source: Gamella et al. 2017 [28]. Adapted with permission of John Wiley and Sons.

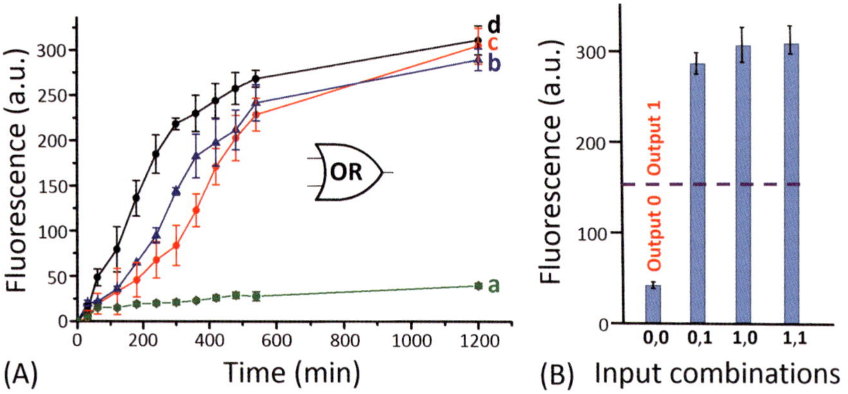

Figure 12.12 (A) Time-dependent fluorescence corresponding to DNA-FAM measured in the solution at $\lambda = 520$ nm upon realization of the OR gate activated with different combinations of the input signals in the alginate film (Input *A* and Input *B* are Glc and Lac, respectively): (a) **0,0**; (b) **0,1**; (c) **1,0**; (d) **1,1**. Note that **0,0** logic combination also corresponds to the uncontrolled leakage of DNA-FAM from the alginate film. (B) The fluorescence output signals measured after 20 hours from the application of the input signals in different combinations. The dashed line shows the threshold separating **0** and **1** logic values of the output signal. Source: Gamella et al. 2017 [28]. Adapted with permission of John Wiley and Sons.

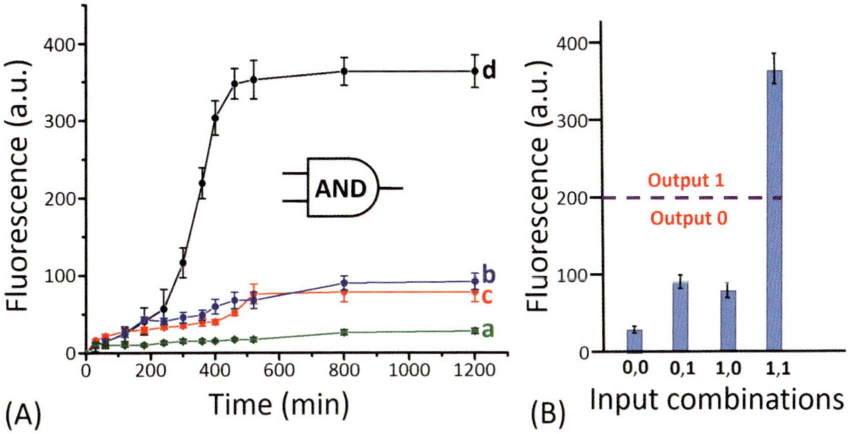

Figure 12.13 (A) Time-dependent fluorescence corresponding to DNA-FAM measured in the solution at $\lambda = 520$ nm upon realization of the AND gate activated with different combinations of the input signals in the alginate film (Input *A* and Input *B* are NADH and Pyr, respectively): (a) **0,0**; (b) **0,1**; (c) **1,0**; (d) **1,1**. Note that **0,0** logic combination also corresponds to the uncontrolled leakage of DNA-FAM from the alginate film. (B) The fluorescence output signals measured after 20 hours from the application of the input signals in different combinations. The dashed line shows the threshold separating **0** and **1** logic values of the output signal. Source: Gamella et al. 2017 [28]. Adapted with permission of John Wiley and Sons.

as expected for the AND logic gate. The delay (lag period) in the fluorescence increase originated from the two-step process involved in the H_2O_2 production. The first biocatalytic reaction (Figure 12.10A) needed time to produce Lac, which then reacted to yield H_2O_2 in the second reaction step. This resulted in nontrivial complex kinetics in the DNA release process, where sigmoidal kinetics has been observed (Figure 12.13A, curve d) with the lag period prior to the rapid increase in the DNA release rate. This lag period particularly originated from very slow diffusion of the intermediate product (Lac) through the alginate hydrogel with high viscosity. This kinetics is different from the kinetics demonstrated by the OR logic operation where each biocatalytic reaction runs independently and the process is not complicated by the formation of an intermediate product. The final fluorescence output corresponded to the expected AND logic function (Figure 12.13B).

While the OR and AND logic gates were analyzed optically in a solution (Figures 12.9 and 12.10), some other logic gates require the whole alginate hydrogel system to operate. Figure 12.14A shows schematically the INHIB (inhibited) logic gate operation in the alginate hydrogel. The hydrogel included GOx-SiO_2-NPs and esterase-(Est)-SiO_2-NPs entrapped. The reaction biocatalyzed by GOx in the presence of Glc (Input *A*) always resulted in the H_2O_2 production. On the other hand, the reaction catalyzed by Est in the presence of ethyl butyrate (Et-O-Bu, Input *B*) resulted in production of ethanol (EtOH). Ethanol operated as a free radical scavenger [125], thus inhibiting the alginate degradation in the presence of H_2O_2. Overall, the signal-triggered DNA-FAM release was observed always when H_2O_2 is produced, but not in the case when

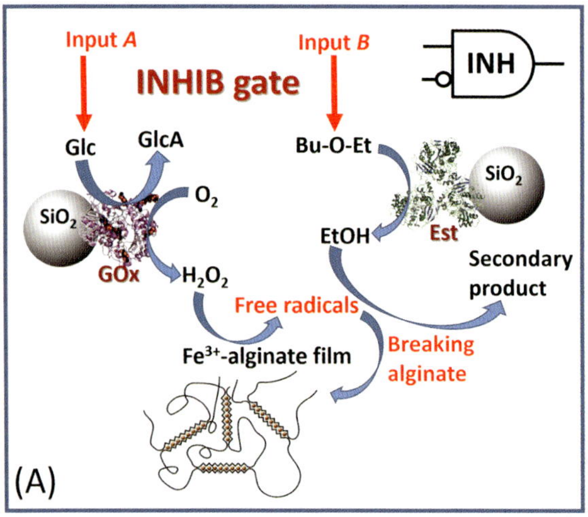

Truth table

Input *A*	Input *B*	Output
0	0	0
0	1	0
1	0	1
1	1	0

(B)

Figure 12.14 (A) The scheme showing operation of the INHIB (inhibition) logic gate realized in the Fe^{3+}-cross-linked alginate film on the electrode surface. The reaction biocatalyzed by GOx results in H_2O_2 when Input *A* (Glc) is applied. The produced H_2O_2 is decomposed to yield free radicals in the Fenton-type reaction catalyzed by iron cations. The produced free radicals damage the alginate film and stimulate the DNA-FAM release. The reaction biocatalyzed by Est in the presence of Input *B* (Et-O-Bu) results in the production of ethanol (EtOH), which operates as a free radical scavenger, thus inhibiting their reaction with alginate and preventing the DNA-FAM release. (B) The truth table of the Boolean INHIB logic gate. Source: Gamella et al. 2017 [28]. Adapted with permission of John Wiley and Sons.

EtOH was formed. This is formulated in the truth table characteristic of the INHIB logic gate (Figure 12.14B).

Figure 12.15A shows the experimentally observed fluorescence increase measured in the solution upon the DNA-FAM release from the alginate hydrogel containing GOx-SiO_2-NPs and Est-SiO_2-NPs activated with Glc and Et-O-Bu, respectively. The input combinations **0,0** and **0,1** do not result in the release process because H_2O_2 is not produced (Glc is not applied). On the other hand, the input combination **1,0** yields H_2O_2 and results in the alginate degradation and DNA-FAM release. Finally, the input combination **1,1** also does not result in the release process – even H_2O_2 is produced. This originates from the effective scavenging of the free radicals by the produced EtOH, thus inhibiting the alginate decomposition and DNA-FAM release. The final result in the form of the fluorescence signal experimentally measured in the solution is shown in Figure 12.15B for different input combinations. This result nicely fits the truth table corresponding to the INHIB logic gate (Figure 12.14B).

In addition to the fluorescence observed in the solution upon the DNA-FAM release, the dissolution of the alginate film from the electrode surface was observed. The process was easily visible by the naked eye (Figure 12.16). The photos shown in Figure 12.16 correspond to the alginate-modified electrodes containing the INHIB gate system entrapped in the alginate hydrogel. The alginate film electrochemically deposited on the graphite electrode has yellow color (mostly due to cross-linking iron cations included in the alginate film).

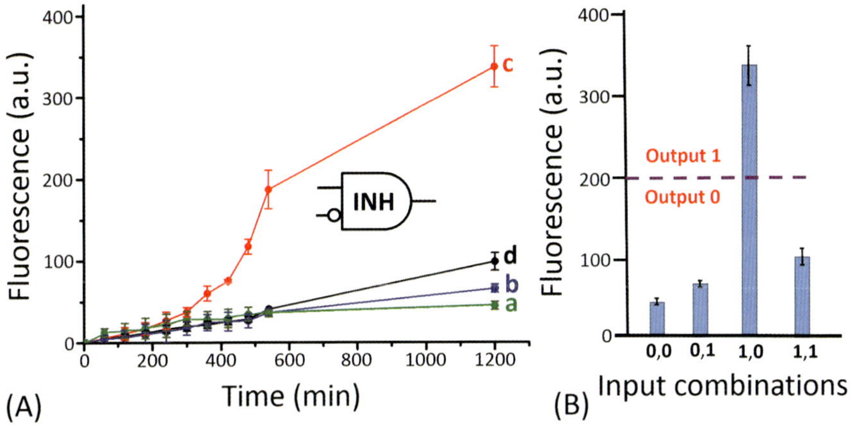

Figure 12.15 (A) Time-dependent fluorescence corresponding to DNA-FAM measured in the solution at $\lambda = 520$ nm upon realization of the INHIB gate activated with different combinations of the input signals in the alginate film (Input *A* and Input *B* are Glc and Et-O-Bu, respectively): (a) **0,0**; (b) **0,1**; (c) **1,0**; (d) **1,1**. Note that **0,0** logic combination also corresponds to the uncontrolled leakage of DNA-FAM from the alginate film. (B) The fluorescence output signals measured after 20 hours from the application of the input signals in different combinations. The dashed line shows the threshold separating **0** and **1** logic values of the output signal. Source: Gamella et al. 2017 [28]. Adapted with permission of John Wiley and Sons.

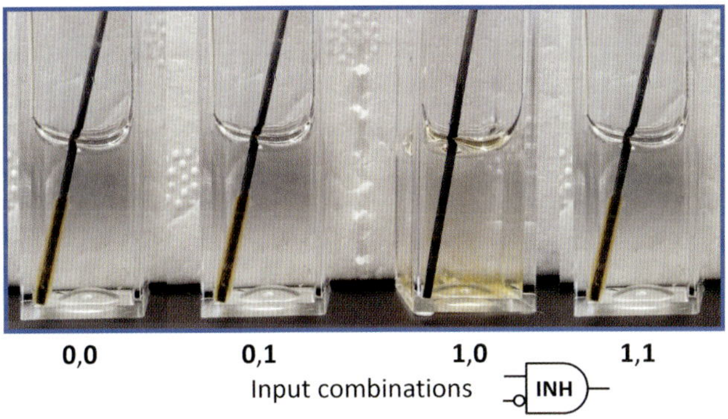

Figure 12.16 Photos showing the alginate-modified electrodes after the INHIB gate was realized for 20 hours in the presence of different combinations of the input signals. The yellow color originates mostly from the cross-linking iron cations in the alginate film. Note that the **1,0** input combination results in the complete dissolution of the alginate film, while all other input combinations preserved the film on the electrode surface. Source: Gamella et al. 2017 [28]. Adapted with permission of John Wiley and Sons.

This yellow colored film completely disappears from the electrode surface; the yellow color moves to the solution when the signal combination **1,0** is applied, while all other input combinations preserve the colored film on the electrode surface. Similar results, but for different input combinations, were observed for the OR and AND logic gates.

In order to address the chemical mechanism responsible for the alginate degradation and the DNA release, the H_2O_2 effect on Fe^{3+}- and Ca^{2+}-cross-linked alginate hydrogels was compared. In this experiment the alginate hydrogel with the entrapped DNA-FAM was produced in the form of small beads (ca. 1 mm diameter) [126]. The alginate beads were used in this experiment instead of a film on the electrode surface because the Ca^{2+}-cross-linked alginate cannot be deposited on an electrode surface electrochemically. The alginate beads also included GOx-SiO_2-NPs and were activated in the presence of glucose added to the solution. The beads composed of the Ca^{2+}-cross-linked alginate hydrogel demonstrated the fluorescence increase in the presence of H_2O_2 comparable with that of small uncontrolled leakage of DNA-FAM. On the other hand, the beads composed of Fe^{3+}-cross-linked alginate hydrogel demonstrated at least 10-fold faster DNA-FAM release compared with the leakage process. This experiment clearly indicated the important role of the iron cations, which are responsible for the Fenton-type catalytic degradation of H_2O_2 yielding free radicals ultimately responsible for the alginate dissolution and DNA-FAM release. The detailed mechanism of these radical reactions was investigated long time ago [121].

It should be noted that the alginate degradation mechanism includes catalytic formation of free radicals from H_2O_2 produced enzymatically. This raises two important questions about the possible toxicity effect of H_2O_2 if the process occurs in a biological environment and about the possible damage to the entrapped DNA molecules (or other biomolecules entrapped and released). These important concerns have been addressed in additional experiments.

After the DNA-FAM release to the solution was completed (the process was stimulated by the OR logic gate and included the H_2O_2 production inside the alginate gel), the H_2O_2 concentration in the solution was measured. The analysis was performed by the standard enzyme assay with the HRP-ABTS system (http://www.sigmaaldrich.com/technical-documents/protocols/biology/enzymatic-assay-of-peroxidase-abts-as-substrate.html) similar to one shown in Figure 12.9B. Maximum detected concentration of H_2O_2 was ca. 1.5 μM in the solution. It should be noted that the H_2O_2 concentration in human plasma is approximately in the range of 1–3 μM [127]. This means that the possible formation of H_2O_2 in the external solution is in the range of its natural concentration in biofluids and it should not produce any toxicity effect if the systems are used in a biological environment.

The possible damage to the DNA species by the free radicals produced catalytically inside the alginate film was studied by its hybridization with other oligonucleotides to yield a catalytically active DNAzyme. The released DNA (in this experiment with a different base sequence and without the fluorescent label) was used to assemble a DNAzyme capable of catalytic split of an oligonucleotide (Q-F) labeled at one end with a fluorescent dye (F) and on the other end with a quencher (Q). Similar DNAzymes have been studied recently, particularly in the systems performing logic operations [81, 106]. The DNAzyme can be assembled only if the oligonucleotide (released DNA) has full complementarity to other DNA molecules being components of the DNAzyme. It should be noted that the DNAzyme assembling and its catalytic activity are very sensitive to the

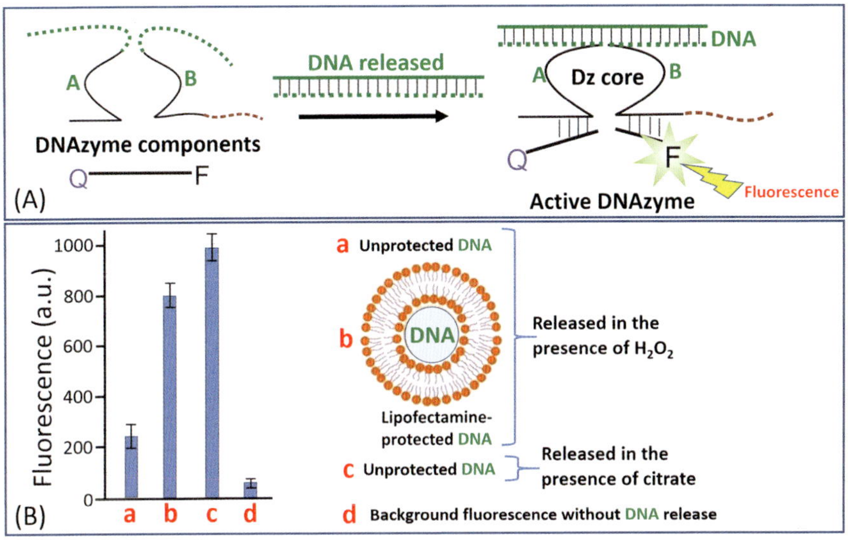

Figure 12.17 (A) The scheme illustrating the DNAzyme operation. The released DNA (in this experiment without the FAM fluorescent label) combines with three other oligonucleotides (A, B, and Q-F) to produce the catalytic Dz core capable of splitting the Q-F oligonucleotide. The split reaction separates the dye and quencher and increases the fluorescence. (B) The fluorescence signal obtained after the DNA release and formation of the DNAzyme: (a) release of the unprotected DNA stimulated by the H_2O_2 formation (similarly to all previously shown results), (b) release of the Lipofectamine-protected DNA stimulated by the H_2O_2 formation, (c) release of the unprotected DNA stimulated by addition of citrate (without formation of H_2O_2), and (d) background fluorescence in the absence of the released DNA. Source: Gamella et al. 2017 [28]. Adapted with permission of John Wiley and Sons.

complementarity of the oligonucleotides. Even a single mismatch in the DNA base sequence affects significantly the DNAzyme assembling and decreases its catalytic activity, thus allowing its use for the DNA analysis [128]. If the released DNA is damaged by the free radicals during the alginate degradation process, it cannot assemble the DNAzyme. When the DNAzyme is assembled, it splits the Q-F oligonucleotide, thus separating the dye and quencher, resulting in the fluorescence signal (Figure 12.17A). Note that this fluorescence signal originates from the catalytic split of Q-F, but not directly from the released DNA. In one experiment the fluorescence signal was measured upon the H_2O_2-triggered DNA release (specifically producing H_2O_2 by the OR logic gate) (Figure 12.17B, bar a). It was compared with the fluorescence signal obtained after the alginate dissolution with the added citrate [43, 44, 126] (note that this process does not include the H_2O_2 production and the alginate dissolution proceeds through a different mechanism without producing free radicals) (Figure 12.17B, bar c). The fluorescence signal measured with the H_2O_2 production was about 25% from the situation when the alginate was dissolved with citrate without formation of H_2O_2. This can be explained by the damage to ca. 75% of the DNA molecules by the free radicals. This is a little bit disappointing but the expected result. In the next experiment to decrease the damage to the entrapped DNA molecules, they were included in the protecting micelles formed by Lipofectamine, which is a known protecting species used for DNA delivering into cells [129]. The Lipofectamine-protected DNA demonstrated much higher activity of the DNAzyme, which means much less damage to the released DNA molecules (Figure 12.17B, bar b). This experiment demonstrated the minimized damage to the entrapped biomolecules (DNA in the present experiment) when the biomolecules are protected with an additional shell (Lipofectamine micelles in the present example).

12.5.2 DNA Release from Fe^{3+}-Cross-Linked Alginate Hydrogel Stimulated by Signals Processed Through Multi-gate Network Composed of Concatenated AND Gates

In the earlier approaches outlined in the previous sections, one- and two-step biocatalytic reactions were realized in alginate films prepared electrochemically at an electrode surface [28]. These processes mimic Boolean logic gates and trigger biomolecular release. However, more complex biocatalytic cascades do not function effectively inside alginate hydrogel presumably because their successive steps involve diffusion-limited transport of intermediate products. In order to achieve a higher throughput multistep biocatalytic reaction sequence, ultimately yielding H_2O_2 that diffuses into and decomposes the alginate hydrogel, which in turn releases entrapped biomolecular species, the biocatalytic system was assembled at external interface of an alginate film prepared at an electrode surface [30]. It transpires that enough of the intermediate reaction products are retained in the vicinity of the hydrogel/water interface, reaching the next catalytic center much more effectively than inside the hydrogel.

The modified electrodes used in this study were prepared by electrochemical deposition of the alginate hydrogel film cross-linked with Fe^{3+} cations. Four enzymes, amyloglucosidase (AMG), glucose dehydrogenase (GDH), lactate dehydrogenase (LDH), and lactate oxidase (LOx), have been covalently attached at external surface of the hydrogel. The catalytic cascade (Figure 12.18A) included the following reactions: (i) each molecule of maltose was hydrolyzed to two molecules of glucose in the reaction catalyzed by AMG; (ii) then, glucose was oxidized and NAD^+ was reduced to yield NADH in the reaction catalyzed by GDH; (iii) then, NADH reduced Pyr to Lac in the reaction catalyzed by LDH; and (iv) finally, Lac was oxidized by O_2 to yield H_2O_2 in the reaction catalyzed by LOx. The produced H_2O_2 can diffuse into the alginate hydrogel, which results in free radicals in a Fenton-type reaction catalyzed by iron cations. These free radicals then decompose the alginate hydrogel and release the entrapped biomolecular species, fluorescent-label-functionalized oligonucleotide (DNA-FAM). Its release was detected by measuring fluorescence in the solution. The four-step biocatalytic cascade can yield H_2O_2 only if all the chemical input signals are applied to activate all the reaction steps. This mimics a network of concatenated AND gates (Figure 12.18B). Indeed, each reaction step (except the very first one) requires two reacting species: one produced *in situ* in the previous catalytic step and another added as the input signal, thus representing the AND Boolean gate. The very first step was activated by a single input

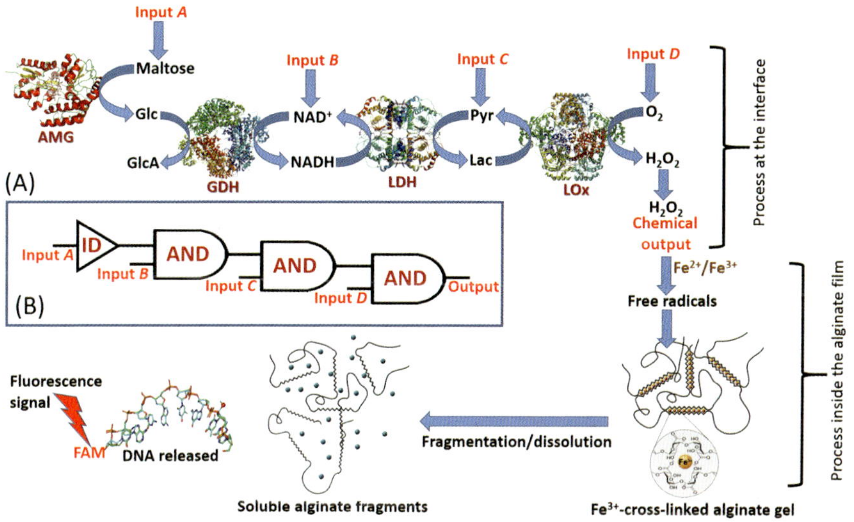

Figure 12.18 (A) The biocatalytic cascade shown schematically. The enzymes, AMG, GDH, LDH, and LOx, were all immobilized at the electrode surface and catalyzed reactions yielding H_2O_2 as the final product only when all the inputs (*A*, *B*, *C*, *D*) were applied at their logic values **1**. The produced H_2O_2 was catalytically decomposed by iron cations present in alginate, and the generated free radicals resulted in the degradation/fragmentation/dissolution of the film, allowing DNA-FAM release. (B) Logic circuit composed of concatenated Identity (ID) and three AND gates corresponding to the biocatalytic cascade performed by the immobilized enzymes shown in (A). Source: Okhokhonin et al. 2018 [30]. Adapted with permission of John Wiley and Sons.

(maltose), mimicking the Identity (ID) gate. Significant degradation/dissolution of the alginate hydrogel was achieved only when all the four input signals were applied (Figure 12.19). It should be noted that the immobilized enzymes bound to the polymer layer remain on the electrode surface (note that the enzymes were bound to a polyethyleneimine layer adsorbed at the external interface) while the alginate film is biocatalytically decomposed/dissolved, thus allowing almost complete disappearance of the alginate film.

As expected for concatenated gates shown in Figure 12.18B, the final result **1** (significant DNA-FAM release) was observed only for the input signal combination **1,1,1,1**, while several other input combinations tested (means some of the input chemicals "added"/kept at zero concentrations) resulted in noticeably smaller fluorescence of the solution, which can be attributed to uncontrolled DNA-FAM leakage (Figure 12.20). It should be noted that the present "model system" realization of the enzyme-based logic system includes not only "binary" processing of the input signals but also chemical actuation (alginate hydrogel degradation/dissolution and DNA-FAM release), offering a prototype not only for the "computing" signal processing but also for the "machinery" actuation, for biomolecular release.

The present example system demonstrated for the first time the multistep process triggered by complex biomolecular signals and resulted in the release from

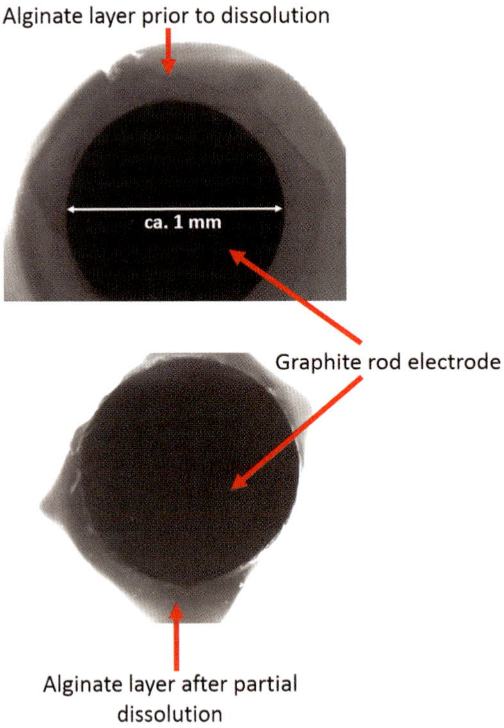

Figure 12.19 Microphotograph (×10 magnification, front view) of typical morphologies at the electrode surface. These electrodes were held in solution for six hours (top image) without application of inputs and (bottom image) with all the inputs applied. The bottom image illustrates partial decomposition/dissolution of the alginate layer. Source: Okhokhonin et al. 2018 [30]. Adapted with permission of John Wiley and Sons.

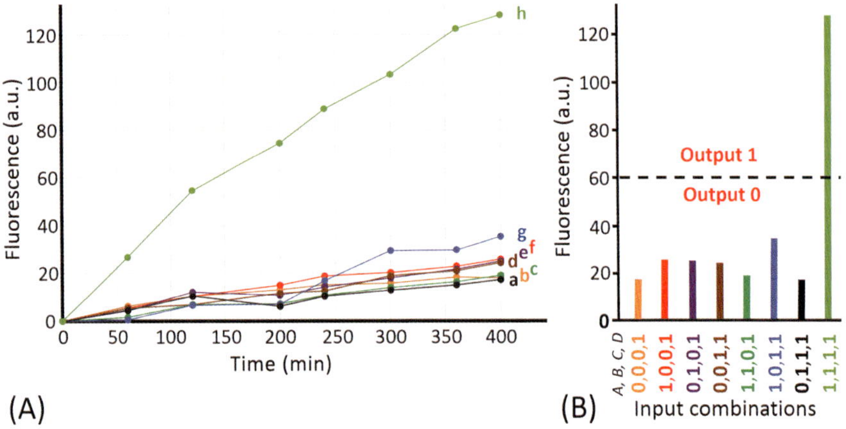

Figure 12.20 (A) Fluorescence response ($\lambda = 517$ nm) corresponding to the DNA-FAM released upon application of the input signals in several selected (out of total 16 possible) different combinations: (a) **0,1,1,1**; (b) **0,0,0,1**; (c) **1,1,0,1**; (d) **0,0,1,1**; (e) **0,1,0,1**; (f) **1,0,0,1**; (g) **1,0,1,1**; (h) **1,1,1,1**. The line segments were added to guide the eye. Only the **1,1,1,1** input combination yields DNA-FAM release, as expected for the concatenated AND gates. (B) The bar chart showing the fluorescence output corresponding to the released DNA-FAM, extracted from part (A) at 400 minutes reaction time. The "binary" digital distinction between **0** and **1** outputs is clearly well defined (as schematically marked by the dashed line). Source: Okhokhonin et al. 2018 [30]. Adapted with permission of John Wiley and Sons.

the polymer-modified electrode surface. The studied approach can be applied to electrode arrays, where each electrode is responding to different biomolecular signals and their complex combinations. In addition to biomolecular signals, the release process can be activated by electrochemical signals as has been shown previously [69, 70]. Overall, the developed approach allows activation of the biomolecular release processes from modified electrodes activated by complex combinations of the electrochemical and biochemical signals through sophisticated logic processes.

12.6 Conclusions and Perspectives

The present chapter illustrates extension of the enzyme logic systems to the downstream molecule release systems, thus operating not only as a logic "device" processing input signals but also as an actuator controlled by the signals. The systems described in this chapter illustrated the versatility of a bioelectronic approach to signal-triggered release of various substances. This approach provides an abundance of inspiration for future research and possible practical applications. Indeed, the broad range of the applied biomolecular signals, also reaching the level of bacterial cells used as input signals [77], would allow a great variety of activating processes. The different biomolecular species and nano-objects released from the alginate matrix in the process triggered by the various signals could be used for many different biomedical and biotechnological applications. The ability of the modified electrodes to work in a biological environment [111–113] is promising for designing implantable bioelectronic devices [130] operating in biofluids and releasing biological substances in response to complex combinations of biomarkers. In future biomedical applications, a similar setup composed of two modified electrodes would allow separation of the potential-generating sensing electrode and drug-releasing electrode; they might be at different locations operating in different environments and under different conditions. Alternatively, Fe^{3+}-cross-linked alginate thin films containing loaded substances could be dissolved, and the substances could be released upon processing biochemical signals directly inside the polymer matrices, bringing even more flexibility to future applications. Overall, this research direction could potentially contribute to future systems for autonomous sensing and actuating in the general frame of a theranostic approach. The systems designed for future practical (possibly biomedical) applications will not be necessary based on alginate hydrogels, but can be composed of various polymeric or nanostructured materials with signal-switchable features to allow release processes or some other actuation functions.

References

1 Katz, E. and Privman, V. (2010). *Chem. Soc. Rev.* 39: 1835–1857.
2 Katz, E. (2015). *Curr. Opin. Biotechnol.* 34: 202–208.
3 Pita, M., Privman, M., and Katz, E. (2012). *Top. Catal.* 55: 1201–1216.

4 Katz, E., Bocharova, V., and Privman, M. (2012). *J. Mater. Chem.* 22: 8171–8178.
5 Privman, M., Tam, T.K., Bocharova, V. et al. (2011). *ACS Appl. Mater. Interfaces* 3: 1620–1623.
6 Privman, M., Tam, T.K., Pita, M., and Katz, E. (2009). *J. Am. Chem. Soc.* 131: 1314–1321.
7 Mailloux, S., MacVittie, K., Privman, M. et al. (2014). *ChemElectroChem* 1: 1822–1827.
8 Katz, E. (2010). *Electroanalysis* 22: 744–756.
9 Katz, E. and Pita, M. (2009). *Chem. Eur. J.* 15: 12554–12564.
10 Tam, T.K., Pita, M., Ornatska, M., and Katz, E. (2009). *Bioelectrochemistry* 76: 4–9.
11 Amir, L., Tam, T.K., Pita, M. et al. (2009). *J. Am. Chem. Soc.* 131: 826–832.
12 Honarvarfard, E., Gamella, M., Poghossian, A. et al. (2017). *Appl. Mater. Today* 9: 266–270.
13 Molinnus, D., Poghossian, A., Keusgen, M. et al. (2017). *Electroanalysis* 29: 1840–1849.
14 Poghossian, A., Katz, E., and Schöning, M.J. (2015). *Chem. Commun.* 51: 6564–6567.
15 Poghossian, A., Malzahn, K., Abouzar, M.H. et al. (2011). *Electrochim. Acta* 56: 9661–9665.
16 Poghossian, A., Krämer, M., Abouzar, M.H. et al. (2009). *Procedia Chemistry* 1: 682–685.
17 Krämer, M., Pita, M., Zhou, J. et al. (2009). *J. Phys. Chem. C* 113: 2573–2579.
18 Minko, S., Katz, E., Motornov, M. et al. (2011). *J. Comput. Theor. Nanoscience* 8: 356–364.
19 Pita, M., Minko, S., and Katz, E. (2009). *J. Mater. Sci.: Materials in Medicine* 20: 457–462.
20 Motornov, M., Zhou, J., Pita, M. et al. (2008). *Nano Lett.* 8: 2993–2997.
21 Motornov, M., Zhou, J., Pita, M. et al. (2009). *Small* 5: 817–820.
22 Tokarev, I., Gopishetty, V., Zhou, J. et al. (2009). *ACS Appl. Mater. Interfaces* 1: 532–536.
23 Tam, T.K., Zhou, J., Pita, M. et al. (2008). *J. Am. Chem. Soc.* 130: 10888–10889.
24 Guz, N., Halámek, J., Rusling, J.F., and Katz, E. (2014). *Anal. Bioanal. Chem.* 406: 3365–3370.
25 Tam, T.K., Pita, M., and Katz, E. (2009). *Sens. Actuat. B* 140: 1–4.
26 Katz, E., Pingarrón, J.M., Mailloux, S. et al. (2015). *J. Phys. Chem. Lett.* 6: 1340–1347.
27 Filipov, Y., Gamella, M., and Katz, E. (2018). *Electroanalysis* 30: 1281–1286.
28 Gamella, M., Privman, M., Bakshi, S. et al. (2017). *ChemPhysChem* 18: 1811–1821.
29 Scheja, S., Domanskyi, S., Gamella, M. et al. (2017). *ChemPhysChem* 18: 1541–1551.
30 Okhokhonin, A.V., Domanskyi, S., Filipov, Y. et al. (2018). *Electroanalysis* 30: 426–435.

31 Skirtach, A.G., Yashchenok, A.M., and Möhwald, H. (2011). *Chem. Commun.* 47: 12736–12746.

32 Coll, C., Bernardos, A., Martínez-Máñez, R., and Sancenón, F. (2013). *Acc. Chem. Res.* 46: 339–349.

33 Myrick, J.M., Vendra, V.K., and Krishnan, S. (2014). *Nanotechnol. Rev.* 3: 319–346.

34 Caldorera-Moore, M.E., Liechty, W.B., and Peppas, N.A. (2011). *Acc. Chem. Res.* 44: 1061–1070.

35 Vallet-Regí, M., Balas, F., and Arcos, D. (2007). *Angew. Chem. Int. Ed.* 46: 7548–7558.

36 Moore, T., Chen, H.Y., Morrison, R. et al. (2014). *Molec. Pharmaceutics* 11: 24–39.

37 Qing, G.Y., Li, M.M., Deng, L.J. et al. (2013). *Mini-Rev. Med. Chem.* 13: 1369–1380.

38 Niikura, K., Iyo, N., Matsuo, Y. et al. (2013). *ACS Appl. Mater. Interfaces* 5: 3900–3907.

39 Bonini, M., Berti, D., and Baglioni, P. (2013). *Cur. Opin. Colloid Interface Sci.* 18: 459–467.

40 Jiao, Y.F., Sun, Y.F., Chang, B.S. et al. (2013). *Chem. Eur. J.* 19: 15410–15420.

41 Michalak, M., Marek, A.A., Zawadiak, J. et al. (2013). *Eur. Polym. J.* 49: 4149–4156.

42 Munkhjargal, M., Matsuura, Y., Hatayama, K. et al. (2013). *Sens. Actuat. B* 188: 831–836.

43 Bocharova, V., Zavalov, O., MacVittie, K. et al. (2012). *J. Mater. Chem.* 22: 19709–19717.

44 Mailloux, S., Zavalov, O., Guz, N. et al. (2014). *Biomater. Sci.* 2: 184–191.

45 Johnson, R.P., John, J.V., and Kim, I. (2014). *J. Appl. Polym. Sci.* 131: 40796.

46 Yang, K., Feng, L.Z., Shi, X.Z., and Liu, Z. (2013). *Chem. Soc. Rev.* 42: 530–547.

47 Zhou, M., Zhou, N.D., Kuralay, F. et al. (2012). *Angew. Chem. Int. Ed.* 51: 2686–2689.

48 Traitel, T., Goldbart, R., and Kost, J. (2008). *J. Biomater. Sci. Polym. Ed.* 19: 755–767.

49 Lim, E.-K., Kim, T., Paik, S. et al. (2015). *Chem. Rev.* 115: 327–394.

50 Shaik, M.R., Korsapati, M., and Panati, D. (2012). *Int. J. Pharma Sci.* 2: 112–116.

51 Kamaly, N., Yameen, B., Wu, J., and Farokhzad, O.C. (2016). *Chem Rev.* 116: 2602–2663.

52 Liechty, W.B., Kryscio, D.R., Slaughter, B.V., and Peppas, N.A. (2010). *Annu. Rev. Chem. Biomol. Eng.* 1: 149–173.

53 Rehm, B.H.A. (ed.) (2009). *Alginates: Biology and Applications, Microbiology Monographs Series*, vol. 13. Dordrecht: Springer.

54 Gombotz, W.R. and Wee, S.F. (2012). *Adv. Drug Delivery Rev.* 64: 194–205.

55 Chan, A.W. and Neufeld, R.J. (2010). *Biomaterials* 31: 9040–9047.

56 Pescosolido, L., Piro, T., Vermonden, T. et al. (2011). *Carbohyd. Polym.* 86: 208–213.

57 Barrias, C.C., Lamghari, M., Granja, P.L. et al. (2005). *J. Biomed. Mater. Res. A* 74A: 545–552.
58 Chaudhari, S.A., Kar, J.R., and Singhal, R.S. (2015). *Curr. Org. Chem.* 19: 1732–1754.
59 Krebs, M.D., Salter, E., Chen, E. et al. (2010). *J. Biomed. Mater. Res. A* 92A: 1131–1138.
60 Jiang, G., Min, S.-H., Oh, E.J., and Hahn, S.K. (2007). *Biotechnol. Bioprocess Eng.* 12: 684–689.
61 Hu, W.W. and Tsou, S.L. (2014). *Carbohydr. Polym.* 101: 240–248.
62 Dey, K. and Roy, P. (2011). *Biotechnol. Lett.* 33: 1101–1105.
63 Hoesli, C.A., Raghuram, K., Kiang, R.L.J. et al. (2011). *Biotechnol. Bioeng.* 108: 424–434.
64 Moebus, K., Siepmann, J., and Bodmeier, R. (2009). *Eur. J. Pharm. Biopharm.* 72: 42–53.
65 Sun, X., Shi, J., Zhang, Z., and Cao, S. (2011). *J. Appl. Polym. Sci.* 122: 729–737.
66 Prasad, K. and Kadokawa, J.-I. (2009). *Alginates: Biology and Applications*, vol. 13 (ed. B.H.A. Rehm), Microbiology monographs series, 175–210. Dordrecht: Springer.
67 Castro, G.R., Kamdar, R.R., Panilaitis, B., and Kaplan, D.L. (2005). *J. Contr. Release* 109: 149–157.
68 Pearson, R.G. (1963). *J. Am. Chem. Soc.* 85: 3533–3539.
69 Jin, Z., Güven, G., Bocharova, V. et al. (2012). *ACS Appl. Mater. Interfaces* 4: 466–475.
70 Jin, Z., Harvey, A.M., Mailloux, S. et al. (2012). *J. Mater. Chem.* 22: 19523–19528.
71 Narayanan, R.P., Melman, G., Letourneau, N.J. et al. (2012). *Biomacromolecules* 13: 2465–2471.
72 Mailloux, S., Halámek, J., Halámková, L. et al. (2013). *Chem. Commun.* 49: 4755–4757.
73 Mailloux, S., Halámek, J., and Katz, E. (2014). *Analyst* 139: 982–986.
74 Mailloux, S., Guz, N., Zakharchenko, A. et al. (2014). *J. Phys. Chem. B* 118: 6775–6784.
75 Mailloux, S., Guz, N., Gamella Carballo, M. et al. (2014). *Anal. Bioanal. Chem.* 406: 4825–4829.
76 Gamella, M., Guz, N., and Katz, E. (2016). *Electroanalysis* 28: 2692–2696.
77 Gamella, M., Guz, N., Mailloux, S. et al. (2014). *Electroanalysis* 26: 2552–2557.
78 Gamella, M., Guz, N., Mailloux, S. et al. (2014). *ACS Appl. Mater. Interfaces* 6: 13349–13354.
79 Gamella Carballo, M., Guz, N., Pingarrón, J.M. et al. (2015). *Chem. Commun.* 51: 7618–7621.
80 Guz, N., Fedotova, T.A., Fratto, B.E. et al. (2016). *ChemPhysChem* 17: 2247–2255.
81 Mailloux, S., Gerasimova, Y.V., Guz, N. et al. (2015). *Angew. Chem. Int. Ed.* 54: 6562–6566.

82 Fratto, B.E., Guz, N., Fallon, T.T., and Katz, E. (2017). *ChemPhysChem* 18: 1721–1725.
83 Pannier, A.K. and Shea, L.D. (2004). *Molec. Therapy* 10: 19–26.
84 Kar, M., Tiwari, N., Tiwari, M. et al. (2013). *Part. Part. Syst. Char.* 30: 166–179.
85 Liu, M., Wang, Z.Y., Zong, S.F. et al. (2015). *J. Biomed. Nanotechnol.* 11: 979–987.
86 Moratz, J., Samanta, A., Voskuhl, J. et al. (2015). *Chemistry – Eur. J.* 21: 3271–3277.
87 Diaz, J.A. and Gibbs-Davis, J.M. (2013). *Small* 9: 2862–2871.
88 Yokoyama, M. (2005). Controlled release of DNA using thermoresponsive polymers. In: *Non-viral Gene Therapy, Gene Design and Delivery* (ed. K. Taira, K. Kataoka and T. Niidome), 165–175. Tokyo: Springer.
89 Little, K.S.R. and Kohane, D.S. (2008). *J. Mater. Chem.* 18: 832–841.
90 Sassolas, A., Leca-Bouvier, B.D., and Blum, L.J. (2008). *Chem. Rev.* 108: 109–139.
91 Xu, C., Taylor, P., Ersoz, M. et al. (2003). *J. Mater. Chem.* 13: 3044–3048.
92 Aytar, B.S., Prausnitz, M.R., and Lynn, D.M. (2012). *ACS Appl. Mater. Interfaces* 4: 2726–2734.
93 Wang, F., Li, D., Li, G. et al. (2008). *Biomacromolecules* 9: 2645–2652.
94 Wang, F., Wang, J., Zhai, Y. et al. (2008). *J. Contr. Release* 132: 65–73.
95 Wang, F., Liu, X., Li, G. et al. (2009). *J. Mater. Chem.* 19: 286–291.
96 Moore, E.J., Curtin, M., Ionita, J. et al. (2007). *Anal. Chem.* 79: 2050–2057.
97 Wang, J., Jiang, M., and Mukherjee, B. (2000). *Bioelectrochemistry* 52: 111–114.
98 Wang, J., Rivas, G., Jiang, M., and Zhang, X. (1999). *Langmuir* 15: 6541–6545.
99 Wang, J., Zhang, X., Parrado, C., and Rivas, G. (1999). *Electrochem. Commun.* 1: 197–202.
100 Jiang, M., Ray, W.W., Mukherjee, B., and Wang, J. (2004). *Electrochem. Commun.* 6: 576–582.
101 Takeishi, S., Rant, U., Fujiwara, T. et al. (2004). *J. Chem. Phys.* 120: 5501–5504.
102 Sempionatto, J.R., Gamella, M., Guz, N. et al. (2015). *Electroanalysis* 27: 2171–2179.
103 Minehan, D.S., Marx, K.A., and Tripathy, S.K. (1994). *Macromolecules* 27: 777–783.
104 Pande, R., Ruben, G.C., Lim, J.O. et al. (1998). *Biomaterials* 19: 1657–1667.
105 Honarvarfard, E., Gamella, M., Guz, N., and Katz, E. (2017). *Electroanalysis* 29: 324–329.
106 Gamella, M., Zakharchenko, A., Guz, N. et al. (2017). *Electroanalysis* 29: 398–408.
107 Masi, M., Gamella, M., Guz, N., and Katz, E. (2016). *Electroanalysis* 28: 2613–2625.
108 Strack, G., Babanova, S., Farrington, K.E. et al. (2013). *J. Electrochem. Soc.* 160: G3178–G3182.

109 Holzinger, M., Le Goff, A., and Cosnier, S. (2012). *Electrochim. Acta* 82: 179–190.
110 Zebda, A., Gondran, C., Le Goff, A. et al. (2011). *Nature Commun.* 2: 370.
111 Halámková, L., Halámek, J., Bocharova, V. et al. (2012). *J. Am. Chem. Soc.* 134: 5040–5043.
112 Szczupak, A., Halámek, J., Halámková, L. et al. (2012). *Energy Environ. Sci.* 5: 8891–8895.
113 MacVittie, K., Halámek, J., Halámková, L. et al. (2013). *Energy Environ. Sci.* 6: 81–86.
114 Katz, E. (ed.) (2012). *Biomolecular Computing – From Logic Systems to Smart Sensors and Actuators.* Weinheim: Wiley-VCH.
115 Katz, E., Lötzbeyer, T., Schlereth, D.D. et al. (1994). *J. Electroanal. Chem.* 373: 189–200.
116 Kyoung, C.S. and Sung, C.Y. (2011). *Polymer – Korea* 35: 444–450.
117 Holme, H.K., Davidsen, L., Kristiansen, A., and Smidsrod, O. (2008). *Carbohydr. Polym.* 73: 656–664.
118 Yang, Z., Li, J.P., and Guan, H.S. (2004). *Carbohydr. Polym.* 58: 115–121.
119 Li, X.X., Xu, A.H., Xie, H.G. et al. (2010). *Carbohydr. Polym.* 79: 660–664.
120 Mao, S.R., Zhang, T.T., Sun, W., and Ren, X.H. (2012). *Pharm. Development Technol.* 17: 763–769.
121 Smidsrød, O., Haug, A., and Larsen, B. (1965). *Acta Chem. Scand.* 19: 143–152.
122 Wilson, R. and Turner, A.P..F. (1992). *Biosens. Bioelectron.* 7: 165–185.
123 Dunford, H.B. (1991). Horseradish peroxidase: structure and kinetic properties. In: *Peroxidases in Chemistry and Biology*, vol. 2 (ed. J. Everse, K.E. Everse and M.B. Grisham), 1–24. Boca Raton, FL: CRC Press.
124 Goldstein, S., Meyerstein, D., and Czapski, G. (1993). *Free Radical Biology and Medicine* 15: 435–445.
125 Phillis, J.W., Estevez, A.Y., and O'Regan, M.H. (1998). *Neuroscience Lett.* 244: 109–111.
126 Privman, V., Domanskyi, S., Luz, R.A.S. et al. (2016). *ChemPhysChem* 17: 976–984.
127 Forman, H.J., Bernardo, A., and Davies, K.J.A. (2016). *Arch. Biochem. Biophys.* 603: 48–53.
128 Kolpashchikov, D.M. (2007). *ChemBioChem* 8: 2039–2042.
129 Cardarelli, F., Digiacomo, L., Marchini, C. et al. (2016). *Sci. Reports* 6: 25879.
130 Katz, E. (ed.) (2014). *Implantable Bioelectronics – Devices, Materials and Applications.* Weinheim: Wiley-VCH.

13

Biofuel Cells Controlled by Biocomputing Systems

13.1 Introduction: Biofuel Cells, Their Applications, and Motivation for Designing Adaptive, Signal-Controlled Devices

Biofuel cells based on purified enzymes [1–10] or whole microbial cells [11–13] are promising future alternative sources of sustainable electrical energy. They include two bioelectrocatalytic electrodes: one for oxidation of organic compounds [14–16] and another for reduction of oxygen [17, 18]. Enzymes or microbial cells are used to biocatalyze redox transformations at the electrodes and/or generate oxidizable substrates from raw organic substances. While microbial biofuel cells might be able to generate substantial electrical power operating with large size fermentation tanks and electrodes, enzyme-based biofuel cells are rather small (sometimes micro-size) devices producing small electrical power [19]. In the last 10–15 years [20], the fundamental problems of coupling between the biochemical and electrochemical processes, mostly related to the efficient interfacial charge transport using mediated [21] or direct (non-mediated) electron transfer [22] between enzymes and electrodes, were solved, and the biofuel cells became feasible. However, their practical applications still require solutions of many difficult engineering problems, particularly related to their long-term operation [23, 24], miniaturization [25–30], and power efficiency [31–33]. Thus, in recent years, most of the papers published on the biofuel cells moved from fundamental chemical problems to chemical and electrical engineering solving practical application issues. The present situation creates an illusion that this topic is no more in the area of basic research. This chapter highlights a new area of fundamental research activity in biofuel cell studies – the coupling of biofuel cells with biocomputing systems to yield "smart" bioelectronic devices releasing energy on-demand upon processing biochemical information.

One of the important potential applications of the biofuel cells is powering of implantable biomedical devices [34–38] (Figure 13.1). In this case a miniaturized biofuel cell must be implanted in a human body and use naturally existing biochemical substances as fuel (e.g., glucose in blood) [39] (Figure 13.2). Adaptive behavior of the implantable biofuel cell self-regulating the power release would be the immense advantage of these bioelectronic devices. Modified

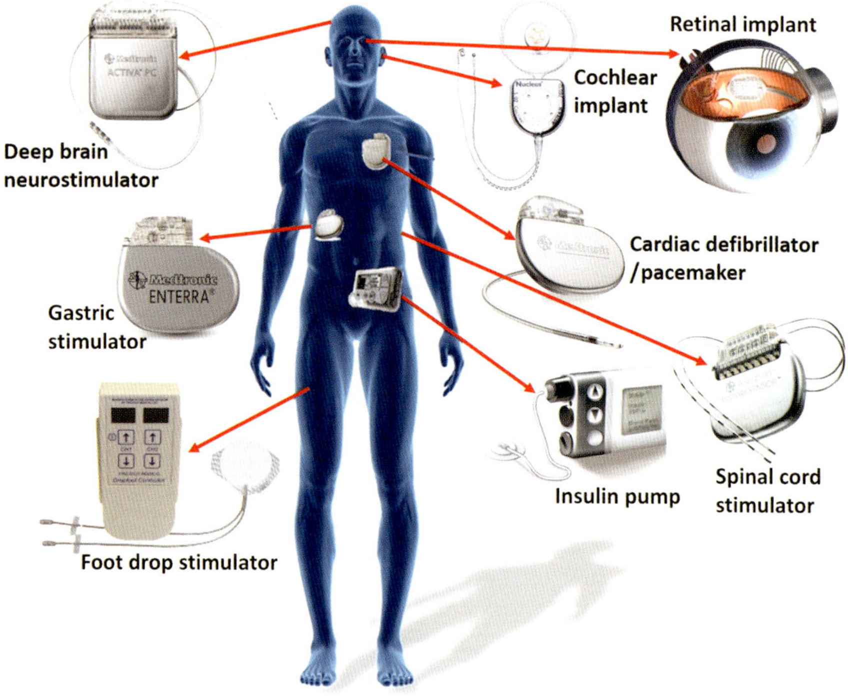

Figure 13.1 Various implantable/wearable medical microelectronic devices – pacemaker, deep brain neurostimulator, spinal cord stimulator, gastric stimulator, foot drop implant, cochlear implant, retinal implant, insulin pump, etc. – all of them need electrical power to operate. Source: Gamella et al. 2018 [19]. Adapted with permission of Elsevier.

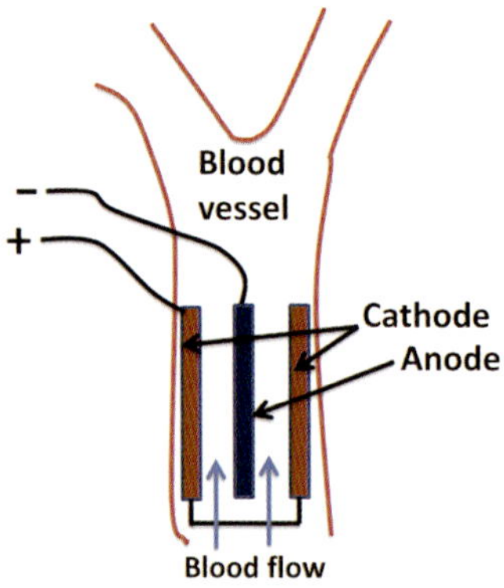

Figure 13.2 Schematic futuristic vision of an enzyme-based biofuel cell implanted in a blood vessel for extracting electrical power by glucose oxidation at an anode and oxygen reduction at a cathode. (Not experimentally realized yet.) This schematics was suggested by Prof. Adam Heller (https://en.wikipedia.org/wiki/Adam_Heller) as early as in 1990s and similar images are shown in numerous websites.

electrodes switchable between electrochemically active and inactive states by various physical and/or chemical signals were actively studied in the last decade [40–45], but very few of them were considered for their use in biofuel cells [46–49]. It should be also noted that most of the developed switchable electrodes were activated/deactivated by artificial signals (usually by light [41], magnetic field [43], temperature changes [42], or electrical potential [49]) applied to the systems, and they do not provide communication with biochemical environment to regulate (switch or tune) the electrode activity according to the presence or absence of biochemical substances.

A new approach became possible when polymer-modified electrodes switchable between ON/OFF states by pH values [50–54] were coupled to biochemical reactions generating pH changes *in situ* [55–58]. This allowed transduction of biochemical input signals (e.g., glucose concentration) to the pH changes governing the electrochemical activity of the switchable electrode [59]. Using a new concept of Boolean logic gates and their biocomputing networks based on enzymatic reactions [60, 61], the complexity of the enzyme systems controlling the electrode activity was scaled up [62]. Finally, the switchable electrodes controlled by complex multienzyme systems, being reversibly activated–inactivated by various patterns of different biochemicals [56, 58, 63–65], were assembled in a biofuel cell producing electrical power depending on the biochemical environment [45, 66–73]. The following sections outline recent advances in this novel research area.

13.2 Biofuel Cells Controlled by Logically Processed Biochemical Signals

Interfaces modified with polyelectrolyte layers (usually in the form of covalently attached polymer brushes) demonstrated pH-switchable features [74–76]. Electrodes functionalized with the polyelectrolyte brushes demonstrated pH-switchable (possibly tunable) electrochemical activity when a neutral state of the polyelectrolyte was impermeable for ionic redox species keeping the electrode mute, while the ionized state of the polymer brush was permeable for ionic redox species of the opposite charge allowing their access to the conducting support and activating the electrochemical process [51, 53, 77]. Protonation–deprotonation of the polymer brush was controlled by the pH value that could be changed in a bulk solution [51, 53, 77] or varied locally at the interface by means of electrochemistry [52, 54]. The pH changes governing the electrode activity can be also produced by enzyme reactions including complex biocatalytic cascades mimicking Boolean logic operation (see also Section 6.2.4 and Figure 6.11).

As soon as the pH-switchable bioelectrocatalytic electrodes and the pH-change-producing enzyme logic systems were formulated, biofuel cells controlled by the logically processed biochemical signals became feasible [67–69]. Only one switchable bioelectrocatalytic electrode is needed in order to control the biofuel cell activity since the power production in the biofuel

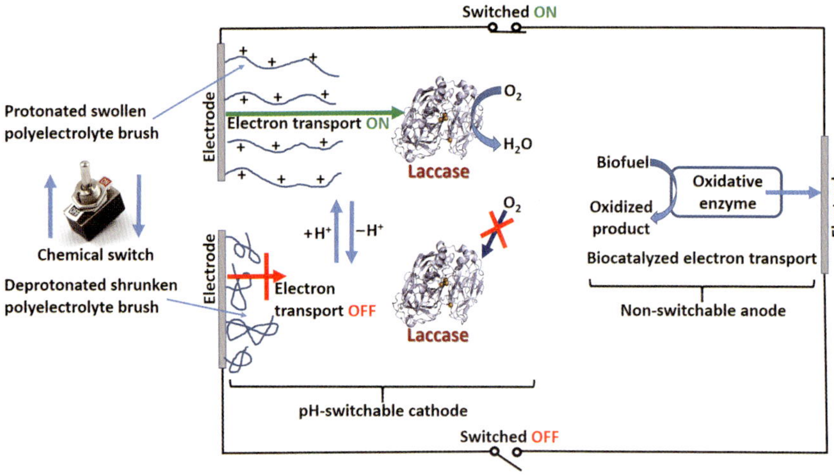

Figure 13.3 Schematics of a signal-switchable biofuel cell. The switchable feature is realized at the O_2-reducing cathode based on the laccase bioelectrocatalyzed reaction at a polyelectrolyte-modified electrode. The polyelectrolyte brush exists in a protonated swollen state at a low pH value, when the polymer layer is hydrophilic and permeable, thus resulting in the electrochemically active ON state. On the other hand, at neutral/basic pH values (above the polyelectrolyte pK_a), the polyelectrolyte is deprotonated, shrank, and not permeable, thus resulting in the electrochemically inactive OFF state. The anode can be based on any bioelectrocatalytic reaction resulting in the oxidation of any biofuel (e.g., glucose) catalyzed by any oxidative enzyme (e.g., glucose oxidase). The biofuel operation (activity) depends on the state of the cathode controlled by the pH value. The present schematic presentation does not specify the mechanism of the charge transfer between the switchable cathode and laccase; more specific example is shown in Figure 13.4.

cells requires simultaneous operation of both electrodes – the cathode and anode for oxygen reduction and biofuel oxidation, respectively. It should be noted that in most of the known biofuel cells the cathode reaction is the same – bioelectrocatalytic oxygen reduction – while the anodic process might differ substantially depending on the used biofuel and the applied biocatalyst. This makes it more practical to control the activity of the bioelectrocatalytic oxygen reduction at the modified cathode being switched between the ON and OFF states by the enzyme-induced pH changes (Figure 13.3).

Laccase is a well-known enzyme frequently used for bioelectrocatalytic reduction of oxygen in biofuel cells [18]; thus it was selected to operate at the modified pH-switchable cathode (Figure 13.3). The pH-controlled laccase-catalyzed oxygen reduction was realized using poly(4-vinylpyridine) (P4VP), functionalized with Os(dmo-bpy)$_2$Cl redox centers (dmo-bpy = 4,4′-dimethoxy-2,2′-bipyridine) (see structure in Figure 13.4A). The redox groups tethered to the polymer brush allowed the mediated electron transfer to laccase active centers only at pH < 4.5 when the polymer was in the protonated and swollen state [50] (Figure 13.4A). In this state the polymer brush was free to move, thus communicating with the conductive support and with the soluble enzyme due to quasi-diffusional translocation of the polymer chains. On the other hand, the mediated electron transfer was fully inhibited at pH > 5.5, when the deprotonated polymer was shrunk.

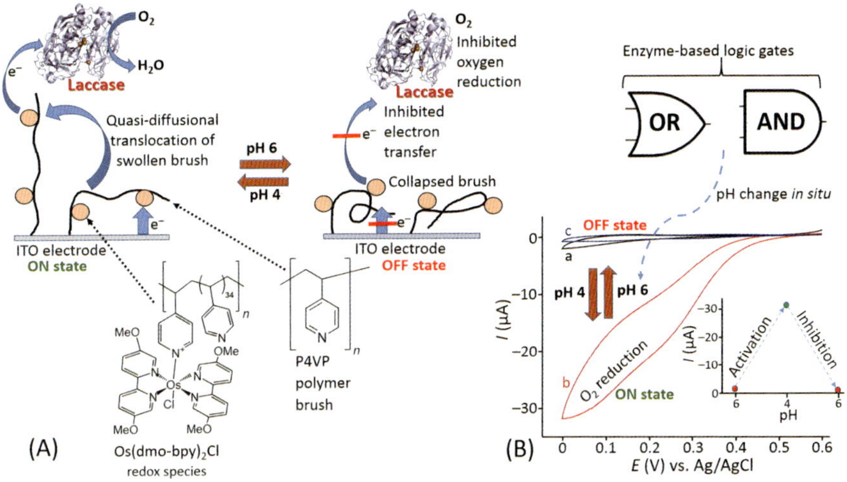

Figure 13.4 (A) Schematics of the pH-switchable laccase-catalyzed O_2 reduction at the electrode modified with a P4VP brush functionalized with Os(dmo-bpy)$_2$Cl redox species operating as an electron transfer mediator for laccase. The redox-functionalized polymer chains are free to move when the polymer is protonated (pH = 4 < pK_a) and swollen. The polymer chains are collapsed when the polymer is deprotonated (pH = 6 > pK_a). The electron transfer function is only realized by the quasi-diffusional translocation of the swollen polymer chains tethered to the electrode surface. The electron transfer through the hydrophobic collapsed polymer film is inhibited because of the long distance between the redox species, which does not allow the electron hopping from one redox site to another. The swollen and collapsed states of the polymer brush represent the ON and OFF states of the electrode. (B) The cyclic voltammograms obtained at the O_2-reducing electrode controlled by the pH value: (a) the inhibited electrode at pH 6, (b) the activated electrode at pH 4, and (c) the electrode inhibited again after pH increase to 6. The pH changes can be produced *in situ* through biocatalytic reactions mimicking various logic gates (e.g., OR/AND logic gates or their networks). Inset: Reversible variation of the bioelectrocatalytic current upon cyclic changes of the pH value.

This did not allow its physical movements for communicating with the electrode support and the enzyme. Notably, the low density of the redox groups attached to the polymer chains did not allow electron hopping in the shrunk state of the polymer layer. The corresponding cyclic voltammograms obtained in the presence of laccase and oxygen demonstrated no electrocatalytic current at pH ca. 6 when the electrode is in the OFF state (Figure 13.4B, curve a) and a well-defined cathodic electrocatalytic current for the O_2 reduction at pH ca. 4 when the electrode is active (Figure 13.4B, curve b). The electrocatalytic current was inhibited again after the pH value returned to the initial value (Figure 13.4B, curve c), thus demonstrating the reversible switching of the electrocatalytic process (Figure 13.4B, inset). This reversible switching of the electrocatalytic process can be coupled with the pH changes generated *in situ* by enzyme logic systems (see Sections 6.2.3 and 6.2.4 and Figures 6.10 and 6.11).

The logically controlled O_2-reducing bioelectrode was coupled with a glucose-oxidizing bioelectrocatalytic system to yield a biofuel cell [67, 68]

(Figure 13.3). The bioelectrochemical oxidation of glucose in the presence of soluble glucose oxidase (GOx) and methylene blue (MB) operating as a diffusional electron transfer mediator [78] was selected as the simplest example of an anodic reaction to demonstrate the logically switchable biofuel cell. The anodic compartment containing the glucose-oxidizing bioelectrocatalytic system in 0.1 M phosphate buffer, pH 7, under argon was separated from the cathodic switchable O_2-reducing electrode by a Nafion® membrane preventing mixing of the cathodic and anodic solutions. The cathodic compartment in addition to the O_2-reducing system (laccase in the solution and the Os-complex electron relay at the electrode surface) also included the enzyme logic system controlling the pH value.

Enzyme systems producing *in situ* pH changes as the logic output signals were developed specifically for controlling the electrochemical activity of the polymer brush-modified electrodes [67]. In a simple example, two parallel biocatalytic reactions were applied to mimic an OR logic gate (Figure 13.5A). Two enzymes – glucose oxidase (GOx) and esterase (Est) – biocatalyzed the oxidation of glucose (Input *A*) and hydrolysis of ethyl butyrate (Input *B*) respectively, mimicking the OR logic gate (Figure 13.5B). The absence of glucose or ethyl butyrate was considered as the digital input signal **0**, while their presence in the operational concentrations was considered as the input signal **1**. The biocatalytic reactions in the presence of glucose and ethyl butyrate resulted in the formation of gluconic acid and butyric acid, respectively, both producing low pH values in the non-buffered solution. Thus, in the presence of any substrate (input signals **0**,**1** or **1**,**0**) or both of them together (input signals **1**,**1**), one of

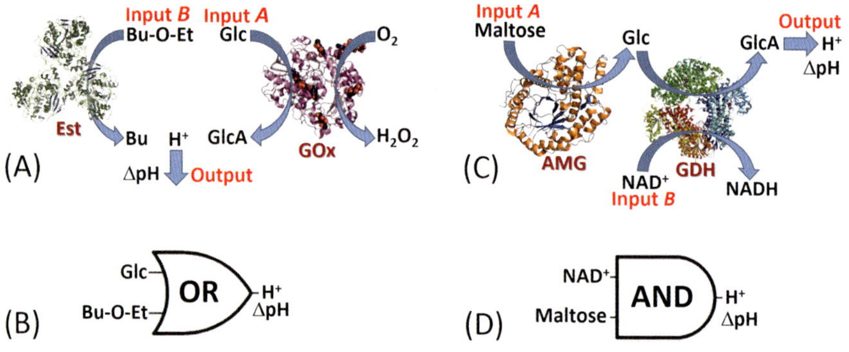

Figure 13.5 (A) The biocatalytic reactions operating in parallel and resulting in pH decrease upon production of butyric acid (Bu) or gluconic acid (GlcA) due to hydrolysis of ethyl butyrate (Et-O-Bu; Input *B*) catalyzed by esterase (Est) or oxidation of glucose (Glc; Input *A*) catalyzed by glucose oxidase (GOx). Notably, the pH decrease can be achieved in the presence of either or both inputs (**0**,**1**; **1**,**0**; **1**,**1** combinations), thus mimicking the OR logic gate. (B) The Boolean OR logic gate realized with the reactions shown in (A). (C) The two-step process catalyzed by amyloglucosidase (AMG) and glucose dehydrogenase (GDH). The reaction catalyzed by AMG results in the conversion of maltose (Input *A*) to Glc, which is further oxidized to GlcA in the reaction step catalyzed by GDH. The second reaction step requires also the presence of NAD^+ (Input *B*). The whole process results in the pH decrease only in the presence of both inputs (**1**,**1** combination), thus mimicking the AND logic gate. (D) The Boolean AND logic gate realized with the reactions shown in (C).

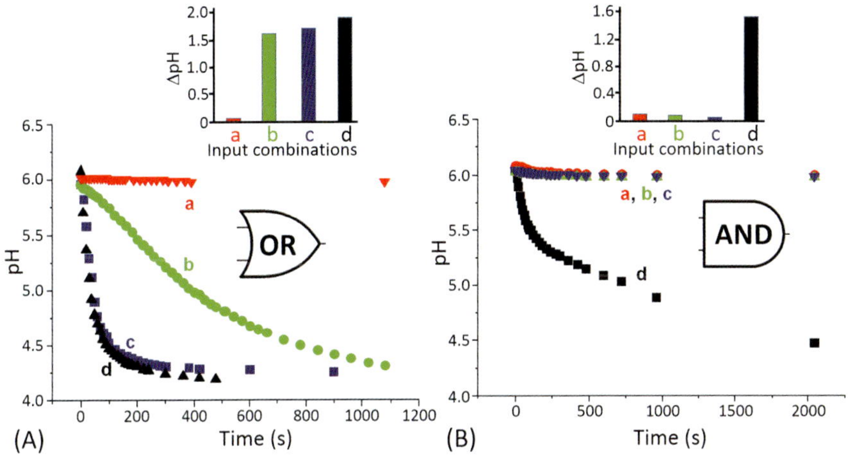

Figure 13.6 (A) The kinetics of the pH changes upon realizing the biocatalytic reactions mimicking the OR logic gate shown in Figure 13.5A while being activated with different input combinations: (a) **0**,**0**; (b) **1**,**0**; (c) **0**,**1**; and (d) **1**,**1**. (B) The kinetics of the pH changes upon realizing the biocatalytic reactions mimicking the AND logic gate shown in Figure 13.5C while being activated with different input combinations: (a) **0**,**0**; (b) **1**,**0**; (c) **0**,**1**; and (d) **1**,**1**. Insets: The pH changes obtained after 1100 seconds from the time when different signal combinations were applied. The exact composition of the solutions and other technical details can be found in Ref. [67]. Source: Amir et al. 2009 [67]. Adapted with permission of American Chemical Society.

the reactions or both of them proceeded and resulted in the acidification of the solution reaching pH ca. 4.2 (Figure 13.6A). The pH value was unchanged keeping the original value of pH ca. 6 only in the absence of both substrates (input signals **0**,**0**). Therefore, the features of the system corresponded to the OR logic operation (Figure 13.6A, inset). The AND logic gate (Figure 13.5D) was composed of two enzymes operating cooperatively: amyloglucosidase (AMG) and glucose dehydrogenase (GDH) (Figure 13.5C). The enzyme system was activated by two biochemical input signals: maltose (Input *A*) and NAD^+ (Input *B*) participating in a two-step chain reaction. In the first step maltose was hydrolyzed by AMG to glucose, and then glucose was oxidized by GDH to gluconic acid. The second reaction required NAD^+ as an electron acceptor, and the chain reaction cannot be completed in the absence of NAD^+ even though glucose was produced in the first step. The absence of maltose or NAD^+ was considered as the digital input signal **0**, while their presence in the operational concentrations was considered as the input signal **1**. In the absence of either or both substrates (input signals **0**,**0**; **0**,**1**; or **1**,**0**), the two-step reaction chain was not completed, and the pH value was not changed. In the presence of both substrates (input signals **1**,**1**), the reaction proceeded till the very end, resulting in the formation of gluconic acid and acidification of the solution reaching pH ca. 4.3 (Figure 13.6B). Therefore, the features of the system corresponded to the AND logic operation (Figure 13.6B, inset).

The V–i (voltage vs. current density) polarization function of the biofuel cell was obtained upon application of variable load resistances and by measuring the

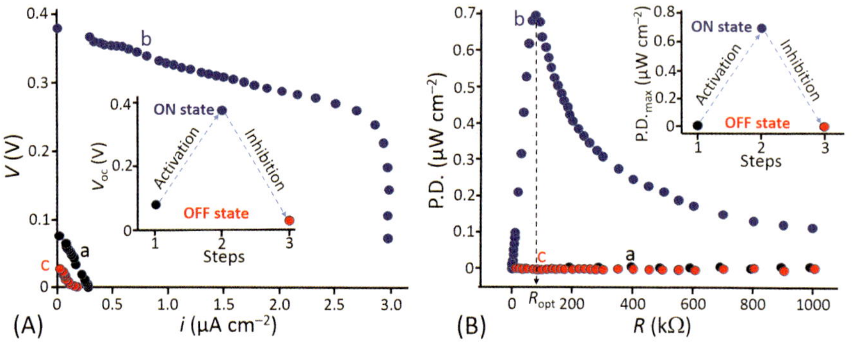

Figure 13.7 The polarization function, *V* vs. *i*, (A) and power density dependence on the external resistance load, P.D. vs. *R*, (B) measured on a variable load resistance (*R*) for the biofuel cell composed of the pH-switchable cathode (shown in Figure 13.4A) operating in the presence of O_2 and a bare ITO-anode operating with soluble GOx, methylene blue mediator, and glucose. The cathode and anode were separated with a Nafion membrane to allow the cathode operation in the presence of air and anode operation under Ar. The biofuel cell operation was studied at different pH values corresponding to the inhibited/activated states of the cathode: (a) pH 6.0 – OFF state; (b) pH ca. 4.2 – ON state; (c) pH ca. 7 – OFF state. The pH decrease (pH 6 → pH ca. 4.2) activating the biofuel cell operation was produced by the biocatalytic reactions mimicking OR/AND logic gates shown in Figure 13.5. The reset to the neutral pH ca. 7 resulting in the inactivation of the biofuel cell was achieved by another biocatalytic reaction (hydrolysis of urea catalyzed by urease). The insets show reversible activation–inhibition of the biofuel cell operation by the stepwise pH changes produced by the biocatalytic reactions. Source: Amir et al. 2009 [67]. Adapted with permission of American Chemical Society.

current and voltage generated on them (Figure 13.7A). The power density produced by the biofuel cell upon connecting to the variable resistances was derived from the V–i measurements (Figure 13.7B). The V–i function of the biofuel cell in its non-active state, pH ca. 6, revealed 80 mV for the open circuit voltage, V_{oc}, and 0.3 μA cm^{-2} for the short circuit current density, i_{sc} (Figure 13.7A, curve a). The maximum released power density was P.D.$_{max}$ = 6 nW cm^{-2} (Figure 13.7B, curve a). The biofuel cell was activated when the OR logic gate was used and **0**,**1**; **1**,**0**; or **1**,**1** input signals were applied (for the AND logic gate the "successful" combination of the input signals was **1**,**1**). When the biofuel cell was activated (through *in situ* pH changes enabling the biocatalytic cathodic process), the V–i function was changed demonstrating ca. 380 mV for the open circuit voltage, V_{oc}, and 3 μA cm^{-2} for the short circuit current density, i_{sc}, at pH ca. 4 (Figure 13.7A, curve b). The maximum power density was significantly increased reaching 700 nW cm^{-2} (Figure 13.7B, curve b). The patterns of the biochemical signals resulting in the biofuel cell switching ON resembled the OR/AND logic functions (Figure 13.8). After the biofuel cell was activated by the set of the biochemical inputs processed by the enzyme logic gates, another Reset signal (urease and urea) was applied to bring the pH value back to its initial value and to inactivate the biofuel cell (Figure 13.7A,B, curves c). Thus, the reversible cyclic activation–inhibition of the biofuel cell was achieved by the application of various biochemical signals (Figure 13.7A,B, insets).

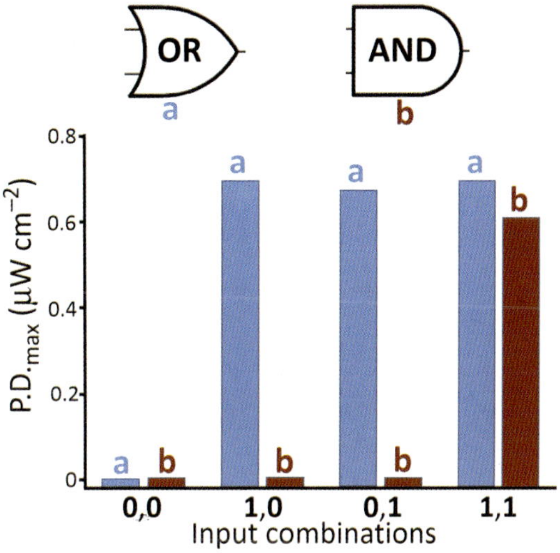

Figure 13.8 The bar chart showing the maximum power density obtained from the biofuel cell controlled by different combinations of the biochemical input signals: (a) OR logic gate and (b) AND logic gate (see Figure 13.5 for the biocatalytic reactions). Source: Amir et al. 2009 [67]. Adapted with permission of American Chemical Society.

A similar biofuel cell (Figure 13.3) was switched ON by much more sophisticated combination of the biochemical input signals when the bioelectrocatalytic cathodic process was controlled by the enzyme logic network composed of several concatenated gates (see Section 6.2.3 and Figure 6.10) [68]. A sequence of biochemical reactions was designed to produce *in situ* pH changes as the final output of the biochemical cascade (Figure 6.10A). The reactions were activated by four chemical input signals – NADH, acetaldehyde, maltose, and sucrose (input signals *A*, *B*, *C*, and *D*, respectively) – and processed by four enzymes, alcohol dehydrogenase (ADH), amyloglucosidase (AMG), invertase (Inv), and glucose dehydrogenase (GDH). The reactions were started from the production of NAD^+ and ethanol biocatalyzed by ADH in the presence of NADH and acetaldehyde being primary input signals. Another reaction chain resulted in the formation of glucose from sucrose and maltose (another couple of the primary inputs) biocatalyzed by Inv and AMG, respectively. The *in situ* produced NAD^+ and glucose were reacted with GDH yielding gluconic acid (GlcA) as the final chemical product, which resulted in the formation of the acidic pH ca. 4.3. The biochemical set of the reactions (Figure 6.10A) can be reformulated in the terms of the logic operations performed by the network composed of three logic gates (Figure 13.9A). Indeed, the production of NAD^+ being an intermediate product in the system was biocatalyzed by ADH only in the presence of NADH and acetaldehyde. The absence of either or both the inputs inhibited the reaction and NAD^+ was not formed. Thus, this part of the system operates as a Boolean AND logic gate. The production of glucose, also being an intermediate product, proceeded in two independent reactions biocatalyzed by AMG and Inv in the presence of maltose and sucrose, respectively. Thus, the glucose formation was activated in the presence of any of the input signal or both of them. This part of the biochemical cascade represented the OR logic gate. The *in situ* produced NAD^+ and glucose reacted to yield gluconic acid and decrease the solution pH value upon the process biocatalyzed by GDH. The product and the respective

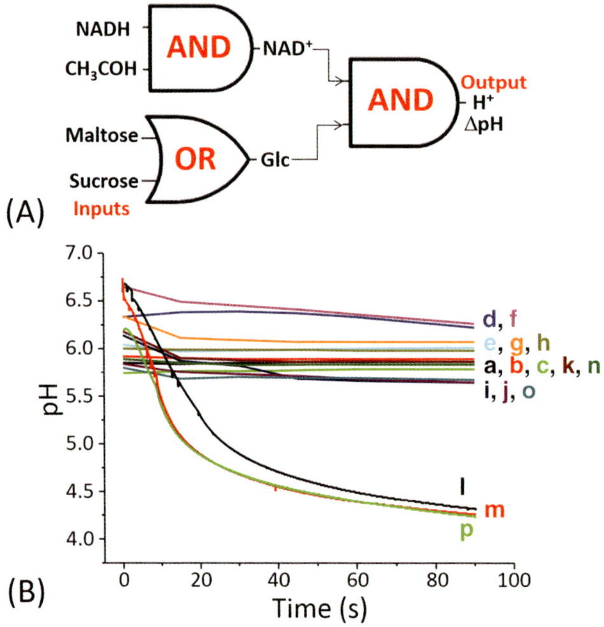

Figure 13.9 (A) The logic network composed of OR/AND gates realized with a cascade of biocatalytic reactions (see Figure 6.10A for the scheme of the biocatalytic cascade), resulting in pH decrease as the final output signal. (B) The kinetics of the pH changes in the course of the biocatalytic reactions activated with different combinations of the input signals: (a) **0,0,0,0**; (b) **1,0,0,0**; (c) **0,1,0,0**; (d) **0,0,1,0**; (e) **0,0,0,1**; (f) **1,1,0,0**; (g) **1,0,1,0**; (h) **1,0,0,1**; (i) **0,1,1,0**; (j) **0,1,0,1**; (k) **0,0,1,1**; (l) **1,1,1,0**; (m) **1,1,0,1**; (n) **1,0,1,1**; (o) **0,1,1,1**; and (p) **1,1,1,1**. The exact composition of the reacting systems and all other technical details can be found in Ref. [68]. Source: Tam et al. 2009 [68]. Adapted with permission of Elsevier.

pH decrease were obtained only in the presence of both reactants, thus featuring the final AND logic gate (Figure 13.9A). The chemical input signals were defined as digital **0** in the absence of the respective chemicals and digital **1** upon addition of their operating concentrations optimized experimentally. All possible 16 combinations of the chemical input signals were examined for changing pH in the system (Figure 6.10C), and only three of them (**1,1,0,1**, **1,1,1,0**, and **1,1,1,1**) resulted in the production of the acidic pH ca. 4.3, which corresponds to ΔpH ca.2.5 (Figure 13.9B). This result was consistent with the expected output of the logic network (Figure 13.9A) as well as with the biochemical scheme (Figure 6.10A). The experimental features of the system followed the Boolean logic function, $O = (A \text{ AND } B) \text{ AND } (C \text{ OR } D)$, where O is the output and A, B, C, and D are the input signals. After completing the enzyme reactions resulting in the logically controlled acidic medium (pH = 4.1–4.5), the pH value was reset to the initial value (pH ca. 6) by the formation of ammonia upon hydrolysis of urea biocatalyzed by urease, thus performing the Reset function.

Similarly to the switchable biofuel cell described above, the biofuel cell was characterized by the polarization curve (Figure 13.10A) and the power release dependence (Figure 13.10B) obtained in the inactive and active states, curves a

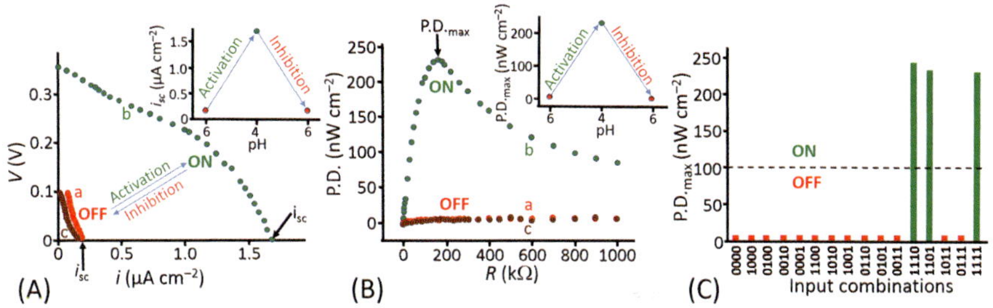

Figure 13.10 *V*–*i* polarization curves obtained for the biofuel cell with different load resistances (A) and the power density as a function of the resistance load (B): (a) in the inactive state prior to the addition of the biochemical input signals (pH value in the cathodic compartment ca. 6), (b) in the active state after the cathode was activated by changing pH to ca. 4.3 by the biochemical signals (input combinations **1,1,1,0**; **1,1,0,1** and **1,1,1,1**), and (c) after the Reset function was activated by the addition of urea to the system containing urease. Insets: Switchable i_{sc} (A) and power density (B) upon transition of the biofuel cell from the mute state to the active state and back activated/inhibited by biochemical signals processed through the enzyme logic network. The bar chart (C) showing the power density produced by the biofuel cell in response to different patterns of the chemical input signals. Dashed lines show thresholds separating digital **0** and undefined and **1** output signals produced by the system. Source: Tam et al. 2009 [68]. Adapted with permission of Elsevier.

and b, respectively. However, the cell operation was controlled by four different biochemical signals applied in 16 combinations; only 3 of them being "successful" for switching the biofuel cell ON according to the built-in logic "program" in the enzyme network (Figure 13.10C). After the biofuel cell activation, another cycle can be started with the reset to the initial "mute" state by the application of the urease/urea signal.

The biochemical networks demonstrate robust error-free processing of biochemical signals upon appropriate optimization of their components and interconnections [79–81] (see also Chapters 3 and 4). However, the limit of the biocomputing network complexity is set by the cross-reactivity of the enzyme-catalyzed reactions. Only enzymes belonging to different biocatalytic classes (oxidases, dehydrogenases, peroxidases, hydrolases, etc.) could operate in a homogeneous system without significant cross-reactivity. If chemical reasons (e.g., for biosensing; see Chapter 11) require the use of cross-reacting enzymes in the system, they must be compartmentalized using pattering on surfaces or applied in microfluidic devices (see Chapter 5). Application of more selective biomolecular interactions would be an advantage to make biocomputing systems more specific to various input signals and less cross-reactive in the chemical signal processing. This goal can be achieved by application of highly selective biorecognition (e.g., immune-specific) interactions for biocomputing [82]. One of the novel immune-based biocomputing systems was already applied for switching the biofuel cell activity by the logically processed antibody signals [69].

A surface functionalized with a mixed monolayer of two different antigens – 2,4-dinitrophenyl (DNP) and 3-nitro-L-tyrosine (NT) loaded on human serum albumin (HSA) and bovine serum albumin (BSA), respectively – was used to analyze the input signals of the corresponding antibodies: anti-DNP (anti-dinitrophenyl IgG polyclonal from goats) and anti-NT (anti-nitrotyrosine IgG polyclonal from rabbits) [69]. After binding to the surface, the primary antibodies were reacted with the secondary antibodies – anti-goat-IgG-HRP and anti-rabbit-IgG-HRP (mouse-origin IgG specific to goat IgG and mouse-origin IgG specific to rabbit IgG, both labeled with horseradish peroxidase, HRP) – to attach the biocatalytic HRP tag to the immune complexes generated on the surfaces (Figure 13.11). The primary anti-DNP and anti-NT antibodies were applied in four different combinations, **0,0**; **0,1**; **1,0**; and **1,1**, where the digital value **0** corresponded to the absence of the antibody and value **1** corresponded to their presence in the experimentally optimized concentrations. The secondary antibody labeled with the HRP biocatalytic tag was bound to the surface only if the respective primary antibody was already there. Since the both secondary antibodies were labeled with HRP, the biocatalytic entity appeared on the surface upon application of **0,1**; **1,0**; and **1,1** signal combinations. Only in the absence of both primary antibodies (signals **0,0**), the secondary antibodies were not bound to the surface, and the HRP biocatalyst did not appear there, thus resembling the OR logic operation. The assembled functional interface was reacted with 2,2′-azino-*bis*(3-ethylbenzothiazoline-6-sulfonic acid) (ABTS) and H_2O_2. The biocatalytic oxidation of ABTS and concomitant reduction of H_2O_2 resulted in

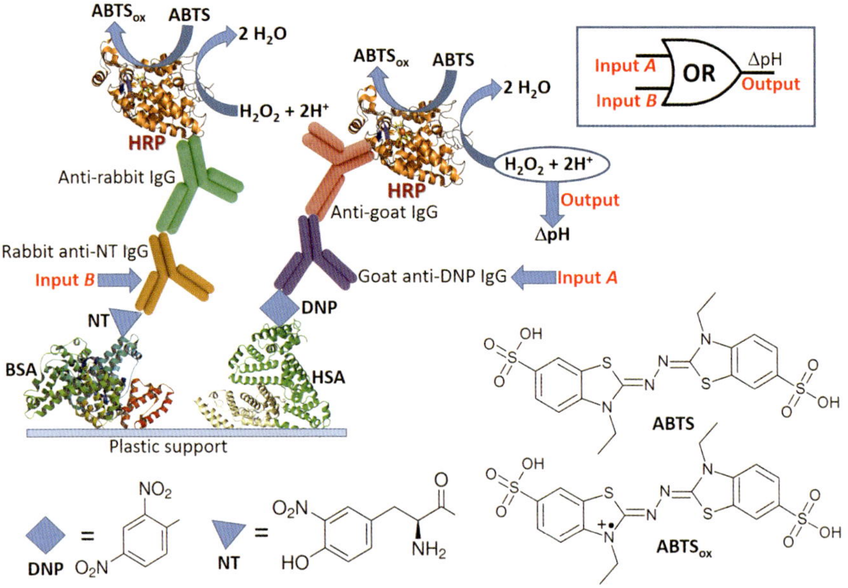

Figure 13.11 The immune-recognition system producing *in situ* pH changes controlling operation of a biofuel cell. The logic input signals are represented by the antibodies: goat anti-DNP IgG (Input *A*) and rabbit anti-NT IgG (Input *B*). Either or both input signals applied result in the assembling of the antibody sequence ended with the HRP biocatalytic tag producing the pH change in the presence of H_2O_2 and ABTS. Note that H^+ cations are consumed in the reaction catalyzed by HRP, thus resulting in the pH increase. The system mimics the Boolean OR logic gate. The structures of the DNP and NT antigens and ABTS/ABTS$_{ox}$ are shown.

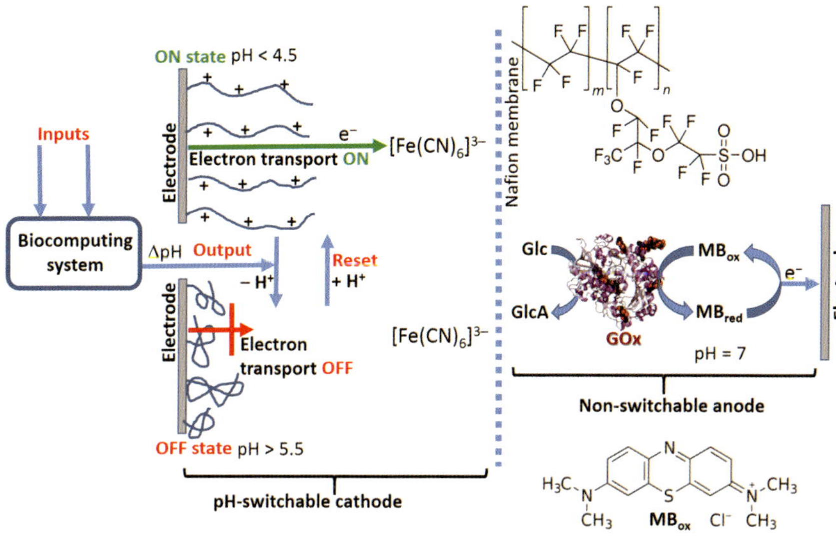

Figure 13.12 Schematics of the pH-switchable biofuel cell with the power release controlled by immune signals (see Figure 13.11 for the signal-generating system). The pH-switchable cathode is represented by a model system with $[Fe(CN)_6]^{3-}$ used as the electron-accepting species reacting at the polymer-brush-modified electrode. The electron acceptor can penetrate through the swollen polymer film protonated at pH < 4.5, thus resulting in the active ON state of the electrode. On the other hand, the electron acceptor cannot penetrate through the collapsed polymer film deprotonated at pH > 5.5, thus resulting in the inhibited OFF state of the electrode. The pH increase resulting in switching off the cathode reaction is produced through the biocomputing process controlled by the immune signals. The anode reaction is represented by the glucose (Glc) oxidation catalyzed by GOx with the electron transport mediated with methylene blue (MB). The cathode and anode solutions are separated with a Nafion membrane.

the increase of the solution pH only when the biocatalytic HRP tag was present on the surface (Figure 13.11). This happened when the primary antibody signals were applied in the combinations **0**,**1**; **1**,**0**; and **1**,**1**. The plastic support functionalized with the antibody species and the HRP biocatalytic tag was placed in the cathodic compartment of a biofuel cell (Figure 13.12). The pH increase generated *in situ* by the enzyme reaction coupled with the immune recognition system yielded the inactive shrunken state of the polymer-brush-modified electrode, thus deactivating the entire biofuel cell. It should be noted that for simplicity, the cathode was represented by a model redox system with a ferricyanide solution instead of the oxygen system (Figure 13.12). The biofuel cell being active at pH 4.5 (Figure 13.13A,B, curves a) was partially inactivated, curves b, by the pH increase up to 5.8 generated by the immune-based logic system. Since the output signal **1** from the logic system resulted in the inactivation of the biofuel cell (operating as the inverter producing **0** output for input **1** and *vice versa*), the system modeled an NOR logic gate (Figure 13.13B, inset). After the biofuel cell inactivation, the next cycle was started by the reset to the initial pH value activating the switchable electrode again. To activate the biofuel cell, GOx and glucose were added to the cathodic compartment, resulting in the pH

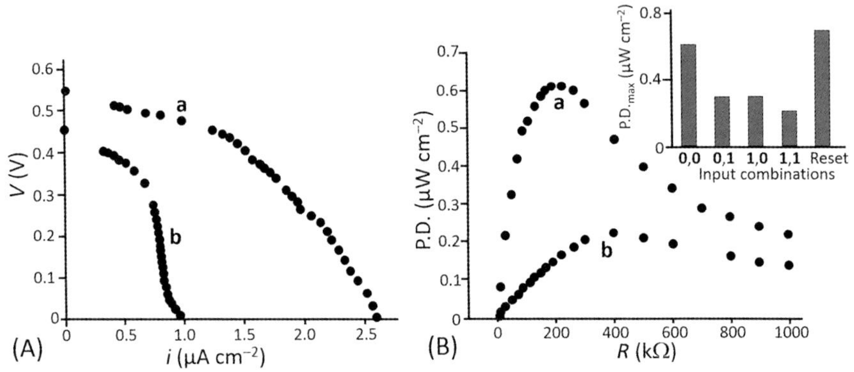

Figure 13.13 (A) Polarization curves of the biofuel cell shown in Figure 13.12 in the ON (a) and OFF (b) states. (B) Power release on different load resistances in the ON (a) and OFF (b) states. Inset: Maximum power produced by the cell upon application of different combinations of the immune input signals. Source: Tam et al. 2009 [69]. Adapted with permission of American Chemical Society.

decrease to ca. 4.2 due to the biocatalytic oxidation of glucose and formation of gluconic acid.

13.3 Biofuel Cells Controlled by Biomolecular Keypad Lock Systems

A multistep binding of complementary antibodies finished with a signal-generating enzyme-labeled antibody was used to engineer a biomolecular keypad lock system [83] (see also Chapter 10). An antigen, 3-nitro-L-tyrosine (NT), bound to a carrier protein (bovine serum albumin, BSA) was physically adsorbed on polystyrene wells of an enzyme-linked immunosorbent assay (ELISA) plate. The surface functionalized with the NT-BSA conjugate was stepwise reacted with different antibodies defined as input signals for the system, each time washing out the non-reacted species. An antibody bound to the surface in a previous step served as an antigen for binding another antibody in the next step. The antibodies were bound to the surface only if they found corresponding complementary antigens attached to the interface in the previous reaction steps; otherwise they were removed upon the washing steps. The correct sequence of the biorecognition steps included the binding of rabbit-origin IgG anti-nitrotyrosine (input signal *A*), goat-origin IgG specific against rabbit IgG (input signal *B*), and mouse-origin IgG specific against goat IgG (input signal *C*). The very final step was the attachment of the enzyme-labeled goat-origin IgG specific against mouse IgG to terminate the assembling process with the catalytic species (Figure 13.14A).

In order to couple the biorecognition events with a switchable biofuel cell yielding a self-powered bioelectronic device, the terminal goat-origin IgG specific against mouse IgG was labeled with urease. This enzyme, when it is bound to the surface through the correct sequence of antibodies, resulted in the

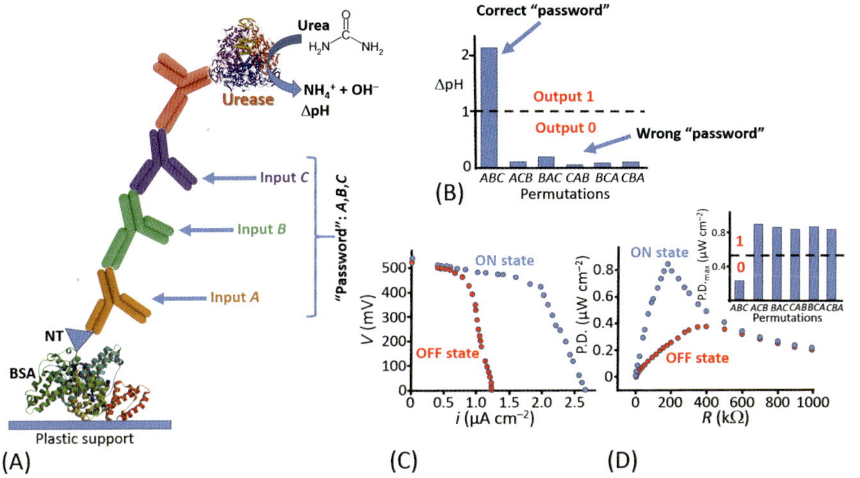

Figure 13.14 (A) The multicomponent immune recognition system assembled upon application of the antibody signals in the correct *A*, *B*, *C* sequence. The last antibody attached is functionalized with the urease biocatalytic tag. Its reaction (hydrolysis of urea) results in the formation of ammonia and pH increase. (B) The pH increase as the function of the order (permutation) of the antibody application. Only the correct order (*A*, *B*, *C* "password") results in the pH change; all other input orders do not allow the urease immobilization and do not result in the pH change. The polarization function (C) and power release (D) measured on a variable resistance in the active ON and inhibited OFF states of the biofuel cell (see Figure 13.12 for the biofuel cell schematics; the solutions with variable pH values were used to change the biofuel cell performance). The inset shows the maximum power release by the biofuel cell dependent on the order (permutation) of the immune signals applied.

pH increase upon the biocatalytic formation of ammonia in the presence of urea. The pH changes generated in the system were measured after application of the antibody signals in different permutations (Figure 13.14B). Only the correct sequence of the antibodies added according to the "password" *A*, *B*, *C* resulted in the attachment of the urease-labeled antibody, thus producing $\Delta pH > 2$ (starting from pH = 4.2), while all incorrect permutations of the antibody signals yielded $\Delta pH < 0.3$.

The solutions generated by the immune recognition system were added to a cathodic compartment of a switchable biofuel cell. A simple model biofuel cell was composed of two indium tin oxide (ITO) electrodes (Figure 13.12). The cathode was modified with a pH-switchable P4VP brush operating in the presence of $K_3[Fe(CN)_6]$ used as a model oxidizer. When pH < 4.5 is applied, the pyridine groups in the polymer brush are protonated yielding the positively charged swollen hydrophilic thin film permeable for anionic redox species, $[Fe(CN)_6]^{3-}$, thus allowing their electrochemical reaction [39]. The polymer brush being in the nonprotonated hydrophobic shrunk state at pH > 5.5 is impermeable for $[Fe(CN)_6]^{3-}$, thus inhibiting the cathodic reaction [63]. The anode was an unmodified ITO-electrode operating in the presence of soluble glucose oxidase (GOx) that oxidized the glucose fuel with the help of a diffusional redox mediator methylene blue, in phosphate buffer, pH 7.0, under Ar. The electrodes were separated with a Nafion membrane to allow the pH change in the cathodic compartment without affecting pH in the anodic solution. The oversimplified design of the biofuel cell was specially selected to demonstrate clearly the power control by the immune-based keypad lock system without any complications from secondary effects related to the bioelectrocatalytic reactions. The experiment was started at pH 4.2 in the cathodic solution when the switchable cathode was in the ON state and the entire biofuel cell was active demonstrating a high current–voltage output (Figure 13.14C) (ON state) and high power release (Figure 13.14D) (ON state). Addition of the solutions with almost unchanged pH produced by the immune system upon "wrong" permutations of the antibody signals did not affect the biofuel cell performance preserving its power output. The solution with the increased pH value of ca. 6.6 obtained upon application of the only correct sequence of the antibody signals (*A*, *B*, *C*) resulted in the restructuring of the pH-switchable polymer brush at the modified cathode, inhibiting the penetration of the $[Fe(CN)_6]^{3-}$ redox species to the cathode surface. This resulted in the decrease of the biofuel cell activity reflected by the smaller current–voltage output (Figure 13.14C) (OFF state) and lower generated power (Figure 13.14D) (OFF state). The power release measured at the external load resistance of 180 kΩ, corresponding to the maximum power produced by the biofuel cell in the active state, was significantly decreased only when the "correct" sequence of the antibodies was applied (*A*, *B*, *C*) and the obtained solution was added to the biofuel cell (Figure 13.14D, inset). This was considered as the "YES" output corresponding to the "opening" of the "lock" upon the correct application of the *A*, *B*, *C*–password.

The designed immune-based keypad lock system demonstrated the IMPLICATION logic function generating the final output "YES" only when the correct order of the antibody input signals is applied (the *A*, *B*, *C* "password"). The "answer" "YES" was obtained in the form of the decreased electrical power produced by the biofuel cell. All other permutations of the antibody inputs did not result in the inhibition of the biofuel cell, preserving it in the initial active state, thus "denying" the access to the locked information. The designed security system operated without the need of any external power source producing electrical power itself. It should be noted that the present example device is aiming at the concept demonstration only, while the real operating system should be based on a microfluidic system (lab-on-a-chip) allowing for its miniaturized design. Conceptually similar biomolecular keypad lock, but based on the enzyme-biocatalyzed reaction cascade instead of the sequence of the immune-recognition reactions, was designed and interfaced with a switchable biofuel cell [84].

13.4 Conclusions and Perspectives

Extensive work performed in the area of switchable electrode interfaces has led to numerous systems controlled by a large variety of physical and chemical signals. Particularly important results are expected when the research advances in signal-responsive materials, modified electrodes, and chemical computing are integrated into a new research area. Further scaling up the complexity of the chemical information processing systems will allow the next level of sophistication when switchable electrodes are controlled by various chemical processes. Particularly interesting would be combination of biomolecular computing systems and switchable modified electrodes. Coupling of "smart" switchable electrodes with sophisticated multistep biochemical pathways could be envisaged in the continuing research. Integration of the signal-responsive electrodes with the information processing systems might be used to develop "smart" multi-signal-responsive biosensors and biofuel cells controlled by complex biochemical environment. The biochemically/physiologically controlled switchable electrodes will operate as an interface between biological and electronic systems in future micro-/nano-robotic devices.

Switchable electrodes controlled by signal-processing enzyme-based logic systems were already integrated in "smart" biofuel cells producing electrical power dependent on complex variations of biochemical signals [48, 66]. Enzyme biocatalytic [67, 68] and immune biorecognition [69] systems have been developed to control performance of switchable biofuel cells. Future implantable biofuel cells producing electrical power on-demand depending on physiological conditions are feasible as the result of the present research [85–87]. Preliminary experiments with a biofuel cell implanted in snails (Figure 13.15A) demonstrated the electrical output dependent on the physiological conditions, e.g., giving food to the snail resulted in the increasing concentration of glucose in the hemolymph, thus increasing the output voltage of the implanted biofuel cell (Figure 13.15B).

Further development of sophisticated enzyme-based biocomputing networks will be an important phase in the development of "smart" bioelectronic devices. Scaling up the complexity of biocomputing system controlling biofuel cell activity will be achieved by networking immune- and enzyme-based logic gates

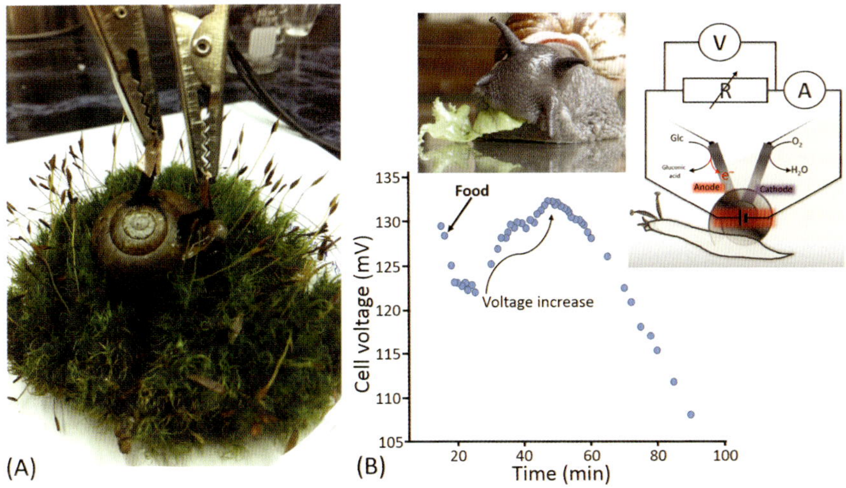

Figure 13.15 (A) The biofuel cell implanted in a snail. (B) The biofuel cell voltage change in time. Note the voltage increase after the snail eats. The insets show the snail eating and the biofuel cell schematics. Source: Halámková et al. 2012 [85]. Adapted with permission of American Chemical Society.

responding to a large variety of biochemical signals. Biofuel cells switchable by enzyme-based [84] or immunosystem-based [83] keypad lock systems have been designed to operate as self-powered biomolecular information security systems. The correct biomolecular "password" introduced into the keypad lock as a sequence of biomolecular input signals resulted in the activation (or inhibition) of the biofuel cell, while all other "wrong" permutations of the molecular inputs preserved the original OFF (or ON) state of the biofuel cell. Further research directed to the increasing stability and robustness of the information processing biocatalytic electrodes could result in many practical applications, including bioelectrocatalytic barcode generation using electrodes characteristic of biofuel cells [88] (see Section 10.3.2). The present developments and future expectations are based on the application of a multidisciplinary approach that will require further collaborative contribution from electrochemists and specialists in materials science and unconventional molecular and biomolecular computing.

References

1 Atanassov, P., Johnson, G., and Luckarift, H. (eds.) (2014). *Enzymatic Fuel Cells: From Fundamentals to Applications*. Weinheim: Wiley-VCH.
2 Cracknell, J.A., Vincent, K.A., and Armstrong, F.A. (2008). *Chem. Rev.* 108: 2439–2461.
3 Luz, R.A.S., Pereira, A.R., de Souza, J.C.P. et al. (2014). *ChemElectroChem* 1: 1751–1777.
4 Yu, E.H. and Scott, K. (2010). *Energies* 3: 23–42.
5 Meredith, M.T. and Minteer, S.D. (2012). *Annu. Rev. Anal. Chem.* 5: 157–179.
6 Leech, D., Kavanagh, P., and Schuhmann, W. (2012). *Electrochim. Acta* 84: 223–234.
7 Koushanpour, A., Gamella, M., and Katz, E. (2017). *Electroanalysis* 29: 1602–1611.
8 Rasmussen, M., Abdellaoui, S., and Minteer, S.D. (2016). *Biosens. Bioelectron.* 76: 91–102.
9 Rincón, R.A., Lau, C., Luckarift, H.R. et al. (2011). *Biosens. Bioelectron.* 27: 132–136.
10 Cosnier, S., Gross, A.J., Le Goff, A., and Holzinger, M. (2016). *J. Power Sources* 325: 252–263.
11 Santoro, C., Arbizzani, C., Erable, B., and Ieropoulos, I. (2017). *J. Power Sources* 356: 225–244.
12 Tharali, A.D., Sain, N., and Osborne, W.J. (2016). *Front. Life Sci.* 9: 252–266.
13 Mathuriya, A.S. and Yakhmi, J.V. (2016). *Crit. Rev. Microbiol.* 9: 127–143.
14 Dudzik, J., Chang, W.C., Kannan, A.M. et al. (2013). *Biofabrication* 5: 035009.
15 Shao, M.L., Poller, S., Sygmund, C. et al. (2013). *Electrochem. Commun.* 29: 59–62.
16 Chen, J., Bamper, D., Glatzhofer, D.T., and Schmidtke, D.W. (2015). *J. Electrochem. Soc.* 162: F258–F264.
17 Mano, N. and de Poulpiquet, A. (2018). *Chem. Rev.* 118: 2392–2468.

18 Le Goff, A., Holzinger, M., and Cosnier, S. (2015). *Cell. Mol. Life Sci.* 72: 941–952.
19 Gamella, M., Koushanpour, A., and Katz, E. (2018). *Bioelectrochemistry* 119: 33–42.
20 Cosnier, S., Gross, A.J., Giroud, F., and Holzinger, M. (2018). *Curr. Opin. Electrochem.* 12: 148–155.
21 Willner, I. and Katz, E. (2000). *Angew. Chem. Int. Ed.* 39: 1180–1218.
22 Ghindilis, A.L., Atanasov, P., and Wilkins, E. (1997). *Electroanalysis* 9: 661–674.
23 Moehlenbrock, M.J. and Minteer, S.D. (2008). *Chem. Soc. Rev.* 37: 1188–1196.
24 Reuillard, B., Abreu, C., Lalaoui, N. et al. (2015). *Bioelectrochemistry* 106, Part A: 73–76.
25 Gonzalez-Guerrero, M.J., del Campo, F.J., Esquivel, J.P. et al. (2017). *Biosens. Bioelectron.* 90: 475–480.
26 Mark, A.G., Suraniti, E., Roche, J. et al. (2017). *Lab Chip* 17: 1761–1768.
27 Heller, A. (2004). *Phys. Chem. Chem. Phys.* 6: 209–216.
28 Chen, T., Barton, S.C., Binyamin, G. et al. (2001). *J. Am. Chem. Soc.* 123: 8630–8631.
29 Gamella, M., Lee, I., Guz, N. et al. (2018). *Int. J. Unconventional Computing* 14: 27–41.
30 Rapoport, B.I., Kedzierski, J.T., and Sarpeshkar, R. (2012). *PLoS One* 7: e38436.
31 Navaee, A. and Salimi, A. (2018). *J. Electroanal. Chem.* 815: 105–113.
32 Sakai, K., Kitazumi, Y., Shirai, O. et al. (2017). *ACS Catal.* 7: 5668–5673.
33 Zhou, M. and Wang, J. (2012). *Electroanalysis* 24: 197–209.
34 Barton, S.C., Gallaway, J., and Atanassov, P. (2004). *Chem. Rev.* 104: 4867–4886.
35 Sales, F.C.P.F., Iost, R.M., Martins, M.V.A. et al. (2013). *Lab Chip* 13: 468–474.
36 Katz, E. (2015). *Bioelectron. Med.* 2: 1–12.
37 Katz, E. and MacVittie, K. (2013). *Energy Environ. Sci.* 6: 2791–2803.
38 Shleev, S. (2017). *ChemPlusChem* 82: 522–539.
39 Heller, A. (2006). *Anal. Bioanal.Chem.* 385: 469–473.
40 Katz, E. (2018). *Signal-Switchable Electrodes and Electrochemical Systems: Materials, Methods, and Applications*. Weinheim: Wiley-VCH.
41 Katz, E. (2018). *Electroanalysis* 30: 759–797.
42 Katz, E. (2016). *Electroanalysis* 28: 1916–1929.
43 Katz, E. (2016). *Electroanalysis* 28: 904–919.
44 Katz, E., Fernández, V.M., and Pita, M. (2015). *Electroanalysis* 27: 2063–2073.
45 Katz, E., Minko, S., Halámek, J. et al. (2013). *Anal. Bioanal.Chem.* 405: 3659–3672.
46 Zang, Y., Zhu, H.M., and Xue, H.G. (2018). *Electrochim. Acta* 259: 676–684.
47 Chen, Y., Gai, P.P., Xue, J.J. et al. (2015). *Biosens. Bioelectron.* 74: 142–149.
48 Katz, E. (2010). *Electroanalysis* 22: 744–756.
49 Katz, E. and Willner, I. (2003). *J. Am. Chem. Soc.* 125: 6803–6813.
50 Tam, T.K., Ornatska, M., Pita, M. et al. (2008). *J. Phys. Chem. C* 112: 8438–8445.
51 Tam, T.K., Pita, M., Motornov, M. et al. (2010). *Electroanalysis* 22: 35–40.

52 Tam, T.K., Pita, M., Trotsenko, O. et al. (2010). *Langmuir* 26: 4506–4513.
53 Motornov, M., Tam, T.K., Pita, M. et al. (2009). *Nanotechnology* 20: 434006.
54 Tam, T.K., Pita, M., Motornov, M. et al. (2010). *Adv. Mater.* 22: 1863–1866.
55 Pita, M., Minko, S., and Katz, E. (2009). *J. Mater. Sci. – Mater. Med.* 20: 457–462.
56 Privman, M., Tam, T.K., Bocharova, V. et al. (2011). *ACS Appl. Mater. Interfaces* 3: 1620–1623.
57 Bocharova, V., Tam, T.K., Halámek, J. et al. (2010). *Chem. Commun.* 46: 2088–2090.
58 Katz, E., Bocharova, V., and Privman, M. (2012). *J. Mater. Chem.* 22: 8171–8178.
59 Tam, T.K., Zhou, J., Pita, M. et al. (2008). *J. Am. Chem. Soc.* 130: 10888–10889.
60 Katz, E. and Privman, V. (2010). *Chem. Soc. Rev.* 39: 1835–1857.
61 Katz, E. (2015). *Curr. Opin. Biotechnol.* 34: 202–208.
62 Pita, M. and Katz, E. (2009). *Electroanalysis* 21: 252–260.
63 Privman, M., Tam, T.K., Pita, M., and Katz, E. (2009). *J. Am. Chem. Soc.* 131: 1314–1321.
64 Wang, X., Zhou, J., Tam, T.K. et al. (2009). *Bioelectrochemistry* 77: 69–73.
65 Pita, M., Privman, M., and Katz, E. (2012). *Top. Catal.* 55: 1201–1216.
66 Katz, E. and Pita, M. (2009). *Chem. Eur. J.* 15: 12554–12564.
67 Amir, L., Tam, T.K., Pita, M. et al. (2009). *J. Am. Chem. Soc.* 131: 826–832.
68 Tam, T.K., Pita, M., Ornatska, M., and Katz, E. (2009). *Bioelectrochemistry* 76: 4–9.
69 Tam, T.K., Strack, G., Pita, M., and Katz, E. (2009). *J. Am. Chem. Soc.* 131: 11670–11671.
70 Katz, E. (2015). Biocatalytic electrodes and biofuel cells controlled by biomolecular signals and implantable biofuel cells operating in vivo – towards bioelectronic devices integrating biological and electronic systems. In: *Electrochemistry*, Specialist Periodical Reports, vol. 13 (ed. C. Banks, R. Mortimer and S. McIntosh), 170–209. Royal Society of Chemistry.
71 Katz, E. (2015). Switchable electrodes and biofuel cells logically controlled by chemical and biochemical signals. In: *Handbook of Bioelectronics: Directly Interfacing Electronics and Biological Systems*, Chapter 17 (ed. K. Iniewski and S. Carrara), 215–238. Cambridge University Press.
72 Mailloux, S., MacVittie, K., Privman, M. et al. (2014). *ChemElectroChem* 1: 1822–1827.
73 Katz, E., Bocharova, V., and Halámek, J. (2014). Switchable electrodes and biofuel cells. In: *Enzymatic Fuel Cells: From Fundamentals to Applications*, Chapter 18 (ed. P. Atanassov, G. Johnson and H. Luckarift), 374–421. Weinheim: Wiley-VCH.
74 Minko, S. (2006). *Polym. Rev.* 46: 397–420.
75 Luzinov, I., Minko, S., and Tsukruk, V. (2004). *Prog. Polym. Sci.* 29: 635–698.
76 Cohen Stuart, M.A., Huck, W.T.S., Genzer, J. et al. (2010). *Nat. Mater.* 9: 101–113.
77 Motornov, M., Sheparovych, R., Katz, E., and Minko, S. (2008). *ACS Nano* 2: 41–52.

78 Bartlett, P.N., Tebbutt, P., and Whitaker, R.C. (1991). *Prog. React. Kinet.* 16: 55–155.

79 Privman, V., Arugula, M.A., Halámek, J. et al. (2009). *J. Phys. Chem. B* 113: 5301–5310.

80 Melnikov, D., Strack, G., Pita, M. et al. (2009). *J. Phys. Chem. B* 113: 10472–10479.

81 Privman, V., Strack, G., Solenov, D. et al. (2008). *J. Phys. Chem. B* 112: 11777–11784.

82 Strack, G., Chinnapareddy, S., Volkov, D. et al. (2009). *J. Phys. Chem. B* 113: 12154–12159.

83 Halámek, J., Tam, T.K., Chinnapareddy, S. et al. (2010). *J. Phys. Chem. Lett.* 1: 973–977.

84 Halámek, J., Tam, T.K., Strack, G. et al. (2010). *Chem. Commun.* 46: 2405–2407.

85 Halámková, L., Halámek, J., Bocharova, V. et al. (2012). *J. Am. Chem. Soc.* 134: 5040–5043.

86 Szczupak, A., Halámek, J., Halámková, L. et al. (2012). *Energy Environ. Sci.* 5: 8891–8895.

87 MacVittie, K., Halámek, J., Halámková, L. et al. (2013). *Energy Environ. Sci.* 6: 81–86.

88 Strack, G., Luckarift, H.R., Nichols, R. et al. (2011). *Chem. Commun.* 47: 7662–7664.

14

Bioelectronic Interface Between Enzyme-Based and DNA-Based Computing Systems

14.1 Introduction: Interfacing Enzyme-Based and DNA-Based Computing Systems Is a Challenging Goal

In the research area of biomolecular computing, DNA computing [1–7] and enzyme-based computing [8, 9] have received exceptional attention representing two most important subareas of unconventional computing. However, these subareas based on the use of different biomolecular species performing different reactions are usually isolated from each other, rarely operating together in one integrated system. They represent different approaches to chemical information processing, having different advantages and disadvantages. DNA computing is believed to be a potential alternative to electronic computers [10, 11] for some computational tasks, due to the advantage of massive parallel data processing [12], a straightforward design of relatively complex circuits [13], and affordability. Among the most obvious applications of DNA-based logic circuits is the analysis of genetic alterations that can be transformed into clinical testing of infectious and genetic diseases [14–18]. Despite the advances in the development of *in vitro* selection, functional DNAs are still limited in the diversity and efficiency of catalytic reactions and are inferior to proteins in terms of affinity and diversity of ligands that DNA can recognize [19, 20]. At the same time, enzymes are proven to be selective and sensitive receptors; they are known as the best catalysts, allowing rate enhancement up to 10^{17}-fold in comparison with uncatalyzed reactions [21]. However, enzyme-based computing is experimentally limited to the systems mimicking operation of only a few concatenated logic gates [8, 9], and the network complexity was restricted by enzymes cross-reactivity and noise build [22–24]. Combining enzyme and DNA computational systems in communicating enzyme–DNA circuits may enable (i) highly selective recognition of a diverse spectrum of biological molecules or disease markers, (ii) catalytic signal amplification, (iii) massive parallel data processing, and (iv) complex computational information processing, particularly for biologically generated signals. So far, mixed enzyme-DNA computational systems have been limited to those that involve enzymes directly acting on DNA, e.g., DNA polymerases, ligases, endonuclease, etc. [25–27]. However, DNA processing enzymes cannot detect such disease biomarkers as small biological molecules, sugars, proteins,

Enzyme-Based Computing Systems, First Edition. Evgeny Katz.

etc. On the other hand, biocomputing systems based on general enzymes (not related to DNA) were successfully used for logic processing and binary sensing of various combinations of physiological biomarkers in the binary YES/NO format [28–31]. Therefore, a universal interface for connecting enzymatic logic gates with DNA information processing circuits is needed. The present chapter introduces the very first approach to enzyme/DNA interfaces, where the primary chemical signals are processed by an enzyme logic system, and then the produced intermediate output signal activates a DNA logic system operating as a downstream computing element. The whole system operates as an integrated enzyme/DNA-based computing device [32]. The enzyme logic system as well as the DNA computing part can be represented by systems of different complexity, including very sophisticated molecular "devices" performing reversible logic operations [33].

14.2 Bioelectronic Interface Transducing Logically Processed Signals from an Enzymatic System to a DNA System

Figure 14.1 shows schematically a bioelectronic interface that recognizes NADH, which is produced as an output of an enzymatic system, and releases a DNA oligonucleotide, which can be processed by a downstream DNA computing system as an input [32]. The interface is based on two modified electrodes. The first electrode communicating with the enzyme computing system (sensing electrode at the left) was coated with adsorbed polyethyleneimine (PEI) and pyrroloquinoline quinone (PQQ) covalently attached to the PEI thin film [34]. The immobilized PQQ served as a catalyst for electrochemical oxidation of the biocatalytically produced NADH [35]. This process resulted in the formation of a negative potential of ca. −60 mV (vs. Ag/AgCl) and the corresponding current sufficient for reduction of Fe^{3+} in the Fe^{3+}-cross-linked alginate film on the second connected electrode (releasing electrode at the right) [36–39]. Note that Fe^{2+} cations are not capable of alginate cross-linking and their formation results in the alginate thin-film dissolution and concomitant release of the entrapped DNA molecules (see Chapter 12 for more explanations). The released DNA was analyzed optically (note that the DNA was labeled with a fluorescent dye) and used for activating the downstream DNA computing process. The final output generated by the DNA computing was also analyzed optically.

In this study [32], two enzyme logic systems have been employed, either of which produced NADH (Figure 14.2). For binary operation of the enzyme systems, digital input **0** was defined as the absence of the corresponding substrates, whereas digital input **1** was defined as experimentally optimized concentrations of the substrates. The first system (Figure 14.2A) operated as a cascade of reactions catalyzed by three enzymes – maltose phosphorylase (MPh), hexokinase (HK), and glucose-6-phosphate dehydrogenase (G6PDH). It mimicked three concatenated Boolean AND logic gates (Figure 14.2C), and the high output signal (production of NADH) was observed only in the

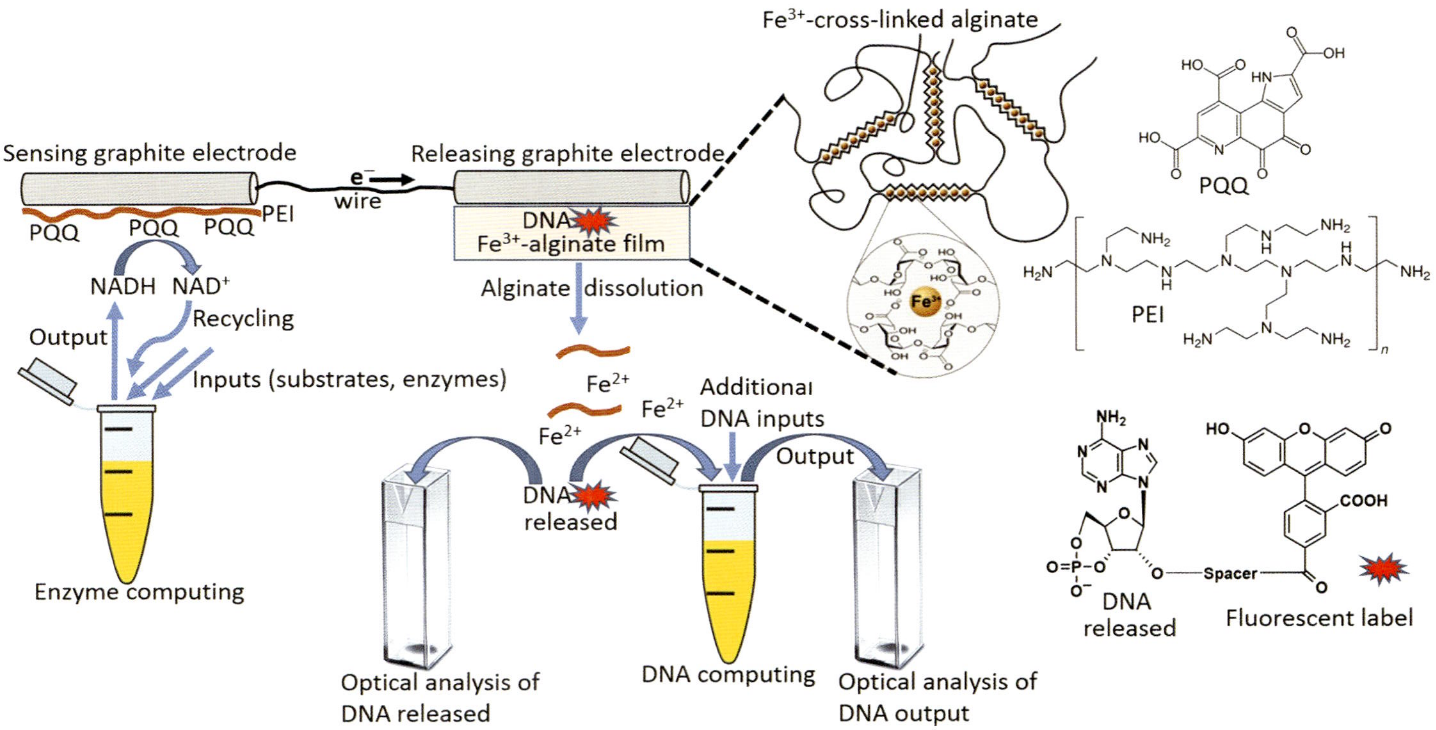

Figure 14.1 General scheme of the system operation: the enzyme computing system produces NADH as an output, which is oxidized on a PQQ-modified electrode generating a negative potential on the electrode. This sensing electrode is electrically connected to an Fe^{3+}-cross-linked alginate-modified electrode. When the negative potential is generated on the sensing electrode, it results in electrochemical reduction of Fe^{3+} to Fe^{2+} at the alginate electrode, thus leading to the alginate layer dissolution and release of the entrapped DNA. The released DNA labeled with a fluorescent dye was optically detected in the solution and used as an input for the downstream DNA computing system.

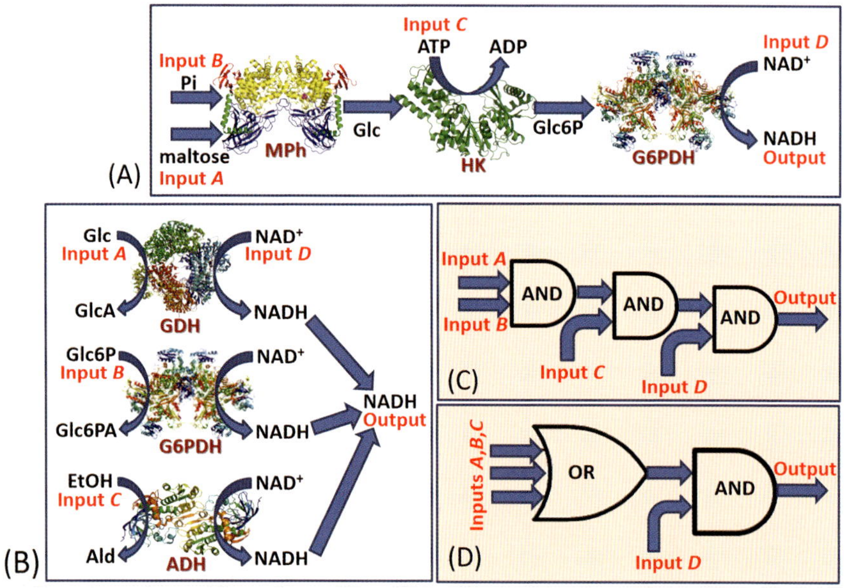

Figure 14.2 Two enzyme systems used in this study and their corresponding logic schemes. (A) A cascade of three AND gates made of maltose phosphorylase (MPh), hexokinase (HK), and glucose-6-phosphate dehydrogenase (G6PDH). The biocatalytic reaction of MPh was activated in the presence of maltose (Input *A*) and inorganic phosphate (Pi, Input *B*) resulting in glucose (Glc) and glucose-1-phosphate by-product formation. In the next reaction step catalyzed by HK, Glc is converted to glucose-6-phosphate (Glc6P) in the presence of ATP (Input *C*). Finally, Glc6P reduces NAD^+ (Input *D*) to NADH in the process biocatalyzed by G6PDH. Overall, the NADH production is only possible in the presence of all 4 input signals activating the enzyme-based system. (B) A combination of three parallel reactions biocatalyzed by three NAD^+-dependent enzymes: glucose dehydrogenase (GDH), G6PDH, and alcohol dehydrogenase (ADH). Each biocatalytic reaction was activated by the corresponding substrate: Glc (Input *A*), Glc6P (Input *B*), and ethanol (EtOH) (Input *C*). The NAD^+ cofactor (Input *D*) was needed for all reactions; thus none of them could proceed in the absence of NAD^+. (C, D) The logic schemes corresponding to the biocatalytic cascades are shown in (A) and (B), respectively. Source: Mailloux et al. 2015 [32]. Adapted with permission of John Wiley & Sons.

presence of all four input substrates (see legend to Figure 14.2). The second system (Figure 14.2B) operated as a 3-input OR gate connected to an AND gate (Figure 14.2D). For this system, the NADH production was activated in the presence of any of substrate inputs *A*, *B*, *C* with the mandatory presence of NAD^+ (input *D*). Upon enzymatic formation of NADH using any of the systems, it reacted with the PQQ electrode producing a negative potential and reoxidized back to the NAD^+ state (Figure 14.1).

Figure 14.3 demonstrates the correct digital behavior of the enzymatic logic gate systems interacting with the PQQ electrode. The PQQ-modified sensing electrode was coupled with a reference electrode to measure the potential biocatalytically produced on the electrode (Figure 14.3A). A negative potential of ca. −60 mV (digital **1**) was achieved when NADH was produced by either of the enzyme logic systems. Otherwise, the potential less negative than −10 mV (digital **0**) was measured. For the enzyme logic system mimicking three concatenated

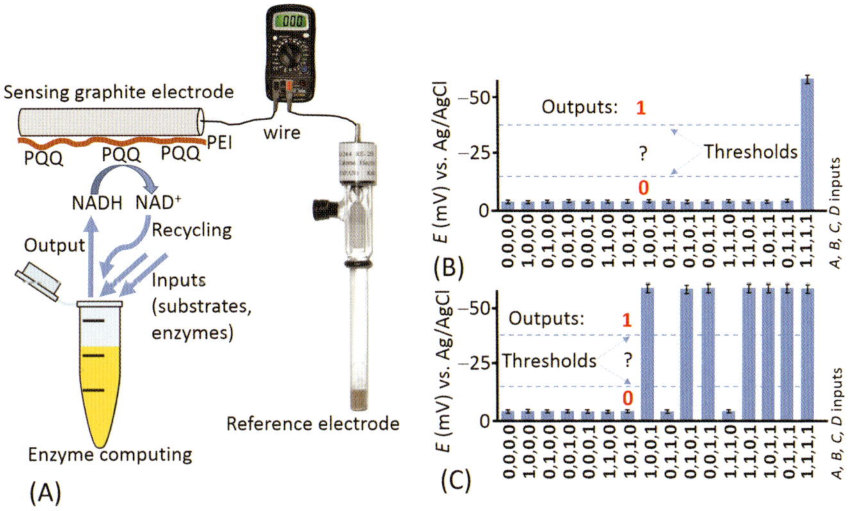

Figure 14.3 (A) Potential measurements on the sensing electrode – general scheme. (B, C) Electric potentials generated on the PQQ-modified electrode interfaced with the biocatalytic systems shown in Figure 14.2A,B, respectively, when different combinations of input signals were applied. The bars show the potential values achieved after 30 minutes of exposing the PQQ-modified electrode to the enzyme systems. The data are average of three independent experiments. The potential produced on the PQQ-modified electrode has a logarithmic dependence on the NADH concentration (according to the Nernst equation), thus resulting in very small variations of the measured potentials. Threshold lines separate logic output **0**, undefined area, and logic output **1**.

AND gates (Figure 14.2A,C), logic output **1** (ca. −60 mV) was measured only in the presence of all reacting input species (input combination **1,1,1,1**). All other input combinations (15 different variants) resulted in the electrode potential less negative than −10 mV, digital **0** (Figure 14.3B). Operation of the enzyme system mimicking a 3-input OR gate followed by an AND gate (Figure 14.2B,D) resulted in output **1** (ca. −60 mV) generated in the following input combinations (*A*, *B*, *C*, *D*) – **0,0,1,1**; **0,1,0,1**; **1,0,0,1**; **0,1,1,1**; **1,0,1,1**; **1,1,0,1**; **1,1,1,1** – while all other input combinations resulted in output signal **0** (Figure 14.3C). The generated potential output patterns (Figure 14.3B,C) correspond to the expected operation of the enzyme logic systems.

To enable transfer of the output signal produced by the enzyme computing systems (Figure 14.2) into a DNA input signal, the PQQ electrode was connected to another electrode (Figure 14.1), which was coated with Fe^{3+}-cross-linked alginate film entrapping a fluorescently labeled DNA oligonucleotide output (DNA1: FITC-5′-TGC AGA CGT TGA AGG ATC CTC). Generation of the negative potential on the PQQ electrode resulted in subsequent reduction of Fe^{3+} into Fe^{2+} at the alginate-coated electrode. It triggered the alginate film dissolution and release of DNA1. It was observed that when the potential of ca. −60 mV (digital **1** output of the enzymatic computing systems) was applied to the second electrode, the alginate film was substantially degraded. At the same time, no visible changes in the film structure were observed at the potential of ca. −5 mV

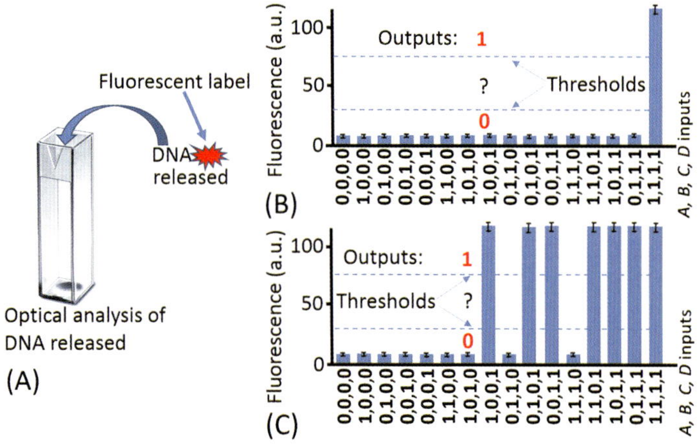

Figure 14.4 (A) Optical analysis of the DNA released (note that the DNA was labeled with a fluorescent dye). (B, C) Fluorescence signal corresponding to the dye-labeled DNA released from the alginate thin film when the PQQ-modified electrode was interfaced with the enzymatic logic gate systems shown in Figure 14.2A,B, respectively, when different combinations of input signals were applied. The bars show the fluorescence measured after 30 minutes of exposing the electrodes to the enzyme systems. The fluorescence is represented by normalized arbitrary values. The data are average of three independent experiments. Threshold lines separate logic output **0**, undefined area, and logic output **1**.

(digital **0** output of the enzyme computing systems), on the same experimental timescale. Fluorescent signal of the solution containing released DNA1 was measured in the presence of different combinations of enzymatic system inputs (Figure 14.4). As expected, high fluorescence (digital **1**) was registered upon the electrochemically stimulated release of DNA1 in the presence of NADH. When no NADH was produced, the fluorescent signal remained low (digital **0**). The concentrations of the released (digital **1**) and leaking (digital **0**) DNA1 were reaching ca. 5 nM and 0.8 nM, respectively, after 30 minutes. This result shows significant discrimination between the leakage and stimulated release of DNA1 entrapped in the alginate-modified interface. It should be noted that there is a perfect correlation between the output signals produced by the enzymatic systems in the form of the potentials (Figure 14.3) with the fluorescence of released DNA1 (Figure 14.4). In other words, the enzyme-generated output was consistently converted into DNA1, which served as an input for DNA computing as detailed below.

Oligonucleotide DNA1 released by the interface was recognized by a 3-input deoxyribozyme AND gate (3-AND) (Figures 14.5 and 14.6). Deoxyribozyme logic gates controlled by DNA oligonucleotide inputs are most well-developed DNA logic constructs up to date [40–47]. Indeed, such gates can be assembled in automaton that plays tic-tac-toe game with a human [41]; they can be organized in multilayer computational cascades [43–46] and a molecular calculator with seven-segment digital display [47]. The design of 3-AND takes advantage of the concept of split (binary) deoxyribozyme sensors [48–51] and consists of two DNA strands folded in the stem-loop structures DNA-a and DNA-b

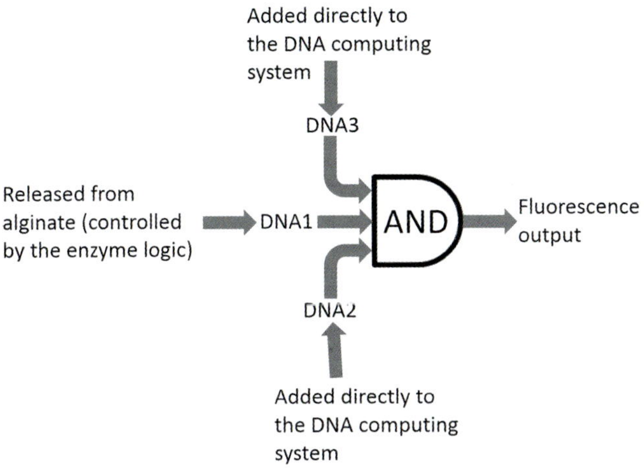

Figure 14.5 The general scheme illustrating the DNA-based 3-AND logic gate operation. More details are shown in Figure 14.6.

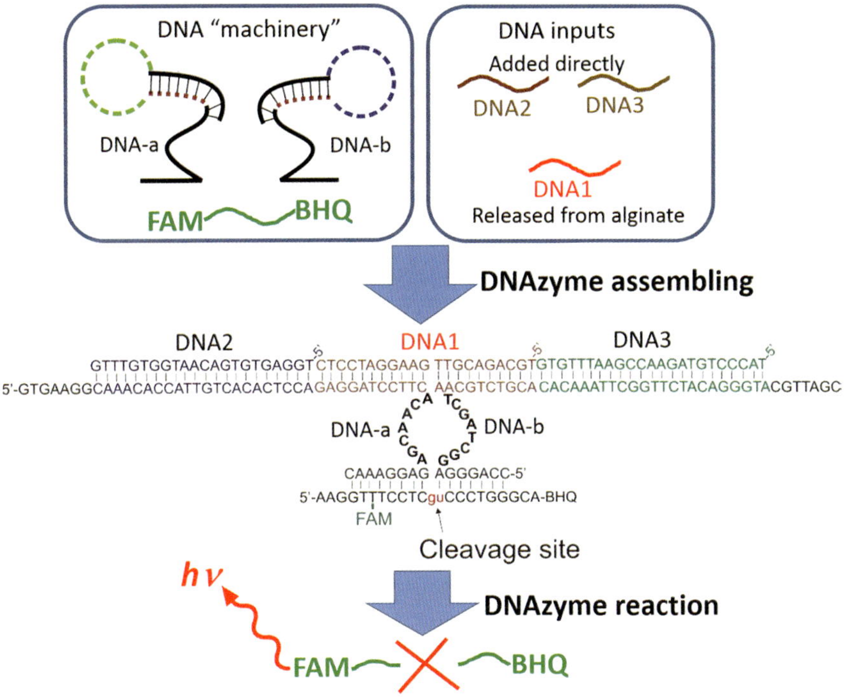

Figure 14.6 Principal scheme of a three-input deoxyribozyme AND gate. Strands DNA-a and DNA-b of the gate are dissociated in the absence of inputs. Dashed lines indicate the input recognition fragments of the strands. The catalytic deoxyribozyme complex was formed in the presence of all three inputs (DNA1, DNA2, and DNA3). Note that DNA1 was released from the alginate electrode in the process stimulated by the enzyme logic system, while two other DNA inputs (DNA2 and DNA3) were added directly to the reacting solution performing the DNA computing operation. The DNAzyme catalytic core cleaves the fluorophore- and quencher-labeled FAM-DNA-BHQ substrate and increases fluorescent signal. Note that in this experiment DNA1 is not fluorescently labeled.

(Figure 14.6). The strands are dissociated in the absence of input oligonucleotides (i.e., in the absence of DNA1, DNA2, and DNA3 inputs). However, hybridization of two oligonucleotide inputs to the loop fragments opens the hairpins and the third input bridges DNA-a and DNA-b strands, which results in the formation of a catalytic core (Figure 14.6). The deoxyribozyme cleaves a fluorophore- and quencher-labeled substrate (FAM-DNA-BHQ in Figure 14.6), thus producing high fluorescence. In this study, DNA1, an output of the enzyme/DNA interface, was used as a bridging input for DNA-a and DNA-b. In experiments with the 3-AND gate, DNA1 without fluorescent label was used (otherwise, the release system was operating in the way detailed previously). Two other inputs (DNA2 and DNA3) mimicked the sequences of microRNAs shown to be promising molecular markers of human cancers [52].

In order to activate the DNA computing system (3-AND gate), DNA1 input was released from the alginate film upon its dissolution stimulated by the enzyme logic system, while two other DNA inputs (DNA2 and DNA3) were added directly to the reacting solution (Figure 14.7A). According to the truth table (Figure 14.7B), the 3-AND gate produces high fluorescence output (digital **1**) only in the presence of all 3 inputs. In the performed experiments, the DNA1 input was produced *in situ* by the stimulated release (digital **1**) or leakage (digital **0**) from the alginate-modified electrode and its concentration was set by the system as a function of logic operation of the enzyme systems. Two other inputs, DNA2 and DNA3, were either used in concentration of 10 nM or absent for digital **1** and **0**, respectively.

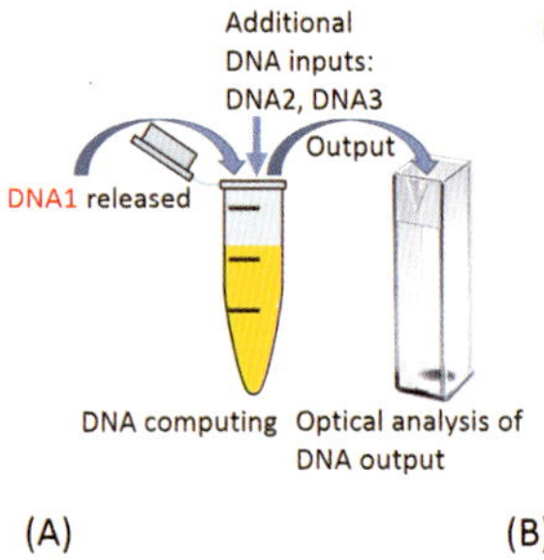

(B)

Added DNA inputs		Released DNA input	
DNA2	DNA3	DNA1	Output
0	0	0	0
1	0	0	0
0	1	0	0
0	0	1	0
1	1	0	0
1	0	1	0
0	1	1	0
1	1	1	1

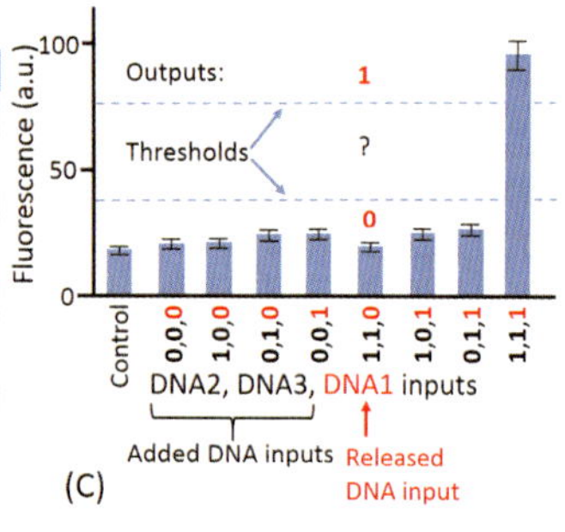

Figure 14.7 Digital performance of the DNA logic gate. (A) General scheme. (B) Truth table. (C) Fluorescent response of DNA logic gate in the presence of all possible DNA input combinations. The concentrations of the inputs were as follows. For low inputs (digital **0**), DNA2, 0 nM, and DNA3, 0 nM; DNA1 was produced *in situ* with the concentration set by the system corresponding to output **0** (when substrate inputs *A*, *B*, *C*, *D* for the enzyme systems were **0,0,0,0**). For high inputs (digital **1**), DNA2, 10 nM, and DNA3, 10 nM; DNA1 was produced *in situ* with the concentration set by the system corresponding to output **1** (when Inputs *A*, *B*, *C*, *D* for the enzyme systems were **1,1,1,1**). Digital values for DNA1 are shown in red. Control bar corresponds to a sample containing only FAM-DNA-BHQ substrate. The bars show the fluorescence measured after 30 minutes of exposing the electrode to the enzyme systems. The fluorescence is represented by normalized arbitrary values. The data are average of three independent experiments. Threshold lines separate logic output **0**, undefined area, and logic output **1**.

The full logic network includes six independent logic inputs, four inputs (*A*, *B*, *C*, *D*) in the enzyme part and two inputs in the DNA part (DNA2 and DNA3; see the oligonucleotide sequences in Figure 14.6), while DNA1 is not an independent input. Therefore, the full truth table includes $2^6 = 64$ variants of logic input combinations. Figure 14.7C shows a simplified representation of the logic process considering only the DNA logic part. Logic value **0** and **1** for the intermediate output/input DNA1 can be realized with various combinations of the enzyme inputs *A*, *B*, *C*, *D*, also being different for two enzyme-based logic systems (Figure 14.2). For simplicity and for minimizing the number of experiments, *A*, *B*, *C*, *D* enzyme inputs were used in combinations **0,0,0,0** and **1,1,1,1** for realizing the DNA1 digital values **0** and **1**, respectively. This simplification is justified by very small signal variations for all combinations of the *A*, *B*, *C*, *D* inputs generating either by output **0** or **1** (Figure 14.4). In other words, the leakage of DNA1 and the release of DNA1 are almost the same regardless of the input combinations resulting in the leakage and release, respectively.

The correct digital response of 3-AND gate was registered at all possible DNA input combinations (Figure 14.7C). Importantly, the high output signal (last bar in Figure 14.7C corresponding to **1,1,1** input combination) could be statistically distinguished from the low output (about fourfold fluorescence increase) obtained for all other input combinations. The fluorescent results were supported by the analysis of the samples by gel electrophoresis [32]. The data proves the expected digital response of 3-AND gate and the possibility of using an electrode-released oligonucleotide for transferring the signal from enzymatic to DNA computational systems.

This study demonstrates the possibility to design an interface that enables communication between enzymatic and DNA-based computing systems. The whole system includes two individual logic subsystems (enzyme based and DNA based) connected electrically to allow the output signal produced by the enzyme logic gates operate as the input signal for the DNA logic gates. The system operated in two distinct steps, first the enzyme logic process and then the DNA logic process, representing the very first realization of the enzyme/DNA logic interface connecting a non-DNA processing enzyme computation system with DNA logic gates. This interface is "universal" because it is compatible with a variety of both enzymatic and DNA molecular logic circuits. NADH communicating between the enzyme system and the interface electrode allows great versatility for the selection of enzymes participating in the biocomputing process, since NADH is produced in a broad variety of reactions. In addition, it is possible to replace NADH with other reducing molecules (e.g., glucose) [53]. The deoxyribozyme gate-based computational systems are also known to show great versatility and complexity [40–47]. The limitations of the interface are the following: (i) The enzyme-based computing system must produce NADH or other reductive species as an output. (ii) The DNA-based computing system must accept nanomolar range concentration of oligonucleotide as an input. However, the amount of the released DNA could be increased if larger electrodes or thicker alginate films are used for the DNA entrapment and release. (iii) In its current design, the signal can be transferred only in one direction: from the enzyme to DNA system. (iv) Only one kind of DNA (or a set of DNA sequences) can be released per electrode pair. More DNA outputs could be released in the controlled way, if a multielectrode array is applied [54]. Despite the limitations, the designed enzyme/DNA system can find some important practical applications. Indeed, the enzymatic and DNA-based computing systems used in this study proved to be relevant to diagnosis of human disorders [1, 17, 18, 55, 56], as well as to very complex information processing [40–47]. The studied system [32] reports on the final outcome results achieved after the system came to the saturation (the end of the process) similarly to most other studies in the field [8, 9, 57–60]. It would be interesting to study the time-dependent output production. The time-dependent outputs were studied experimentally and modeled theoretically for some multistep biocatalytic reactions applied for logic operations [61, 62]. Also time-dependent dissolution of alginate thin film and concomitant release of loaded substances (particularly DNA species) were reported recently [63–65]. The present system includes a number of processes with complex kinetics (biocatalytic cascades, electric potential formation, reductive dissolution of alginate, oligonucleotide release, and finally DNA reactions). Study of the combination of these time-dependent processes and their kinetics could become the subject of challenging investigation. It should be also noted that the alginate film dissolution and subsequent DNA release could be achieved in non-electrochemical systems using direct chemical communication with enzyme logic systems. Similar release processes have been studied using enzyme systems producing H_2O_2 [63–65] or citrate [66, 67] as the final output, thus excluding the electrical interface between enzyme logic gates and DNA computing systems.

14.3 The Bioelectronic Interface Connecting Enzyme-Based Reversible Logic Gates and DNA-Based Reversible Logic Gates: Realization in a Flow Device

While the previous section explains integration of the enzyme and DNA computing systems represented by relatively simple Boolean logic gates (OR, AND) and their simple combinations, the present section demonstrates scaling up the complexity of the computing processes represented by multicomponent/multistep reversible logic operations. The biomolecular realization of reversible logic processes performed in flow devices functionalized with various enzymes is detailed in Chapter 5. Also, Chapter 5 offers discussion on the reversible logic definition. Notably, the reversible systems are only logically reversible, but not physically reversible. The present section is aimed at demonstrating the concerted operation of logically reversible processes performed first by an enzyme-based system and then continued in a DNA-based system activated by enzyme-generated outputs through the bioelectronic interface similar to the one illustrated earlier. The system discussed here integrates *reversible* enzyme logic gate (represented by the Fredkin gate [68] with three-input/three-output information channels) and *reversible* DNA gate (represented by the Feynman gate [69, 70] with two-input/two-output information channels).

The (bio)molecular realizations of the Fredkin [71–74] and Feynman [75–80] gates have been already reported by several research groups, being designed and studied as single (stand-alone) computing units. However, their integration in complex circuits, particularly when one of the units operates with enzymes and the second with DNA molecules, is very challenging, and the present system is a unique biomolecular device performing reversible logic operations with enzymes and DNA molecules in a concerted way. It is particularly important to note that the modularity of the used flow system allows for easy substitution of the used logic gates with alternative logic systems that have different logic functionalities. Overall, the designed system in specific, as well as the approach in general, represents one more step advancing the complexity and operational flexibility of biomolecular systems while working to mimic their electronic computing counterparts. The following sections explain the processes used to mimic the Fredkin gate with enzyme reaction and then Feynman gate with DNA reactions, as well as their integration in the continuous process with the help of a bioelectronic interface.

14.3.1 Enzyme-Based Fredkin Gate Processing Biomolecular Signals Prior to the Bioelectronic Interface

While many sophisticated reversible logic gates (e.g., Toffoli, Peres, and Feynman gates) can be assembled by combining together several basic Boolean operations, such as AND, XOR, etc., the Fredkin gate represents a special kind of logic processes, which cannot be easily realized by combinations of trivial Boolean functions. When the Fredkin gate is assembled from XOR, NOT, and AND logic operations, their unique logic operations translate into a very complicated set of logic combinations, which transpires into the need for complicated circuitry that is nearly impossible to directly mirror enzymatically [81]. In the Fredkin gate, the logic value of Input A controls the directions of Inputs B and C. When the Control Input A has **0** value, Inputs B and C are directly copied to the corresponding Outputs Q and R; otherwise when Input A has **1** value, Inputs B and C switch their directions, and they are copied to Outputs R and Q, respectively. Thus, the Fredkin gate represents a Controlled-Swap (CSWAP) gate with the swapping (exchanging) of the output channels depending on the Control channel. It should be noted, that the logic value read at Output P indicates whether Inputs B and C are directly transposed to Outputs Q and R or switched between these channels (see additional explanations on the Fredkin gate operation in Section 5.6.2 and Figure 5.23 for the truth table and schematics of the Fredkin gate).

The Fredkin gate was recently realized in an enzyme-based system that was assembled from flow cells modified with various enzymes (see Section 5.6.2 and Figures 5.23–5.26). In the present system the same Fredkin gate skeleton was reused [33], yet the input concentrations were re-optimized to benefit the interface operation. The major difference compared to the previous design is in the extension of the Fredkin gate to the electrochemical interface that is connected to the DNA computing system. Control Input A was defined as the presence of glucose (Glc) for its logic value **1**, while the presence of lactate (Lac) was used as the logic value **0**. Data Inputs *B* and *C* were both defined as the presence of NAD^+ for their logic value **1**. The absence of NAD^+ in the input channels (filled with the background solutions only) was used as the logic value **0** for the Inputs *B* and *C*. All three inputs (*A*, *B*, and *C*) were applied in eight possible combinations. The output signals generated by the enzyme-based Fredkin gate are represented by the biocatalytically produced NADH, which appears in the output channels *Q* or *R* depending on the logic value of Input *A*. Output *P* is represented by $ABTS_{ox}$, which is produced upon biocatalytic oxidation of ABTS (the operation of this gate is explained in Section 5.6.2) (Figure 14.8A). The output signals in channels *P*, *Q*, and *R* were analyzed optically (Figure 14.8B).

The real difference from the previously explained Fredkin gate (Section 5.6.2) starts at the next reaction (information processing) step where the enzymatically produced NADH activates the electrochemical system (Figure 14.8C). This system includes two identical electrochemical flow cells, which have inlets connected to output channels *Q* and *R* of the Fredkin gate. Each cell has two graphite electrodes, one modified with PQQ and another coated with Fe^{3+}-cross-linked alginate with the loaded DNA. The only difference was in the sequence of the DNA loaded in the thin alginate film: one alginate electrode was loaded with DNA1 (labeled with a fluorescent dye FAM) and another with DNA2, which are input signals for the next information processing step – DNA-based Feynman gate.

PQQ is a well-known electrocatalyst for oxidation of NADH [35]. When Outputs *Q* or *P* appear at the logic value **1**, thus containing in the solution NADH at concentration of ca. 160 μM, the PQQ-modified electrodes start NADH oxidation, and a negative potential of ca. −70 mV (vs. Ag/AgCl reference) is produced on the active electrode. On the other hand, the PQQ electrodes have ca. 0 mV potential when the solution pumped to the electrochemical cells from the Fredkin gate does not contain NADH (logic value **0**). Since the appearance of NADH in the output solutions is directly translated into the potential formation on the PQQ electrodes, it is obvious that the pattern of the potential outputs measured for different combinations of Inputs *A*, *B*, and *C* repeats the same pattern observed optically for the NADH formation in Outputs *Q* and *R* (compare the bar diagrams in Figure 14.8B,C). In simpler terms, the potentials measured on the PQQ electrodes repeat the same logic as Outputs *Q* and *R* after the Fredkin gate, thus conserving the logic of the outputs but being measured differently. The PQQ electrodes in both electrochemical cells were connected electrically to the alginate-modified electrodes loaded with DNA. Anytime when the PQQ electrode becomes negatively polarized in the presence of NADH, it produces a current resulting in Fe^{3+} reduction in the alginate film on the connected electrode. When Fe^{3+} is reduced to yield Fe^{2+} cations, the alginate film is dissolved, and the loaded DNA is released to the solution (Figure 14.9). This process repeats the pattern of Outputs *Q* and *R* generated by the Fredkin gate for different combinations of primary Inputs *A*, *B*, and *C*, thus translating the biocatalytically produced output signal in the form of NADH to the released DNA operating as an input signal to the downstream DNA computing system. Overall, the negative potential formation on the PQQ electrode and the corresponding alginate dissolution and concomitant DNA

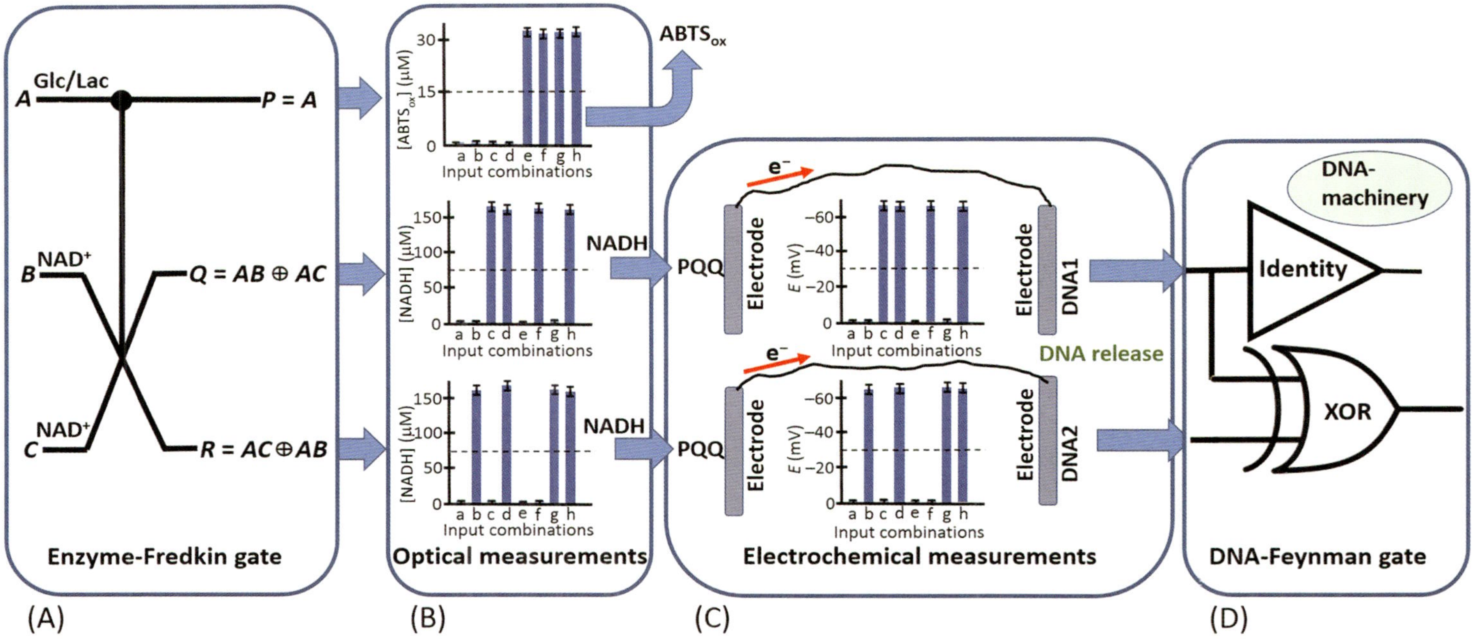

Figure 14.8 The block scheme of the entire system including (A) the enzyme-based Fredkin gate, (C) the electrochemical cells operating as an interface between the enzyme system and DNA system, and (D) the DNA-based Feynman gate composed of the ID and XOR gates operating in parallel. (B) The optically measured outputs (*P*, *Q*, and *R*) generated by the Fredkin gate upon application of various combinations of Inputs *A*, *B*, and *C*. Output *P* corresponding to the formation of $ABTS_{ox}$ was measured at $\lambda = 415$ nm, while Outputs *Q* and *R* corresponding to the production of NADH were measured at $\lambda = 340$ nm. The bars in the diagrams correspond to the following combinations of the *A*, *B*, *C* inputs: (a) **0,0,0**; (b) **0,0,1**; (c) **0,1,0**; (d) **0,1,1**; (e) **1,0,0**; (f) **1,1,0**; (g) **1,0,1**; and (h) **1,1,1**. The optically measured outputs were recalculated to the corresponding concentrations of $ABTS_{ox}$ and NADH. (C) In addition to the schematically shown electrochemical cells composed of the PQQ-modified and alginate/DNA-modified electrodes, the electrochemical outputs are shown. The bars in the diagrams correspond to the following combinations of the *A*, *B*, *C* inputs: (a) **0,0,0**; (b) **0,0,1**; (c) **0,1,0**; (d) **0,1,1**; (e) **1,0,0**; (f) **1,1,0**; (g) **1,0,1**; and (h) **1,1,1**. These outputs were measured as electrical potentials produced on the PQQ electrodes (measured vs. Ag/AgCl reference), and they were consistent with the optically measured production of NADH in Outputs *Q* and *R* of the Fredkin gate. The dash lines in all bar diagrams represent the thresholds separating **0** and **1** output signals. Source: Guz et al. 2016 [33]. Adapted with permission of John Wiley & Sons.

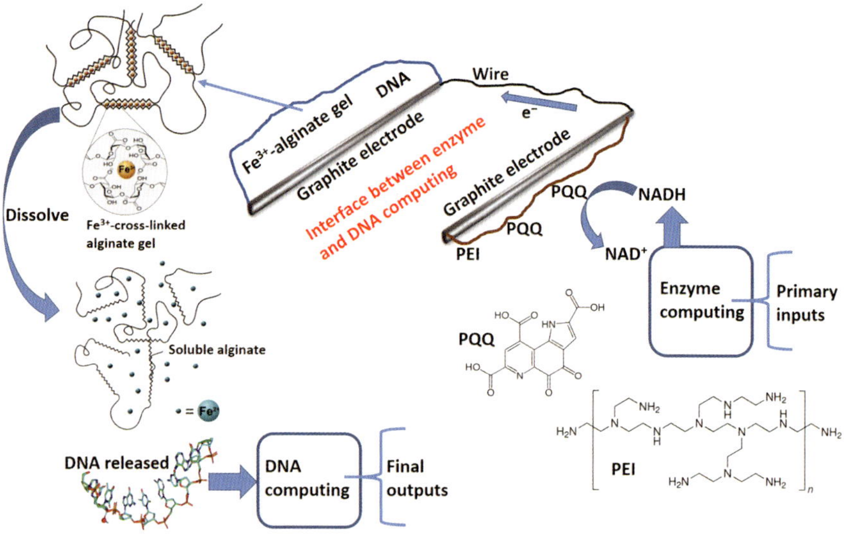

Figure 14.9 Operation of the electrochemical interface between the enzyme and DNA computing systems. The PQQ-modified electrode generated negative potential in the presence of enzymatically produced NADH. This potential resulted in the current passing from the PQQ electrode to the alginate electrode, thus resulting in the reduction of Fe^{3+} ions and dissolution of alginate hydrogel and release of the entrapped DNA. The released DNA was applied as a signal to the DNA computing system. Note that two electrochemical cells activated by Outputs *Q* and *R* generated by the enzyme system released two different DNA samples, DNA1 and DNA2, both operating as new input signals to the DNA system. Structures of PQQ and PEI are shown. Source: Guz et al. 2016 [33]. Adapted with permission of John Wiley & Sons.

release repeat the pattern of the logic outputs from the Fredkin gate, thus being extensions of Outputs *Q* and *R* (Figure 14.8A,B,C).

14.3.2 Reversible DNA-Based Feynman Gate Activated by Signals Produced by the Enzyme-Based Fredkin Gate

The DNA molecules released from the alginate-modified electrodes activated by logic output signals produced by the Fredkin gate were used to activate the next step in the information processing, which was represented by the DNA-based Feynman logic gate [33]. In other words, the downstream DNA-based Feynman gate was activated by the chemical outputs (represented by NADH) produced *in situ* by the upstream enzyme-based Fredkin gate, which was activated in its turn by the primary input signals *A*, *B*, and *C*. The logically reversible Feynman gate can be represented by a combination of ID and XOR gates operating in parallel (Figures 14.8D and 14.10A). The Feynman gate (also known as a Controlled NOT [CNOT] gate) includes two-input and two-output channels [82]. The truth table for the Feynman gate (Figure 14.10D) corresponds to the experimentally observed output signals generated by the system mimicking the Feynman gate (Figure 14.10B,C). DNA1 and DNA2 released from the alginate electrodes served as input signals for the Feynman gate. A number of other oligonucleotides

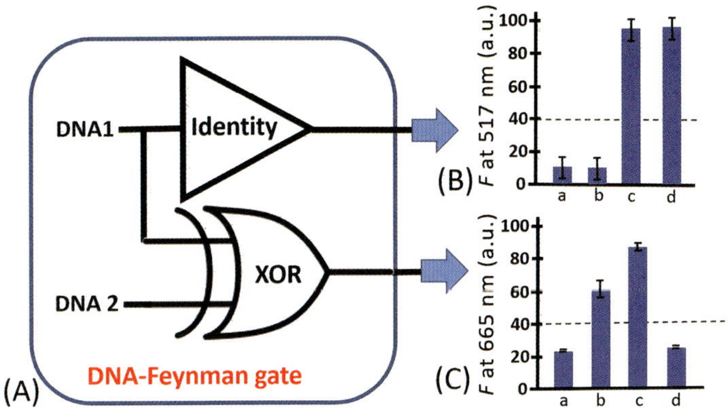

(D)

Input (A′)	Input (B′)	Output (P′)	Output (Q′)
0	0	0	0
0	1	0	1
1	0	1	1
1	1	1	0

Figure 14.10 (A) Logic scheme (including the ID and XOR gates operating in parallel) of the Feynman gate and the output signals produced by the DNA system. (B) The bar diagram showing the output signals ($\lambda_{max} = 517$ nm) produced by the DNA-based ID gate with the following DNA inputs: (a) **0**,**0**; (b) **0**,**1**; (c) **1**,**0**; and (d) **1**,**1**. Note that four input combinations are shown for consistency with the truth table (D), while the input signals processed by the ID gate are only **0** and **1**. The output signals were measured as fluorescence produced by the FAM-labeled DNA released from the alginate electrode. (C) The bar diagram showing the fluorescent output signals produced by the DNA-based XOR gate with the following combinations of the DNA inputs: (a) **0**,**0**; (b) **0**,**1**; (c) **1**,**0**; and (d) **1**,**1**. The output signals were measured as fluorescence ($\lambda_{max} = 665$ nm) generated by the DNA reactions. The dashed lines in both bar diagrams represent the thresholds separating **0** and **1** output signals. (D) The truth table for the Feynman (CNOT) gate. Inputs *A′* and *B′* and Outputs *P′* and *Q′* correspond to the DNA-based part of the system. Note that Inputs *A′* and *B′* in the DNA system correspond to the Outputs *Q* and *R* of the enzyme system. Source: Guz et al. 2016 [33]. Adapted with permission of John Wiley & Sons.

(A1, A2, B1, B2, Q–F) (Figure 14.11) (see the list of oligonucleotide sequences in Appendix 14.A) served as "machinery" (meaning non-variable components of the logic system) for the Feynman gate. One of the oligonucleotides, Q–F, included a quencher (BHQ2) and a fluorescent dye (Qz6) (Q and F, respectively) covalently bound to different ends of the oligonucleotide. The distance separating the fluorophore and quencher was short enough to allow the effective quenching of the photo-excited state of the dye, thus inhibiting its fluorescence. The DNA information processing system (Feynman gate) was realized in a single solution fed by the flows coming from both electrochemical cells releasing DNA inputs DNA1 and DNA2. When no inputs were added to the DNA "machinery" system, meaning **0**,**0** input combination for DNA1 and DNA2, the fluorescence observed in the system was only negligibly higher than the control (Q–F alone) due to a background reaction corresponding to the logic output **0**. When one of the DNA inputs, DNA1 or DNA2 (input combinations **0**,**1**; **1**,**0**), was added to the "machinery" system, it hybridized with the oligonucleotides present in the solution and produced a deoxyribozyme catalytic core that cuts the Q–F

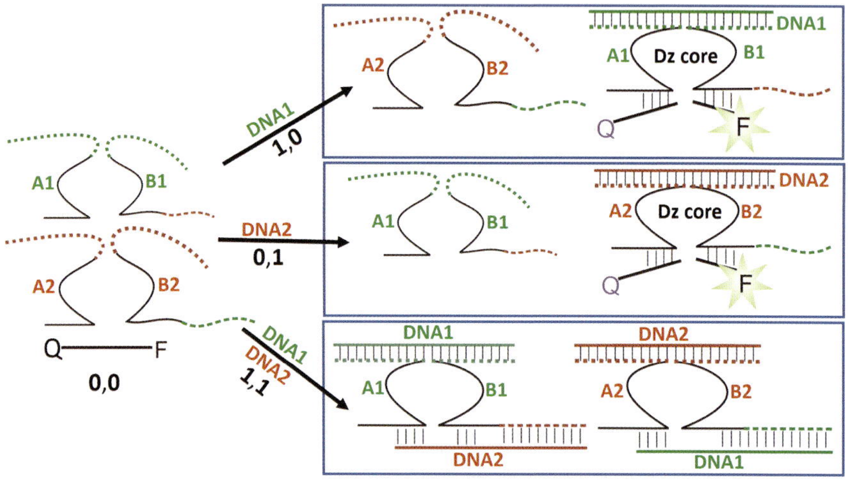

Figure 14.11 Schematic representation of the DNA reactions mimicking XOR function of the Feynman gate realized for different combinations of the DNA1 and DNA2 inputs. The produced fluorescence was used as the final output signal. The sequences of the used DNAs are given in Appendix 14.A. All DNAs except DNA1 and DNA2 were the constant (non-variable) parts of the solution ("machinery"). Source: Guz et al. 2016 [33]. Adapted with permission of John Wiley & Sons.

sequence, thus separating the fluorophore and quencher. The fluorophore separated from the quencher produced a high fluorescence signal considered as logic output **1**. Upon addition of the both DNA inputs (input combination **1,1**), the DNA inputs were both hybridized to the oligonucleotides present in the solution resulting in catalytically mute structures not capable of cutting the Q–F sequence. The fluorescence was quenched, thus resulting in the logic output **0**. Minor fluorescence observed upon application of **0,0** and **1,1** input combinations originates from the incomplete quenching of the fluorophore attached to the oligonucleotide. The realized DNA reactions and the obtained logic outputs mimic the XOR gate operation, which is a part of the Feynman gate (Figure 14.10A). The second part of the Feynman gate represented by the ID gate was realized in a very simple way. One of the DNA inputs (DNA1) was labeled with a fluorescent dye (FAM). When it was released from the alginate electrode, its fluorescence was observed and considered as the output value **1**. Obviously, the absence of this fluorescence, in case DNA1 was not released from alginate electrode, was considered as the output **0**. It should be noted that the fluorescence observed in the XOR gate (λ_{max} = 665 nm) and fluorescence measured in the ID gate (λ_{max} = 517 nm) were produced by the different dyes (Qz6 and FAM, respectively) and they did not interfere with each other. The experimental results measured with the DNA logic system for the ID and XOR channels are shown as bar diagrams demonstrating the fluorescent outputs for different combinations of the DNA inputs (Figure 14.10B,C). The experimental result is shown for the simplified logic process when only four combinations of the DNA inputs were used to activate the DNA-based Feynman gate. The full set of the data was represented with eight input combinations, starting from the primary Inputs *A*, *B*, and *C* used for the activation of the enzyme-based Fredkin gate. The number of shown input combinations was reduced due to their redundancy. In experimental practice, some of the logically triggered releases produce the same effect on the DNA computing system; therefore to avoid redundancy and unnecessary complexity in the shown data some input combinations and the corresponding output signals are not shown. Indeed, any combination of the primary inputs that produces NADH will yield a negative potential on the PQQ electrode, resulting in the release of DNA from the affected alginate electrode. Another reason for the reduced number of logic variants illustrated in Figure 14.10C is the clear demonstration of the XOR logic of the DNA computing system. If the whole set of the inputs was to be shown, the output would not be a clear representation of an XOR function, it would in fact be much more complex. It should be noted that XOR functions have been already realized with DNA systems [83–88]; however, they have never been included in the complex network similar to the present study.

14.4 Conclusions and Perspectives

Further developments in the area of biomolecular computing will certainly be motivated by practical applications. While several potential applications have already emerged (e.g., using genetically encoded computing circuits [89, 90] or simply borrowing ideas from DNA computing systems, such as their use as gene regulators in living cells [91]), novel applications are needed to justify the continuing research in the genre of biomolecular computing. In turn, the quest for novel applications will require new structural and functional features to be integrated into the biocomputing systems to yield increased complexity and enhanced functionality. The present study demonstrated a possibility to increase the complexity of biomolecular computing systems, particularly by integration of parts based on the enzyme- and DNA-based reactions. While the present design does not pretend to be practically useful for any specific application, it demonstrated advantages of the modular design, where complex information processing steps can be realized by combinations of individual cells performing simple functions. These individual cells performing enzyme reactions that can electrochemically stimulate release processes and DNA reactions can be easily exchanged to other functional elements, thus allowing different functions depending on specific needs. Additionally, the number of input and output channels can be increased, thus increasing the system complexity. Considering that the present systems were realized as a combination of relatively large flow cells, the next experimental step will include miniaturization of the systems to a microfluidic lab-on-a-chip system. This lab-on-a-chip system would retain the ability to perform complex computational functions and allow for biomolecular release processes that are logically controlled by complex combinations of variable input signals. The bioelectronic (bioelectrochemical) interface connecting enzyme and DNA processes could also be miniaturized using electrode microarrays. Overall, the present approach, while already demonstrating a significant advancement in sophisticated biocomputing processes, can foster novel and complex systems that show functionality similar to microelectronic circuitries. It can be noted that it is hard to expect a miniaturization of the system that is comparable with modern electronics. Yet, when considering that the advantage of the biomolecular circuitries belies in their unique activation by metabolites representing physiological changes in living organisms, the importance cannot be overlooked. Even more challenging approach to the reverse operation, where a DNA computing system activates, an enzyme logic system should be considered as the extension of the developed system. The first step in this direction has been already done with a sensing electrode activated with a DNA input signal [65].

14.A Oligonucleotides Used in the System Mimicking Feynman Gate (See Section 14.3.2)

DNA1:	6-FAM-5′-TAG AGT AACC T CAC ACG GAA TGT TTC
DNA2:	5′-GCTTA CAA CCAAT GAA GGA TCC TC
A1:	GAA ACA TTC CGT A CAA CGA GGTTGTGC
B1:	GAG GAT CCT TC GTTG GTG A GG CTA GCT GTG AGG TTA CTC TA
A2:	GAG GAT CCT TC A CAA CGA GG TTG TGC
B2:	GAA ACA TTC CGT GTT GGT GA GG CTA GCT ATT GG TTG TAAGC
Q–F:	Qz6-CAGCACAACCguCACCAACCG-BHQ2

Nucleotides complementary to inputs DNA1 or DNA2 are underlined.

The following abbreviations are used: BHQ2 (Black Hole Quencher®) is a trademark registered in the United States for a fluorescence quencher; 6-FAM attached to the DNA is a fluorescein derivative; and Qz6 is a Quasar 670 fluorescent dye.

References

1 Gerasimova, Y.V. and Kolpashchikov, D. (2015). *Chem. Commun.* 51: 870–872.
2 Pei, R., Matamoros, E., Liu, M. et al. (2010). *Nat. Nanotechnol.* 5: 773–777.
3 Stojanovic, M.N., Stefanovic, D., and Rudchenko, S. (2014). *Acc. Chem. Res.* 47: 1845–1852.
4 Stojanovic, M.N. and Stefanovic, D. (2011). *J. Comput. Theor. Nanosci.* 8: 434–440.
5 Benenson, Y. (2012). *Nat. Rev. Genet.* 13: 455–468.
6 Kahan, M., Gil, B., Adar, R., and Shapiro, E. (2008). *Physica D* 237: 1165–1172.
7 Ezziane, Z. (2006). *Nanotechnology* 17: R27–R39.
8 Katz, E. and Privman, V. (2010). *Chem. Soc. Rev.* 39: 1835–1857.
9 Katz, E. (2015). *Curr. Opin. Biotechnol.* 34: 202–208.
10 Benenson, Y. (2009). *Mol. BioSyst.* 5: 675–685.

11 Stojanovic, M.N. (2011). *Isr. J. Chem.* 51: 99–105.
12 Adleman, L.M. (1994). *Science* 266: 1021–1024.
13 Santini, C.C., Bath, J., Turberfield, A.J., and Tyrrell, A.M. (2012). *Int. J. Mol. Sci.* 13: 5125–5137.
14 Wang, D., Fu, Y., Yan, J. et al. (2014). *Anal. Chem.* 86: 1932–1936.
15 Gil, B., Kahan-Hanum, M., Skirtenko, N. et al. (2011). *Nano Lett.* 11: 2989–2996.
16 Hemphill, J. and Deiters, A. (2013). *J. Am. Chem. Soc.* 135: 10512–10518.
17 Konry, T. and Walt, D.R. (2009). *J. Am. Chem. Soc.* 131: 13232–11323.
18 Cornett, E.M., Campbell, E.A., Gulenay, G. et al. (2012). *Angew. Chem. Int. Ed.* 51: 9075–9077.
19 Schlosser, K. and Li, Y. (2009). *Chem. Biol.* 16: 311–322.
20 Toh, S.Y., Citartan, M., Gopinath, S.C.B., and Tang, T.-H. (2014). *Biosens. Bioelectron.* 64: 392–403.
21 Radzicka, A. and Wolfenden, R. (1995). *Science* 267: 90–93.
22 Privman, V., Zavalov, O., Halámková, L. et al. (2013). *J. Phys. Chem. B* 117: 14928–14939.
23 Privman, V., Arugula, M.A., Halámek, J. et al. (2009). *J. Phys. Chem. B* 113: 5301–5310.
24 Privman, V., Strack, G., Solenov, D. et al. (2008). *J. Phys. Chem. B* 112: 11777–11784.
25 Benenson, Y., Paz-Elizur, T., Adar, R. et al. (2001). *Nature* 414: 430–434.
26 Kolpashchikov, D.M. and Stojanovic, M.N. (2005). *J. Am. Chem. Soc.* 127: 11348–11351.
27 Beyer, S. and Simmel, F.C. (2006). *Nucleic Acids Res.* 34: 1581–1587.
28 Katz, E., Wang, J., Privman, M., and Halámek, J. (2012). *Anal. Chem.* 84: 5463–5469.
29 Wang, J. and Katz, E. (2011). *Isr. J. Chem.* 51: 141–150.
30 Halámková, L., Halámek, J., Bocharova, V. et al. (2012). *Analyst* 137: 1768–1770.
31 Halámek, J., Bocharova, V., Chinnapareddy, S. et al. (2010). *Mol. Biosyst.* 6: 2554–2560.
32 Mailloux, S., Gerasimova, Y.V., Guz, N. et al. (2015). *Angew. Chem. Int. Ed.* 54: 6562–6566.
33 Guz, N., Fedotova, T.A., Fratto, B.E. et al. (2016). *ChemPhysChem* 17: 2247–2255.
34 Moore, A.N.J., Katz, E., and Willner, I. (1996). *Electroanalysis* 8: 1092–1094.
35 Katz, E., Lötzbeyer, T., Schlereth, D.D. et al. (1994). *J. Electroanal. Chem.* 373: 189–200.
36 Mailloux, S., Halámek, J., and Katz, E. (2014). *Analyst* 139: 982–986.
37 Mailloux, S., Guz, N., Zakharchenko, A. et al. (2014). *J. Phys. Chem. B* 118: 6775–6784.
38 Mailloux, S., Guz, N., Gamella Carballo, M. et al. (2014). *Anal. Bioanal. Chem.* 406: 4825–4829.
39 Jin, Z., Güven, G., Bocharova, V. et al. (2012). *ACS Appl. Mater. Interfaces* 4: 466–475.

40 Stojanovic, M.N., Mitchell, T.E., and Stefanovic, D. (2002). *J. Am. Chem. Soc.* 124: 3555–3561.
41 Stojanovic, M.N. and Stefanovic, D. (2003). *Nat. Biotechnol.* 21: 1069–1074.
42 Macdonald, J., Li, Y., Sutovic, M. et al. (2006). *Nano Lett.* 6: 2598–2603.
43 Stojanovic, M.N., Semova, S., Kolpashchikov, D. et al. (2005). *J. Am. Chem. Soc.* 127: 6914–6915.
44 Yashin, R., Rudchenko, S., and Stojanovic, M.N. (2007). *J. Am. Chem. Soc.* 129: 15581–15584.
45 Brown, C.W. III, Lakin, M.R., Horwitz, E.K. et al. (2014). *Angew. Chem. Int. Ed.* 53: 7183–7187.
46 Brown, C.W. III, Lakin, M.R., Stefanovic, D., and Graves, S.W. (2014). *ChemBioChem* 15: 950–954.
47 Poje, J.E., Kastratovic, T., Macdonald, A.R. et al. (2014). *Angew. Chem. Int. Ed* 53: 9222–9225.
48 Kolpashchikov, D.M. (2007). *ChemBioChem* 8: 2039–2042.
49 Gerasimova, Y.V., Cornett, E., and Kolpashchikov, D.M. (2010). *ChemBioChem* 11: 811–817.
50 Mokany, E., Bone, S.M., Young, P.E. et al. (2010). *J. Am. Chem. Soc.* 132: 1051–1059.
51 Gerasimova, Y.V. and Kolpashchikov, D.M. (2010). *Chem. Biol.* 17: 104–106.
52 Liu, Z., Sall, A., and Yang, D. (2008). *Int. J. Mol. Sci.* 9: 978–999.
53 Mailloux, S., Halámek, J., Halámková, L. et al. (2013). *Chem. Commun.* 49: 4755–4757.
54 Gamella, M., Zakharchenko, A., Guz, N. et al. (2017). *Electroanalysis* 29: 398–408.
55 Gerasimova, Y.V., Cornett, E.M., Edwards, E. et al. (2013). *ChemBioChem* 14: 2087–2090.
56 Gerasimova, Y.V. and Kolpashchikov, D.M. (2013). *Angew. Chem. Int. Ed.* 52: 10586–10588.
57 Katz, E. (ed.) (2012). *Molecular and Supramolecular Information Processing: From Molecular Switches to Logic Systems.* Weinheim: Wiley-VCH.
58 Katz, E. (ed.) (2012). *Biomolecular Information Processing – From Logic Systems to Smart Sensors and Actuators.* Weinheim: Wiley-VCH.
59 Szacilowski, K. (2012). *Infochemistry – Information Processing at the Nanoscale.* Chichester: Wiley.
60 de Silva, A.P. (2013). *Molecular Logic-Based Computation.* Cambridge: Royal Society of Chemistry.
61 Privman, V., Domanskyi, S., Mailloux, S. et al. (2014). *J. Phys. Chem. B* 118: 12435–12443.
62 Halámek, J., Bocharova, V., Arugula, M.A. et al. (2011). *J. Phys. Chem. B* 115: 9838–9845.
63 Okhokhonin, A.V., Domanskyi, S., Filipov, Y. et al. (2018). *Electroanalysis* 30: 426–435.
64 Gamella, M., Privman, M., Bakshi, S. et al. (2017). *ChemPhysChem* 18: 1811–1821.
65 Gamella, M., Guz, N., and Katz, E. (2016). *Electroanalysis* 28: 2692–2696.
66 Mailloux, S., Zavalov, O., Guz, N. et al. (2014). *Biomater. Sci.* 2: 184–191.

67 Bocharova, V., Zavalov, O., MacVittie, K. et al. (2012). *J. Mater. Chem.* 22: 19709–19717.

68 Fredkin, E. and Toffoli, T. (1982). *Int. J. Theor. Phys.* 21: 219–253. (First published in 1981 as a DARPA report: http://www.dtic.mil/dtic/tr/fulltext/u2/a101383.pdf).

69 O'Brien, J.L., Pryde, G.J., White, A.G. et al. (2003). *Nature* 426: 264–267.

70 Monroe, C., Meekhof, D.M., King, B.E. et al. (1995). *Phys. Rev. Lett.* 75: 4714–4717.

71 Fratto, B.E. and Katz, E. (2016). *ChemPhysChem* 17: 1046–1053.

72 Orbach, R., Remacle, F., Levine, R.D., and Willner, I. (2012). *Proc. Natl. Acad. Sci. U.S.A.* 109: 21228–21233.

73 Roy, S. and Prasad, M. (2010). *Opt. Eng.* 49: 065201.

74 Klein, J.P., Leete, T.H., and Rubin, H. (1999). *Biosystems* 52: 15–23.

75 Fratto, B.E. and Katz, E. (2015). *ChemPhysChem* 16: 1405–1415.

76 Katz, E. and Fratto, B.E. (2017). Enzyme-based reversible logic gates operated in flow cells. In: *Advances in Unconventional Computing*, Chapter 2, vol. 2 (ed. A. Adamatzky), 29–59. Springer.

77 Cervera, J. and Mafé, S. (2010). *ChemPhysChem* 11: 1654–1658.

78 Remón, P., Ferreira, R., Montenegro, J.-M. et al. (2009). *ChemPhysChem* 10: 2004–2007.

79 Moseley, F., Halámek, J., Kramer, F. et al. (2014). *Analyst* 139: 1839–1842.

80 Zhou, C., Wang, K., Fan, D. et al. (2015). *Chem. Commun.* 51: 10284–10286.

81 De Vos, A. (2010). *Reversible Computing: Fundamentals, Quantum Computing, and Applications.* Weinheim: Wiley-VCH.

82 Garipelly, R., Kiran, P.M., and Kumar, A.S. (2013). *Int. J. Emerg. Technol. Adv. Eng* 3: 417–423.

83 Stojanovic, M.N. and Stefanovic, D. (2003). *J. Am. Chem. Soc.* 125: 6673–6676.

84 Gao, W., Zhang, L., Liang, R.P., and Qiu, J.D. (2015). *Chem. Eur. J.* 21: 15272–15279.

85 Gao, W., Zhang, L., Zhang, Y.M. et al. (2014). *J. Phys. Chem. C* 118: 14410–14417.

86 Zadegan, R.M., Jepsen, M.D.E., Hildebrandt, L.L. et al. (2015). *Small* 11: 1811–1817.

87 Liu, Y.Z., Dong, B.R., Wu, Z.T. et al. (2014). *Chem. Commun.* 50: 12026–12029.

88 Yang, C.N., Chen, Y.L., Lin, H.Y., and Hsu, C.-Y. (2013). *Chem. Commun.* 49: 8860–8862.

89 Roybal, K.T., Rupp, L.J., Morsut, L. et al. (2016). *Cell* 164: 770–779.

90 Haefliger, B., Prochazka, L., Angelici, B., and Benenson, Y. (2016). *Nat. Commun.* 7: 10709.

91 Green, A.A., Silver, P.A., Collins, J.J., and Yin, P. (2014). *Cell* 159: 925–939.

15

What Is Next? Mimicking Natural Biological Information Processes

15.1 Motivation and Goals

The unconventional computing [1] based on molecular [2–5] and biomolecular [6] processes has been mostly following the concept of binary information processing used in electronics [7]. However, the conceptually novel approaches adopted from natural biological processes are needed to bring the (bio)molecular computing to the next complexity level and to provide tools for new applications [8].

The information processing paradigm of nature has been very successful. We do not fully understand it, but advances have been made in systems biology to explore aspects of the functioning of the nature's information processing [9–12]. Presently, we cannot even remotely mimic the complexity of the natural processes by making "artificial life" systems starting with biomolecules/biochemistry. However, an interesting avenue of research has been to consider specific processes – memory, learning, etc. – as "network elements" that could offer new functionalities to our otherwise more conventionally manufactured systems of Si-electronics.

Here we take up a new challenge: we consider how to actually realize with biochemical processes certain basic bio-inspired information processing steps. Ultimately, such systems can offer tools for experimenting with information processing networks based on synthetic autonomous biochemical processes, to allow a new avenue for understanding the nature's information processing paradigm. Our emphasis here will be on feed-forward loops, as an example of possible concept transfer from biological systems to artificial enzyme-based information processing systems. Indeed, it is the most common network motif in information processing in natural systems, and the challenge will be to carry it out with few coupled biochemical reactions, which will be an immensely simpler realization than that in nature.

Both general and specific designs are presented and theoretically substantiated, with the bulk of the required experimental work to be carried out in the future. It is hoped that the described designs would initiate a new research direction of synthetic information processing not mimicking the full scope of what the nature does, as "artificial life," but rather taking up a more limited and therefore hopefully more tractable goal of mimicking only the nature's information processing

Enzyme-Based Computing Systems, First Edition. Evgeny Katz.

design, with biomolecular/biochemical reaction processes vastly simpler than those that evolved naturally.

15.2 Example and Discussion of Feed Forward Loops

In order to make the presentation specific, let us first consider an example, which is later revisited and described in detail in Section 15.3, of a possible design of a feed-forward loop function with a cascade of enzyme-catalyzed reactions. This section offers a general introduction, whereas the details of the actual biochemistry of this and other systems' functioning are explained in later sections. Unlike the various earlier realized "biochemical gates," feed forward is in most cases not a binary function [13–15].

Another important difference in attempting actual biochemical realizations is that feed forward has two "signal transduction" steps, each involving the input signal, X, usually not being directly converted into the output. Rather, in the primary (direct) step, the input acts as the activator (promoter), denoted by →, or repressor (inhibitor), denoted by ⊣, of the ongoing process(es) that generate the output signal, Z (see the schematic in Figure 15.1A). The feed-forward loop is completed by adding the secondary (indirect) step in which X activates or represses the ongoing process(es) that generate another, intermediate signal, Y, which in turn activates or represses the process(es) of the production of Z.

Here we will consider the simpler (for biochemical realizations) situation when X or Y, rather than X and Y together – which is another feed-forward

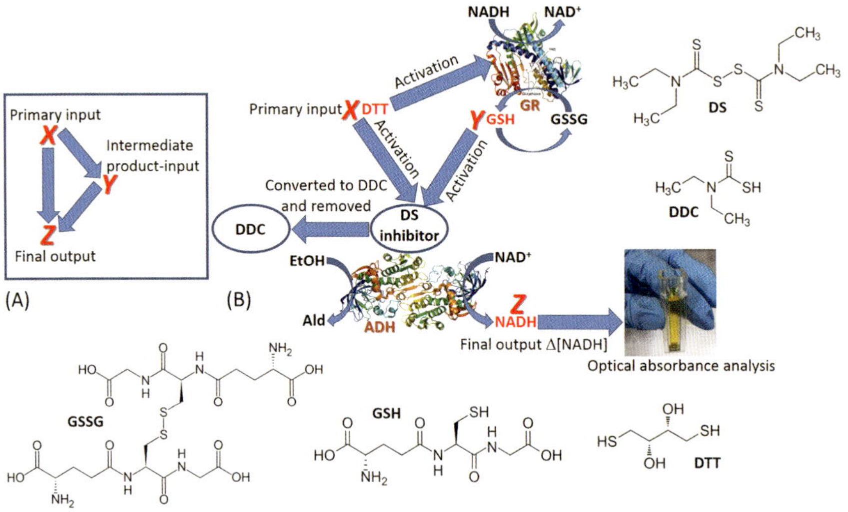

Figure 15.1 Feed-forward system with *activation* in all the signal transduction steps. (A) The schematic shows the activations involved. (B) The biochemical processes are explained in detail in Section 15.3. Two enzyme-biocatalyzed processes continuously produce chemicals that are signals Y and Z. Chemical input X activates (promotes) the production of both Z and Y, whereas Y promotes the production of Z. The chemical notations and abbreviations are explained in Section 15.3.

option – activate the output signal Z production. Then in the simplest classification [15, 16] there can be six different loops corresponding to choosing activation or repression in each of the steps:

$$X \rightarrow Z \text{ or } X \dashv Z, \quad X \rightarrow Y \text{ or } X \dashv Y, \quad Y \rightarrow Z \text{ or } Y \dashv Z \tag{15.1}$$

The feed-forward loop is then called "coherent" or "incoherent" depending on whether the net effect of X on the production of Z in the secondary step is the same as in the primary.

The most abundant in nature [16–18] feed forward involves three activations. A potential realization of such a process with an enzymatic cascade is shown in Figure 15.1B, for which the notation, details of the biochemical processes, and the system's functioning are explained later, in Section 15.3, as part of a discussion of experimental realizations, including also systems involving repression.

Here we would like to address several important features expected of such systems. First, we need at least two enzymatic (or other biochemical) processes that yield signals Y and Z, and these processes' rates at time t should be affected (controlled) by the value of the input at that time, $X(t)$. In some situations activation can be made rather sharp as a function of parameters, and inhibition can also be made sharp. Thus, various responses of the feed-forward loop can be made quite steep. Feed forward can therefore on its own in some regimes approximate binary gates. Therefore, ideas have been put forth theoretically [19], but to our knowledge never actually experimentally realized, for multi-gate logic with DNA structure oligonucleotide systems made of binary feed-forward functions.

However, in a general setting the feed forward's role in nature is obviously not binary. Rather, the feed-forward loop, specifically the one with three promotions shown in the schematic in Figure 15.1A, plays a stabilizing role in nature's networks: it delays [17, 20–22] the changes in the response, $Z(t)$, to avoid erratic swings and "waste of resources" in natural pathway responses to environmental variations/fluctuations, specifically, those in $X(t)$. The secondary step – with the X to Y to Z transduction – takes a fraction of the input signal and processes it in parallel to the primary transduction channel. It should be designed to act to time delay (as compared with the direct X to Z transduction in the primary step), the effect of a part of the changes in the input signal, X, as far as its net impact on Z goes.

Therefore, in order to experimentally accomplish proper feed-forward realizations, one needs to go beyond the "input at $t = 0$ to output at gate time $t_{\text{gate}} > 0$" response paradigm of the digital-gate biocomputing. The availability of $X(t)$ has to be experimentally controlled by adding/removing (inputting/deactivating) this compound by physical or (bio)chemical means at the externally controllable rate (which can be negative) R_{ext}, such that

$$\frac{\mathrm{d}X}{\mathrm{d}t} = R_{\text{ext}}(t) + \text{reaction terms} \tag{15.2}$$

where the "reaction terms" describe the kinetics of a possible consumption of X by the biochemical processes of the feed-forward loop itself. The quantification of the feed-forward effect will consist of observing how the resulting time dependence of $Z(t)$ is affected by the presence of the secondary transduction step, $X \rightarrow Y \rightarrow Z$, which can be enabled at various degrees of activity controlled by chemicals needed for that step's functioning.

One expected effect is that sharp variations in X will not cause an immediate response by changes in Z. Rather, with the second step active at proper levels, the response will only occur when the input signal, X, changes (up or down) by certain threshold amounts, and furthermore, the system might have its own timescales for response rather than be driven by the input's variation.

The "stabilizing/resource-conserving" effects expected of feed-forward functions have never been realized in simple "synthetic" biochemical systems. The challenges and possible approaches to accomplish feed-forward realization in enzyme-based systems are discussed in the following sections. For these the experimental realizations will have to be designed guided by the needs of quantifying the anticipated characteristics of their response. The chemical kinetic modeling of the expected systems is also commented.

15.3 Enzymatic Feed-Forward Loops

In this section the potential enzymatic cascade-based feed-forward designs are presented. First, the system shown in Figure 15.1B is described in detail, its design is outlined, and preliminary experimental observations are reported. The system involves only activations in all its signal transduction steps corresponding to the options in Eq. (15.1). The cascade includes the functioning of two enzymes as biocatalysts. Glutathione reductase (GR) biocatalytically converts glutathione from its oxidized form (GSSG) to reduced form (GSH) (their structures are shown in Figure 15.1). The latter, GSH, acts as an intermediate signal, Y, in the feed-forward functioning. Concomitantly, β-nicotinamide adenine dinucleotide is converted [23] from its reduced form (NADH) to the oxidized form (NAD^+). Alcohol dehydrogenase (ADH) biocatalytically oxidizes ethanol (EtOH) to yield acetaldehyde (Ald), while β-nicotinamide adenine dinucleotide is converted [24] from its oxidized form, NAD^+, to the reduced state NADH. These two processes can yield the net increase in the amount of NADH that can be measured optically by changes in absorption and that will be designated as the output signal, Z. Thus, as expected for feed forward, signals Z and Y are generated continuously once the reactions are started. In fact, the net rate of production of NADH in this system must be kept in check, to avoid rapid buildup of the signal Z. This can be done [25] by initially largely inhibiting the activity of enzyme ADH by adding disulfiram (DS; see its structure in Figure 15.1), which forms an ADH-DS complex that has low biocatalytic activity.

The input signal, X, can then be dithiothreitol (DTT; see its structure in Figure 15.1) that when added to the system has the following effects on the process rates:

(i) Its promotion of GR results in a substantial increase [26] of the GR enzyme activity, thus increasing the rate of the signal Y generation. This corresponds to $X \rightarrow Y$ in Eq. (15.1).
(ii) In addition, DTT chemically converts DS from its original disulfide form to the thiol form, diethyldithiocarbamate (DDC; see its structure in

Figure 15.1), shifting the kinetics to result in the breakup of the inhibited enzyme complex ADH-DS and restoring ADH to high activity, thus increasing the rate of the signal Z generation: $X \rightarrow Z$ in Eq. (15.1).

(iii) Importantly, GSH (signal Y), which is the product of the reaction biocatalyzed by GR, also chemically removes DS from the ADH-DS inhibited complex, acting to increase the rate of the signal Z generation. Therefore, the step $Y \rightarrow Z$ in Eq. (15.1) is built into the biochemical system's functioning.

The excess concentration of NADH,

$$\Delta \mathrm{NADH}(t) = \mathrm{NADH}(t) - \mathrm{NADH}(0) \tag{15.3}$$

which is our signal $Z(t)$, is generated at a rate increased by the direct effect of the input signal X. And it is also increased through the indirect step of X accelerating the signal Y production, while the latter in turn contributes to increasing the output of Z.

The full realization and characterization of the proposed enzymatic cascade will require addressing several challenges, even though some of the processes have already been studied in the literature. The latter include the process of removing the DS inhibitor from ADH-DS complex by DTT (signal X). Indeed, DTT is an established reactant for converting disulfide chemical species to their thiol derivatives [27]. The inhibition of ADH by the disulfide form of DS has also been studied, including the property that its thiol derivative, DDC, is removed from the enzyme complex and does not inhibit ADH [25, 28]. Therefore, the primary step, $X \rightarrow Z$, should be realizable in a controllable fashion for experiments to probe the time dependence properties of the process.

Regarding the secondary step, for $X \rightarrow Y$ it has been already mentioned that DTT is known to promote the activity of GR [26]. As a test of feasibility of realizing the $Y \rightarrow Z$ step, preliminary experiments were performed (Figure 15.2) to demonstrate the effect of GSH (signal Y) on the rate of NADH production (contributing to signal Z). First, ADH ($0.63\,\mathrm{units\,ml^{-1}}$) was prepared in the inhibited form by incubation with the added optimized concentration of disulfiram (0.8 mM). This disulfiram concentration caused a significant inhibition of ADH, by forming the ADH-DS complex. ADH could then be reactivated by the thiol–disulfide exchange with the reduced glutathione produced *in situ* by the GR reaction. Lower disulfiram concentrations did not result in a substantial inhibition of ADH, while higher concentrations did not allow the enzyme reactivation. The inhibited ADH was tested for the production of NADH in the presence of NAD^+ (1.0 mM) and ethanol (1.7 M). Despite its inhibition, the enzyme ADH demonstrated some activity producing NADH (Figure 15.2A, curve a). Addition of NADH (50 μM) initially to the solution in the absence of the disulfide-reducing system (GR and GSSG) did not noticeably affect the rate of the biocatalytic production of the excess NADH (Figure 15.2A, curve b). The same experiment performed in the presence of GR ($2\,\mathrm{units\,ml^{-1}}$), GSSG (3 μM), and NADH (the same added amount, 50 μM) resulted in an enhanced production of NADH, witnessing the inhibitor removal and ADH reactivation (Figure 15.2A, curve c). The rate of the NADH production was increased approximately twofold (Figure 15.2B), thus, demonstrating the feasibility of the $Y \rightarrow Z$ step.

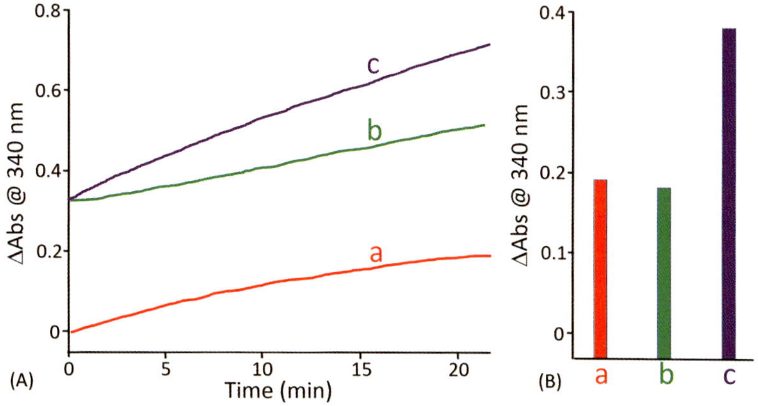

Figure 15.2 Experimental probe of the feasibility of processes needed for the $Y \to Z$ step of the feed-forward loop shown in Figure 15.1. (A) Absorbance at $\lambda_{max} = 340$ nm, as a function of time, measuring the amount of NADH: (a) produced only with the inhibited ADH-DS enzyme complex; (b) the same with some initially added NADH, but in the absence of the other chemicals (GR and GSSG) needed for the signal Y production; and (c) in the presence of GR and GSSG (and the same quantity of the initially added NADH). (B) The bar chart showing the output signal (Z) measured at 20 minutes as the optical absorbance change corresponding to NADH concentration increase: $\Delta\text{Abs} = \text{Abs}(t) - \text{Abs}(0)$, at $t = 20$ minutes.

Regarding the interconnectivity of the system just described (Figure 15.1B) with other biocatalytic processes for attempting networking, it should be noted that the output, NADH, is a substrate (one of the input chemicals) common for many enzymes, mostly for dehydrogenases. The input, dithiothreitol, is less common in purely enzymatic reactions, but it can be produced by enzyme-catalyzed processes [29]. For example, enzyme, protein disulfide-isomerase (PDI), can produce DTT in its reduced form that is needed to initiate the processes shown in Figure 15.1B. It should be also mentioned that the cascade in Figure 15.1B can be realized with the initial NADH replaced by a different chemical, NADPH (which is converted to $NADP^+$, nicotinamide adenine dinucleotide phosphate, by the enzyme GR). Then the whole amount of the NADPH produced by ADH (or ADH-DS) at time $t > 0$ can be identified as the output signal $Z(t)$.

Generally, for nearly any enzyme there are compounds that can promote or inhibit that enzyme's activity. The latter effect can be used to devise systems that involve feed forward with some or all of the processes in Eq. (15.1) corresponding to repression instead of activation. As an illustration, Figure 15.3 shows such a design of a system with all three processes being repressions. Two enzymatic reactions are ongoing: glutathione oxidase (GlutOx) biocatalytically converts glutathione from its reduced form (GSH) to the oxidized form (GSSG), while concomitantly oxygen is converted to hydrogen peroxide [29] that is taken as signal Y. Another enzyme, pyruvate kinase (PK), is converting adenosine triphosphate (ATP) to adenosine diphosphate (ADP) with the concomitant conversion of pyruvate (Pyr) to phosphor (enol)pyruvate (PEP). The output can be defined as the produced amount of ADP. However, to actually measure it, the ATP consumption (the decrease in its concentration) should be used to yield the output signal Z,

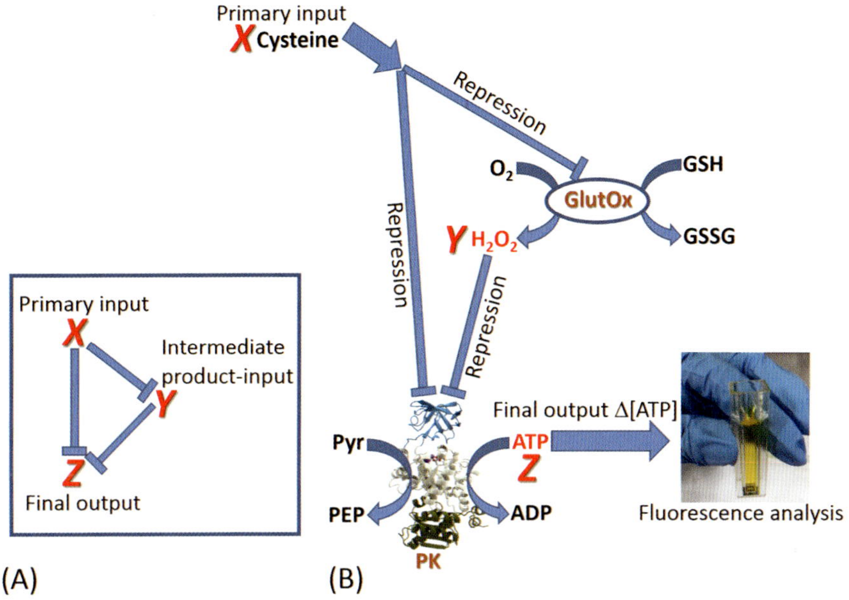

Figure 15.3 Feed-forward system with *repression* of all the signal processing steps. (A) The general schematics. (B) The biocatalytic realization discussed in the text. Abbreviations are explained in the text.

since it can be measured by the standard optical fluorescent assay for ATP – light emission generated by luciferin–luciferase system in the presence of ATP [30]. Note that ADP/ATP are produced/consumed stoichiometrically. Hydrogen peroxide, H_2O_2, produced in the first biocatalytic reaction is known as an inhibitor for PK [29], and therefore the first biocatalytic process is repressing the second one; this realizes $Y \dashv Z$. In addition, cysteine, defined as signal X, when added to the system as the primary input will repress both biocatalytic processes. Cysteine is a known inhibitor [29] of both GlutOx, yielding $X \dashv Y$, and PK, which corresponds to $X \dashv Z$.

Regarding its interconnectivity with other biocatalytic processes, the designed system (Figure 15.3) offers a substantial flexibility. Its output, ADP (or ATP), is utilized by more than 1700 ATP/ADP-dependent enzymes catalogued in the standard database [29]. The input, cysteine, is also compatible with enzymatic processes and can be produced, for example, by cysteine reductase [29].

For the two considered feed-forward systems (Figures 15.1B and 15.3B), with all activation or repression, as well as for other possible systems with different $\rightarrow/\dashv$ combinations, modeling/design and careful selection of parameters are required for proper functioning to achieve compatible reaction conditions. Their actual experimental realization, kinetic modeling, and then the first attempts at networking with other biocatalytic steps are not straightforward and will require a substantial research effort. This expectation is based on what was learned in earlier work with digital biocatalytic gates. Indeed, to our knowledge, feed-forward loops, while being extensively modeled in the literature, have never been actually experimentally realized as autonomously functioning synthetic rather than natural biomolecular systems. Importantly for the experimental work, all the enzymes and their substrates (input chemicals) needed for the experimental realization of the above formulated processes are commercially available.

15.4 Process Design and Kinetic Modeling

In modeling feed-forward loops, one can set up [11] coupled phenomenological rate equations describing signal variations in the ongoing process steps. This approach can yield the expected features, including the delayed response of the output to the input's variations/fluctuations and other properties [15]. As explained shortly, rate equations arise naturally in the discussed systems because of the nature of the biochemical processes involved. However, they will be more complicated and contain different terms than those considered in purely phenomenological formulations. The schematic in Figure 15.4 outlines one of the possible structures generic to the proposed processes considered in Section 15.3, with random selection of notations for chemicals (some not shown, cf. Figures 15.1 and 15.3) and identifications of the signals.

The discussed systems involve the catalytic functions of enzymes, e.g., E_1 (Figure 15.4). Most enzymes have several functional pathways, but for the discussed purposes it will generally suffice [31] to use the standard Michaelis–Menten-type model that focuses on the dominant mechanism, described by the following process sequence. The enzyme first binds a chemical called a substrate, say, S_1 to form a complex C. This complex can either on its own or by binding another substrate, S_2 – this option is common for the considered system – produce the product(s) of the biocatalytic reaction, here $P_{1,2}$ (Figure 15.4), restoring the enzyme to its original form. In the chemical reaction notation, we have

$$S_1 + E_1 \underset{k_{-1}}{\overset{k_1}{\rightleftarrows}} C, \quad S_2 + C \overset{k_2}{\rightarrow} E_1 + P_1 + P_2 \tag{15.4}$$

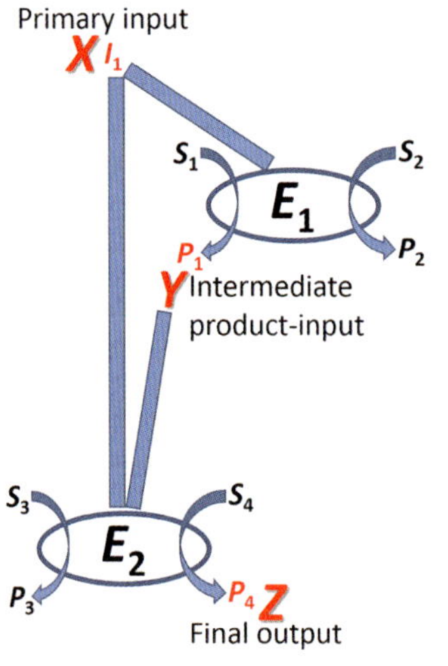

Figure 15.4 Illustration of the described feed-forward designs in terms of the constituent enzymatic processes. Activations or repressions (promotions or inhibitions) are shown by blue lines (cf. Figures 15.1 and 15.3).

This is of course just one of the possible reactant labelings (cf. Figure 15.4) as far as their role in the cascade goes. Here the arrows with rate constants above/below represent chemical processes rather than activation. The second step can usually be assumed irreversible, but the first one requires two rate constants. These process parameters, here $k_{\pm 1}$, k_2, are generally not known individually and have to be fitted from experiments. Activation/repression can involve several mechanisms, one of which can be a complex formation, for example,

$$I_1 + E_1 \underset{r_{-1}}{\overset{r_1}{\rightleftarrows}} \overline{E_1} + W \tag{15.5}$$

Here the "complex" is the modified enzyme $\overline{E_1}$ with a different activity (with larger or smaller rate constants, $\overline{k_{\pm 1}}$, $\overline{k_2}$, in processes similar to those in Eq. (15.4)), and it can be restored to the original form, E_1, by reacting with some other chemical, here denoted W.

If, for instance, I_1 is our input, X, and then the added "reaction terms" in Eq. (15.2) enter via such chemical processes,

$$\frac{dI_1}{dt} = R_{\text{ext}}(t) - r_1 I_1(t) E_1(t) + r_{-1} W(t)\,\overline{E_1}(t) \tag{15.6}$$

whereas the time dependence of the entering quantities is in turn set by their own rate equations, for example,

$$\frac{dE_1}{dt} = -r_1 I_1 E_1 + r_{-1} W \overline{E_1} - k_1 S_1 E_1 + k_{-1} C \tag{15.7}$$

Note that in the next stage, when writing the rate equation for S_1, for instance, terms resulting from its reaction with both the original and modified enzymes will enter, with their respective rates,

$$\frac{dS_1}{dt} = -k_1 S_1 E_1 + k_{-1} C - \overline{k_1} S_1 \overline{E_1} + \overline{k_{-1}}\,\overline{C} \tag{15.8}$$

where the notation (such as for $\overline{C}$) is self-explanatory.

Even within a relatively simple chemical kinetics description outlined here, enzymatic cascades thus lead to systems of numerous coupled chemical rate equations, with parameters that depend on the physical and chemical conditions of the experiment and are documented only to a very limited extent (typically, at most a single parameter, calculated in our notation from the quantities $k_{\pm 1}$ and $S_2(0)k_2$, call the Michaelis–Menten constant, is uniformly tabulated).

The described approach should enable modeling to select adjustable quantities (concentrations of those chemicals that are not designated as input/output signals) as needed to achieve expected feed-forward responses to various protocols $R_{\text{ext}}(t)$ of controlling the input signal availability, thus guiding the experimental work to achieve proper functioning of the synthetic systems, stand-alone and ultimately "wired" (connected via chemical process steps) with other enzymatic processes to attempt simple networking.

The main difference of the considered designs when compared to the much simpler earlier studied binary gate systems is the need for activation/repression as separate ongoing processes, i.e., in that enzymes are not just biocatalysts of fixed activity, but their activity changes ($E_1 \leftrightarrow \overline{E_1}$). As mentioned earlier, the need to have one of the products of the first enzymatic process that activates/represses the second process is also a significant experimental design challenge.

15.5 Simpler Biocatalytic Systems: Not a Feed-Forward Loop Yet

The complexity of the feed-forward systems realization based on the activation/inhibition of the enzymes results in temptation to consider easier ways to achieve similar results. Indeed, the final output signal can be produced through simple biocatalytic reactions (without alteration of the enzyme activity) processing inputs through different pathways with different kinetics [32]. Let us consider, as an example, the cascade sketched in Figure 15.5A, where X (primary input signal) is one of the substrates for enzyme E_1 as well as for enzyme E_2. The former enzyme outputs Z as one of its products, whereas the latter outputs Y. However, Y is a substrate for enzyme E_3, which also outputs Z. Overall, the final output Z is produced through the direct pathway catalyzed by enzyme E_1 and also through two-step process catalyzed by two enzymes E_2 and E_3, which results in the delay for the Z formation (Figure 15.5B). The biocatalytic system considered here is different from the feed-forward systems discussed

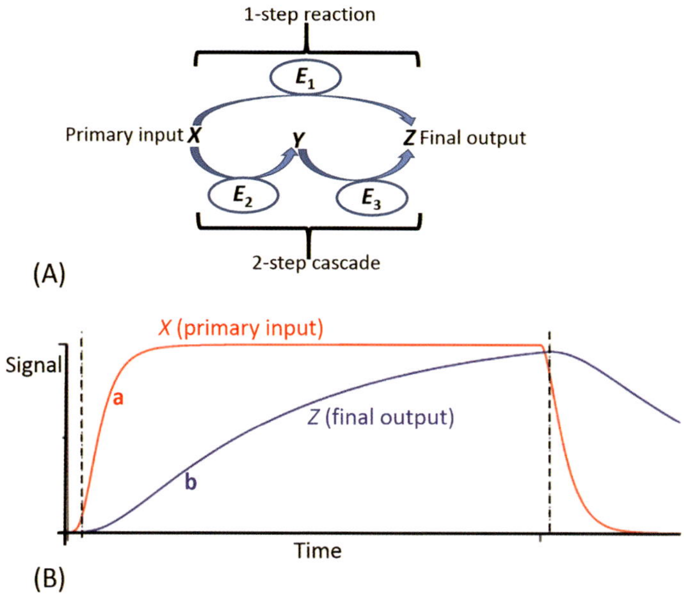

Figure 15.5 (A) The biocatalytic process realized through two different pathways composed of direct (1-step reaction) and indirect (2-step cascade) processes. (B) The time-dependent variation of the primary input signal X (curve a) and output signal Z (curve b) with the delayed changes.

earlier – it does not rely on activation/inhibition of enzymes and only uses two different pathways (a short one and a long one) for producing the final output signal. Such systems are easier to design for experimental realizations because they involve process cascades of the type already realized in the binary biocomputing gate research [31]. The "activation" by X and by Y consists of them simply being the actual inputs enabling the reactions, but not being promoters or inhibitors of the enzymes used. The availability of the primary input substrate, $X(t)$, was controlled by adjusting $R_{ext}(t)$ to yield the shown time dependence (Figure 15.5B, curve a). In this model, adjustable quantities (concentrations of those chemicals that are not designated as input/output signals) can be selected as needed to achieve variants of delayed response (Figure 15.5B, curve b) to various protocols, $R_{ext}(t)$, of controlling the input signal availability. However, this simplified approach cannot provide the full-featured feed-forward realization even for the all-activations case. It should be noted that the simple delay in the formation of the output signal can be produced in flow cell devices without any biocatalytic process [33].

More complicated cascades, such as that shown in Figure 15.1 and 15.3, will have to be explored in future work. These systems actually imitate the nature's design: instead of X being the actual direct input for the production of Y and Z, it activates (or, if needed, represses) the already ongoing processes that produce Y and Z and similarly for the mechanism by which Y affects the production of Z. The simplified approach discussed in this section and outlined in Figure 15.5 should be avoided, when the feed-forward loops are the real goal. This is why the simplified approach does not work and the challenges in realization of the feed-forward systems are still high. The full realization of feed-forward systems through enzyme-catalyzed reactions proposed in this section is still waiting for their experimental realization. The present discussion is serving as a motivation, providing guidelines for this future work.

15.6 Conclusion

This chapter is mostly formulating goals and challenges, rather than reporting on obtained experimental results. The conceptual setup for cascades of enzyme-catalyzed biochemical reactions that realize the feed-forward response was outlined. The new aspects of the devised systems, specifically for feed forward involve the aggressive use of chemicals, which are not enzymes' substrates, but rather are compounds (called ligands) typically binding to other than substrate reaction sites of the enzyme molecules and causing promotion (activation) or slowdown (repression) of the enzymatic activity. The offered discussion highlights the experimental challenges and the required modeling involved in designing, realizing, and characterizing the basic steps, with feed forward as a process with a single controlled input. However, at least some degree of modeling and experimentation with networking cannot be avoided. In order to probe the full range of the control by the input(s) variation, the external input rate(s) must be varied not only for positive values (which can be accomplished by physically adding the input reactant, as well as chemically) but also for negative values that in most cases will require an additional chemical or biochemical process for deactivation. The discussed non-Boolean/nonbinary processes are coming to the area of the unconventional computing from the examples given by nature but realized in much simpler biocatalytic systems. The important question is how far we can go with simple biomolecular (e.g., enzyme-based systems) in mimicking complex biological process without the use of all sophisticated biological mechanisms. The answer remains open and it is awaiting new research efforts.

References

1 Calude, C.S., Costa, J.F., Dershowitz, N. et al. (eds.) (2009). *Unconventional Computation*, Lecture Notes in Computer Science, vol. 5715. Berlin: Springer.
2 Szacilowski, K. (2012). *Infochemistry – Information Processing at the Nanoscale.* Chichester: Wiley.
3 de Silva, A.P. (2013). *Molecular Logic-Based Computation.* Cambridge: Royal Society of Chemistry.
4 Katz, E. (ed.) (2012). *Molecular and Supramolecular Information Processing – From Molecular Switches to Logic Systems.* Weinheim: Wiley-VCH.
5 Sienko, T. (ed.), Adamatzky, A., Conrad, M., and Rambidi, N.G. (Series eds.) (2003). *Molecular Computing.* Cambridge, MA: MIT Press.
6 Katz, E. (ed.) (2012). *Biomolecular Computing – From Logic Systems to Smart Sensors and Actuators.* Weinheim: Wiley-VCH.
7 Kumar Sarkar, S., Kumar De, A., and Sarkar, S. (2015). *Foundation of Digital Electronics and Logic Design.* Singapore: Pan Stanford.
8 Privman, V. and Katz, E. (2015). *Phys. Status Solidi A* 212: 219–228.
9 Melham, T. (2013). *Prog. Biophys. Mol. Biol.* 111: 129–136.
10 Navlakha, S. and Bar-Joseph, Z. (2011). *Mol. Syst. Biol.* 7: 546.
11 Seaton, D.D. and Krishnan, J. (2012). *Phys. Biol.* 9: 045009.
12 Alon, U. (2007). *An Introduction to Systems Biology. Design Principles of Biological Circuits.* Boca Raton, FL: Chapman & Hall/CRC Press.
13 Georgiopoulos, M., Li, C., and Kocak, T. (2011). *Perform. Eval.* 68: 361–384.
14 Annema, A.-J. (1996). *Feed-Forward Neural Networks: Vector Decomposition Analysis, Modelling and Analog Implementation.* New York, NY: Springer Science.
15 Alon, U. (2007). *Nat. Rev.* 8: 450–461.
16 Mangan, S. and Alon, U. (2003). *Proc. Natl. Acad. Sci. U.S.A.* 100: 11980–11985.
17 Kashtan, N. and Alon, U. (2005). *Proc. Natl. Acad. Sci. U.S.A.* 102: 13773–13778.
18 Mattick, J.S. (2007). *J. Exp. Biol.* 210: 1526–1547.
19 Qian, L. and Winfree, E. (2011). *J. R. Soc. Interface* 8: 1281–1297.
20 Dekel, E., Mangan, S., and Alon, U. (2005). *Phys. Biol.* 2: 81–88.
21 Shen-Orr, S.S., Milo, R., Mangan, S., and Alon, U. (2002). *Nat. Genet.* 31: 64–68.

22 Tostevin, F. and ten Wolde, P.R. (2009). *Phys. Rev. Lett.* 102: 218101.
23 Schirmer, R.H., Krauth-Siegel, R.L., and Schulz, G.E. (1989). Glutathione reductase. In: *Coenzymes and Cofactors, Vol. 3: Glutathione,* Chem. Biochem. Med. Aspects, Part A, Chapter 14 (ed. D. Dolphin, R. Poulson and O. Avromonic), 553–596. New York, NY: Wiley.
24 Leskovac, V., Trivic, S., and Pericin, D. (2002). *FEMS Yeast Res.* 2: 481–494.
25 Carper, W.R., Dorey, R.C., and Beber, J.H. (1987). *Clin. Chem.* 33: 1906–1908.
26 Rachdan, D., Lou, M.F., and Harding, J.J. (2005). *Curr. Eye Res.* 30: 919–925.
27 Cleland, W.W. (1964). *Biochemistry* 3: 480–482.
28 Zhou, J., Arugula, M.A., Halámek, J. et al. (2009). *J. Phys. Chem. B* 113: 16065–16070.
29 BRENDA (BRaunschweig ENzyme DAtabase). The Comprehensive Enzyme Information System. http://www.brenda-enzymes.org (accessed 19 March 2019).
30 Roda, A. (ed.) (2010). *Chemiluminescence and Bioluminescence: Past, Present and Future.* Cambridge: Royal Society of Chemistry Press.
31 Privman, V., Zavalov, O., Halámková, L. et al. (2013). *J. Phys. Chem. B* 117: 14928–14939.
32 Katz, E., Privman, V., and Zavalov, O. (2014). Structure of feed-forward realizations with enzymatic processes. In: *Proceedings of The Eighth International Conference on Quantum, Nano/Bio, and Micro Technologies (ICQNM 2014),* 22–27. Wilmington, DE: ThinkMind Online Publishing.
33 Verma, A., Fratto, B.E., Privman, V., and Katz, E. (2016). *Sensors (MDPI)* 16: 1042.

16

Conclusions and Perspectives: Where Are We Going?

In summary, the book reviews various aspects of enzyme-based logic gates, their networking, and their interfacing with systems that offer transduction of the biochemical output into electronic or other signals. The present chapter briefly comments on future challenges for this field of research, sometimes balancing on the border between real science and science fiction.

16.1 Conclusions

While the very original motivation for the molecular unconventional computing originated from the belief that this direction will bring computation to the new high efficiency, which is much above the limits set by the traditional solid-state electronics, this expectation has not been fulfilled in more than 20 years long research efforts. Indeed, even the original promise [1] was only for DNA computing applied to very specific combinatorial problems [2], which can benefit from massive parallel computing performed by numerous individual molecules. Many following scientific reports that have been repeating statements on "potentially high efficiency of molecular computing" do not really have any solid reason for such statements. The expectations for high-speed calculations are hardly preserved for the DNA-based computing but do have any background for molecular systems with synthetic isomerizable molecules and enzyme systems overviewed in this book. Therefore, another motivation should be found for this kind of computing/logic systems rather than building a molecular computer in the future. One of the possibilities, particularly emphasized in the book, is biosensing application of the enzyme logic systems [3, 4]. This application does not require superfast operation, and it is not based on too many components working together – relatively low complexity systems can be sufficient for logic processing of biomolecular signals, particularly for biomedical applications.

An important development in enzyme-based logic has been initiated with the introduction of the novel biosensor concepts [3, 4], with multiple input signals processed biochemically before transduction to the output. Such systems would require utilization of realistic [5], rather than model, biochemical inputs, with proper ranges for the "digital" **0** and **1** signal cutoffs, as well as "gate machinery" compatible with the environment of the sensing application. Furthermore,

Enzyme-Based Computing Systems, First Edition. Evgeny Katz.

new demands for the biochemical gate networking will require a strategy to systematize the modeling, utilizing modular network analysis and detailed network optimization approaches [6, 7], supplemented with single-gate optimization of key elements in the network [8–11]. As the biocomputing networks become larger and more complex, new gates and non-Boolean network elements will have to be realized, characterized, and modeled. The latter, non-Boolean elements, should include filters, functioning, for instance, by diverting some of the output or input(s). It has been argued [8–11] that this can induce sigmoid behavior. More long-term projects will also address issues of signal splitting/balancing and signal amplification. Ultimately, for larger networks the issues of network design for digital error correction will also come into play. Enzyme properties not presently explored for information processing might offer interesting avenues for optimization. For example, allosteric enzymes [12] frequently have the self-promoter (sigmoid-response) property with respect to their substrate concentrations.

Biochemical networks can offer robust error-free operation upon appropriate optimization of their components and interconnections. However, the limit of the biocomputing network complexity is set by the cross-reactivity of enzyme-catalyzed reactions. Only enzymes belonging to different biocatalytic classes (oxidases, dehydrogenases, peroxidises, hydrolases, etc.) could operate in a single "soup" without significant cross-reactivity. If chemical reasons require the use of cross-reacting enzymes in the system, they must be space-separated to allow their operation. This might be achieved by the use of reaction–diffusion chemical processors separating the reacting species in time and space [13]. Another approach would include compartmentalization of the reacting species [14] using pattering on surfaces or in microfluidic devices. Microfluidic hardware required for digital processing of chemical signals is already being developed [15] and has been applied for some biomolecular, e.g., DNA computing, systems [16]. Further development of this research area requires cooperative work of engineers, biochemists, and computer specialists. The ultimate goal of this work will be the design of a microfluidic lab-on-a-chip performing multienzyme-catalyzed cascades and operating similarly to an electronic chip by being able to integrate relatively large networks for processing biochemical signals.

Another way to resolving the problem of the enzyme cross-reactivity in networks is by utilizing other-than-enzymatic biomolecular components with higher specificity. Combination of enzymes with other biomolecules, specifically, with immune-recognition biomolecules (antigens–antibodies), offers interesting avenues for new information processing designs and applications. The use of immune species (antigens–antibodies), which offer a large variety of biorecognition combinations, could substantially increase the variability of biocomputing steps with minimum cross-reactivity. Some of the immune reactants could be labeled with enzymes, thus providing interconnections with the enzyme-based networks and resulting in integrated biocomputing systems based on biorecognition–biocatalysis processes. Preliminary work has already demonstrated logic operations performed by antigen–antibody interactions coupled with enzyme logic gates [17]. The logically processed immune signals were utilized to control the operation of a biofuel cell [18].

A variety of biomedical and biotechnological applications could be envisaged for the developed hybrid systems. These applications can range from biochemical processes, e.g., biomolecular oxidative damage [19], to bioelectronic devices, e.g., biofuel cells [20–22], controlled by enzyme logic systems. As mentioned, the biocomputing approach reviewed here paves the way to novel digital biosensors [3, 4] processing multiple biochemical signals and producing a combination of outputs dependent on a complex pattern of inputs. These biosensors range from medical devices reporting on physiological dysfunctions, e.g., injuries [5, 23], to sensors applied in forensic investigations [24] and alert-type devices for homeland security and military use [25, 26]. The biochemical signals can be processed by chemical means based on the enzyme logic systems and the difference between different physiological scenarios can be directly derived from the chemically processed information, hence obviating the need for computer analysis of the biosensing information. In addition to analysis of the data, the output signals might be directed to chemical actuators (e.g., signal-responsive membranes or polymer films), leading to on-demand drug release. A diverse range of "smart" (stimuli-responsive) materials, with switchable physical properties, has been developed for *in vivo* drug delivery [27]. The new hybrid materials with built-in Boolean logic will be capable of switching physical properties in response to the output of an enzyme logic system toward autonomous on-demand drug delivery [28, 29]. The output signals generated by enzyme logic networks will activate "smart" chemical actuators, resulting, for example, in opening of a membrane or dissolving a polymer film releasing a drug, and could lead to a novel unconventional approach to a decision-making (sense-and-treat) biosensor/bioactuator. One can anticipate that such devices controlled by biochemical logic networks will facilitate decision-making in connection to autonomous feedback-loop drug delivery systems and will revolutionize the monitoring and treatment of patients and many other biomedical applications. The demonstrated approaches to interfacing of biomolecular computing systems with signal-responsive materials enable the use of various biocatalytic reactions to control the properties of responsive materials and systems with built-in Boolean logic. This approach would be an efficient way to fabricate "smart" multi-signal-responsive drug delivery systems, sensors, miniaturized switches, microfluidic devices, etc., which can function without communication with an external electronic computer revolutionizing many biotechnological applications.

16.2 Perspectives

It is always difficult to predict novel results and concepts. The most important and interesting results appear being not expected and not predicted. On the other hand, most predictions are based on extrapolations of already existing knowledge. Still, let us try to highlight some possible research directions in the study of the enzyme logic systems in their broad definition. Some of these suggestions are based on solid experimental background, but some of them are mostly speculative.

16.2.1 Information Processing Through Complex Biological Pathways in Cells

It has been already shown that biological processes in cells can mimic some basic Boolean logic [30, 31]. For example, bacterial cells (*Pseudomonas aeruginosa*) *lasI*/*rhlI* double mutant were activated for production of phenazine by two input signals – 3-oxododecanoyl homoserine lactone (3-oxo-C12-HSL) and *N*-butanoyl-l-homoserine lactone (C4-HSL) – processed through the AND logic gate (Figure 16.1A). The *las* and *rhl* cascades important for quorum sensing (QS) were inactive in the mutant cells. One of the important gene systems under the control of the *las*/*rhl* regulatory system is the *phz* operon, which codes for the enzymes required for phenazine biosynthesis. Thus, the mutant cells were not producing phenazine, which is a redox mediator in electrochemical processes characteristic of wild type of the cells. When two input signals, 3-oxo-C12-HSL and C4-HSL, were applied, the phenazine production was restored, and the mutant cells, which are normally electrochemically inactive, have demonstrated the electrochemical activity. Notably, the applied input signals have to be used together, thus reproducing the AND logic gate. Obviously, the internal operation of this logic gate inside the cells was controlled by very sophisticated (not fully known) biological process with multilevel regulation. This result can be translated to the novel concept of the logic processes performed by living cells or even organisms. The biological systems can accept many different signals producing multiple output signals with various logic applied (Figure 16.1B). At this time it seems to be very difficult to predict in which way this could be useful and how far in using biological substances for logic/computing operations we can go. In the most trivial way, this can lead to the signal-controlled microbial biofuel cells or fermentation reactors. The challenge in this research direction should be given to biologists, but their cooperation with chemists and particularly with computer experts is needed.

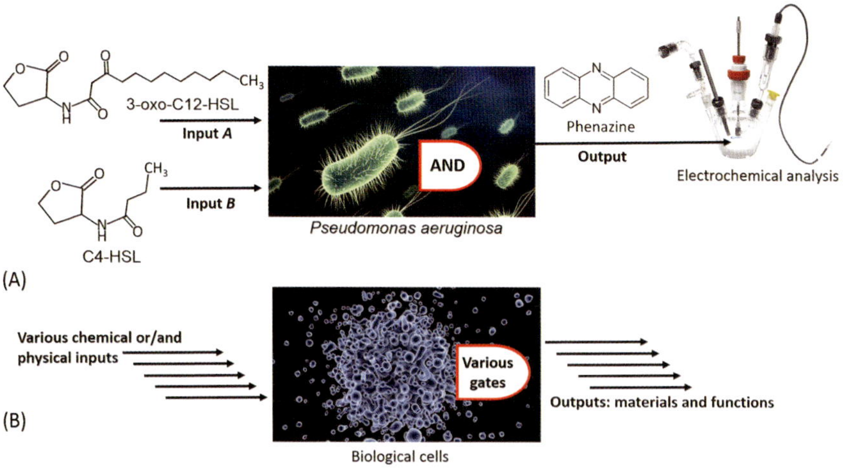

Figure 16.1 (A) The Boolean AND logic gate realized in *P. aeruginosa* bacterial cells. (B) A perspective system based on biological cells performing complex logic operations on many input signals. The logic operations can be programmed in the cells.

16.2.2 Signal-Controlled Bioelectronic Devices and Signal-Triggered Molecular Release

Numerous enzyme logic systems controlling various bioelectronic devices (particularly biofuel cells [20–22]), biocatalytic/biosensing electrodes [32], and electronic chips [33–35] have been reported in the last decade. Some of these systems have been used for the signal-triggered molecular (e.g., drug) release. Particularly, self-powered systems for the molecular release based on biofuel cells are very promising for future biomedical applications. Figure 16.2 shows schematically a biofuel cell activated with two biomarkers for the release process, while the operation proceeds according to the AND logic gate "programmed" in the system composition [36]. This research direction seems to be less fantastic and much more experimentally proved; however, it is still far for practical applications. The challenge is still in moving from the test tube experiments to real physiological/biological systems. This will certainly need contribution from materials science experts, biologists, and, then in later steps, experts in medicine.

16.2.3 Allosteric and Hybrid Enzymes

Allosteric enzymes [12], being activated or inhibited by binding of an effector molecule (which is not a substrate molecule), might be particularly useful for constructing signal-controlled biocatalytic systems. In case a substrate molecule operates as an effector (homotropic allosteric modulator), the enzyme can demonstrate sigmoidal activity dependence on the substrate concentration.

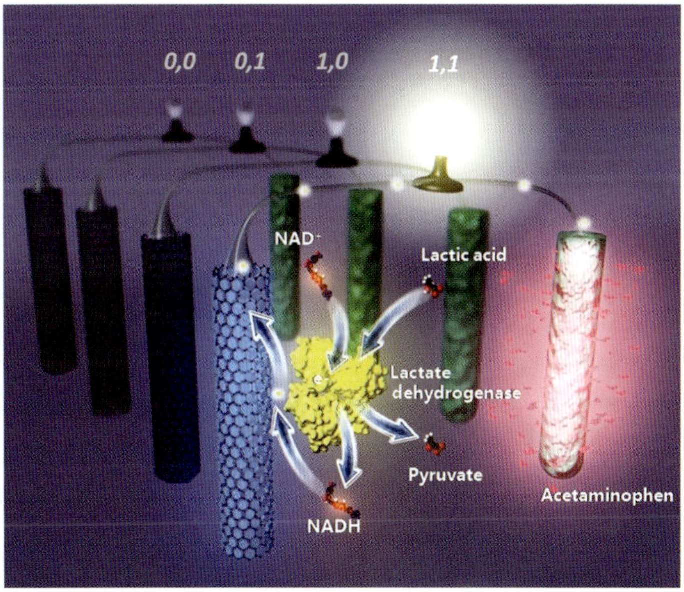

Figure 16.2 Logic-controlled self-powered "sense–act–treat" system that is based on a biofuel cell: toward logic-activated therapeutic intervention. Source: Zhou et al. 2012 [36]. Adapted with permission of John Wiley & Sons.

It has been shown that the sigmoidal concentration dependence (obtained by different means; see Chapter 3) is important for reducing noise in binary operating logic gates [10, 11, 37–39]. However, naturally available allosteric enzymes are not available for all required logic operations, particularly related to biosensing of various biomarkers [5]. This problem can be solved with artificial hybrid molecules composed of a biocatalytic enzyme part and a covalently bound reporter domain. Typically, this involves construction of chimeric enzymes via insertion of a regulatory receptor domain into the biocatalytic domain. Construction of such chimeric enzymes (recombinant fusion proteins) [40] is based on recombinant DNA technology in protein engineering [41]. Conformational changes of the receptor domain in response to complex formation with the ligand induce structural changes in the biocatalytic part of the chimeric enzyme, thereby modulating its activity. This approach has been applied recently to construct pyrroloquinoline quinone-dependent glucose dehydrogenase (PQQ-GDH) [42] (typical quinoprotein [43]) fused with calmodulin (CaM) [44] unit (Figure 16.3A) responsive to Ca^{2+} cations and a special M13 peptide, both important for the CaM conformational changes (Figure 16.3B) and then translated to the conformational changes and activity modulation of the biocatalytic PQQ-GDH domain [45]. Immobilization of the switchable chimeric PQQ-GDH-CaM enzyme on a graphene-functionalized electrode resulted in its electrocatalytic operation reversibly controlled by addition/removal of Ca^{2+}

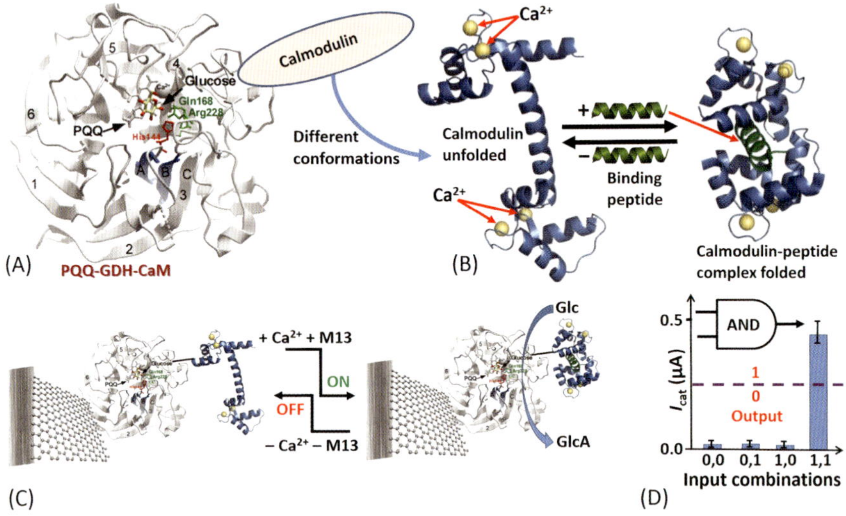

Figure 16.3 (A) The fusion protein composed of a biocatalytic PQQ-GDH unit connected to calmodulin receptor changing its conformation upon complex formation with Ca^{2+} cations and M13 peptide. (B) Different conformations of calmodulin upon reversible binding of M13 peptide (note that Ca^{2+} cations are already in the complex). (C) Reversible activation/inhibition of the bioelectrocatalytic glucose oxidation upon PQQ-GDH-CaM complex formation/dissociation with Ca^{2+} cations and M13 peptide. The PQQ-GDH-CaM chimeric enzyme was immobilized on a carbon fiber electrode modified with graphene nanosheets. (D) The electrocatalytic current produced when Ca^{2+} cations and M13 peptide were added in different combinations mimicking the AND logic gate.

cations and M13 peptide (Figure 16.3C). Since both effectors, Ca^{2+} cations and M13 peptide, should be present together to have the cooperative effect on the conformational changes in the CaM domain, the system responds to them according to the AND logic gate, where the output signal corresponds to the electrocatalytic current (Figure 16.3D). The designed switchable electrocatalytic system was also applied to the construction of a switchable biofuel cell [45]. This approach based on chimeric fusion proteins is very promising for constructing various types of the signal-controlled enzymes useful for the design of different logic systems. Further research in this direction requires collaborative efforts of molecular biologists and experts in bioelectronics and computer science.

16.2.4 Enzyme System Controlled by Various Chemical and Physical Signals

The previous section highlighted a new approach to construction of signal-switchable enzymes using a well-developed fusion technology, resulting in hybrid structures composed of biocatalytic and receptor domains responding to variety of (bio)molecular signals (e.g., combination of Ca^{2+} cations and a special M13 peptide [45]). Many more approaches to the construction of signal-switchable enzymes have been explored in the last two decades. This can include covalent binding of photoisomerizable species to a backbone of enzymes (e.g., spiropyran/merocyanine) [46] and association of enzyme molecules with magnetic nanoparticles [47, 48], thus allowing to modulate the enzyme activity by light signals and by application of external magnetic field. Obviously, if the enzyme activity is followed by electrochemical means, the bioelectrochemical activity is always controlled by the potential applied at the sensing electrode. When a competitive enzyme is present (e.g., GOx competing with PQQ-GDH for glucose oxidation), the enzyme responsible for the bioelectrochemical process (PQQ-GDH in the present example) can be mute because the required substrate (glucose in this example) can be fully consumed in the side reaction, which is not producing any current at the electrode. All these approaches applied separately are not novel, and they are supported by numerous examples documented in many scientific reports. They can be used for constructing signal-controlled biocatalytic cascades and have been already used for realization of different logic gates, if the reactions are considered as binary processes. However, applying several switchable mechanisms together (Figure 16.4) could result in very interesting and novel results. Some of the mechanisms schematically shown in Figure 16.4 can drive the process in different directions (some for activation and some for inhibition). For example, the presence of a competitive enzyme and application of magnetic field can cancel effect of each other [48]. Some of the signals need to be applied together (e.g., Ca^{2+} and M13 in the example of PQQ-GDH-CaM [45]) in order to activate the enzyme, thus performing the AND logic operation. Overall, many different signals applied to one biocatalytic species can allow sophisticated logic operation programmed through the selected signal-affected units. This can bring the complexity and flexibility of the logic operation to the next level.

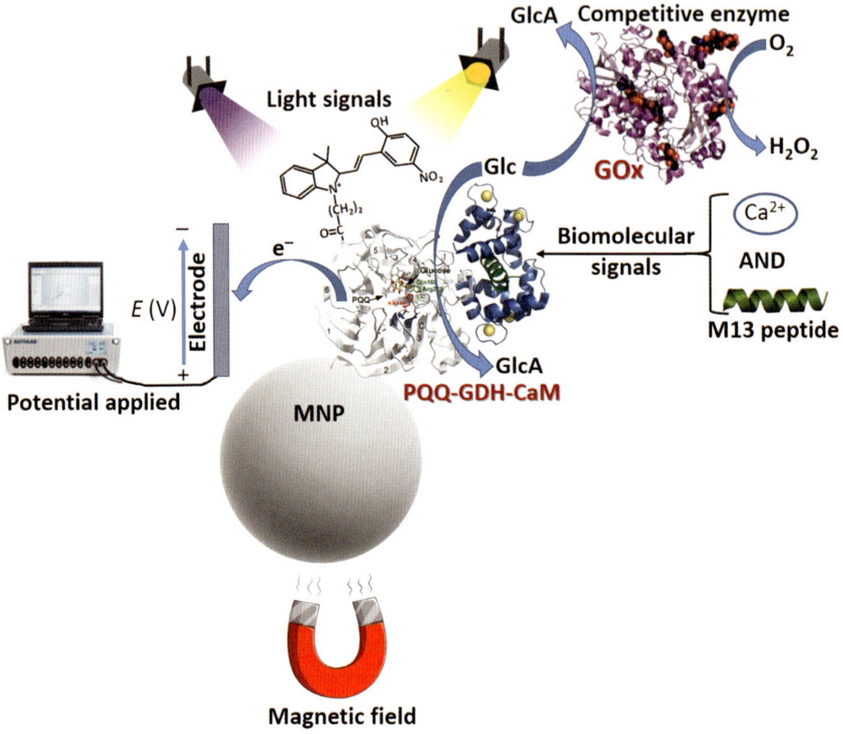

Figure 16.4 The enzyme operation controlled by many different chemical/physical input signals. The PQQ-GDH-CaM chimeric enzyme bound to a magnetic nanoparticle and functionalized with spiropyran/merocyanine photoisomerizable groups is reversibly switched between ON and OFF states by light signals, the presence of Ca^{2+} cations and M13 peptide, and by application of magnetic field. The electron transfer from the enzyme to an electrode is controlled by a potential applied. Also, the presence of GOx can activate a competitive redox reaction, thus inhibiting the PQQ-GDH-CaM operation. The scheme does not reflect any experimental result; however, most of the signal functions have been studied separately.

16.2.5 Molecular and Nanomachines for Self-Propulsion and Logic Operation

The most recent research resulted in construction of various micro- and nano-size motors self-moving and using chemical/catalytic processes to drive them through solutions containing molecular "fuel" [49–51]. One kind of this nano-motors can be activated for self-propulsion by (bio)catalytic decomposition of H_2O_2, resulting in formation of O_2 bubbles that result in driving the nanorods in the opposite direction (Figure 16.5A). While in many systems the H_2O_2 decomposition was catalyzed by inorganic catalysts [52], the use of catalase enzyme has been shown as well [53]. Hydrogen peroxide needed as "fuel" for the self-propulsion can be produced *in situ* through the GOx-catalyzed reaction [54]. Nano-species nonsymmetrically functionalized with different reductive and oxidative enzymes (e.g., glucose oxidase oxidizing glucose and bilirubin oxidase reducing oxygen at both sides of a carbon fiber [55]) can be moving due to enzymatically induced self-electrophoresis. One can

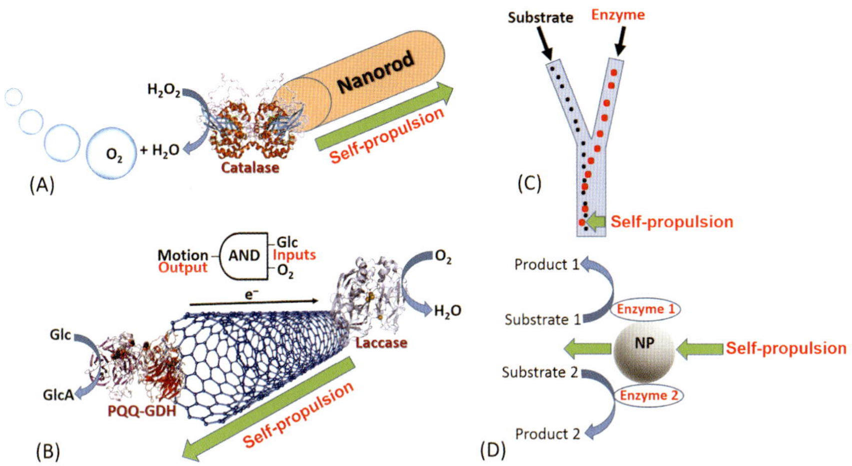

Figure 16.5 (A) Catalase-modified nanorod self-propulsion in the presence of H_2O_2 in the solution. (B) The system proposed for self-propulsion based on enzymatically induced self-electrophoresis. (C) Self-propulsion of individual enzyme molecules while reacting with a corresponding substrate being mixed in laminar flow. (D) A multienzyme-functionalized NP moving and performing logic operations in enzymatic reactions.

speculate on the possibility to bring this mechanism to the nanoscale, where a carbon nanotube connects electrically different oxidative–reductive enzymes (Figure 16.5B). Different, sometimes not fully known, mechanisms can be responsible for moving individual enzyme molecules reacting with substrates. This can be exemplified by reacting enzymes with a substrate in a laminar flow in Y-shaped microfluidic channel where the enzyme molecules are clearly moving in the direction of the substrate [56] (Figure 16.5C). This unique phenomenon provides background for thinking on nano-species (different kinds of nanoparticles, NPs) functionalized with various enzymes capable to move in the direction of higher concentration of the corresponding substrates, whatever mechanism is responsible for that. More than that, we can think on the systems where the NPs are functionalized with several different enzymes performing reactions on mixed substrate compositions (Figure 16.5D). Overall, all possible self-moving enzyme-modified species can perform various Boolean logic operations [57], also performing actuation function at the nanoscale.

The example systems and possible research directions collected here are not the most promising and may not be mandatory continuations in the research field of the enzyme logic/computing systems. In the most probability the development will go in different directions, or the directions mentioned briefly here will not be connected to the enzyme logic and will be continued with different motivation.

16.3 Final Comments

While in a long perspective, a molecular computer might become a reality, particularly for some specific applications, for example, for solving complex combinatorial problems, the present state of the art does not allow any practical use of biomolecular, particularly enzyme-based, systems for real computational applications. However, some other applications of biomolecular information processing systems are not that far from immediate use. One of them is binary sensing where multiple input signals are logically processed, resulting in YES/NO decisions in the binary (**0**,**1**) format. In this subarea of molecular logic, successful implementation of systems at the present level of complexity is at reach, particularly for biomedical applications. As a general trend, the research in biomolecular information processing, which has been motivated originally to progress unconventional computing applications, is broadly developing to areas not directly related to computing in its narrow definition. This research opens up novel concepts in sensing/biosensing, switchable smart materials controlled by logically processed signals, bioelectronic devices (e.g., biofuel cells) controlled by external signals, signal-controlled release processes, user authentication, and so forth.

References

1 Rambidi, N.G. (1997). *BioSystems* 44: 1–15.
2 Adleman, L.M. (1994). *Science* 266: 1021–1024.
3 Katz, E., Wang, J., Privman, M., and Halámek, J. (2012). *Anal. Chem.* 84: 5463–5469.
4 Wang, J. and Katz, E. (2011). *Isr. J. Chem.* 51: 141–150.
5 Halámek, J., Windmiller, J.R., Zhou, J. et al. (2010). *Analyst* 135: 2249–2259.
6 Privman, V., Zavalov, O., Halámková, L. et al. (2013). *J. Phys. Chem. B* 117: 14928–14939.
7 Privman, V., Arugula, M.A., Halámek, J. et al. (2009). *J. Phys. Chem. B* 113: 5301–5310.
8 Filipov, Y., Domanskyi, S., Wood, M.L. et al. (2017). *ChemPhysChem* 18: 2908–2915.
9 Privman, V., Fratto, B.E., Zavalov, O. et al. (2013). *J. Phys. Chem. B* 117: 7559–7568.
10 Zavalov, O., Bocharova, V., Privman, V., and Katz, E. (2012). *J. Phys. Chem. B* 116: 9683–9689.
11 Halámek, J., Zavalov, O., Halámková, L. et al. (2012). *J. Phys. Chem. B* 116: 4457–4464.
12 Traut, T.W. (2008). *Allosteric Regulatory Enzymes*. New York, NY: Springer.
13 Adamatzky, A., De Lacy Costello, B., and Asai, T. (2005). *Reaction-Diffusion Computers*. New York, NY: Elsevier.
14 Fratto, B.E. and Katz, E. (2015). *ChemPhysChem* 16: 1405–1415.
15 Fair, R.B., Khlystov, A., Tailor, T.D. et al. (2007). *IEEE Des. Test Comput.* 24: 10–24.
16 Zhang, Y., Yu, H., Qin, J., and Lin, B. (2009). *Biomicrofluidics* 3: 044105.
17 Strack, G., Chinnapareddy, S., Volkov, D. et al. (2009). *J. Phys. Chem. B* 113: 12154–12159.
18 Tam, T.K., Strack, G., Pita, M., and Katz, E. (2009). *J. Am. Chem. Soc.* 131: 11670–11671.

19 Zhou, J., Melman, G., Pita, M. et al. (2009). *ChemBioChem* 10: 1084–1090.
20 Amir, L., Tam, T.K., Pita, M. et al. (2009). *J. Am. Chem. Soc.* 131: 826–832.
21 Tam, T.K., Pita, M., Ornatska, M., and Katz, E. (2009). *Bioelectrochemistry* 76: 4–9.
22 Katz, E. and Pita, M. (2009). *Chem. Eur. J.* 15: 12554–12564.
23 Zhou, J., Halámek, J., Bocharova, V. et al. (2011). *Talanta* 83: 955–959.
24 Kramer, F., Halámková, L., Poghossian, A. et al. (2013). *Analyst* 138: 6251–6257.
25 Bocharova, V., Halámek, J., Zhou, J. et al. (2011). *Talanta* 85: 800–803.
26 Chuang, M.-C., Windmiller, J.R., Santhosh, P. et al. (2011). *Chem. Commun.* 47: 3087–3089.
27 LaVan, D.A., McGuire, T., and Langer, R. (2003). *Nat. Biotechnol.* 21: 1184–1191.
28 Gamella, M., Privman, M., Bakshi, S. et al. (2017). *ChemPhysChem* 18: 1811–1821.
29 Okhokhonin, A.V., Domanskyi, S., Filipov, Y. et al. (2018). *Electroanalysis* 30: 426–435.
30 Li, Z., Rosenbaum, M.A., Venkataraman, A. et al. (2011). *Chem. Commun.* 47: 3060–3062.
31 Arugula, M.A., Shroff, N., Katz, E., and He, Z. (2012). *Chem. Commun.* 48: 10174–10176.
32 Privman, M., Tam, T.K., Bocharova, V. et al. (2011). *ACS Appl. Mater. Interfaces* 3: 1620–1623.
33 Honarvarfard, E., Gamella, M., Poghossian, A. et al. (2017). *Appl. Mater. Today* 9: 266–270.
34 Molinnus, D., Poghossian, A., Keusgen, M. et al. (2017). *Electroanalysis* 29: 1840–1849.
35 Poghossian, A., Katz, E., and Schöning, M.J. (2015). *Chem. Commun.* 51: 6564–6567.
36 Zhou, M., Zhou, N., Kuralay, F. et al. (2012). *Angew. Chem. Int. Ed.* 51: 2686–2689.
37 Privman, V., Domanskyi, S., Mailloux, S. et al. (2014). *J. Phys. Chem. B* 118: 12435–12443.
38 Mailloux, S., Zavalov, O., Guz, N. et al. (2014). *Biomater. Sci.* 2: 184–191.
39 Bakshi, S., Zavalov, O., Halámek, J. et al. (2013). *J. Phys. Chem. B* 117: 9857–9865.
40 Schmidt, S.R. (ed.) (2013). *Fusion Protein Technologies for Biopharmaceuticals – Applications and Challenges.* Hoboken, NJ: Wiley.
41 Chaudhuri, K. (2013). *Recombinant DNA Technology.* New Delhi: The Energy and Resources Institute, TERI.
42 Duine, J.A., Frank, J., and Van Zeeland, J.K. (1979). *FEBS Lett.* 108: 443–446.
43 Duine, J.A. (1991). *Eur. J. Biochem.* 200: 271–284.
44 Villalobo, A., Ishida, H., Vogel, H.J., and Berchtold, M.W. (2018). *Biochim. Biophys. Acta* 1865: 507–521.
45 Gamella, M., Guo, Z., Alexandrov, K., and Katz, E. (2019). *ChemElectroChem* 6: 638–645.
46 Katz, E. (2018). *Electroanalysis* 30: 759–797.

47 Katz, E. (2016). *Electroanalysis* 28: 904–919.
48 Filipov, Y., Zakharchenko, A., Minko, S., and Katz, E. (2018). *ChemPhysChem* 19: 3035–3043.
49 Gáspár, S. (2014). *Nanoscale* 6: 7757–7763.
50 Chen, C., Soto, F., Karshalev, E. et al. (2018). *Adv. Funct. Mater.* 28: 1806290.
51 Gao, W. and Wang, J. (2014). *Nanoscale* 6: 10486–10494.
52 Wang, J. (2013). *Faraday Discuss.* 164: 9–18.
53 Sanchez, S., Solovev, A.A., Mei, Y., and Schmidt, O.G. (2010). *J. Am. Chem. Soc.* 132: 13144–13145.
54 Pantarotto, D., Browne, W.R., and Feringa, B.L. (2008). *Chem. Commun.* 1533–1535.
55 Mano, N. and Heller, A. (2005). *J. Am. Chem. Soc.* 127: 11574–11575.
56 Sengupta, S., Dey, K.K., Muddana, H.S. et al. (2013). *J. Am. Chem. Soc.* 135: 1406–1414.
57 Zhang, L., Zhang, H., Liu, M., and Dong, B. (2016). *ACS Appl. Mater. Interfaces* 8: 15654–15660.

Index

f

m

n